This book is to be returned on or before
the last date stamped below.
Fine 50c per day

The Neurobiology of Taste and Smell

Second Edition

The Neurobiology of Taste and Smell

Second Edition

Edited by

Thomas E. Finger

University of Colorado Health Sciences Center
Denver, Colorado

Wayne L. Silver

Wake Forest University
Winston-Salem, North Carolina

Diego Restrepo

University of Colorado Health Sciences Center
Denver, Colorado

 **WILEY-LISS**

A JOHN WILEY & SONS, INC., PUBLICATION
New York • Chichester • Weinheim • Brisbane • Singapore • Toronto

Library of Congress Cataloging-in-Publication Data:

The neurobiology of taste and smell / edited by Thomas E. Finger, Wayne L. Silver, Diego Restrepo.—2nd ed.
    p. ; cm.
    Includes bibliographical references and index.
    ISBN 0-471-25721-4 (alk. paper)
    1. Taste. 2. Smell. 3. Neurophysiology. I. Finger, Thomas E. II. Silver, Wayne L. III. Restrepo, Diego.
    [DNLM: 1. Olfactory Receptor Neurons. 2. Taste Buds. 3. Smell. 4. Taste. WL 700 N4937 2000]
    QP455 .N47 2000
    573.8'77—dc21   00-020619

CONTENTS

PREFACE

In the concluding chapter in the first edition of this book in 1987, Lloyd Beidler speculated on the future directions of research in the chemical senses. Some of the speculation, for example, the use of genetics to study the chemical senses, proved to be accurate (see Chapter 3 in the current volume). Yet, it is doubtful that anyone could have predicted the explosion of research and interest in the chemical senses that has occurred in the past 10–12 years. The use of molecular biological and biophysical techniques has provided a wealth of new information that has, in many ways, transformed the field. It is this new information that has prompted publication of a second edition.

The objectives of the second edition remain the same as the first. We provide a broad survey of the current state of research and knowledge about the chemical senses. The book is intended especially for the students and investigators new to the field. Accordingly, we asked the authors to provide an overview of their areas rather than a comprehensive literature review. Results and interpretations were stressed over technical details.

The format of the second edition has been changed somewhat to reflect the direction the field has taken. The book retains the three primary sections: Part I, "Chemical Sensitivity and Sensibility"; Part II, "Olfaction"; and Part III, "Gustation." The separate chapter on chemoreception in invertebrates has been eliminated, since information about invertebrate chemoreception now is included in each of the remaining chapters where appropriate. A chapter on genetic models of chemoreception has been added to Part I. While this chapter focuses on three different species (namatodes, fruit flies, and mice), it describes how modern genetic techniques have been used to identify important molecular components of chemosensory systems. Rather than include a separate chapter on the vomeronasal system, we have included this information on the chapter on chemosensory signaling (Chapter 5) in Part I as well as in Part II (Chapters 6–9). This second edition also emphasizes transduction and molecular biology of olfaction and taste to reflect the flood of chemosensory research in these areas.

All of the chapters of this second edition have been entirely rewritten; most of the authors are new. They come from a variety of backgrounds and have contributed significantly to their respective fields. We believe they have provided a comprehensive description of the crucial issues and important developments within each area. If a particular area of chemical senses research was omitted, it was done so unintentionally. The editors apologize to any investigators who feel that their work is neglected or given short shrift. The chapters are not meant to be comprehensive reviews of an area but are designed to provide an overview and to offer appropriate primary and secondary sources for a student or investigator wishing to delve deeper into a particular area.

Thomas E. Finger

Wayne L. Silver

Diego Restrepo

Denver, Colorado and

Winston-Salem, North Carolina

CONTRIBUTORS

Barry W. Ache Whitney Laboratory, University of Florida, 9505 Ocean Shore Blvd. St. Augustine, FL 32086

Linda A. Barlow Department of Biological Sciences, University of Denver, Denver, CO 80208

Paul A.S. Breslin Monell Chemical Senses Center, 3500 Market Street, Philadelphia, PA 19104-3308

Bruce Bryant Monell Chemical Senses Center, 3500 Market Street, Philadelphia, PA 19104-3308

Gail D. Burd Department of Molecular and Cellular Biology, University of Arizona, Life Sciences Building 444, Tucson, AZ 85721

John R. Carlson Department of Molecular, Cellular, and Developmental Biology, Yale University, New Haven CT 06520-8103

Nirupa Chaudhari Department of Physiology/Biophysics, University of Miami School of Medicine Miami FL 33101

Thomas A. Christensen Arizona Research Laboratories, Division of Neurobiology, University of Arizona, 611 Gould-Simpson Building, Tucson, AZ 85721

Barry J. Davis Department of Anatomy and Neurobiology, University of Maryland School of Medicine, Baltimore MD 21201

Albert I. Farbman Department of Neurobiology and Physiology, Northwestern University, 2153 Sheridan Road, Evanston, IL 60208

Thomas E. Finger Department of Cellular and Structural Biology, University of Colorado Health Sciences Center, 4200 East 9th Avenue, Denver, CO 80262

John I. Glendinning Department of Biological Science, Barnard College, 3009 Broadway, Columbia University, New York, NY 10027

Robert E. Johnston Department of Psychology, 286 Uris Hall, Cornell University, Ithaca, NY 14853

Sue C. Kinnamon Department of Anatomy and Neurobiology, Colorado State University, Ft. Collins, CO 80523

Timothy S. McClintock Department of Physiology and Biophysics, University of Kentucky College of Medicine, Chandler Medical Center, Lexington, KY 40536-0084

Nancy E. Rawson Monell Chemical Senses Center, 3500 Market Street, Philadelphia, PA 19104-3308

Diego Restrepo Department of Cellular and Structural Biology, University of Colorado Health Sciences Center, 4200 East 9th Avenue, Denver, CO 80262

Piali Sengupta Department of Biology, Brandeis University, Waltham, MA 02254

Wayne L. Silver Department of Biology, Wake Forest University, Winston Salem, NC 27109

Sidney A. Simon Department of Neuorobiology and Anesthesiology, Duke University Medical Center, Durham NC 27710

David V. Smith Department of Anatomy and Neurobiology, University of Maryland School of Medicine, 685 West Baltimore Street, Baltimore, MD 21201-1509

Leslie P. Tolbert Arizona Research Laboratories, Division of Neurobiology, University of Arizona, Tucson AZ 85721

Judith Van Houten Department of Biology, University of Vermont, Burlington, VT 05405-0086

Joel White Department of Neuroscience, Tufts University School of Medicine, 136 Harrison Avenue, Boston MA 02111

1

CHEMICAL SENSITIVITY & SENSIBILITY

1

Overview and Introduction

THOMAS E. FINGER and DIEGO RESTREPO
Rocky Mountain Taste and Smell Center,
Department of Cellular and Structural Biology,
University of Colorado Health Sciences Center, Denver, CO

WAYNE L. SILVER
Department of Biology, Wake Forest University, Winston-Salem, NC

1. CHEMICAL SENSITIVITY AND CHEMICAL SENSES

All organisms respond to chemicals in their external environment. This capability is exhibited by simple, single-celled organisms, for example, bacteria, and protista, as well as complex metazoans, for example, mammals, arthropods, and molluscs. Even plants respond to chemicals in their surroundings (Rhoades, 1985). Chemoreceptors in these diverse organisms detect nutrients, conspecifics, predators, or potentially harmful conditions. A plethora of receptor sites has evolved to detect the great diversity of natural chemical stimuli.

Most animals have chemoreceptors for monitoring internal as well as external chemical conditions. Internal sensors monitor physiologically important variables such as blood glucose, CO_2, or pH. The exteroceptive chemical senses of humans are taste, smell, and chemesthesis (common chemical sense). This book focuses on these exteroceptive senses, although many principles of coding and transduction may be shared by the interoceptive modalities.

The Neurobiology of Taste and Smell, Second Edition, Edited by Thomas E. Finger, Wayne L. Silver, and Diego Restrepo.
ISBN 0-471-25721-4   Copyright © 2000 Wiley-Liss, Inc.

Compared to hearing and vision, the chemical senses provide little of the total salient sensory experience for humans. Concomitantly, our lexicon and understanding of these modalities is relatively poor. Humans, being air-breathing vertebrates, have a passageway connecting the nose and mouth. Substances placed in the mouth can stimulate chemoreceptor cells in both organs, and, although information enters the central nervous system (CNS) via separate routes, taste and smell information come together in cortical association areas giving rise to the sensation of flavor. Thus, our language does not always make a clear distinction between stimulation of the gustatory and the olfactory receptor cells. The word *taste* is used colloquially to describe any chemical sensation produced by food in the mouth. Yet this percept results from combined stimulation of various chemoreceptors including free nerve endings in the epithelium (chemesthesis) (See Chapter 4, Bryant and Silver) as well as gustatory and olfactory receptor cells. Likewise, our use of the word *smell* embraces sensations arising from both olfactory and trigeminal chemoreceptors of the nasal cavity. In scientific usage, therefore, it is important to define carefully terms such as *smell, taste, gustation,* and *olfaction.*

Although such definitions are vitally important to a discussion of the neurobiology of chemical senses (Moulton, 1967), attempts to extend these definitions to animals other than humans can be problematic. Taste buds occur in all vertebrates, so the term *gustation* can be applied relatively directly to this group of animals. But what about invertebrates? Invertebrates have no receptor endorgans that look like taste buds, but many do have mouthparts containing chemoreceptors responsive to food stimuli. Despite the similarities in function and position, these mouth chemoreceptors and taste buds are clearly not homologous in terms of having evolutionary continuity. Yet, should the invertebrate oral chemoreceptors be considered gustatory endorgans and do they mediate a sense equivalent to taste? Most chemosensory scientists would answer "yes," but caution must be exercised in presuming commonality of mechanism or function because these systems are evolutionarily convergent rather than being derived from a common ancestor. Similar questions can be raised regarding the equivalence of vertebrate olfaction and numerous chemoreceptor systems in invertebrates, for example, antennal chemoreception in arthropods. Further complicating the situation are chemoreceptors in single cell organisms, which mediate behaviors akin to eating or social activity.

Even in vertebrates, identification of exteroceptive chemosensory modalities is not simple. For example, the nasal cavity contains three well-recognized chemosensory modalities: the main olfactory system, the accessory olfactory (vomeronasal) system, and chemesthetic capabilities. Two other possible chemosensory systems, the terminal nerve and the septal organ, have also been reported in the nasal cavity of vertebrates. The terminal nerve is a conserved feature found in all vertebrates and comprises a population of bipolar neurons whose axons form a distinct nerve fascicle separate from the olfactory (and vomeronasal) nerves. Although some investigators have suggested that the terminal nerve system is chemosensory (Demski and Northcutt, 1983), no experimental evidence supports this conjecture. The distal processes of these cells do not reach an epithelial surface and the system appears to convey information *from* rather than *to* the brain. Thus, the terminal nerve may serve a modulatory rather than a primary sensory role (White and Meredith, 1995). The septal organ is a patch of epithelium found in many mammals and is located on the septum ventral to the olfactory epithelium. The neurons located in this patch of epithelium express mRNA for putative olfactory receptors. The significance of its location ventral to the olfactory epithelium is not known.

2. CHEMOSENSORY-MEDIATED BEHAVIOR

Throughout the animal kingdom, chemical stimuli play an important role in behaviors such as feeding, territorial recognition, and sexual or social activities. Single-celled organisms may move toward a food source, move away from potentially toxic conditions, and find a potential sexual partner (see Chapter 2; van Houten). Thus, the diverse chemosensory functions all occur in a single cell. In metazoan forms, the different functions are usually divided into distinct sensory modalities. Taste receptor organs are always utilized in the ultimate decision of ingestion or rejection of a potential food item. In contrast, olfactory information drives a multitude of behaviors including kin recognition, social behaviors, homing, and alerting to the presence of predators or prey. Some chemical signals elicit stereotyped behaviors in conspecifics. The term *pheromone* was originally defined by Karlson and Luscher (Karlson and Luscher, 1959) as "substances which are secreted to the outside by an individual and received by a second individual of the same species in which they release a *specific* reaction" (our italics). This narrow definition of pheromones implies deterministic effects mediated by a chain of neurons that principally encode this one type of information, or "labeled lines." Although some compounds in insects fit within this narrow definition, more complicated chemical signal blends, whose odor quality may not be transmitted along a labeled line system, can elicit pheromonal responses (see Chapter 5, Johnston). The vomeronasal organ has been implicated in the reception of pheromone signals. However, stating that pheromone detection occurs exclusively through the vomeronasal system is an oversimplification since pheromone-like signals are also detected by the main olfactory system (see Chapter 5, Johnston).

3. TRANSDUCTION OF CHEMICAL SIGNALS

In order to produce an ultimate behavioral response, stimuli must produce changes in cellular activity. External chemicals often interact with specific membrane receptors or ion channels to effect a cellular response. Some stimuli, however, may cross the cell membrane to interact directly with second messenger systems or intracellular receptors (see Chapter 4, Bryant and Silver; Chapter 9, Christensen and White; and Chapter 13, Glendinning et al.).

The concept that molecules (atoms) of different sizes and shapes give rise to different qualities of sensation dates back to Democritus in the fifth century BCE (as described by Theophraetus, *De sensu* cited in Guthrie, 1965). In contemporary usage, we recognize molecules and receptors. Given the enormous diversity in chemical structure and size of chemical signals, the corresponding diversity of cellular receptor mechanisms is not surprising. Simple ionic stimuli may either directly gate or pass through particular ion channels. For example, a rodent's sensitivity to sodium is mediated by direct passage of these ions through amiloride-sensitive epithelial sodium channels. Similarly, in amphibians, detection of pH may be due to gating of an apical potassium channel by protons (see Chapter 13, Glendinning et al.). In contrast, detection of many other chemical stimuli in both single-celled and metazoan organisms involves binding of the stimulus molecule to a transmembrane receptor molecule coupled to an intracellular signaling cascade.

In terms of diversity of receptors, the olfactory organ of mammals may be unparalleled. Rodents may have upwards of 500 to 1000 distinct genes coding for separate olfactory receptor proteins (see Chapter 8, McClintock). These genes represent the largest functionally related family in the genome, comprising 0.8–1.6% of the approximately 60,000 mammalian genes (Mombaerts, 1999). In the mouse, olfactory receptor genes lie in at least 14 clusters spread over 9 chromosomes. Even humans, with our relatively poor olfactory capabilities, have several hundred olfactory coding sequences in our genome, although roughly a third of these are likely to be pseudogenes (genes that do not encode proteins because of deleterious deletion or insertion events resulting in nonsense mutations or frameshifts) (Sharon et al., 1998). The diversity in detection of potential odorants by the olfactory system has been likened to the immune system's capability of producing millions of immunoglobulins, each with a different antigen binding site (Thomas, 1974). However, the heterogeneity of receptor sites in the olfactory system, a result of differences in genomic sequence, is unlike the heterogencity in the immune system, which results from recombination of a smaller number of coding elements. Thus, heterogeneity of olfactory receptors occurred over an evolutionary time frame of millions of years, whereas diversity in the immune recognition elements is generated over a much faster time frame of days to weeks.

4. RECEPTOR SPECIFICITY

Chemoreceptor cells can be very specific or quite broad in their requirements for an appropriate stimulus. The classical example of a specific receptor cell is the pheromone receptive cells on the antennae of certain insects. In this case, particular antennal chemoreceptor neurons may respond to only one chemical component of a phermonal blend of related compounds (Kaissling, 1996). This specificity of response suggests high specificity in molecular recognition and expression. In contrast, other chemoreceptor cells, for example, amino-acid receptors in catfish (Caprio et al., 1993) appear to respond to a family of structurally related amino acids. What is not clear is whether the broader specificity of the catfish taste receptor cells is attributable to expression of multiple, specific receptor molecules in a single cell, or whether the receptor site of a single receptor molecule is less specific.

Traditionally, chemoreceptors have been divided into two classes: generalists and specialists. With the advent of molecular and single-cell methodologies, this distinction is not as clear-cut as was previously thought. Molecular studies indicate that each olfactory receptor cell expresses one, or at most very few, olfactory receptor molecules and that these molecules are narrowly tuned (Sengupta et al., 1996; Zhao et al., 1998). However, physiological studies of individual olfactory receptor neurons show responsiveness to a fairly broad spectrum of chemical stimuli (Firestein, et al. 1993). Likewise, taste receptor cells may respond to stimuli from several different taste qualities. However, this broad responsiveness of the receptor cell does not necessarily mean that taste receptor molecules are broadly "tuned." Each receptor cell may possess multiple receptor proteins or transduction mechanisms (see Chapter 13, Glendinning et al.). For the sense of smell the resolution of this question will have to take into account the statistical problem inherent in odorant/receptor (or receptor cell) studies, due to the current technical limitations in testing a significant fraction of all possible stimulus molecules. The use of a subset of stimulus molecules can erroneously give the impression that receptors are narrowly tuned (see Chapter 9, Christensen and White).

Specific genetically linked or engineered losses of chemoreceptive capability may be useful for determining receptor specificity in various organisms, for example, nematodes (Sengupta et al., 1996), mice (Mombaerts et al., 1996) (see Chapter 3, Sengupta and Carlson). Even humans (Amoore, 1977) exhibit specific anosmias and "taste blindness" apparently related to genetic variability.

5. ENCODING OF CHEMOSENSORY INFORMATION

In single-celled organisms, different chemical signals can activate different second messenger systems to produce different effector responses (see Chapter 2; van Houten). In more complex organisms with nervous systems, the different chemical stimuli may activate different receptor cells. Information from these cells then is transmitted to the central nervous system, which must interpret the signal in order to produce an appropriate behavioral response. The exact nature of the signal between the receptor cells and the central nervous system is not entirely understood. In some systems there appears to be a tight linkage between stimulation of a class of receptor neurons and the resulting behavior, for example, specific amphid receptor neurons of *Caenorhabditis elegans* (Sengupta et al., 1996), or the pheromone-encoding system of moths. In such cases, the information appears transmitted over a "labeled line." In contrast, in order to interpret information from other chemosensory systems, the central nervous system may compare the pattern of signals arriving "across fibers" that are connected to receptor cells of differing specificity. For example, in the main olfactory system of rodents, the nature of an odorant can be discerned only by examining the spatiotemporal combinatorial pattern of activity of the glomeruli in the olfactory bulb (see Chapter 9, Christensen and White). For the taste system, it is arguable whether some tastes are encoded on labeled lines or whether comparisons across fibers are necessary to discern the nature of a chemical stimulus (see Chapter 14, Smith and Davis).

6. SUMMARY

Chemical sensitivity is a nearly ubiquitous phenomenon of living systems. The ability to detect and respond to chemicals is present in single-celled as well as multicellular organisms. In more complex animals, the ability to detect and respond to various chemical stimuli involves particular chemosensory cells that connect to unique areas within the central nervous system, thereby permitting complex, differential responsiveness to different stimuli.

Vertebrates possess three main chemosensory modalities: olfaction, gustation, and the common chemical sense, recently named "chemesthesis." Although functional equivalents of these senses can be identified in invertebrates, similarities in these systems must be viewed as convergence rather than homology. Thus, functional similarities should be interpreted more in terms of information theory and constraints on encoding chemosensory information rather than as common evolutionary heritage.

REFERENCES

Amoore, J. (1977). Specific anosmia and the concept of primary odors. *Chem Senses Flavor* 2:267–281.

Caprio, J., J. G. Brand, J. H. Teeter, T. Valentincic, D. L. Kalinoski, J. Kohbara, T. Kumazawa, and S. Wegert (1993). The taste system of the channel catfish: from biophysics to behavior. *Trends Neurosci* 16:192–197.

Demski, L. S. and R. G. Northcutt (1983). The terminal nerve: a new chemosensory system in vertebrates? *Science* 220:435–437.

Firestein, S., C. Picco, and A. Menini (1993). The relation between stimulus and response in olfactory receptor cells of the tiger salamander. *J Physiol (Lond)* 468:1–10.

Guthrie, W. K. C. (1965). *A History of Greek Philosophy*. Cambridge: Cambridge University Press.

Kaissling, K. E. (1996). Peripheral mechanisms of pheromone reception in moths. *Chem Senses* 21:257–268.

Karlson, P. and M. Luscher (1959). Pheromones: a new term for a class of biologically active substances. *Nature* 183:55–56.

Mombaerts, P. (1999). Molecular biology of odorant receptors in vertebrates. *Annu Rev Neurosci* 22:487–509.

Mombaerts, P., F. Wang, C. Dulac, R. Vassar, S. K. Chao, A. Nemes, M. Mendelsohn, J. Edmondson, and R. Axel (1996). The molecular biology of olfactory perception. *Cold Spring Harb Symp Quant Biol* **61**:135–145.

Moulton, D. (1967). The interrelations of the chemical senses. In: M.R. Kare and O. Maller, eds. *The Chemical Senses and Nutrition*. Baltimore: Johns Hopkins Press, 249–261.

Rhoades, D. F. (1985). Pheromonal communication between plants. In: G.A. Cooper-Driver and T. Swain, eds. *Chemically Mediated Interactions Between Plants and Other Organisms*. New York: Plenum, 195–218.

Sengupta, P., J. H. Chou, and C. I. Bargmann (1996). odr-10 encodes a seven transmembrane domain olfactory receptor required for responses to the odorant diacetyl. *Cell* **84**:899–909.

Sharon, D., G. Glusman, Y. Pilpel, S. Horn-Saban, and D. Lancet (1998). Genome dynamics, evolution, and protein modeling in the olfactory receptor gene superfamily. *Ann N Y Acad Sci* **855**:182–193.

Thomas, L. (1974). *The Lives of a Cell: Notes of a Biology Watcher*. New York: The Viking Press. 153 pp.

White, J. and M. Meredith (1995). Nervus terminalis ganglion of the bonnethead shark (*Sphyrna tiburo*): evidence for cholinergic, and catecholaminergic influence on two cell types distinguished by peptide immunocytochemistry. *J Comp Neurol* **351**:385–403.

Zhao, H., L. Ivic, J. M. Otaki, K. Hashimoto, K. Mikoshiba, and S. Firestein (1998). Functional expression of a mammalian odorant receptor [see comments]. *Science* **279**:237–242.

2

Chemoreception in Microorganisms

JUDITH VAN HOUTEN
Department of Biology, University of Vermont, Burlington, VT

1. INTRODUCTION

Why begin a survey of the neurobiology of taste and smell with an examination of microorganisms? One could consider the evolution of chemical sensing by looking for hints of a primordial system in a comparison of living unicellular and multicellular animals. While this is interesting, it should prove even more useful to look for at microorganisms for general principles of chemical sensing across phyla. It is not uncommon for a discovery about how microorganisms sense to be considered novel and later for it to be found widespread among the metazoa. The interesting challenge is to tease apart the mechanisms that a single cell needs uniquely for its lifestyle from the mechanisms that they hold in common with some or all signaling cells.

Microorganisms use chemical sensing to locate food and mates and to escape predators and toxic conditions. Generally, the end response of the chemoresponse pathway in microorganisms is a change in crawling or swimming that leads the cells to accumulate in or escape from a stimulus. There are exceptions, such as yeast, that while not motile, can grow or change morphology to change orientation in response to chemical cues. In this chapter, however, we will stop short of the motor response end of chemoresponse, and we will discuss the upstream elements of signal transduction pathways of chemical sensing in bacteria, yeast, slime molds, and ciliates.

The Neurobiology of Taste and Smell, Second Edition, Edited by Thomas E. Finger, Wayne L. Silver, and Diego Restrepo
ISBN 0-471-25721-4   Copyright © 2000 Wiley-Liss, Inc.

11

2. THE BACTERIA

Perhaps the best understood chemosensory system of all is the chemotaxis of *Escherichia coli*. Other bacteria, *Bacillus subtilis* and *Salmonella typhimurium*, come in as close seconds and still others have been scrutinized in considerable detail. Nonetheless, we will consider primarily *E. coli* here, keeping in mind that most of its signal transduction components are in common with other bacteria and, as we will see, with eukaryotes. The studies of *E. coli* chemosensory transduction have moved to an exceedingly fine molecular and structural level, its detail surpassing that in studies of eukaryotic systems.

    E. coli cells have multiple receptors for attractant and repellent ligands and all of these receptors feed into a common signal transduction pathway that has become known as the two-component system, consisting of a sensor and a response regulator. The ultimate endpoint of this system is to affect the direction of rotation of the randomly distributed flagella and cell movement. A counter-clockwise rotation leads to smooth swimming of the cells, a clockwise rotation leads to a tumble and an abrupt turn in the cell's swimming path. This is a common aspect of attraction or repulsion of little organisms. That is, the cells accumulate or disperse by the modulation of turns in a swimming path that leads to attraction or repulsion through a biased random walk (Falke et al., 1997; see Van Houten, 1990 for discussion). Therefore, ultimately the binding of attractant ligand to its receptor suppresses tumbles and allows the cell to swim in relatively long, smooth runs, while the binding of a repellent increases tumbles

$\longrightarrow$

FIG. 2.1. (a) The chemosensory two-component pathway of *E. coli* and *S. typhimurium*. Arrows indicate action of one component on another. Attractants and repellents in the periplasm bind to specific transmembrane receptors or to soluble binding proteins that in turn bind to transmembrane receptors. The transmembrane receptors are coupled by a scaffolding protein (CheW) to a cytoplasmic histidine kinase (CheA), which in turn regulates two response regulators (CheB and CheY). Phosphorylation of CheB modulates the adaptation system in which CheR methylates specific regulatory glutamate side chains on the cyotplasmic surface of the receptor, and phospho-CheB hydrolyzes these modifications. The steady-state level of receptor methylation provided by opposing CheR and CheB reactions enables the pathway to adapt to background stimuli and also provides a simple chemical memory. Phosphorylation of CheY modulates the rotary flagellar motor as phospho-CheY docks to the motor switch apparatus, thereby controlling the direction of motor rotation and the swimming behavior of the cell. Although CheY can catalyze its own dephosphorylation, the rate of phospho-signal inactivation is enhanced by a phosphatase activity (CheZ).

    (b) Domain organization of chemosensory pathway components. (TM, transmembrane; MG, methylation). The receptor is the aspartate receptor. Che A is the transmitter histidine kinase, composed of four functionally distinct domains involved in phoshotransfer (P1), response regulatory docking (P2), dimerization, and histidine autophosphorylation and receptor coupling. CheY and CheB share homologous aspartate kinase receiver domains; CheB also possesses a separate methylesterase domain. Residues shown in bold indicate phosphorylation sites on CheA, CheY, and CheB CheZ is a phosphatase that dephosphorylates CheY. (From Falke et al., 1997, with permission.)

and makes the cell turn frequently to interrupt swimming runs. The changes in swimming are transient since the receptors and behavior adapt, resetting the level of turning and making the cell ready to respond to the next change it detects. This adaptation also serves as a form of memory for the temporal sensing that the bacteria carry out. In other words, the cells detect change and when the amount of ligand is not changing, the cells adapt, even though ligand may still be present.

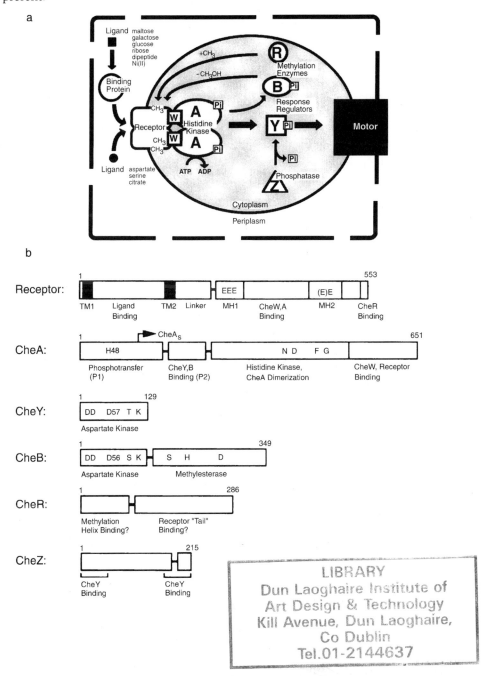

2.1. Transmembrane Chemoreceptors

Figure 2.1A shows the general scheme of the *E. coli* pathway. There are five transmembrane receptors with ligand binding domains in the periplasmic space between the cell membrane and wall. Three of these receptors specifically bind only aspartate, serine, or citrate, but all five also bind protein-ligand complexes from the periplasmic space. These complexes present to their respective transmembrane receptors sugars or dipeptides or nickel. For example, the transmembrane receptor that binds aspartate directly (Tar) also binds the maltose periplasmic space binding protein. The soluble binding proteins are often referred to as soluble receptors, and they are somewhat analogous to the olfactory binding proteins in mucous, but the bacterial proteins are less promiscuous. Thus, the bacterial transmembrane receptors have the capacity to respond to two ligands even simultaneously (see Falke et al. 1997 and Taylor and Zhulin, 1998 for reviews; also Parkinson and Kofoid, 1992; Hazelbauer et al., 1993; Stock and Mowbray, 1995).

The bacterial transmembrane receptors all have common structural features: an extracellular binding domain for one or more ligand, a transmembrane region, a cytoplasmic linker, and methylation and signaling regions. The receptors form homodimers and tend to cluster together on the cell. When ligand binds to one or both of the specific sites, the protein tilts and the signaling helix makes a piston-like movement helix toward the cytoplasm. The detail about these conformational changes has been gathered through multiple and elegant biochemical, mutational, and crystallographic experiments (Falke et al., 1997). The piston movement appears to be key to the propagation of the signal to the next component in the pathway, Che A (the transmembrane kinase), a histidine kinase and the first part of the two-component system. You will note in Figure 2.2 that the receptor (sensory domain) and Che A (methylation region) are part of a large complex. Che A and the components Che R, W, and Y are part of the complex for all of the five transmembrane receptors, thus all signals from the receptors activate the same signal transduction pathway.

Che A histidine kinase autophosphorylation and kinase activation are inhibited upon attractant ligand binding to receptor and activated upon repellent binding to receptor. When active, Che A phosphorylates Che Y and the phosphorylated form of Che Y binds to a component of the flagellar motor, Fli M, (Fig 2.1), causing clockwise rotation of the flagella and tumbling. Hence, repellents or the removal of attractants cause tumbling and repulsion by activating Che A; attractants or removal of repellents cause smooth swimming by down-regulating Che A. Che Z is a constitutive phosphatase that dephosphorylates Che Y, but it does not affect Che A. (See Fig. 2.1B for a summary of these proteins.)

2.2. Receptor Adaptation

Bacteria respond, as do most microorganisms, to changes in stimulus and can detect small changes in stimulus level even in a high-stimulus background. This is accomplished by adaptation of the receptor through covalent modification. A

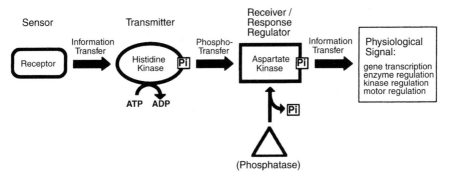

FIG. 2.2. Information flow through a two-component signaling pathway. Shown are the standard prokaryotic and eukaryotic signaling components including (a) the sensor module, typically a transmembrane receptor with two putative membrane-spanning helices; (b) a transmitter histidine kinase that is regulated by the receptor and catalyzes autophosphorylation on histidine; and (c) a receiver or response regulator whose active site catalyzes phosphotransfer from the transmitter, thereby yielding autophosphorylation on aspartate. The response regulator can catalyze its own dephosphorylation, but some pathways require a separate phosphatase to generate more rapid dephosphorylation, or to provide additional pathway regulation. Different pathways display highly specializes assemblages of the modular elements; for example, the sensor, transmitter, and response regulator modules can be separate proteins or can be fused together in various combinations. (From Falke et al., 1997, with permission.)

methyltransferase (Che R) and an opposing methylesterase (Che B) control adaptation of the receptor through modulating the receptor methylation domain (Fig. 2.1). Che R is part of the receptor complex (Fig. 2.3) and is well positioned to methylate its resident receptor. Levels of methylation depend primarily on the opposition by Che B. The effects of adaptation are perhaps easiest to see with an example of an attractant binding to its receptor, after which Che A kinase is down-regulated and phosphorylation levels of Che Y and also Che B decrease (Fig. 2.1). The unphosphorylated form of Che Y does not bind to FliM of the motor and the cell swims smoothly. The unphosphorylated and inactive Che B is not able to oppose Che R methyltransferase and methyl groups are added to the receptor. The increasing methylation in turn activates Che A, which activates Che B and Che Y through phosphorylation. The behavior of the cell returns to a basal level of turning and the methylation level also balances the activity of Che A.

The methylation apparatus is not specific for a receptor, and, while Che R is part of each receptor complex, Che B is free to act on any and all of the receptors clustered together, usually at one pole or the other of the cell, in what is termed the "nose" (Parkinson, 1993). (Such clustering probably acts as a mechanism to increase sensitivity (Bray et al., 1998).) Therefore, methylation occurs globally on all the receptors and reflects the net binding of attractant and repellent ligands on all the receptors at the time. The methylation level is also responsible for resetting the cell's Che A activity, bringing behavior back to a basal level,

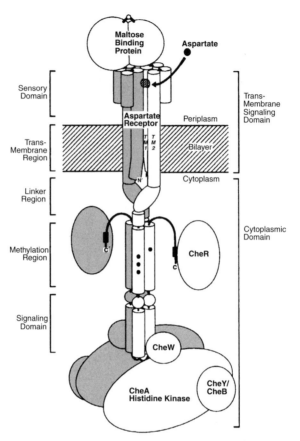

FIG. 2.3. A typical receptor-kinase signaling complex illustrated by the aspartate receptor. The transmembrane receptor provides the architectural framework of the supermolecular signaling complex. Most of the chemosensory pathway components are associated with this complex, either stably or transiently. The kinetically stable core ternary complex is composed of the dimeric receptor (illustrated as a collection of helices), the coupling protein CheW, and the dimeric histidine kinase CheA. Other components are believed to be in rapid equilibrium between bound and soluble forms, including periplasmic binding proteins, the methyltransferase CheR, the methylesterase CheB, and the motor response regulator CheY. (From Falke et al., 1997, with permission.)

despite the persistence of stimulus, and making the cell ready for the next change in stimulus concentration.

2.3. The Two-Component System

Perego and Hoch (1996) consider the two-component systems to be sensory recognition and signal transduction systems that allow the bacteria to cope with a large number of environmental signals and cope quickly. The two components are (1) a transmitter, which is a sensor histidine kinase that is activated by autophosphorylation of a histidine and (2) a receiver, which is a response regulator that specifically recognizes the sensor kinase and receives the phosphate that is transferred from the histidine kinase to an aspartate residue on the response regulator (Fig. 2.2). (The histidine kinase and aspartate kinase can be separate proteins or separate domains of the same protein.) Often the sensor kinase also has phosphatase activity and can dephosphorylate the response regulator as well as phosphorylate it. Thus, the activation of the response regulator depends upon the relative levels of the histidine kinase versus phosphatase activities in the first component. In the *E. coli* system, Che A is the histidine kinase, but it has no phosphatase activity. Che A integrates information about the extent of receptor activation by stimulus with the state of receptor methylation to modulate the level of histidine kinase activity accordingly. One of the response regulators is Che Y, which binds to Che A until the transfer of the phosphate from histidine to aspartate reduces its affinity for Che A. The constitutive Che Z ultimately removes the acyl-phosphate from Che Y. Che Y needs no transmitter output histidine kinase domain, just a receiver aspartate kinase domain for the transfer of phosphate from Che A and also the ability to bind to FliM in its phosphorylated form. However, another response regulator, Che B, has both a receiver domain and a response domain since its methylesterase becomes active upon receipt of the acyl-phosphate. The Che B aspartate kinase receiver domain catalyzes the phosphotransfer from the histidine of Che A to an aspartate of Che B. Che Z does not consider Che B as a substrate for dephosphorylation, but Che B has an intrinsic dephosphorylase activity, which makes the acyl-phosphate very short-lived.

2.4. Other Receptors

In addition to the five transmembrane receptors for attractants and repellents, there are still other receptors like the AER receptor that detects redox energy levels, and feeds into the same two component signaling system with Che A and Che Y (Taylor and Zhulin, 1998). Thus, the bacteria make extensive use of two-component systems for responding to their environment and physiological conditions, sometimes with modification of the number of transmitters or receivers (Parkinson and Kofoid, 1992; Perego and Hoch, 1996; Chang and Stewart, 1998). An especially interesting aspect of bacterial sensing is the common use of Che A, R, B, and Y by all of these receptors. In this scheme, Che A integrates information from all the cellular inputs and the cells respond to the net effects of attractants, repellents, and physiological status. Other important features of the bacterial chemosensory systems that remain to be understood are that cells can sense and adapt to ligand concentrations that range over five orders of magnitude, and that the cells can also respond to changes in receptor occupancy of only 0.1% per second even in high background levels of stimulus (see Spiro et al., 1997, for description of the challenges). Thus work remains to be done to fully understand the implications of the two-component system combined with the methylation system of adaptation.

The bacteria are not alone in employing this rather efficient and rapid means of signaling and integrating signals through transmitters and receivers. Yeast (both *Saccharomyces cerevisiae* and *Schizosaccharomyces pombe*), and plants make use of two-component systems in a variety of signaling tasks (Morgan et al., 1995; Chang and Stewart, 1998). Interestingly, in yeast the two-component system for osmotic regulation feeds into one of the MAP kinase pathways, which are known to be fairly universal in signaling in eukaryotic cells. While slime molds (see sect. 4) use G protein coupled receptors for some sensory signaling, they also use histidine kinase two-component systems to control cAMP degradation during development and ammonium and osmo-sensing (Thomason et al., 1999). Therefore, it is most likely that the two-component systems of bacteria with their interesting phosphorelay and tight layers of control and adaptation will be found to be widespread in the signaling world of eukaryotes.

With the discovery of eukaryotic-like protein kinases and SH3 domains in bacteria, it is possible that the two-component systems and the "eukaryotic" phosphorylation signal mechanisms not only merge within one cell but interact with each other in both prokaryotes and eukaryotes. This is especially possible when one considers that a CheA protein of *Streptomyces* has SH3 domains and likely shows mixed functions (Bakal and Davies, 2000).

2.5. Quorum Sensing

Before leaving the interesting world of bacteria, it is essential to mention yet one more application of two-component signaling pathways: quorum sensing, that is, monitoring their own population densities and, indeed, counting their own populations. Many bacterial responses, such as symbiotic bioluminescence, antibiotic production, fruiting body development, virulence in pathogenesis, and gene transcription, among others (Kleerebezem et al., 1997), occur only when the population has risen to sufficient levels. Cells communicate and coordinate their activities by sending out into the extracellular medium signal molecules, like pheromones, which the cells can respond to when signal reaches sufficient levels. For the gram positive bacteria *B. subtilis* and *Staphlococcus aureus*, the signals are peptides. The peptide pheromones bind to transmembrane sensors, which have histidine kinase activity and which pass on their information about the presence of the peptides to response regulators through a phosphorelay. Ultimately, transcription factors are affected and the gene products needed for development or virulent infection are produced.

The extracellular signals for gram negative bacteria primarily are N-acyl-L-homoserine lactones (AHL) (Swift et al., 1996), AHLs activate two-component systems in *Vibrio harveyi*, *Agrobacterium tumefaciens*, *Erwinia* sp., *Photobacterium fischeri*, *Pseudomonas aeruginosa*, *Serratia liquefaciens*, and *Rhizobium legumino-sarum* for bioluminescence, antibiotic synthesis, conjugation, enzyme synthesis, nodulation, and swarming. However, it has been difficult to detect extracellular signals used in quorum sensing from gram negative enteric bacteria like *E. coli* and *S. typhimurium*. Certainly no AHL signal is evident. Recently, however, there have been suggestions that *E. coli* and *S. typhimurium* can at least secrete a signal (yet uncharacterized) that can be used by *V. harveyi* to activate its quorum-sensing system. These recent studies suggest that most all bacteria use some form of quorum sensing to respond in a step-wise fashion to extracellular environmental information (Fuqua and Greenberg, 1998; Surette and Bassler, 1998).

3. YEAST

Many elegant genetic and biochemical studies have provided details about the signal transduction pathways involved in yeast mating and in the responses of yeast cells to their environment. The study of yeast can provide new insights into how a single cell keeps its signal transduction pathways separate or comingles them as appropriate.

Haploid yeast cells pay close attention to their environment and respond to the presence of mates, low nitrogen availability, high or low osmotic pressure, carbon and nitrogen deprivation, and glucose levels using specific signal transduction pathways that have very different outcomes (Thievelein, 1994; Madhani and Fink, 1998). At least four of these signaling systems involve MAP kinase cascades; the fifth involves another familiar signal transduction pathway including protein kinase A. We will focus first on two MAP kinase pathways and, after briefly reviewing the other pathways, we will discuss how such similar pathways can be kept separate and also interact without chaos for the cell.

3.1. G Protein Cascades and MAP Kinases

In the presence of cells of complementary mating type and the pheromone that they secrete, the cells will become mating reactive and fuse with cells to become diploid. In the absence of the mating pheromones, the cells will form filaments and become invasive. Figure 2.4 shows the mating pathway in which pheromone binds to a seven transmembrane receptor and activates the complex of trimeric G proteins to exchange GDP for GTP and to dissociate from the receptor. Pheromones are peptides called **a** or α and they are secreted from **a** or α cells, respectively. Yeast cells express one of the two complementary receptors designed to respond to pheromones of the complementary mating types: receptors for **a** pheromone are on cells that secrete α pheromone, for example. Once released from the receptor, the $\beta\gamma$ subunits of the trimeric G proteins carry on the signal transduction, with the G_α coming into play in adaptation and not in direct signal transmission (see Bardwell et al., 1994; Ferguson et al., 1994 for reviews). The demonstration of the activity of the $\beta\gamma$ subunits in yeast mating was the first example of an active role for these subunits in signal transduction; now there are numerous examples of involvement of the $\beta\gamma$ subunits in effector enzyme and channel function (Ford et al., 1998). (See section 4 on *Dictyostelium*.)

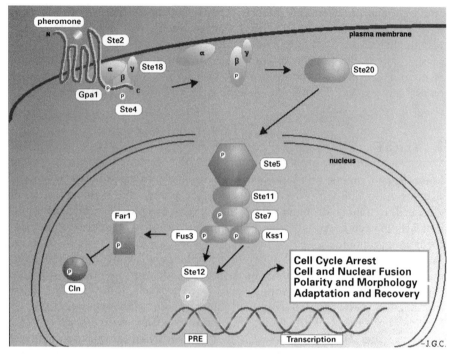

FIG. 2.4. Signal transmission during the pheromone response of *S. cerivisiae*. The initial flow of information is illustrated schematically. The precise order of function and cellular location of all of the components shown, as well as the specific stage at which the signal moves to the nucleus, are not known. Components that are known to be phosphorylated in response to pheromone are indicated with a "P." (From Bardwell, et al., 1994, with permission.)

The target of the βγ subunits is thought to be the Ste20 protein (Leeuw et al., 1998), which in turn activates the MAP kinase cascade, possibly through its own serine/threonine kinase activity (Fig. 2.4). The enzymes of the cascade are held together on a scaffold of Ste5 protein (Ste5p to distinguish it from the gene), and the cascade begins with the member of the MAP kinase kinase kinase family, Ste11p, and moves on to the MAP kinase kinase and MAP kinase counterparts Ste7p and Fus3p. Fus3p, once active, will phosphorylate targets like Far1p, an inhibitor of cell cycling, and Ste12p, a transcription factor that activates transcription of the genes needed for the morphological and other changes involved in mating. Figure 2.4 also includes MAP kinase Kss1p (see below).

Figure 2.5 shows five of six MAP kinases coded for in the yeast genome and that each MAP kinase pathway leads to a unique response: filamentous invasive growth, coping with osmotic pressure or hypotonic shock (two different pathways), and nutrient starvation. This is particularly curious when you consider Figure 2.6 A. It appears that downstream from the receptor level, much of the MAP kinase cascade is shared between the mating pheromone and filamentation/invasion pathways, except for the MAP kinases Fus3p and Kss1p, which are specific to their pathways. Even the target of the MAP kinases, transcription factor Ste12p, is common to both pathways. How do yeast keep the mating pheromone from inducing filamentous invasive growth? Why are the signals for invasive growth not confused with mating signals?

3.2. Transcriptional Control

The first puzzle part to fall into place to explain pathway specificity is the combinatorial control of transcription. Genes transcribed for mating are controlled by the response element called PRE (pheromone response element), which binds dimers of Ste12p or Ste12p/Mcm 1p. The genes needed for filamentation and invasion are under the control of FREs (filamentation response elements), which consist of a PRE along with a TCS (Tec1p binding site) and which bind Ste12p in combination with Tec1p transcription factor. The problem now reduces to how the cell knows to activate Tec1p or Ste12p alone or possibly Ste12p with Mcm1p.

The solution to avoiding crosstalk must include the MAP kinases Fus3p and Kss1p that are specific to the two pathways. Some mutants in the gene for Fus3p are sterile, and mutants in the gene for Kss1p are not able to grow invasively as haploids. However, when the phosphorylated Fus3p gene product is totally absent as in a null mutant, the Kss1p MAP kinase can inappropriately take its place and some weak mating reactivity as well as filamentation result from pheromone stimulation (Fig. 2.6B). Normally this cross talk is avoided, probably by some inhibitory activity of Fus3p that keeps Kss1p from posing as an "imposter" in the mating MAP kinase pathway. Likewise, Kss1p has an inhibitory

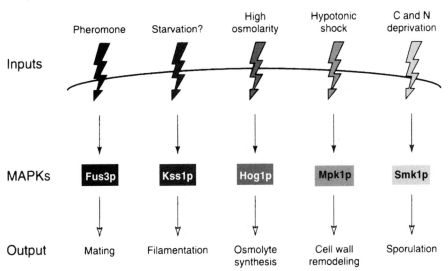

FIG. 2.5. MAPK signaling pathways of *S. cerivisiae*. Physiological signals (inputs) and developmental/cellular responses (outputs) are indicated. Sporulation is induced by carbon and nitrogen starvation ("C and N deprivation"). The question mark after "starvation" indicates that there is uncertainty about the stimulus for the filamentation pathway. (From Madhani and Fink, 1998, with permission.)

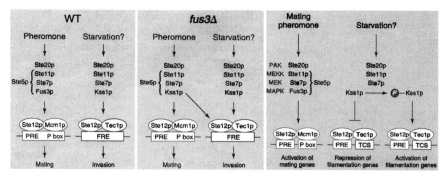

FIG. 2.6. Model for the function of Kss1p in wild-type and in *fus3* mutant haploid cells. (a) Wild-type (WT): Fus3p functions as the MAPK for the mating pathway; Kss1p functions as the MAPK for the invasion pathway. For simplicity, the inhibitory activity of Kss1p is not depicted. (b) *fus3Δ* : Kss1p impersonates Fus3p. As a consequence, mating can occur but the pheromone pathway inappropriately activates haploid invasion and FREs via Kss1p. (c) Mating and filamentation signaling pathways share components but use different MAPKs. The mating pathway is depicted on the left and the filamentation pathway on the right. Ste5p binds Ste7p and Fus3p, and is required for pheromone-induced signaling. Note that the combination of a PRE and TCS is referred to as a FRE. (From Madhani and Fink, 1998, with permission.) ·

effect on its own pathway if signals for filamentation and invasion are not present (i.e., Kss1p is not phosphorylated). As shown in Figure 2.6C, unphosphorylated Kss1p inhibits the transcription activity of Ste2p and Tec1p on the FRE, but this inhibition is relieved when the filamentation/invasion pathway is stimulated and Kss1p is phosphorylated (Cook et al., 1997).

Another mechanism for preventing inappropriate cross talk among pathways is the formation of complexes of the proteins of the MAP kinase cascade and subsequent sequestration of the complex with its activated kinases and their target MAP kinase. Ste5p is thought to assemble the mating pathway kinases Ste11p, Ste7p, and Fus3p (Printen and Sprague, 1994) and somehow orchestrate shuttling of activated kinases from the region of the receptor to the nucleus where target transcription factors are contacted. The sequestration in large complexes could prevent exchange of activated enzymes that are in common with other pathways. The complex could also explain the inhibitory effects of Fus3p that keep Kss1p from being activated by the pheromone system. Perhaps the complex as a whole is selective for the correct MAP kinase such as Fus3p and the complex will bind to Fus3p even if Fus3p is mutated and inactive, keeping Kss1p from being activated by pheromone. It would be predicted that Kss1p would indeed cross over and be activated by pheromone in a Fus3p null mutant, which has no Fus3p to hold its place in the complex occupied and to keep Kss1p from reaching the complex.

3.3. Downstream Events

Yeast manage to keep signaling pathways separate (Madhani and Fink, 1998), even though some pathways may actually share a few targets or effects, even though the grand response of the pathway is separate (e.g., mating and cell cycle arrest rather than filamentous growth). For example, the polarized deposition of cell wall in response to pheromone and the early events of filament formation resemble each other and Madhani and Fink (1998) speculate that the reason for sharing most of the same MAP kinase cascade and the same transcription factors and risking confusion is that two pathways do, indeed, require some common gene expression. Thus, limited amounts of cross talk to activate some common transcription factors could be efficacious.

An interesting connection between the bacterial and yeast systems is in the yeast pathway that responds to hyperosmotic shock (Morgan et al., 1995; Posas et al., 1996). While this pathway is shown in Figure 2.5 as a MAP kinase cascade pathway, it begins with a two-component system with low osmolarity activating a transmembrane protein that has histidine kinase activity. This transmitter in turn transfers a phosphate to the response regulator, which stimulates the HOG1 kinase cascade. This cascade shares Ste11p with the pheromone and filamentation/invasion pathways, and, like these pathways, it is not activated by inappropriate signals.

4. DICTYOSTELIUM

When the amoebae of the slime mold *Dictyostelium discoideum* are grazing on bacteria and food is plentiful, the cells grow and divide and generally pay attention to gradients of folic acid from the bacteria. Following folic acid should lead them to food. However, when the bacteria are depleted, the cells undergo a remarkable developmental program, which includes changing of receptors for extracellular chemical cues. The folate binding sites are down regulated and replaced with the a series of four cyclic AMP receptors, beginning with cAR1 and a little later cAR3. Focal cells begin to emit cyclic AMP in pulses about 5 min apart and surrounding cells begin an oriented movement toward the source of this signal. The receptors adapt and the stimulated cells themselves secrete cyclic AMP in pulses, thereby propagating the signal wave to outlying cells. When the extracellular cyclic AMP is degraded, the receptors again become sensitive and the cells once more respond to a wave of cyclic AMP as it passes by. Cells stream in toward the focal cell and eventually form a multicellular slug. However, the need to respond to cyclic AMP is not over; the developing cells will continue to use cyclic AMP at different concentrations and through different receptors (cAR2 and cAR4) to stimulate the differentiation of cells into stalk and spores of the fruiting body. The formation of spores should allow the slime mold to weather the bad nutritional times, until bacteria are available and once again amoebae can safely graze. (See Gross, 1994; Ginsburg et al., 1995; Chen et al., 1996; Van Haastert and Kuwayama, 1997 for reviews.)

4.1. cAMP Receptor Cascade

Perhaps the easiest way to review the pathways of *Dictyostelium* chemotaxis is to follow the events in Figure 2.7, which are displayed in chronological order. (Note that in Figure 2.7, a cell can be stimulated through uniform application of cAMP

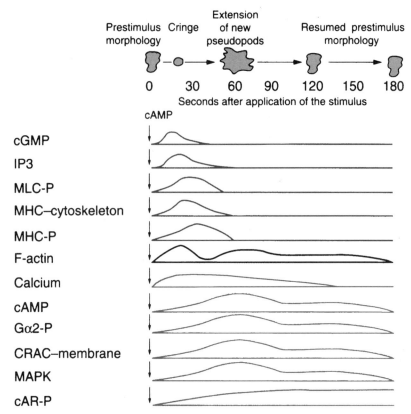

FIG. 2.7. Responses to the application of cAMP. The morphological changes that cells undergo at given times after application of the stimulus are shown at the top. The profiles below the different morphological stages show the approximate time courses of different response. Responses with similar kinetics are indicated by the same color. Abbreviations: cGMP, accumulation of cGMP; IP3, accumulation of IP$_3$; MLC-P, phosphorylation of myosin light chains; MHC-cytoskeleton, association of myosin heavy chain with the cytoskeleton; MHC-P, phosphorylation of myosin heavy chain; F-actin, the fraction of actin found as F-actin; calcium, intracellular calcium levels; cAMP, accumulation of cAMP or activation of adenylyl cyclase; Gα2-P, phosphorylation of Gα2 subunit; CRAC-membrane association of CRAC with the membrane fraction; MAPK, activation of MAP kinase; cAR-P, phosphorylation of cAR1. (From Mei-Yu et al., 1996, with permission.)

as opposed to a gradient. The cells quickly change their cytoskeleton and "cringe" before putting out pseudopodia as though they were going to migrate. Cells in a gradient of cAMP extend pseudopodia asymmetrically and crawl up the gradient. While the gradient is necessary for asymmetric response of the cell and oriented motility, the cringing demonstrates that signal transduction can occur as in uniform application of a stimulus. The cells adapt and recover their original motility even in the continued presence of stimulus.) Extracellular cAMP binds to the seven transmembrane receptor cAR1 and activates the trimeric G protein complex that includes Gα2. Deletion of either Gβ or Gα2 eliminates the rapid increase in intracellular cGMP, the slower increases in IP$_3$, the changes in cytoskeletal elements such as phosphorylation of myosin I and II and translocation of myosin II to the cytoskeleton, polymerization of actin needed for oriented motility, and the slower rise in intracellular cAMP. The Gβγ subunits activate the adenylyl cyclase (reviewed by Chen et al., 1996), which produces the cyclic AMP that is secreted as part of the signal relay system. The Gβγ subunits, which are membane tethered, appear to recruit a cytoplasmic activator of the adenylyl cyclase (CRAC) to the membrane where it activates the enzyme. Alternatively the activation of PI3K by the Gβγ subunits creates the lipid binding sites for proteins with plekstrin homology (PH) domains such as CRAC, PKB (required for chemotaxis, Meili et al., 1999), and VAV (a Rho exchange factor), which rapidly concentrate at the membrane where the stimulus is applied (Parent and Devreotes, 1999; Jin et al., 2000).

The rise in cGMP is essential in the chemotaxis response, especially for the oriented motility through the cell's actomyosin complex in particular (reviewed in Kuwayama and Van Haastert, 1996; Van Haastert and Kuwayama, 1997). The mechanism of regulation of the particulate guanylyl cyclase is being elucidated in part through two chemotaxis mutants with opposite effects on the cyclase. The mutants have different defects in the same cytoplasmic protein that appears to have the capacity to both activate and inhibit the cyclase, which could explain the different phenotypes of mutants with alterations in the same gene (Kuwayama and Van Haastert, 1996).

The activation of PLC in response to extracellular cAMP requires Gα2 and a receptor other than cAR1 (perhaps cAR3), however, disruption of PLC and the concomitant loss of the IP$_3$ increase upon stimulation has no effect on the developmental pathway (Van Haastert and Dijken, 1997). Nonetheless, there remains the possibility for a role in chemotaxis for a PKC that is activated by PLC activity and diacyl glycerol (reviewed in Phillips et al., 1997).

Green fluorescent protein (GFP) labeled receptors, Gβ, CRAC and others have shown us that while the receptors remain evenly distributed at the membrane and G proteins are distributed in a shallow gradient from leading edge to rear of the cell, PH domain proteins are found only at the site of stimulation (Jin et al., 2000). Thus, *Dictyostelium* can enormously sharpen and localize a signal transduction cascade, which can account for how the cell detects a < 2% difference in stimulus from front to back, even in the presence of a high background.

4.2. G Protein-Independent Signaling

The use of mutants in studying *Dictyostelium* has made it possible to uncover some interesting receptor-dependent but also G protein independent signal transduction pathways in chemotaxis. For example, as shown in Figure 2.7 there is an influx of extracellular calcium that begins at 5 sec (Nebl and Fisher, 1997).

Mutants with no functional Gα2, which normally couples with cAR1, the receptor for chemotaxis, still show this calcium influx. Likewise, an activation of a MAP kinase pathway persists in these mutants and also in cells lacking Gβ. Therefore, it appears that the receptors that cannot signal to the rest of the chemotaxis pathway for lack of G proteins can still activate calcium influx and MAP kinase, perhaps by direct coupling of receptor to effectors (Van Haastert and Van Dijken, 1997; reviewed in Chen et al., 1996).

The development of restriction-enzyme-mediated integration mutagenesis (REMI) made it possible not only to generate mutants, but also to clone the affected gene rather rapidly (Kuspa and Loomis, 1992). Chemotaxis mutants with impaired adenylyl cyclase activation have been mapped to the genes for ERK2, a member of the MAP kinase family, and for a ras exchange factor similar to Cdc25p in yeast, among others (Segall et al., 1995; Insall et al., 1996; Knetsch et al., 1996; reviewed in Chen et al., 1996). Characterization of these new mutants suggests that the control of adenylyl cyclase must be by at least two pathways, one of which is G protein-independent, involving members of the MAP kinase family. A MAP kinase kinase (DdMEK1) is required for chemotaxis in *Dictyostelium*, but its target is the guanylyl cyclase (Ma et al., 1997). The MAP kinase kinase, does not regulate the MAP kinase ERK2. Therefore, at least two MAP kinase pathways are involved in chemotaxis and aggregation; one is essential for chemoattractant activation of adenylyl cyclase, and the other for chemoattractant activation of the guanylyl cyclase. Both are mediated by the cAMP receptor, but at least one pathway can be independent of G proteins.

4.3. Quorum Sensing in Slime Molds

There are some parallels between the bacterial and slime molds systems that should be noted. *Dictyostelium* can count their cell numbers as can baceria. The slime mold amoebae monitor their density by detecting a conditioned medium factor (CMF) that is secreted into the surrounding medium and is detected by a receptor-mediated mechanism (Brazill et al., 1997, 1998). The apparent function of CMF is to signal the density of starving cells, that is, amoebae that are ready for chemotaxis, aggregation and development. Without CMF, the aggregation process does not begin avoiding fruiting bodies made of small aggregates of cells, and the cAMP dependent guanylyl cyclase, adenylyl cyclase, and calcium fluxes are all inhibited.

The mechanism by which CMF works is intriguing. First, CMF binds to its receptor, which couples with Gα1 in a trimeric G protein complex. The active receptor liberates Gβγ subunits, which activate PLC. PLC, in turn, inhibits the GTPase activity of Gα2, the Gα that couples with the cAR receptors. Thus, Gα2 remains uncoupled (i.e., GTP bound) from the Gβγ subunits longer, and the free Gβγ subunits activate the guanylyl cyclase and the adenylyl cyclase. Thus, CMF acts through a receptor-mediated system to facilitate the activity of cAR1 and cAMP signaling. This interplay between receptors generally is beyond the current state of knowledge of olfactory or gustatory receptors. However, interplay among receptors through G protein modulation and also the interplay among G protein coupled receptors and the MAP kinase pathways should be considered (see Daaka et al., 1997).

Perhaps the most important lesson for olfaction and taste from the *Dictyostelium* system is the very subtle amino acid differences among the cAR receptors that create an array of receptors that all respond to the same ligand but with different kinetic properties. Apparently a variety of these different properties is needed over developmental time. cAR1 and cAR3 are expressed early in development and cAR1 is expressed first and is the highest affinity receptor. Cells lacking cAR1 will eventually aggregate at low efficiency probably due to the transduction by way of cAR3; deletion of both receptors arrests aggregation and development (Chen et al., 1996). Cells engineered to be missing both cAR1 and cAR3 serve as a clean slate for development. The constitutive expression of cAR 1, 2, or 3 and chimera receptors in a $cAR1^-$ $cAR3^-$ genetic background restores the developmental program in response to cAMP, although concessions have to be made to account for the different affinities of the receptor expressed. Therefore, all of the receptors are capable of signaling in the chemotaxis pathway if forced to be present during early stages of development. They all signal through $G\alpha2$, but normally they appear at different times in development and probably are tailored to respond to the level of stimulus appropriate at that time. (There is some evidence for a role for $G\alpha3$ in regulation of signaling, but the receptor is not defined and this would be in addition to $G\alpha2$ function (Brandon and Podgorski, 1997).) Also of interest for olfaction and taste is that these 7TM receptors do not have to be phosphorylated to desensitize (Kim et al., 1997).

5. CILIATES

With the ciliates, our study of chemosensation enters the realm of "swimming neurons," that is, eukaryotic microbes that are excitable cells. Studying membrane biophysical phenomena in these organisms, especially those underlying mechanical and chemical stimulation, has certain advantages. Ciliates can be penetrated by electrodes for electrophysiological recording or voltage clamping, grown in quantity for biochemistry, mutated, and analyzed by molecular genetic techniques. Most advantageous of all is their behavior, which can be analyzed to provide a window into the physiology of the cell.

Ciliated protozoa use chemical sensing to respond to the presence of potential mates, find food, and avoid predators and toxic conditions (reviewed in Van Houten et al., 1981; Van Houten, 1994). They even respond to some unlikely stimuli like opiates, PDGF, β-endorphin, and lysozyme (Chiesa et al., 1993; Marino and Wood, 1993; Koppelhus et al., 1994; Hennessey et al., 1995; Renaud et al., 1995). Here we will review two very different ciliates, *Euplotes* and *Paramecium*, that make very different uses of chemical stimuli.

5.1. *Euplotes*

Euplotes are very large cells (>100 μm) that creep along the substratum on fused cilia called cirri. When two cells of different mating type cross paths, they influence each other with secreted pheromones to become mating reactive. Once mating reactive, the cells will form pairs, even among cells of the same mating type, and exchange meiotic nuclei. Each cell constitutively produces a pheromone that is coded by the *mat* locus, and since *Euplotes* are diploid, each cell can express up to two different pheromones. The products of the *mat* loci are codominant, creating complicated mating reactions among the heterozygotes. (See Van Houte, 1994; Luporini et al., 1995 for reviews.)

Euplotes raikovi is a marine ciliate with as many as 12 *mat* locus alleles. The pheromone gene products are homodimers of 38 to 40 amino acid peptides, which are processed from larger prepropeptides. (See Brown et al., 1993; Luginbuhl et al., 1994; Mronga et al., 1994; Ottiger et al., 1994; for structures.) The pheromones bind tightly (Kd = 10^{-9} M) and specifically to the cells. At this point, the reader probably is expecting a discussion of complementary receptors for pheromones on the cell to be stimulated, not unlike the yeast mating system in which **a** cells secrete **a** pheromone but express α receptors. In the yeast system, a cell cannot confuse itself with binding of its own pheromone; only complementary cells can respond. However, in *Euplotes raikovi* the pheromone and receptor are very closely related, and a 55 amino acid longer form of the pheromone serves as the receptor (Luporini et al., 1995). The extra sequence serves as the membrane anchor. The secreted pheromone binds to the membrane-bound version on the secreting cell's surface, possibly acting in an autocrine fashion to stimulate vegetative growth. With the appearance of an *E. raikovi* cell of another mating type there will be other pheromones in the medium from the newcomer, and these pheromones will bind to the first cell's receptors, eliciting not the autocrine effects but the mating response. Thus, there will be two outcomes of pheromone binding to receptor: autocrine growth response when binding is to its own receptor and mating reactivity when binding with the same affinity is to the receptor of a different mating type.

Euplotes octocarinatus goes about mating in a different way. This fresh water ciliate has multiple (10) mating types, which are determined by combinations of only four codominant alleles. The larger peptide pheromones of *E. octocarinatus* are thought to bind to receptors on cells of complementary mating type and not to serve an autocrine function by binding to self. The cDNA sequences for all four pheromone genes are known and at least one is expressed in bacteria (Brunen-Nieweler et al., 1994).

The contrast of the two *Euplotes* species is very interesting for their common use of peptide pheromones for mating reactivity, and while using very different approaches to sort self from nonself for mating. Like peptide growth factors, pheromones are used for autocrine stimulation in *E. raikovi*, while like the yeast mating pheromones, they are used exclusively to distinguish nonself and initiate mating in both species.

Euplotes vannus shows an unusual behavior to keep it hovering in the area of bacteria, their food. The cells walk back and forth in a slow forward movement and backward jerk, which keeps the cells trapped in the area of stimulus. These behaviors, called side-stepping motions (Ricci et al., 1987), correlate with long depolarizations (Stock et al., 1997). This strategy for chemoattraction contrasts with that of paramecia, which take more of the chemokinesis approach of bacteria and swim smoothly up gradients of attractants.

5.2. Paramecium

Paramecia spend most of their time swimming forward by beating their cilia; only occasionally do they turn in their swimming path. The abrupt turn is caused by an action potential that transiently increases calcium in the axonemal space and changes the power stroke of the cilia. While calcium remains high, the changed power stroke sends the cell backward, but calcium is quickly sequestered or removed so the power stroke and swimming return to normal. Thus, the observer sees only an abrupt turn in the swimming path. As the negative resting membrane potential of the cell is modulated up or down with changing extracellular K^+ or injected current, the cilia beat more quickly with hyperpolarization and more slowly with depolarization. Sufficient depolarization triggers the calcium action potential. Thus the cells are slaves to their membrane potential, and their swimming behavior (speed and frequency of turning) gives insight into their physiological state. The corollary is that since cells respond to environmental stimuli by altering swimming patterns, the signal transduction pathways by which the environmental stimuli act must ultimately affect membrane electrical properties. (See review by Machemer, 1989.)

The stimuli for *P. tetraurelia* include compounds like acetate, folate, lactate, NH_4Cl, biotin, glutamate, cAMP (Bell et al., 1998), many of which are signals shared by *Caenorhabditis elegans* and *Dictyostelium* (Sengupta, 1997; Bell et al., 1998). Interestingly, all three organisms graze or feed on bacteria and these stimuli should lead them to their food organisms, which are the sources of the stimuli.

Paramecia in attractant stimuli swim fast and smoothly with few turns; in repellents they swim more slowly and turn often (Van Houten, 1978). The cells adapt to uniform concentrations of stimuli, by mechanisms that we do not yet understand. However, the adaptation is very important for the ultimate accumulation of the population of cells (Van Houten, 1990). Adaptation, as in bacterial chemotaxis, seems to reset the level of response so that cells beginning to swim down a gradient of attractant or up a gradient of repellent make an immediate turn, which often sends the cell back into the relative attractant environment. We developed a computer simulation (Van Houten and Van Houten, 1982), which very closely matches the empirical results in assays of population response in T-mazes (Van Houten et al., 1982). Especially important in quality of the simulation are the immediate off-response upon leaving an attractant stimulus and adaptation.

The behavior of paramecia moving up and down gradients of attractants and repellents resembles that in bacterial "chemotaxis." However, the paramecia modulate speed as well as frequency of turning, while bacteria seem to concentrate their efforts in tumbling or suppressing tumbling. The smooth fast swimming of *Paramecium* in attractants is characteristic of hyperpolarization and we have confirmed that attractant stimuli do, indeed, hyperpolarize membrane potential (Van Houten, 1979; Preston and Van Houten, 1987).

5.3. Three Signal Transduction Pathways for Attractants

The chemosensory signal transduction pathways in *Paramecium* all ultimately must result in changes in swimming and therefore in membrane potential. We will focus here on three attractant stimulus pathways and how hyperpolarization is accomplished (reviewed in Van Houten, 1998).

Figure 2.8 shows examples of the three different pathways. Not all known variations on the theme are represented here, but the purpose of Figure 2.8 is to show the variety of signaling pathways, within one cell, that all converge in different ways on membrane potential. In pathway 1, the stimulus could be acetate, folate, cyclic AMP, or biotin. There are specific surface binding sites for each stimulus and each has been shown to hyperpolarize the cells about 8–10 mV. No second messenger has been identified, despite efforts to measure cAMP, cGMP, and IP$_3$. However, there are measurable transient K$^+$ and sustained conductances upon stimulation, and we have examined the possibility that the plasma membrane Ca^{2+} pump carries this sustained current (Wright and Van Houten, 1990; Wright et al., 1992, 1993; Yano et al., 1996). We have cloned the *Paramecium* plasma membrane pump gene, and found it to be 43% identical to the human Ca^{2+} pump (Elwess and Van Houten, 1997). The pump is an interesting molecule with many opportunities for regulation through its own inhibitory domain, calmodulin, PKA, and PKC among other mechanisms. We calculate that as little as a 1% increase in its activity could account for the conductances that we observe (Wright et al., 1992 see below).

In pathway 2, glutamate binds to its receptor (Yang, 1995) and rapidly activates adenylyl cyclase: intracellular cyclic AMP increases are noticeable by 30 msec and increases seven fold by 150 msec (Yang et al., 1997). The newly formed cyclic AMP likely activates protein kinase A, which in turn activates the pump to maintain the hyperpolarizing current for the length of time that glutamate remains in the environment. Consistent with this model are the observations that kinase inhibitors such as H7 and H8 specifically inhibit chemoresponse to glutamate but not to other stimuli (Yang et al., 1997).

Pathway 3 requires no receptor for its lone stimulus NH$_4$Cl, which probably crosses the membrane as NH$_3$ and alkalinizes the inside of the cell. We used ion sensitive fluorescent dyes to observe rapid alkalinization (Davis et al., 1998), but we do not understand the coupling of pH to hyperpolarizations. The pump by all our tests is not involved in this pathway.

For the study of ion channels in the development and persistence of the stimulus-induced hyperpolarizations, we are fortunate to have in *Paramecium*

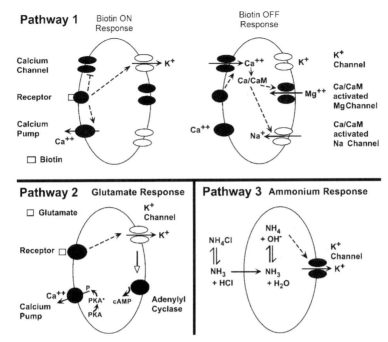

FIG. 2.8. Three pathways for chemosensory signal transduction in *Paramecium tetraurelia*. Pathway 1: For stimuli such as acetate, cAMP, folate, biotin that activate a small K conductance to hyperpolarize the cell and subsequently activate the plasma membrane calcium pump to sustain the hyperpolarization. All involve calcium/calmodulin, probably at the level of the pump. For biotin, there is a calcium influx upon removal of the stimulus, which is not present or not as noticeable with the other stimuli. Pathway 2: Glutamate appears to activate a K conductance to hyperpolarize the cell, and also rapidly activates the adenylyl cyclase. The rapid rise in cAMP activates PKA, which has direct effects on the axoneme and ciliary beating and also on the calcium pump. Pathway 3: Ammonium chloride rapidly alkalinizes the cell, probably by crossing the membrane as ammonia. The alkalinization hyperpolarizes the cell by a yet undefined mechanism.

tetraurelia a zoo of mutants, each of which has a specific single site mutation and many of which are defective in specific ion channel functions. There are at least 10 known channel conductances in *P. tetraurelia*, and there are mutants in most all (Saimi and Kung, 1987). Our computer-based motion analysis of mutants unresponsive to stimuli representative of the three pathways (see Davis et al., 1998; Van Houten et al., 1998) have led us to the following models for the receptor-mediated pathways 1 and 2 (Fig. 2.7).

We are currently testing these models with voltage clamp and molecular genetic approaches. The computer behavior simulation studies that we carried out years ago confirmed the importance of the obvious on-response, that is, the immediate response of simulated cells to attractant stimulus (Van Houten and Van Houten, 1982). However, upon re-examination the simulation results remind us that the off-response is essential for a robust population response and maintenance of cells in the attractant area, and now we have a physiological correlate of the off-response (Preston et al., 1998; Van Houten et al., 1998).

Receptor proteins have remained somewhat elusive. We have purified one receptor for cyclic AMP and partially cloned its gene (Van Houten et al., 1991) and partially purified the receptor for folate. It is important to keep an open mind about receptor structure since there are many variations on the theme: guanylyl cyclases as receptors (Yu et al., 1997); tyrosine kinases; glycosyl phosphatidylinositol (GPI) anchored proteins as well as G protein coupled receptors. While the cAMP receptor is an integral membrane protein, the *Paramecium* folate receptor is a particularly good candidate for GPI anchoring as opposed to integration into the membrane (see below). GPI anchored proteins congregate with other signal transduction proteins in detergent insoluble microdomains on the cell surface (rafts) or in caveolae (lipid microdomain invaginations of the cell membrane, associated with caveolin protein and the site of internalization by potocytosis) (Brown and Rose, 1992; Garcia et al., 1993; Lisanti et al., 1994; Casey, 1995; Parton and Simons, 1995; Weimbs et al., 1997). Caveolae and rafts of special lipids raise the effective concentrations of signaling proteins as much as 1000 fold, thereby increasing apparent affinity and affecting rates of "reaction limited" processes (Kholodenkio et al., 2000). Rafts also provide answers for how proteins with no transmembrane sequences signal to kinase cascades inside the cell (Cary and Cooper, 2000). These structures are highly characteristic of polarized cells, with the caveolae at the apical surface. Considering their morphology, taste and olfactory cells seem good candidates for caveolae and concentration of signaling proteins in membrane domains. There may be a particularly good reason to examine taste cells for caveolae or detergent insoluble microdomains since a GPI anchored component of an amiloride sensitive sodium channel has been found in kidney epithelial cells and amiloride sensitive sodium channels are implicated in taste transduction (Vallet et al., 1997; see Chapter 13 by Glendinning et al.).

Paramecia have abundant GPI anchored surface proteins. We have gathered preliminary evidence using blocking antibodies that chemoreceptors including the folate receptor are among them. (Paquette et al., 1996, 1997, 2000). The *Paramecium* folate receptor of 30 kD fits nicely with most mammalian folate receptors, which are small GPI anchored proteins (Antony, 1996; Yano et al., 2000).

We have employed a multidisciplinary approach to paramecium attractant chemoresponse using biochemistry, electrophysiology, immunocytochemistry, and behavioral assays (motion analysis for individual cells and T-mazes for populations) of both wild-type and mutant cells (reviewed in Van Houten, 1998; see also Preston and Hammond, 1997). Mutants have given us important insights that we would otherwise overlook (Van Houten, 1979; Schulz et al., 1984; Smith et al., 1987; Bell et al., 1998; Davis et al., 1998). Now we are adding the new approaches of antisense technology and stable transformation with expression plasmids to our arsenal and we can ask new questions and test our hypotheses in new ways. We are, for example, using an antisense expression vector for the gene for the first enzyme of GPI anchor synthesis, confirmed that we can manipulate levels of GPI anchored proteins on the cell surface and thereby selectively perturbing chemoresponse (Murakami et al., 1999; Yano et al., 2000).

5.4. Paramecium Repellent Responses

Paramecia are repelled from some stimuli by frequent turns in the swimming paths, characteristic of depolarization. Quinine, quinidine, and other bitter substances depolarize the cells and elicit action potentials, which cause the turns (Van Houten, 1979). The conductances elicited by quinine and other bitter substances are being teased apart and there appears to be a complex behavior mediated by multiple pathways (Oami, 1996, 1998).

5' ribonucleotides and glutamate act synergistically in amino acid or umami taste (Yamaguchi, 1991; see Chapter 13 by Glendinning et al.); IMP and GMP are especially good in this interaction with glutamate. Paramecia are attracted to glutamate, which acts synergistically with cGMP to elicit larger than additive behavioral responses to glutamate alone, and to this extent paramecia seem to have an umami taste. However, IMP does not behave as expected for an umami substance. It shares a glutamate binding moiety on the cell surface, but this is not the glutamate site for attraction to glutamate (Van Houten et al., 2000). More curiously, IMP does not synergize with glutamate and actually is a repellent for paramecia (Yang, 1995). IMP depolarizes the cells and causes frequent turns and slow swimming, characteristic of a repellent. We speculate that the cells use IMP, a downstream degradation product of ATP, as an indicator that bacterial action and decay have gone too far and conditions are toxic.

Yet another unusual repellent is GTP, which at low concentrations causes dispersal by eliciting turns (Clark et al., 1993). However, the signal transduction system for GTP is different from that for quinine or IMP. GTP elicits slow oscillations of membrane potential and repeated bouts of turning. Oscillations of calcium are thought to underlie the bouts of turning and the GTP effects can be antagonized by intracellular cyclic AMP (presumably through a PKA mechanism) and sarco/endoplasmic reticulum calcium (SERCA) pump inhibitors (Wassenberg et al., 1997; Mimikakis et al., 1998). These results suggest that some of the potentially oscillating calcium could come from internal stores. Voltage clamp studies show that both Mg^{2+} and Na^+ conductances are activated upon GTP stimulation; both of these are calcium dependent (Clark et al., 1997), perhaps utilizing the calcium released from intracellular stores for their activation in response to GTP.

A mutant defective in the GTP response has been isolated, and should be most useful in untangling complex behavioral response to GTP (Mimikakis et al., 1998b). It also will be interesting to see how the unusual repulsion from lysozyme (Hennessey et al., 1995) relates to the GTP response.

5.5. Insights from Paramecium

Stein and Meredith (1994) comment on the commonality of mechanisms of excitation among eukaryotes, including *Paramecium*. Ion channels in particular can be recognized across phyla and studies of *Paramecium* channels can be expected to provide insights into first principles of ion channel and sensory function (Hille, 1984). Similarly, because *Paramecium* is an excitable cell with neuronal properties, studies of its chemosensory signal transduction will provide some new insights into chemical sensing in higher organisms. In particular, it will be interesting to see how GPI anchored proteins, off-responses, and pump conductances figure into signal transduction in other systems. *Paramecium* also provides a model for multiple chemosensory signal pathways acting concurrently in a single cell. Students of taste cell function in particular might find useful parallels in how *Paramecium* integrates at the level of the membrane potential and responds to the net effects of multiple inputs.

6. SUMMARY

An important purpose of this chapter is to make students of the chemical senses in vertebrates aware of the progress in studies of chemosensory physiology, biochemistry, genetics, and behavior in microorganisms. Clearly there is common ground between microorganisms and vertebrates in the conservation of signal transduction through receptors to second messengers and even to modulation of cellular electrical activity. There also are areas of apparent divergence between these groups in the employment of MAP kinase pathways, two-component signaling, cross talk among 7-TMS and MAP kinase pathways, quorum sensing, GPI anchored proteins, and more in the repertoire of microorganisms. However, the microorganisms are likely only pointing the way to mechanisms and pathways that soon will be found to be employed by vertebrates as well. Perhaps Arkowitz (1999) demonstrates this best in his review article comparing *Dictyostelium* and yeast systems of chemotaxis and chemotropism as models for polarized cell growth and movement that are essential for organization of cells and development of complex structures. The oriented movement and growth of slime mold amoebae and yeast in response to external chemical cues has important insights for mechanisms in more complex systems.

REFERENCES

Antony, A. (1996). Folate receptors. *Ann Rev Nutr* **16**: 501–521.

Arkowitz, R. A. (1999). Responding to attraction: chemotaxis and chemotropism in *Dictyostelium* and yeast. *Trends Cell Biol* **9**: 20–27.

Bakal, C. J. and J. E. Davies (2000). No longer an exclusive club: eukaryotic signaling domains in bacteria. *Trends Cell Biol* **10**: 32–37.

Bardwell, L., J. G. Cook, C. J. Inouye, and J. Thorner (1994). Signal propagation and regulation in the mating pheromone response pathway of the yeast *Saccharomyces cerevisiae*. *Dev Biol* **166**: 363–379.

Bell, W. E., W. Karstens, Y. Sun, and J. L. Van Houten (1998). Biotin chemoresponse in *Paramecium*. *J Comp Physiol A* **183**: 361–366.

Brandon, M. A. and G. J. Podgorski (1997). Gα3 regulates the cAMP signaling system in *Dictyostelium. Mol Biol Cell* **9**: 1677–1685.

Bray, D., M. D. Levin, and C. J. Morton-Firth (1998). Receptor clustering as a cellular mechanism to control sensitivity. *Nature* **393**: 85–88.

Brazill, D. T., R. Gundersen, and R. H. Gomer (1997). A cell-density sensing factor regulates the lifetime of a chemoattractant-induced Gα-GTP conformation. *FEBS Lett* **404**: 100–104.

Brazill, D. T., D. F. Lindsey, J. D. Bishop, and R. H. Gomer (1998). Cell density sensing mediated by a G protein-coupled receptor activating phospholipase C. *J Biol Chem* **273**: 8161–8168.

Brown, D. A. and J. K. Rose (1992). Sorting of GPI-anchored proteins to glycolipid-enriched membrane subdomains during transport to the apical cell surface. *Cell* **68**: 533–544.

Brown, L. R., S. Mronga, R. A. Bradshaw, C. Ortenzi, P. Luporini, and K. Wüthrich (1993). Nuclear magnetic resonance solution structure of the pheromone Er-10 from the ciliated protozoan *Euplotes raikovi. J Mol Biol* **231**: 800–816.

Brunen-Nieweler, C., F. Meyer, and K. Heckmann (1994). Expression of the pheromone 3-encoding gene of *Euplotes octocarinatus* using a novel bacterial secretion vector. *Gene* **150**: 187–192.

Cary, L. A. and J. A. Cooper. (2000). Molecular switches in lipid rafts. *Nature* **404**: 945–947.

Casey, P. J. (1995). Protein lipidation in cell signalling. *Science* **268**: 802–805.

Chang, C. and R. C. Stewart (1998). The two-component system. *Plant Physiol* **117**: 723–731.

Chen, M-Y., R. H. Insall, and P. N. Devreotes (1996). Signaling through chemoattractant receptors in *Dictyostelium. Trends Genet* **12**: 52–57.

Chiesa, R., W. I. Silva, and F. L. Renaud (1993). Pharmacological characterization of an opiod receptor in the ciliate Tetrahymena. *J Euk Microbiol* **40**: 800–804.

Clark, K. D., T. M. Hennessey, and D. L. Nelson (1993). External GTP alters the motility and elicits an oscillating membrane depolarization in *Paramecium tetraurelia. Proc Natl Acad Sci U S A* **90**: 3782–3786.

Clark, K., T. M. Hennessey, D. L. Nelson, and R. R. Preston (1997). Extracellular GTP causes membrane-potential oscillations through the parallel activation of Mg^{2+} and Na^{+} currents in *Paramecium tetraurelia. J Membr Biol* **157**: 159–167.

Cook, J. G., L. Bardwell, and J. Thorner (1997). Inhibitory and activating functions for MAPK Kss1 in the *S. cerevisiae* filamentous-growth signalling pathway. *Nature* **390**: 85–90.

Daaka, Y., L. M. Luttrell, and R. J. Lefkowitz (1997). Switching of the coupling of the β_2-adrenergic receptor to different G proteins by protein kinase A. *Nature* **390**: 88–91.

Davis, D., J. Fiekers, and J. L. Van Houten (1998). Intracellular pH and chemoresponse to NH_4^+ in *Paramecium. Cell Motil Cytoskel* **40**: 107–115.

Elwess, N. and J. L. Van Houten (1997) Cloning and molecular analysis of the plasma membrane Ca^{2+} ATPase in *P. tetraurelia. J Euk Microbiol* **44**: 250–257.

Falke, J. J., R. B. Bass, S. L. Butler, S. A. Chervitz, and M. A. Danielson (1997). The two-component signaling pathway of bacterial chemotaxis: a molecular view of signal transduction by receptors, kinases and adaptation enzymes. *Annu Rev Cell Dev Biol* **13**: 457–512.

Ferguson, B., J. Horeka, J. Printen, J. Schultz, B. J. Stevenson, and G. F. Sprague (1994). The yeast pheromone response pathway: new insights into signal transmission. *Cell Molec Biol Res* **40**: 223–228.

Ford, C. E., N. P. Skiba, H. Bae, Y. Daaka, E. Reuveny, L. R. Shekter, R. Rosal, G. Weng, C-S. Yang, R. Iyengar, R. J. Miller, L. Y. Jan, R. J. Lefkowitz, and H. E. Hamm (1998). Molecular basis for interactions of G protein $\beta\gamma$ subunits with effectors. *Science* **280**: 1271–1274.

Fuqua, C. and E. P. Greenberg (1998). Cell-to-cell communication in *Escherichia coli* and *Salmonella typhimurium*: they may be talking but who's listening? *Proc Natl Acad Sci U S A* **95**: 6571–6572.

Garcia, M., C. Mirre, A. Quaroni, H. Reggio, and A. LeBivic (1993). GPI-anchored proteins associate to form microdomains during their intracellular transport in Caco-2 cells. *J Cell Sci* **104**: 1281–1290.

Ginsburg, G. T., R. Gollop, Y. Yu, J. M. Louis, C. L. Saxe, and A. R. Kimmel (1995). The regulation of *Dictyostelium* development by transmembrane signalling. *J Euk Microbiol* **42**: 200–204.

Gross, J. D. (1994). Developmental decisions in *Dictyostelium discoideum*. *Microbiol Rev* **58**: 330–351.

Hazelbauer, G. L., H. C. Berg, and P. Matsumura (1993). Bacterial motility and signal transduction. *Cell* **73**: 15–22.

Hennessey, T. M., M. Y. Kim, and B. H. Satir (1995). Lysozyme acts as a chemorepellent and secretagogue in *Paramecium* by activating a novel receptor-operated Ca^{2+} conductance. *J Membr Biol* **148**: 13–25.

Hille, B. (1984). *Ion Channels of Excitable Membranes*. Sunderland, MA.: Sinauer Assoc. Inc.

Insall, R., J. Borleis, and P. N. Devreotes (1996). The aimless RasGEF is required for processing of chemotactic signals through G-protein-coupled receptors in *Dictyostelium*. *Curr Biol* **6**: 719–729.

Iwadate, Y., K. Katoh, H. Asai, and M. Kikuyama (1997). Simultaneous recording of cytosolic Ca^{2+} levels in *Didinium* and *Paramecium* during a *Didinium* attack on *Paramecium* *Protoplasma* **200**: 117–127.

Jin, T., N. Zhang, Y. Long, C. A. Parent, and P. N. Devreotes (2000). Localization of the G protein $\beta\gamma$ complex in living cells during chemotaxis. *Science* **287**: 1034–1036.

Kim, J. Y., R. D. Soede, P. Schaap, R. Valkema, J. A. Borleis, P. J. Van Haastert, P. N. Devreotes, and D. Hereld (1997). Phosphorylation of chemoattractant receptors is not essential for chemotaxis or termination of G-protein-mediated responses. *J Biol Chem* **272**: 27313–27318.

Kholodenkio, B., J. B. Joek, and H. V. Westerhoff (2000). Why cytoplasmic signaling proteins should be recruited to cell membranes. *Trends Cell Biol* **10**: 173–178.

Kleerebezem, M., E. N. Quadri, O. P. Kuipers, and W. M. de Vos (1997). Quorum sensing by peptide pheromones and two-component signal-transduction systems in gram-positive bacteria. *Molec Microbiol* **24**: 895–904.

Knetsch, M. L. W., S. J. P. Epskamp, P. W. Schenk, Y. Wang, J. E. Segall, and B. E. Snaar-Jagalska (1996). Dual role of cAMP and involvement of both G-proteins and ras in regulation of ERK2 in *Dictyostelium discoideum*. *EMBO J* **15**: 3361–3368.

Koppelhus, U., P. Hellung-Larsen, and V. Leick (1994). An improved quantitative assay for chemokinesis in *Tetrahymena*. *Biol Bull* **187**: 8–15.

Kurvilla, H. G., M. Y. Kim, and T. M. Hennessey (1997). Chemosensory adaptation to lysozyme and GTP involves independently regulated receptors in Tetrahymena thermophila. J Euk Microbiol **44**: 263–268.

Kuspa, A. and W. F. Loomis (1992). Tagging developmental genes in Dictyostelium by restriction enzyme-mediated integration of plasmid DNA. Proc Natl Acad Sci U S A **89**: 8803–8807.

Kuwayama, H. and P. J. M. Van Haastert (1996). Regulation of guanylyl cyclase by a cGMP-binding protein during chemotaxis in Dictyostelium discoideum. J Biol Chem **271**: 23178–23724.

Leeuw, T., C. Wu, J. D. Schrag, M. Whiteway, D. Y. Thomas, and E. Leberer (1998). Interaction of a G-protein β subunit with a conserved sequence in Ste20/PAK family protein kinases. Nature **391**: 191–194.

Lisanti, M. P., P. E. Scherer, Z-T. Tang, and M. Sargiacomo (1994). Caveolae, caveolin and caveolin-rich membrane domains: a signalling hypothesis. Trends Cell Biol **4**: 231–235.

Luginbuhl, P., M. Ottiger, S. Mronga, and K. Wüthrich (1994). Structure comparison of the pheromones Er-1, Er-10 and Er-2 from Euplotes raikovi. Protein Sci **3**: 1537–1546.

Luporini, P., A. Vallesi, C. Miceli, and R. A. Bradshaw (1995). Chemical signaling in ciliates. J Euk Microbiol **42**: 208–212.

Ma, H., M. Gamper, C. Parent, and R. A. Firtel (1997). The Dictyostelium MAP kinase kinase DdMEK1 regulates chemotaxis and is essential for chemoattractant-mediated activation of guanylyl cyclase. EMBO J **16**: 4317–4332.

Machemer, H. (1989). Cellular behaviour modulated by ions: electrophysiological implications. J Protozool **36**: 463–487.

Madhani, H. D. and G. R. Fink (1998). The riddle of MAP kinase signaling specificity. Trends Genet **14**: 151–155.

Marino, M. J. and D. C. Wood (1993). β-endorphin modulates a mechanoreceptor channel in the protozoan Stentor. J Comp Physiol A **173**: 233–240.

Meili, R., C. Ellsworth, S. Lee, T. B. Reddy, H. Ma, and R. A. Firtel (1999). Chemoattractant-mediated transient activation and membrane localization of Akt/PKB is required for efficient chemotaxis to camp in Dictyostelium. EMBO J **18**: 2092–2105.

Mimikakis, J. L., K. Clark, and D. L. Nelson (1998a). The purinergic response in Paramecium tetraurelia is inhibited by intracellular cAMP. J Membr Biol **163**: 19–23.

Mimikakis, J. L., D. L. Nelson, and R. R. Preston (1998b) Oscillating response to a purine nucleotide disrupted by mutation in Paramecium tetraurelia. Biochem J **330**: 139–147.

Morgan, B. A., N. Bouquin, and L. H. Johnston (1995). Two-component signal-transduction systems in budding yeast MAP a different pathway? Trends Cell Biol **5**: 453–457.

Mronga, S., P. Luginbuhl, L. R. Brown, C. Ortenzi, P. Luporini, R. A. Bradshaw, and K. Wüthrich (1994). The NMR solution structure of the pheromone Er-1 from the ciliated protozoan Euplotes raikovi. Protein Sci **3**: 1527–1536.

Murakami, L. G., J. Yano, R. R. Preston, and J. L. Van Houten (1999). Antisense approach to GPI anchored chemoreceptors in Paramecium. Soc for Neurosci Abstr **25**: 127.

Nebl, T. and P. R. Fisher (1997). Intracellular Ca^{2+} signals in Dictyostelium chemotaxis are mediated exclusively by Ca^{2+} influx. J Cell Sci **110**: 2845–2853.

Nern, A. and R. A. Arkowitz (1998). A GTP-exchange factor required for cell orientation. Nature **391**: 195–198.

Oami, K. (1996). Distribution of chemoreceptors to quinine of the cell surface of *Paramecium caudatum*. *J Comp Physiol* **A 179**: 345–352.

Oami, K. (1998). Membrane potential responses of *Paramecium caudatum* to bitter substances: existence of multiple pathways for bitter responses. *J Exp Biol* **201**: 13–20.

Ottiger, M., T. Szyperski, P. Luginbuhl, C. Ortenzi, P. Luporini, R. A. Bradshaw, and K. Wüthrich (1994). The NMR solution structure of the pheromone Er-2 from the cilaited protozoan *Euplotes raikovi*. *Protein Sci* **3**: 1515–1526.

Paquette, C. A., A. Bush, and J. L. Van Houten (1996). GPI anchored proteins and chemoresponse in *Paramecium*. *Molec Biol Cell* Supp. **7**: 269.

Paquette, C. A., A. Bush, and J. L. Van Houten (1997). GPI anchored receptors in *Paramecium*. *Chem Senses* **22**: 766.

Paquette, C. A., V. Rakochy, A. Bush, and J. L. Van Houten (2000). GPI anchored proteins in *Paramecium tetraurelia*: Possible role in chemoresponse. *J Exp Biol*. In press.

Parent, C. A. and P. N. Devreotes (1999). A cell's sense of direction. *Science* **284**: 765–770.

Parkinson, J. S. (1993). Signal transduction schemes of bacteria. *Cell* **73**: 857–871.

Parkinson, J. S. and D. F. Blair (1993). Does *E. coli* have a nose? *Science* **259**: 1701–1702.

Parkinson, J. S. and E. C. Kofoid (1992). Communication modules in bacterial signaling proteins. *Annu Rev Genet* **26**: 71–112.

Parton, R. G. and K. Simons (1995). Digging into caveolae. *Science* **269**: 1398–1399.

Perego, M. and J. A. Hoch (1996). Protein aspartate phosphatases control the output of two-component signal transduction systems. *Trends Genet* **12**: 97–101.

Phillips, P., M. Thio, and C. Pears (1997). A protein kinase C-like activity involved in the chemotactic response of *Dictyostelium*. *Biochim Biophys Acta* **1349**: 72–80.

Posas, F., S. M. Wurgler-Murphy, T. Maeda, E. A. Witten, T. C. Thai, and H. Saito (1996). Yeast HOG1 MAP kinase cascade is regulated by a multistep phosphorelay mechanism in the SLN1-YPD1-SSK1 "two-component" osmosensor. *Cell* **86**: 865–876.

Preston, R. R. and J. L. Van Houten (1987). Chemoreception in *Paramecium*: acetate- and folate-induced membrane hyperpolarization. *J Comp Physiol* **60**: 525–536.

Preston, R. R. and J. A. Hammond (1997). Phenotypic and genetic analysis of "chamelcon," a *Paramecium* mutant with an enhanced sensitivity to magnesium. *Genetics* **146**: 871–880.

Preston R. R., W. E. Bell, and J. L. Van Houten (1998). Genetic dissection of the chemoresponse hyperpolarization in *Paramecium*. *Chem Senses* **21**: 2100.

Printen, J. A. and G. F. Sprague (1994). Protein-protein interactions in the yeast pheromone response pathway: Ste5p interacts with all members of the MAP kinase cascade. *Genetics* **138**: 609–619.

Renaud, F. L., I. Colon, J. Lebron, N. Ortiz, F. Rodriguez, and C. Cadilla (1995). A novel opiod mechanism seems to modulate phagocytosis in *Tetrahymena*. *J Euk Microbiol* **42**: 205–208.

Ricci, N., R. Gianetti and C. Miceli (1987). The ethogramm of *Euplotes crassus I*. The wild type. *Eur J Protistol* **23**: 129–140.

Saimi, Y. and C. Kung (1987). Behavioral genetics of paramecium. *Annu Rev Genet* **21**: 47–65.

Schulz, S., M. Denaro and J. Van Houten (1984). Relationship of folate binding and uptake to chemoreception in *Paramecium*. *J Comp Physiol* **155**: 113–119.

Segall, J. E., A. Kuspa, G. Shaulsky, M. Ecke, M. Maeda, C. Gaskins, R. A. Firtel, and W. F. Loomis (1995). A MAP kinase necessary for receptor-mediated activation of adenylyl cyclase in *Dictyostelium*. *J Cell Biol* **128**: 405–413.

Sengupta, P. (1997). Cellular and molecular analyses of olfactory behavior in C. *elegans. Cell Dev Biol* **8**: 153–161.

Smith, R., R. R. Preston, S. Schulz, and J. Van Houten (1987). Correlation of cyclic adenosine monophosphate binding and chemoresponse in *Paramecium. Biochim Biophys Acta* **928**: 171–178.

Spiro, P. A., J. S. Parkinson, and H. G. Othmer (1997). A model of excitation and adaptation in bacterial chemotaxis. *Proc Natl Acad Sci U S A* **94**: 7263–7268.

Stein, B. E. and M. A. Meredith (1994). *The Merging of the Senses.* Cambridge, MA: MIT Press.

Stock, A. M. and S. L. Mowbray (1995). Bacterial chemotaxis: a field in motion. *Curr Opin Struct Biol* **5**: 744–751.

Stock, C., T. Kruppel, and W. Leuken (1997). Kinesis in *Euplotes vannus*—ethological and electrophysiological characteristics of chemosensory behavior. *J Euk Microbiol* **44**: 427–433.

Surette, M. G. and B. L. Bassler (1998). Quorum sensing in *Escherichia coli* and *Salmonella typhimurium. Proc Natl Acad Sci U S A* **95**: 7046–7050.

Swanson, R. V., L. A. Alex, and M. I. Simon (1994). Histidine and aspartate phosphorylation: two-component systems and the limits of homology. *Trends Biochem Sci* **11**: 485–490.

Swift, S., J. P. Throup, P. Williams, G. P. C. Salmond, and G. S. A. B. Stewart (1996). Quorum sensing: a population-density component in the determination of bacterial phenotype. *Trends Biochem Sci* **21**: 214–219.

Taylor, B. L. and I. B. Zhulin, I.B. (1998). In search of higher energy: metabolism-dependent behaviour in bacteria. *Molec Microbiol* **28**: 683–690.

Thievelein, J. M. (1994). Signal transduction in yeast. *Yeast* **10**: 1753–1790.

Thomason, P., D. Traynor, and R. Kay (1999). Taking the plunge: terminal differentiation in *Dictyostelium. Trends Genet* **15**: 15–19.

Vallet, V., A. Chiralbi, H-P. Gaeggeler, J-D. Horisberger, and B. C. Rossier (1997). An epithelial serine protease activates the amiloride-sensitive sodium channel. *Nature* **389**: 607–610.

Van Haastert, P. J. M. and H. Kuwayama (1997). cGMP as second messenger during *Dictyostelium* chemotaxis. *FEBS Lett* **410**: 25–28.

Van Haastert, P. J. M. and P. van Dijken (1997). Biochemistry and genetics of inositol phosphate metabolism in *Dictyostelium. FEBS Lett* **410**: 39–43.

Van Houten, J. (1978). Two mechanisms of chemotaxis in *Paramecium. J Comp Physiol* **127**: 167–174.

Van Houten, J. (1979). Membrane potential changes during chemokinesis in *Paramecium. Science* **204**: 1100–1103.

Van Houten, J. L. (1990). Chemosensory transduction in *Paramecium.* In: J. P. Armitage and J. M. Lackie, eds. *Biology of the Chemotactic Response.* Cambridge: Cambridge University Press, pp 297–322.

Van Houten, J. (1994). Chemosensory transduction in eukaryotic microorganisms: trends for neuroscience? *Trends Neurosci* **17**: 62–71.

Van Houten, J. L. (1998). Chemosensory transduction in *Paramecium. Eur J Protistol* **34**: 301–307.

Van Houten, J. and R. R. Preston (1987). Chemoreception in single-celled organisms. In: T. E. Finger and W. L. Silver, eds. *Neurobiology of Taste and Smell.* New York: Wiley, pp 11–38.

Van Houten, J. L., W. Q. Yang, and A. Bergeron (2000). Glutamate chemosensory transduction in *Paramecium*. *J Nutr* **130**: 946S–949S.

Van Houten, J. L. and J. C. Van Houten (1982). Computer analysis of *Paramecium* chemokinesis behavior. *J Theor Biol* **98**: 453–468.

Van Houten, J., D. C. R. Hauser, and M. Levandowsky (1981). Chemosensory behavior in protozoa. In: M. Levandowsky and S. H. Hutner, eds. Biochemistry and Physiology of Protozoa, Vol. 4. New York: Academic Press, pp 67–124.

Van Houten, J., E. Martel, and T. Kasch (1982). Kinetic analysis of chemokinesis in *Paramecium*. *J Protozool* **29**: 226–230.

Van Houten, J., B. Cote, J. Zhang, J. Baez, M. L. Gagnon (1991). Studies of the cyclic AMP chemoreceptor of *Paramecium*. *J Membr Biol* **119**: 15–24.

Van Houten, J. L., W. E. Bell, and R. R. Preston (1998). Genetic dissection of the attraction chemoresponse hyperpolarization of *Paramecium*. *Soc Neurosci Abstr* **24**: 2100.

Wang, N., G. Shaulsky, R. Escalante, and W. F. Loomis. (1996). A two-component histidine kinase gene that functions in *Dictyostelium* development. *EMBO J* **15**: 3890–3898.

Ward, M. J. and D. R. Zusman (1997). Regulation of directed motility in *Myxococcus xanthus*. *Mol Microgiol* **24**: 885–893.

Wassenberg, J. J., K. D. Clar, and D. L. Nelson (1997). Effect of SERCA pump inhibitors on chemoresponses in *Paramecium*. *J Euk Microbiol* **44**: 574–581.

Weimbs, T., S. H. Low, S. J. Chapin, and K. E. Mostov (1997). Apical targeting in polarized epithelial cells: there is more afloat than rafts. *Trends Cell Biol* **7**: 393–399.

Wright, M. V., and J. L. Van Houten (1990). Characterization of a putative Ca^{2+} transporting ATPase in the pellicles of *Paramecium*. *Biochim Biophys Acta* **1029**: 241–251.

Wright, M. V., M. Frantz, and J. L. Van Houten (1992). Lithium fluxes in *Paramecium* and their relationship to chemoresponse. *Biochim Biophys Acta* **1107**: 223–230.

Wright, M., N. L. Elwess, and J. L. Van Houten (1993). Calcium transport and chemoresponse in *Paramecium*. *J Comp Physiol B* **163**: 288–296.

Xiao, Z. and P. N. Devreotes (1997). Identification of detergent-resistant plasma membrane microdomains in *Dictyostelium*: enrichment of signal transduction proteins. *Mol Biol Cell* **8**: 855–869.

Yamaguchi, S. (1991). Basic properties of umami and effects on humans. *Physiol Behav* **49**: 833–841.

Yang, W. Q. (1995). Dissertation. University of Vermont, Burlington, VT.

Yang, W. Q., C. Braun, H. Plattner, J. Purvee, and J. L. Van Houten (1997). Cyclic nucleotides in glutamate chemosensory signal transduction of *Paramecium*. *J Cell Sci* **110**: 2567–2572.

Yano, J., D. Fraga, R. Hinrichsen, and J. L. Van Houten (1996). Effects of calmodulin antisense oligonucleotides on chemoresponse in *Paramecium*. *Chem Senses* **21**: 55–58.

Yano, Y., K. Garner, J.-D. Herlihy, V. Rakochy, W. N. White, and J. L. Van Houten (2000). GPI anchored proteins in the chemoresponse of *Paramecium* to folate. *Chemical Senses*. In press.

Yu, S., L. Avery, E. Baude, and D. L. Garbers (1997). Guanylyl cyclase expression in specific sensory neurons: a new family of chemosensory receptors. *Proc Natl Acad Sci U S A* **94**: 3384–3387.

3

Genetic Models of Chemoreception

PIALI SENGUPTA
Department of Biology and Volen Center for Complex Systems,
Brandeis University, Waltham, MA

JOHN R. CARLSON
Department of Molecular, Cellular, and Developmental Biology,
Yale University, New Haven, CT

1. INTRODUCTION

Genetic analysis offers a means of identifying molecular components essential for chemosensory function. Classical genetic investigation usually begins with the identification of either an induced mutant or a naturally occurring variant that exhibits a phenotype distinct from that of a reference strain. The phenotypic difference is then mapped to a genetic locus, which then can be characterized in both genetic and molecular detail to yield information about the product of the gene. This process has been facilitated enormously in recent years by a variety of technological advances, including enhancer trapping, positional cloning techniques, and large-scale genomic sequencing.

Reverse genetics, by contrast, takes as its point of departure a DNA sequence whose product is suspected to play a role in chemosensory function. Mutations affecting the product are then sought, and the phenotypic consequences of such mutations are examined. For example, if animals carrying mutations in the DNA sequence encoding a particular transduction enzyme are

The Neurobiology of Taste and Smell, Second Edition, Edited by Thomas E. Finger, Wayne L. Silver, and Diego Restrepo.
ISBN 0-471-25721-4   Copyright © 2000 Wiley-Liss, Inc.

isolated and found to exhibit olfactory abnormalities, then that enzyme is inferred to play a role in olfaction. Reverse genetics also has been greatly facilitated by recent technical advances such as new methods of gene-targeting.

Genetic analysis is most conveniently performed in model systems, such as the worm, the fruit fly, and the mouse. In many instances, the use of genetic models has paved the way for identifying and characterizing conserved processes in other organisms. While the molecular underpinnings of many processes are surprisingly well conserved between vertebrates and invertebrates, significant differences also have been identified. Genetic models have thus not only allowed us to dissect conserved pathways in detail, but also have been important in identifying the diversity of mechanisms used by animals to attain similar goals.

Chemosensation is used by almost all members of the animal kingdom to detect predators and prey, and to communicate with members of their own species. In the last decade, there has been an explosion of information regarding the molecular components of neurons that function to transduce chemical signals and allow the animal to respond appropriately. In addition, an interesting finding has been that at least some of the basic signal transduction mechanisms used for chemoreception are conserved between vertebrates and invertebrates. This chapter addresses the use of genetic models in the study of chemoreception, and discusses how the information obtained from such studies has complemented and added to knowledge obtained from molecular and physiological investigation.

2. *CAENORHABDITIS ELEGANS* AS A MODEL SYSTEM FOR THE STUDY OF CHEMOSENSATION

2.1. Why Worms?

The freeliving soil nematode C. *elegans* is an excellent model organism for studying chemosensory behavior. First, the animals possess a complex, highly discriminatory and powerful chemosensory response to large numbers of chemicals. Second, the compact nervous system of the animal has been extensively studied, such that each individual neuron in the animal can be characterized. Third, since the animals exist primarily as hermaphrodites (although they can also be cross-fertilized by males), large-scale genetic screens and subsequent genetic manipulations of behavioral mutants are easy to perform. Fourth, transgenes can be introduced easily and the animals' three-day lifecycle allows for rapid completion of experiments. Finally, C. *elegans* is the first metazoan organism whose genome has been completely sequenced. Projections from current data indicate that a large percentage of the predicted ∼19,000 genes share similarities with vertebrate counterparts (Sonnhammer and Durbin, 1997; Bargmann, 1998; Ruvkun and Hobert, 1998). Thus, reverse genetic analyses of sequenced genes provide a further tool for investigation of the chemosensory system in vivo.

2.2. Chemosensory Behavior

The lifestyle of C. elegans requires that worms be able to respond to many chemicals. Worms live in the soil and feed on several different bacteria. C. elegans detect bacteria by sensing bacterial metabolites such as esters and ketones (Dainty et al., 1985; Zechman and Labows, 1985). Since different bacteria produce different metabolites, the ability of worms to sense multiple chemicals maximizes detection of food sources. This chemosensory behavior can be examined in the laboratory by assaying their responses to pure chemicals. Worms respond to a gradient of a chemical by either moving up the gradient (attraction) or by moving down the gradient (repulsion). Several chemicals elicit neither response.

Worms respond to both water-soluble and volatile chemicals. Water-soluble attractants include the ions Na^+ and Cl^- as well as small inorganic molecules such as cAMP, lysine, and the neurotransmitter serotonin (Ward, 1973; Dusenbery, 1974; Bargmann and Horvitz, 1991a). Water-soluble repellents include D-tryptophan and extract of garlic (Dusenbery, 1975; Bargmann et al., 1990). Volatile attractants include molecules of several chemical classes such as ketones, esters, aldehydes, alcohols, pyrazines, and thiazoles (Bargmann et al., 1993). The response elicited by a particular volatile compound depends primarily on its chain length. For example, short-chain alcohols such as isoamyl alcohol elicit attractive responses, while longer chain alcohols such as octanol elicit strong repellent responses (Bargmann et al., 1993). Behavioral responses are also sensitive to the concentrations of the odorants used. Benzaldehyde, a chemical that is strongly attractive at lower concentrations (.05 nM), is strongly repulsive at higher concentrations (10 nM).

C. elegans is capable of discriminating between remarkably similar compounds in behavioral assays. For instance, worms are able to discriminate between the two attractive chemicals 2,3-butanedione and 2,3-pentanedione behaviorally. Experiments have shown that these chemicals are sensed by different neurons, and the responses can be separated genetically (P. Sengupta and C. I. Bargmann, unpub. obs.; Sengupta et al., 1996). Independent categories of chemical attractants have been defined based on cross-saturation and cross-adaptation experiments in behavioral assays (Ward, 1973; Bargmann et al., 1993). These experiments defined more than 4 categories of water-soluble molecules (Ward, 1973) and at least eight different classes of volatile attractants (Bargmann et al., 1993).

In addition to bacterial metabolites, worms also respond to a complex mixture of pheromones produced by other worms. C. elegans constitutively secretes a complex mixture of fatty and bile acids collectively called dauer pheromone (Golden and Riddle, 1982, 1984). High levels of pheromone trigger entry into an alternative larval state called the dauer state. Dauer larvae do not reproduce, but animals can recover from this state and resume normal development when conditions are favorable. Recovery from the dauer stage is mediated primarily by the chemosensory system, detecting the presence of food and lower levels of pheromone.

2.3. Neuroanatomy of the Chemosensory System

Neurons mediating chemosensory responses have dendrites in the amphid sensory organs, the so-called "nose" of this animal (See Chapter 6, Farbman). C. *elegans* hermaphrodites have a compact nervous system consisting of 302 neurons. Examination of serial sections of electron micrographs has allowed the prediction of most presynaptic and postsynaptic partners of each neuron (White et al., 1986). These studies have also described the axonal, dendritic, and ciliary morphologies of many neurons, and have shown that the positions of the cell bodies of most neurons are relatively invariant from animal to animal (Ward et al., 1975; Sulston and Horvitz, 1977; Sulston et al., 1983; White et al., 1986). These data coupled with expression patterns of recently identifed cell-specific genes (see Table 3.1) allow the easy identification of single neuron types.

The neurons responsible for particular chemosensory responses have been defined in experiments in which single neuron types were killed using a laser microbeam and the responses of the operated animals were assessed (Bargmann and Horvitz, 1991a; Bargmann et al., 1993) (Table 3.2). All neurons that sense water-soluble chemicals have dendrites ending in a single or double cilia directly exposed to the environment. The neuron types AWA, AWB, and AWC, which sense only volatile chemicals, each have unique characteristic ciliary endings that are indirectly exposed to the environment (Ward et al., 1975; Ware et al., 1975; Perkins et al., 1986; see Chapter 6, Farbman). Genetic and structural data implicate the cilia as the location of the primary signal transduction events in sensory signaling. Animals with structurally defective cilia due to mutations in genes such as *che-2*, *osm-3* (encoding a kinesin-like heavy chain) (Shakir et al.,

TABLE 3.1 Summary of described genes

Gene	Predicted Protein	Chemosensory Neurons Where Expressed*
odr-10	Olfactory receptor for diacetyl	AWA
odr-3	G_α subunit	AWA, AWB, AWC, ADF, ASH
gpa-2	G_α subunit	AWC**
gpa-3	G_α subunit	ASE, ADF, ASG, ASH, ASI, ASJ, ASK, ADL**
odr-4	Novel	AWA, AWB, AWC, ADF, ASG, ASH, ASI, ASJ, ASK, ADL**
tax-4	α subunit of cyclic nucleotide-gated channel	AWC, ASE, ASG, ASI, ASJ, ASK, AFD (thermosensory)**
tax-2	β subunit of cyclic nucleotide-gated channel	AWB, AWC, ASE, ASG, ASI, ASJ, ASK, AFD (thermosensory)**
osm-9	Predicted calcium channel	AWA, AWC, ASE, ADF, ASG, ASH, ASI, ASJ, ASK, ADL**

*Expression pattern inferred from expression of GFP fusion genes.
**Also expressed in additional neurons.

See text for references.

TABLE 3.2 Functions of chemosensory neurons defined by laser killing experiments

Sensory Neuron	Chemicals Sensed
AWA	Volatile attractants: diacetyl, pyrazine, thiazole
AWB	Volatile repellents: 2-nonanone (long-range avoidance)
AWC	Volatile attractants: benzaldehyde, butanone, isoamyl alcohol, thiazole, 2,3-pentanedione
ASE	Water-soluble attractants: Na^+, Cl^-, cAMP, biotin, lysine
ADF	Water-soluble attractants: Na^+, Cl^-, cAMP, biotin (minor), dauer pheromone
ASG	Water-soluble attractants: Na^+, Cl^-, cAMP, biotin, lysine (minor), dauer pheromone (minor)
ASH	Water-soluble repellents: osmotic shock Volatile repellents: 1-octanol, benzaldehyde (short range avoidance) Nose touch avoidance
ASI	Water-soluble attractants: Na^+, Cl^-, cAMP, biotin, lysine (minor) dauer pheromone
ASJ	Dauer pheromone
ASK	Water-soluble attractant: lysine
ADL	Water-soluble repellents Volatile repellents: 1-octanol (short- and long-range avoidance)

1993) and *osm-6* (a novel cilium protein) (Collet et al., 1998), all exhibit defects in a broad range of chemosensory responses (Perkins et al., 1986).

The laser killing experiments showed that distinct, nonoverlapping sets of neurons sense attractants and repellents (Bargmann and Horvitz, 1991a; Bargmann et al., 1993; Troemel et al., 1995, 1997) (Table 3.2). These experiments also showed that each chemosensory neuron in *C. elegans* responds to multiple molecules of diverse chemical structures. For example, the AWA neuron type responds to the structurally dissimilar compounds diacetyl, pyrazine, and trimethylthiazole (Bargmann et al., 1993). This enables the animal to maximize its ability to respond to chemicals using only a few neurons. A given chemical may be sensed by multiple neurons, thus increasing the fidelity of the response. Na^+ ions are sensed primarily by the ASE neurons, but the ASI, ASG, and ADF neurons also contribute to the response (Bargmann and Horvitz, 1991a). The water-soluble dauer pheromone is sensed by the ASI and ADF neurons with lesser contributions from the ASG neurons (Bargmann and Horvitz, 1991b).

2.4. Olfactory Receptors

Behavioral and genetic screens for mutants with defects in responses to one or more odorants resulted in the identification of genes such as the *odr-10* olfactory receptor (Sengupta et al., 1996). *odr-10* null mutants fail to respond only to the

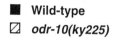

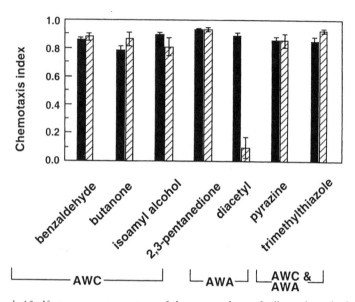

FIG. 3.1. *odr-10* olfactory receptor mutants fail to respond specifically to the volatile odorant diacetyl (2,3-butanedione). Responses of wild-type animals (black bars) and *odr-10* null mutants (hatched bars) were examined to the chemicals indicated. Neurons required for sensing the chemicals are marked. Chemotaxis index = (number of animals at the odorant−number of animals at the control)/total number of animals on the plate. A chemotaxis index of 1.0 indicates perfect attraction. 1 μl of ethanol was used as the control; 1 μl of diluted odorant was used as the attractant. Dilutions were as follows: benzaldehyde—1:1000, butanone—1:1000, isoamyl alcohol—1:200, diacetyl—1:1000, pyrazine—10 mg/ml, 2,4,5-trimethylthiazole—1:1000, 2,3-pentanedione—1:1000. (Adapted from Sengupta, 1996.)

attractive volatile chemical 2,3-butanedione (diacetyl) while responding normally to all other chemicals tested, including chemicals of related structure (Fig. 3.1). *odr-10* encodes a protein predicted to contain seven transmembrane domains, characteristic of G protein-coupled receptors. ODR-10 is localized specifically to the cilia of the AWA neurons, which had been shown in laser killing experiments to be required to sense diacetyl. When *odr-10* is eliminated from the AWA neurons, but misexpressed in the AWB neurons, which normally sense volatile repellents, the animals are now repelled by the normally attractive odorant diacetyl (Troemel et al., 1997). These genetic experiments demonstrated the in vivo ligand specificity of an odorant receptor. Heterologous expression experiments where *odr-10* was expressed in HEK293 cells have further confirmed that ODR-10 confers selective responses to diacetyl (Wellerdieck et al., 1997; Zhang et al., 1997).

Examination of the genome sequence of *C. elegans* has revealed a plethora of olfactory receptors. The predicted number of receptors currently stands at about 800, of which about 550 are predicted to encode functional receptors (Troemel et al., 1997; Robertson, 1998). The remainder are most likely pseudogenes, based on the presence of mutations in the genomic sequence and lack of expression (see below). Based on primary sequence, these receptors can be subdivided into several families called *sra*, *srb*, *srg*, *srd*, *sre*, *stl*, and *str* (Troemel et al., 1995, 1997; Robertson, 1998). *str* family members are most closely related to *odr-10* and number approximately 200 (Troemel et al., 1997; Robertson, 1998). Olfactory receptors in worms share limited homology to olfactory receptors from rodents, frogs or fish (See Chapter 8, McClintock). They also do not exhibit similarities to the two families of putative pheromone receptors that have been identified in rodents. Thus, although these receptors may have arisen from the same branch of the G protein-coupled receptor superfamily, they appear to have diverged enormously during evolution.

Examination of the expression patterns of the predicted receptor genes in *C. elegans* provides further evidence that these are chemosensory receptors. The expression patterns of predicted genes can be inferred to a first degree by creating transgenic worms expressing a fusion gene between the reporter GFP and the promoter of the gene being studied. These experiments with a subset of the olfactory receptors showed that many (although not all) of the predicted olfactory receptors are expressed in small subsets of chemosensory neurons (Troemel et al., 1995, 1997; Sengupta et al., 1996). Analyses of the expression patterns of these genes have suggested a molecular basis for the behavioral data described earlier. Each neuron expresses multiple receptors from different families, thus suggesting a mechanism to explain how each neuron responds to multiple chemicals. Also, some receptors are expressed in multiple neuron types, suggesting a mechanism for how some chemicals are sensed by multiple neurons. It should be noted that some of the predicted receptors may be involved in processes other than chemosensation, as demonstrated by their expression in other types of neurons. However, this could also reflect an artifact of the transgenic expression experiments. Interestingly, a subset of these receptors shows sexually dimorphic expression patterns and are expressed in male-specific neurons required during mating (Troemel et al., 1995).

Examination of animals carrying fusion genes where GFP was fused to the C-terminus of an olfactory receptor showed that several of these receptor proteins are localized to the sensory cilia (Sengupta et al., 1996; Dwyer et al., 1998). Behavioral screens have identified genes required for the localization of specific olfactory receptors (Dwyer et al., 1998). *odr-4* (encoding a novel protein) and *odr-8* mutants exhibit a restricted set of chemosensory defects. In these animals, ciliary localization of a subset of olfactory receptors was found to be specifically affected, accounting in part for their behavioral defects. Interestingly, not all functions of these neurons were defective, suggesting that other receptors were localized properly in these mutants. Furthermore, ciliary localization of other signal transduction genes was not affected (Dwyer et al., 1998).

Analysis of the genomic organization of the receptor genes revealed that many of these genes are clustered in the genome in clusters of two to ten genes. As in vertebrates, some of the clusters contain genes from different families and are expressed in different neurons (Ben-Arie et al., 1994; Troemel et al., 1995; Sullivan et al., 1996). However, a subset of these clusters in C. *elegans* are unique in that they may represent polycistronic clusters. Polycistronic clusters consist of genes that are transcribed coordinately from a common promoter. Such clusters of related genes have been demonstrated in C. *elegans* and may represent a mechanism of coordinate regulation (Zorio et al., 1994). It is yet to be shown that the expression of all the receptors from such clusters is spatially or temporally coordinated in C. *elegans*.

2.5. G Proteins

Seven transmembrane domain receptors signal intracellularly by interaction with heterotrimeric G proteins. The genome sequence reveals the presence of several predicted G-protein α subunits in C. *elegans*. Functions have been defined for several of these gene products (Mendel et al., 1995; Segalat et al., 1995; Brundage et al., 1996; Korswagen et al., 1997; Zwaal et al., 1997; Berger et al., 1998; Roayaie et al., 1998, Jensen et al., 1999). Several Gα protein genes are expressed in chemosensory neurons and a few have been implicated in regulating chemosensory behaviors. These include the genes *odr-3*, *gpa-2*, and *gpa-3* (Zwaal et al., 1997; Roayaie et al., 1998; Jensen et al., 1999).

The gene *odr-3* was identified in genetic screens similar to those in which mutations in *odr-10* were identified (Bargmann et al., 1993). *odr-3* mutants fail to respond to most chemicals sensed by the olfactory neurons AWA, AWB, and AWC. In addition, they exhibit defects in sensing chemicals of high osmotic strength and in the response to light mechanical touch (behaviors mediated by the ASH neurons), implicating *odr-3* in the sensory transduction pathways for multiple modalities (Bargmann et al., 1993; Roayaie et al., 1998). *odr-3* encodes a member of the Gα protein family with greater homology to the *gpa* class of Gα proteins in C. *elegans* than to vertebrate G proteins. Among vertebrate G proteins, ODR-3 shares the most similarity to the G_i and G_o family of G proteins. The ODR-3 protein is expressed specifically in the olfactory neurons whose functions are affected in the mutant, with strongest expression observed in the AWC neurons. The protein is localized to the sensory cilia of these neurons (Roayaie et al., 1998).

Overexpression of either wild-type copies of *odr-3* or constitutively activated alleles results in strong defects in chemosensation similar to those observed in loss-of-function alleles, providing further evidence of the direct role of ODR-3 in transducing sensory signals (Roayaie et al., 1998). Interestingly, further analysis of the loss-of-function and the overexpressing phenotypes indicated that *odr-3* also plays a role in the morphogenesis of the sensory cilia. However, the role of *odr-3* in ciliary morphogenesis appears to be distinct from its role in odorant sensation and the alteration in ciliary structure does not appear to cause the odorant defects. These genetic experiments showed that *odr-3* plays a role in both sensory transduction and in ciliary morphogenesis.

In addition to *odr-3*, 13 other Gα subunit genes are expressed almost exclusively in different and partly overlapping subsets of chemosensory neurons. Analysis of the behavioral phenotypes of animals carrying null mutations in these *gpa* genes as well as gain-of-function alleles indicated that several of these genes play a role in normal chemotaxis towards dauer pheromone, other water-soluble attractants, water-soluble repellents, and volatile chemicals. (Fino Silva and Plasterk, 1990; Lochrie et al., 1991; Zwaal et al., 1997; Jensen et al., 1999).

These experiments indicate that multiple G proteins may function in parallel to transduce and integrate chemosensory signals in *C. elegans*. Since each chemosensory neuron expresses many different chemosensory receptors, it is likely that different subsets of them couple with distinct G proteins. The final behavioral output of the animal is then a result of the integration of the signals transduced through several pathways.

2.6. Second Messengers

Genetic and physiological evidence in vertebrates suggests that the signal from most odorants is transduced via a rise in intracellular cAMP levels and the opening of a cyclic nucleotide gated channel (Bakalyar and Reed, 1990; Dhallan et al., 1990; Bradley et al., 1991; Goulding et al., 1992; Liman and Buck, 1994; Brunet et al., 1996; Liu et al., 1996). Experiments in *C. elegans* implicate cGMP and a cGMP-gated channel in the transduction of odorant signals in a subset of chemosensory neurons, while signal transduction in other chemosensory neurons may be mediated by calcium.

A family of membrane receptor guanylyl cyclases comprising about 30 members has been identified in the genome sequence (Yu et al., 1997). Some of these genes appear to be expressed in subsets of sensory neurons including chemosensory neurons, and one (GCY-10) appears to be localized to the sensory cilia of a few neurons. These gene products may direct the generation of cGMP in response to odorant signals. However, it is also possible that a subset of these molecules may directly bind odorants.

Behavioral screens have resulted in the identification of the genes encoding the two subunits of a cyclic nucleotide gated channel. The gene *tax-2* encodes the β subunit and the gene *tax-4* encodes the α subunit (Coburn and Bargmann, 1996; Komatsu et al., 1996). Heterologous expression studies have shown that TAX-4 alone can form a functional channel and that this channel is preferentially sensitive to cGMP (Komatsu et al., 1996). Among the neurons whose sensory functions are mediated by the TAX-2/TAX-4 channel are the AWC olfactory neurons, the chemosensory ASE neurons, the volatile repellent-sensing AWB neurons, and the thermosensory AFD neurons. The functions of other chemosensory neurons including the AWA olfactory neurons and ASH polymodal neurons are mediated by a channel encoded by the *osm-9* gene (Colbert et al., 1997). *osm-9* encodes a predicted calcium-selective channel with limited homology to the recently identified capsaicin receptor in nociceptive neurons and the *Drosophila* phototransduction channels *trp* and *trp*-like (Montell and Rubin, 1989; Phillips et al., 1992; Caterina et al., 1997).

Thus, in *C. elegans*, at least two second messenger systems appear to function in chemosensory signal transduction (Fig. 3.2). However, as mentioned above, the G-protein encoding gene *odr-3* affects the functions of both neurons such as AWC that uses cyclic nucleotides, as well as AWA that is predicted to use Ca^{2+} as second messengers (Roayaie et al., 1998). *odr-3* then must be involved in both these pathways in different cell types. In addition, these cells express different odorant receptors. Are receptors expressed in one neuron type constrained to act

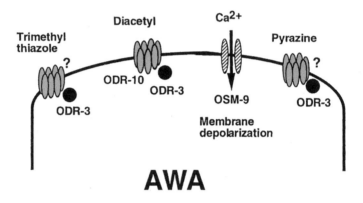

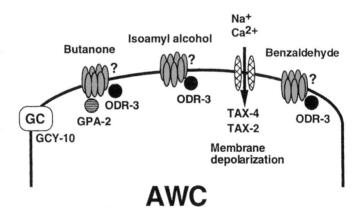

FIG. 3.2. Signaling mechanisms used by the AWA and AWC olfactory neurons. Volatile chemicals are recognized by seven transmembrane domain receptors such as ODR-10. Receptors for other chemicals have not yet been identified. Signals are transduced by G protein α subunits such as ODR-3 and GPA-2. In the AWA neurons, the putative calcium channel encoding gene *osm-9* is required for sensory responses. In the AWC neurons, guanylyl cyclases such as GCY-10 may play a role in the generation of cyclic nucleotides in response to ligand(s). A cyclic nucleotide-gated channel encoded by the *tax-2* and *tax-4* genes is required for AWC-mediated sensory responses. See text for additional details.

only through the type of channel normally found in that cell type? The olfactory receptor ODR-10 can signal via different second messenger systems depending on the cell where it is expressed. In the AWA neurons where ODR-10 is expressed, ODR-10 responds via the *osm-9* encoded channel. However, when misexpressed in the AWB neurons (which do not express *osm-9*), ODR-10 can signal through the *tax-2/tax-4* encoded channel (Troemel et al., 1997).

3. *DROSOPHILA* AS A MODEL SYSTEM FOR THE STUDY OF CHEMOSENSATION

3.1. Why Flies?

The fruit fly *Drosophila melanogaster*, like the worm, offers several important advantages for the study of chemosensation (Siddiqi, 1987; Carlson, 1991, 1996). The number of its chemosensory neurons is small relative to vertebrates, yet it is capable of sophisticated chemosensory function. A variety of powerful genetic and molecular methods allow convenient study of both smell and taste. Perhaps most important is that chemosensory function can be measured easily in vivo, either in simple behavioral paradigms or in tests of physiology. Both olfactory and gustatory function can be conveniently measured physiologically, either at low resolution by recording from an entire chemosensory organ or at higher resolution through measurements from an individual chemosensory neuron.

The chemical senses are ancient sensory modalities, and many principles of their function and organization are well conserved between insects and vertebrates (Hildebrand, 1995). There are many striking parallels between flies and mammals, ranging from molecular components of the periphery, such as odorant-binding proteins (OBPs), to the architectural organization of olfactory processing centers, which show a highly conserved glomerular structure (see Chapter 9, Christensen and White).

3.2. Chemosensory Behavior of *Drosophila*

Drosophila is sensitive to a wide array of odors. Most tested odors elicit an attractive response in a Y-maze paradigm, although many that act as attractants at low concentrations are repellents at high concentrations (Ayyub et al., 1990). Some odors, such as benzaldehyde, repel flies at all concentrations tested. Larvae are attracted to most odors that have been tested, although they are repelled by a long-chain alcohol, nonanol (Cobb et al., 1992).

Pairwise combinations of odors have been tested in odor-masking experiments (Rodrigues, 1980; Siddiqi, 1983), similar to those with worms. Larvae were tested for their ability to respond to one odor in a saturating background of a second odor. These experiments have allowed odors to be grouped into five classes, with odors in each class reducing response to other odors in the same class. Odors within a class might compete for the same binding site; alternatively, signals elicited by two odors within a class might converge at a limiting step subsequent to binding. The classes thus defined consisted of molecules with the same functional group, that is, alcohols, aldehydes, ketones, organic acids, and acetate esters, with few exceptions.

Sudden changes in the concentration of odors in a fly's environment can elicit a jump response (McKenna et al., 1989). This behavior has been interpreted as an escape response, and is elicited by some, but not all, odors. The response is dose-dependent and is mediated primarily by the antenna. Flies also are capable of osmotropotaxis, that is, orienting toward an odor (Borst and Heisenberg, 1982).

When a gustatory sensillum on either the leg or the labellum is stimulated, a fly extends its mouthparts, in a behavior known as the proboscis extension response (PER). This response has been characterized in great detail, both behaviorally and physiologically, in species of large flies (Dethier, 1976). In *Drosophila* the PER is elicited by sugar, but inhibited when NaCl is added to the sugar (Arora et al., 1987).

Flies demonstrate strong preferences when presented with two taste stimuli (Tompkins et al., 1979). Flies manifest preferences for different sugar solutions, as shown in an elegant paradigm in which animals are allowed to feed from the wells of a microtiter dish (Tanimura et al., 1982). Alternate wells of the dish contain agar with either of two sugars. Wells containing one sugar are marked with red dye; those containing the other sugar are marked with blue dye. After feeding in the dark, flies are classified according to the color of their abdomen, which provides an indication of their feeding preferences. Larvae also show a strong contact chemosensory response (Tompkins, 1979; Miyakawa et al., 1980; Miyakawa, 1981; Lilly and Carlson, 1990).

Flies show a variety of sophisticated reproductive behaviors. For example, male flies perform a stereotyped sequence of behaviors toward virgin females, including orientation toward the female, following, tapping of her abdomen with his foreleg, wing vibration, licking of her genitalia, and then mounting (Hall, 1994). These behaviors are stimulated by various cues, of which some are chemosensory (Jallon, 1984; Ferveur, 1997; Ferveur et al., 1997).

3.3. Chemosensory System Development

The fly antenna and maxillary palp develop from the eye-antennal imaginal disc. The development of an olfactory sensillum appears to follow a logic distinct from that of embryonic sense organs, which are largely lineage dependent, and retinal ommatidia, which are dependent on cellular interactions. Rather, the development of an olfactory sensillum on the antenna depends on both lineage mechanisms and cellular interactions (Reddy et al., 1997). The process begins with the selection of a founder cell in the antennal imaginal disc (Ray and Rodrigues, 1995). A few hours later, several adjacent cells not related to the founder cell by lineage begin to express sensory markers, and, along with the founder cell, form a presensillum cluster, which divides as a group to give rise to the cells of a sensillum. The final number of cells in a sensillum is influenced by programmed cell death (see also Chapter 10, Burd and Tolbert).

The specification of one class of sensilla, the sensilla coeloconica, depends on a basic helix-loop-helix (bHLH) transcription factor, *atonal* (Reddy et al., 1997). The genes responsible for specifying the other two classes of antennal sensilla have not yet been identified, but the process does not require the *achaete-scute* genes, which specify most of the other external sense organs studied to date. The *lozenge* gene, which also encodes a transcription factor (Daga et al., 1996), is required in the early development of basiconic sensilla (Lienhard and Stocker, 1991). Antennae of *lozenge*[3] mutants lack all basiconic sensilla (and some trichoid sensilla) (Stocker et al., 1993), and many of the founder cells fail to form (Ray and Rodrigues, 1995). These defects lead to profound defects in olfactory physiology (Riesgo-Escovar et al., 1997a, 1997b).

The larval antennal nerve appears to be the precursor of the adult antennal nerve, and is used as a pathway for the adult projections (Tissot et al., 1997). Afferents from the antennal disc join the antennal nerve beginning at 6–9 h after puparium formation (APF). Adult antennal afferents arrive in the antennal lobe 24 h APF and in the contralateral lobe at 30 h APF (Stocker et al., 1995; Tissot et al., 1997). Projections from the maxillary palp are believed to have entered the antennal lobe by 36 h APF. Larval afferents appear to persist until the mid-pupal stage. The antennal lobe derives from a larval precursor. It increases in volume by 50- to 100-fold, and changes from an apparently homogeneous structure into a glomerular structure, with glomeruli becoming visible at 48 h APF (Stocker et al., 1995).

Taste bristles on the labellum develop from the labial imaginal discs during the first 30 h of pupation (Ray et al., 1993). The sensory progenitors are specified in three waves that occur at 0, 6, and 16 h after pupation. The mitotic activity in those bristles specified in the first wave is completed before the onset of the second wave. A single bristle is composed of eight cells that share a common lineage. The trichogen cell, which forms the shaft of the bristle, and the tormogen cell, which forms the socket, are siblings from a common precursor; the thecogen cell has a common ancestry with the five neurons that it surrounds.

3.4. Odorant-Binding Proteins

Insects, like vertebrates, contain small, abundant soluble proteins in the aqueous medium surrounding olfactory receptor neurons (Vogt and Riddiford, 1981; Pelosi, 1994). Some of these proteins have been shown to bind odorants, or pheromones, and are called odorant-binding proteins (OBPs). They are believed to shuttle hydrophobic odorants through the aqueous medium to and/or from odorant receptors.

Drosophila contains a family of genes encoding OBP homologs (McKenna et al., 1994; Pikielny et al., 1994; see also Chapter 7, Ache and Restrepo). These genes exhibit sequence similarity to moth OBP genes, including the presence of six cysteines at stereotyped positions. Different members of the *Drosophila* OBP family are expressed in different subregions of the third antennal segment, with at least some genes also expressed in the maxillary palps and proboscis (Pikielny et al., 1994; Ozaki et al., 1995). The spatial segregation of OBP gene expression within the olfactory system is consistent with physiological evidence for spatial segregation of different functional types of sensilla (Siddiqi, 1987; Clyne et al., 1997).

Although OBP homologs are dispersed in several chromosomal regions, at least two of the genes are clustered in a pair, less than 1 kb apart (Hekmat-Scafe et al., 1996). These two genes, OS-E and OS-F, were found by immunoelectron microscopy to be co-expressed in the same sensilla. These results showed that an individual sensillum can express more than one OBP, which may have implications for olfactory coding. Different moth OBPs have been shown to bind different odorants preferentially (De Kramer and Hemberger, 1987; Du and Prestwich, 1995). Olfactory receptor neurons in insects are compartmentalized into sensilla, and the expression of different OBPs in a sensillum could influence to which odors the neurons within a given sensillum have access. Expression of multiple OBPs in a sensillum could increase the range of odors that evoke responses, and different combinations of OBPs could allow responses to different combinations of odors. Thus, overlapping expression patterns of OBPs might provide a molecular basis for one means of combinatorial coding in the olfactory system.

3.5. Olfactory Mutants

A diverse collection of olfactory mutants has been identified, reflecting the diversity of approaches used to isolate them. Most mutants have been isolated in behavioral screens, using a variety of paradigms. Some of these behavioral paradigms are striking in their simplicity: for example, the *smell impaired (smi)* mutants were isolated by virtue of their reduced tendency to walk away from a Q-tip soaked in the repellent benzaldehyde (Anholt et al., 1996). Other paradigms, such as the Y-maze paradigm used to isolate the *odA* mutant (Alcorta, 1991), have employed apparatus remarkable for the sophistication of its design and engineering. A number of olfactory mutants have been identified by testing the olfactory response of mutants originally isolated by other criteria. These range from phototransduction mutants (Riesgo-Escovar et al., 1995), to the *anachronism (ana)* mutant, which is defective in a glycoprotein repressor of neuroblast proliferation secreted by glia in the developing larval CNS and which has a strong defect in larval olfaction (Park et al., 1997).

Many olfactory mutants, although not all, are defective in both larval and adult olfaction. Although the larval and adult olfactory organs are distinct in morphology and developmental origins, the genetic foundation of olfaction clearly is largely shared between larvae and adults. This conclusion is also supported by molecular evidence from a large-scale enhancer trap screen, which produced a set of lines showing reporter gene expression in both larval and adult olfactory organs, with little expression elsewhere (Riesgo-Escovar et al., 1992).

3.5.1. Mutants with Odor-Specific Defects

Among the earliest olfactory mutants isolated were the *olf* mutants (Rodrigues and Siddiqi, 1978; Ayyub et al., 1990; Hasan, 1990), isolated in a Y-maze behavioral paradigm. Interestingly, most of these mutants showed partial anosmias, with defective response to only a subset of odors. For example, *olfA* mutants are defective in response to aldehydes, including benzaldehyde, salicylaldehyde, and formaldehyde, but their responses to ethyl acetate, acetone, acetic acid, and ethanol are normal. Mutants of *olfB* and *olfE* were also isolated as benzaldehyde-defective mutants with normal behavior to ethyl acetate and certain other odors, whereas mutants of *olfC* were isolated by virtue of defective response to acetate esters, with responses to benzaldehyde normal in almost all cases. Table 3.3 gives a listing of the mutants discussed below.

A mutant defective in response to benzaldehyde (odor of almond), but normal to all other odors tested, was isolated in a T-maze behavioral paradigm (Helfand and Carlson, 1989). The mutation also produces female sterility and hyperpigmentation of the dorsal thorax; subsequent testing showed that it is a null allele of *pentagon* (*ptg*), a gene identified by Bridges in 1922 and named for its pentagonal patch of thoracic pigmentation. The olfactory specificity of the *ptg*

TABLE 3.3 **Drosophilla olfactory mutants discussed in this review and brief description of salient phenotypes**

Mutant	Salient Phenotypes*
acj6 (abnormal chemosensory jump 6)	Reduced jump response to several odors; abnormal physiological response in antenna and palp
ana (anachronism)	Reduced larval response
Indf *(Indifferent)*	Abnormal larval responses to odors with 8–10 carbons
norpA (no receptor potential A)	Reduced physiological response in maxillary palp
odA	Abnormal behavioral and physiological responses to ethyl acetate
olfA	Reduced response to aldehydes
olfB	Reduced response to benzaldehyde
olfC	Reduced response to acetate esters
olfD (allelic to smellblind*)*	Reduced larval and adult responses to all odors
olfE	Reduced response to benzaldehyde
ptg (pentagon)	Reduced response to benzaldehyde
rdgB (retinal degeneration B)	Abnormal behavioral and physiological responses
Sco *(Scutoid)*	Reduced physiological response to ethyl acetate and acetone
smi (smell impaired); multiple mutants of this class	Reduced avoidance of benzaldehyde

*Behavioral responses at the adult stage, unless otherwise specified.

defect is especially interesting in light of the pleiotropism of the mutant phenotype: it indicates that a mutation can be odorant-specific without being olfactory-specific. Thus the development or physiology of specific odor pathways may require more general components that are not required by other odor pathways.

The *acj6* (*abnormal chemosensory jump*) mutant was isolated in a screen for mutants defective in the olfactory jump paradigm, which has been used to isolate a number of mutants (McKenna et al., 1989; Anand et al., 1990). *acj6* has a severe defect in the electrical response of the antenna and maxillary palp to most odors (Ayer and Carlson, 1991, 1992). However, not all odor responses are equally affected, with some, such as EPG response to benzaldehyde, altered little if at all. Visual system physiology of *acj6* was normal, but the mutant shows reduced mobility, again consistent with an additional role for an odor-specific gene outside the olfactory system.

The *Scutoid* (*Sco*) mutant has a different type of odor-specific physiological defect. *Sco* is defective in its response to ethyl acetate and acetone, but is normal in its response to benzaldehyde, propionic acid, butanol, isobutanol, and acetic acid. Its visual physiology also appears normal. *Sco* is a commonly used genetic marker easily identifiable by its loss of thoracic bristles, and was identified as an olfactory mutant independently in two laboratories that were each using it as a marker in screens for olfactory mutants (Dubin et al., 1995). The molecular basis of its olfactory defect is unknown.

Particularly interesting phenotypes are shown by the *Indifferent* (*Indf*) mutants, which show defects in larval responses to odorants between 8 and 10 carbons in length (Cobb et al., 1992; Cobb, 1996). Two alleles have been isolated, and the locus has been mapped to position 96A2-7 on the third chromosome. Remarkably, depending on the odorant and the allele, *Indifferent* mutations may eliminate an attraction response, eliminate a repulsion response, cause an attractive response to a stimulus that elicits no response from wild-type, or cause a repulsion response to a stimulus which elicits none in wild-type. In the case of octanol, which elicits no response from wild-type, one allele gives an attractive response, whereas the other allele gives a repellent response.

3.5.2. Mutants Associated with a Na^+ Channel

Some mutants, such as the *olfD* mutant (Ayyub et al., 1990), are abnormal in response to all odors tested. Subsequent analysis of *olfD* mutations revealed them to be allelic to the *smellblind* gene (*sbl*) (Lilly and Carlson, 1990). *sbl* was identified in a screen for mutants defective in an olfactory-driven learning paradigm, and subsequently was found to be an olfactory mutant (Aceves-Piña and Quinn, 1979). Further analysis of *olfD* and *sbl* mutations showed that they are alleles of the *para* voltage-gated Na^+ channel (Lilly et al., 1994a). Interestingly, the viable *olfD* and *sbl* mutations also showed both heat- and cold-sensitive developmental lethality (Lilly et al., 1994b). The cold-sensitivity is reminiscent of a cold-sensitive human sodium-channel disease, paramyotonia congetica (Ptacek et al., 1992).

3.5.3. Mutants Associated with the IP$_3$ Transduction Pathway

ota1 (olfactory trap abnormal) was isolated as a mutant that failed to enter an olfactory trap constructed of a microfuge tube and pipette tips (Woodard et al., 1989). The mutant has a physiological defect in odorant response of its antenna and maxillary palp, as well as a physiological defect in its eye (Woodard et al., 1992; Riesgo-Escovar et al., 1994). Genetic analysis showed that it was allelic to *rdgB (retinal degeneration B)*, a well-studied visual system mutant (see Chapter 7, Ache and Restrepo). These results showed that at least one gene is required for normal physiology in two sensory systems, olfaction and vision.

rdgB encodes a phosphatidyl inositol transfer protein, and immunocytochemistry confirmed that it is present in both the antenna and maxillary palp (Vihtelic et al., 1993; Riesgo-Escovar et al., 1994). What role does *rdgB* play in olfactory physiology? There is strong evidence that phototransduction in the *Drosophila* eye is mediated by the IP$_3$ signal transduction cascade, which is implicated in olfactory transduction in larger invertebrates (Breer et al., 1990; Fadool and Ache, 1992). In this pathway, phosphatidyl inositol bisphosphate, or PIP$_2$, is cleaved by phospholipase C into diacylglycerol, which activates a protein kinase, and IP$_3$, which acts to increase Ca^{+2} levels. After PIP$_2$ is cleaved, there is a need to replenish it in the membrane. One model for the role of *rdgB* is that it acts in replenishing membranes with PI, which is then phosphorylated to PIP$_2$.

These results suggested the possibility that the *norpA*-encoded phospholipase C might be required for olfactory physiology in *Drosophila*. *norpA* mutants were found to be defective in the amplitude of response to all tested odors in the maxillary palp (Riesgo-Escovar et al., 1995). Interestingly, the decrease was severe, but not complete, and the mutants showed no reduction whatsoever in the antenna, consistent with the possibility that one of the other phospholipase C genes identified in *Drosophila* might also play a role in olfactory physiology. Molecular analysis confirmed that *norpA* is expressed in the maxillary palp, but revealed no expression in the antenna, consistent with the physiological evidence.

Further evidence for a role of the IP$_3$ pathway in olfactory transduction comes from the finding that an IP$_3$ receptor homolog is expressed in the third antennal segment and the maxillary palp, as well as the proboscis (Raghu and Hasan, 1995). Moreover, a particular splice form of a gene encoding a G-protein α subunit is expressed in certain olfactory and gustatory neurons (Talluri et al., 1995). However, expression of a cyclic nucleotide-gated channel gene also has been detected in the antenna (Baumann et al., 1994), and it is clear that further work is required to elucidate the role of different signaling pathways in *Drosophila* olfaction.

3.5.4. Taste Mutants

Among the first taste mutants isolated in *Drosophila* were the *Lot* mutants, named for their high tolerance to salt (Falk and Atidia, 1975). They were isolated by screening for mutants that fed upon a sucrose solution containing high concentrations of salt and a red dye. Wild-type flies are inhibited by the salt; the mutants fed and could be distinguished by their externally visible red intestines.

A behavioral countercurrent paradigm, in which flies make a series of binary choices between two different substrates, was used to isolate 12 *gus* (*gustatory*) mutants, which defined six X-linked genes (Tompkins et al., 1979). Interestingly, these mutants include four that showed mistactic responses, that is, they partitioned preferentially onto substrates containing stimuli that the wild-type avoided, and some phenotypes were stimulus-specific. *gusA* showed abnormal behavioral response to quinine sulfate, but normal behavior to NaCl. By contrast, *gusC* behaved normally to quinine sulfate but abnormally to NaCl. *gusD* and *gusE* behaved abnormally to both stimuli. *gusA* and *gusE* showed defects in the proboscis extension assay when sugar was used as stimulus, whereas *gusC* and *gusD* responded normally. *gusA* affected the behavior of both larvae and adults tested with quinine sulfate, and analysis of a temperature-sensitive allele has shown that its temperature-sensitive period is during embryogenesis, consistent with a developmental role for the gene. *gusE* affected the response of adults to quinine sulfate without affecting larval behavior, and analysis of a temperature-sensitive allele has also provided evidence for a developmental basis of the phenotype (Tompkins, 1979).

A set of *gust* mutants was isolated independently using the proboscis extension assay and a Y-maze in which one arm was spread with quinine sulfate (Rodrigues and Siddiqi, 1978). These mutants included two allelic mutants that do not respond to sucrose, although they respond to quinine and NaCl, a mutant (*gustB*) with abnormal response to NaCl but normal response to quinine and sucrose, and a mutant that appears blind to all three stimuli. In all of these mutants, responses were abnormal whether the labellum or the legs were stimulated.

The *gustB* mutants have a particularly interesting phenotype (Arora et al., 1987). Behaviorally they show a greater preference for NaCl than wild-type in feeding preference tests, and the proboscis extension response evoked by stimulation of the tarsi with sugar is much more tolerant to NaCl than in wild-type. Physiologically, the *gustB* gene appears to alter the specificity of the S neuron: analysis of the *gustB* mutant by single-unit recording has shown that the S cell, which in wild-type responds to sugars, is altered in the mutant such that it is excited by salts. This alteration in specificity may explain the increased preference of the mutant for NaCl. The gene has been mapped to region 10E1 on the X chromosome, and several alleles have been isolated.

The *Tre* gene has been suggested to encode a taste receptor capable of binding the sugar trehalose (Tanimura et al., 1982, 1988). The gene was identified through a natural variation in trehalose sensitivity found between two wild-type strains. The differences in trehalose sensitivity were specific, in that sensitivity to fructose, glucose, or sucrose was the same between the two strains. The natural variation maps to a locus called *Tre* in region 5AB of the X chromosome. Two alleles of *Tre* have been identified, with the high-sensitivity allele designated Tre^+ and the low-sensitivity allele designated *Tre*.

The *malvolio* mutant was identified in a screen for taste behavior (Rodrigues et al., 1995); no defects in the physiology of sensory neurons were detected. The *malvolio* gene shows sequence similarity to certain transporter genes.

The *east* gene shows defects in both taste and olfaction; its defects, however, are specific to the adult stage (Vijay Raghavan et al., 1992). Its gustatory responses are normal to sugar and to low concentrations of salt, as measured in feeding preference tests, but the mutants are more tolerant of high salt concentrations.

Analysis of certain mutants defined by virtue of nongustatory phenotypes has shown that the *Shaker* potassium channel gene is required for behavioral response to sucrose, NaCl, and KCl (Balakrishnan and Rodrigues, 1991). The physiology of labellar chemosensory neurons to these stimuli was normal, suggesting a role for the channel in central processing. In the same study, mutants of the *shakingB* gene, whose function is associated with gap junctions, were also found to show defects in taste behavior.

4. THE MOUSE AS A GENETIC MODEL

4.1. Why Mice?

The mouse has become the mammalian genetic model of choice. Inbred strains of mice were established as early as the second decade of the twentieth century. A large number of inbred strains and of recombinant inbred strains derived by systematic inbreeding from the cross of two existing progenitor strains have been established (Green and Witham, 1991). The original motivation for establishing these lines was the study of the genetic determinants of cancer, but these strains have been useful in studying scientific questions as varied as neurodegenerative diseases, obesity, and sensory mechanisms. The importance of mice as a genetic model has been enhanced in the last decade with the advent of a variety of techniques allowing the generation, identification, and recovery of mouse mutations (Flaherty, 1998) and by the development of methods for high-speed positional cloning (Okazaki and Hayashizaki, 1997). In addition, the National Institutes of Health has announced a major effort to map the mouse genome, and to produce large numbers of behaviorally identified mouse mutants.

4.2. Genetics of Taste in Mice

In 1970, two-choice tests first demonstrated the existence of genetic variation in the ability of mice to avoid sucrose octaacetate (SOA) (Warren and Lewis,

1970). Since then a number of reports have confirmed that this avoidance is a monogenic polymorphism (Whitney and Harder, 1986; Lush, 1981) at a single autosomal locus, *Soa*, mapped to a position on chromosome 6, approximately 62 cM from the centromere (Capeless et al., 1992). Three phenotypes have been described for SOA sensitivity (Harder et al., 1992): taster, nontaster, and demitaster. They are determined by three alleles (Soaa, Soab, Soac) respectively. Tasters avoid concentrations of SOA as low as 0.001 nM. Nontasters do not avoid SOA up to nearly saturated concentrations (1 mM) and demitasters fall somewhere in between. Shingai and Beidler (1985) reported that taste nerve responses in demitaster strains were less than those in taster strains, while Miller and Whitney (1989) demonstrated that taster mice have more circumvallate taste buds than nontasters.

Whitney and colleagues have developed congenic strains of mice (Harder et al., 1996), genetically identical except for the *Soa* locus. Behavioral testing of these strains revealed that differences in the *Soa* locus had a major effect on the sensitivity to a subset of bitter compounds. Target gene effects were found for strychnine, brucine, denatonium, PROP and quinine, but not for caffeine, cyclohexamide, thiamine and the nonbitter compounds, NaCl and calcium hydroxide (Boughter and Whitney, 1998). Additional studies have suggested a polygenic inheritance for quinine and PROP avoidance (Harder and Whitney, 1998).

More recent studies have resulted in determination of other loci involved in taste. Two loci on chromosome 4 account for over 50% of variability in sucrose intake between C57BL/6ByJ and 129/J. Peripheral nerve recordings show that one of these loci affects the response threshold, while the other affects response magnitude (Bachmanov et al., 1997). Studies with a series of inbred strains of mouse show that separate genes control the variability of consumption of NaCl (Bachmanov et al., 1999). Similarly, variability in consumption of MSG among mouse strains appears to depend on different genes (Beauchamp et al., 1998).

The advent of molecular techniques has resulted in identification of genes encoding proteins important in taste transduction. Transgenic mice deficient for the G-protein gustducin have been produced (Wong et al., 1996). These mice are less sensitive than controls to some bitter substances as well as to sweet substances. The most parsimonious explanation for these experiments is that gustducin plays a role in taste transduction for both bitter and sweet (See Chapter 13, Glendinning et al.).

4.3. Genetics of Olfaction in Mice

A number of mouse strains have been shown to be anosmic or hyposmic to certain compounds (Reed, 1996). Perhaps the best studied of these strains are mice anosmic for androstenone and isovaleric acid (Wysocki et al., 1977). Diminished EOG responses in androstenone-anosmic mice demonstrate that the anosmia is due to the lack of responsiveness of the olfactory receptor neurons as opposed to a deficit elsewhere in the central olfactory pathways. Interestingly, anosmic mice can be induced to become sensitive to androstenone by repeated exposure to the substance (Wang et al., 1993). The gene responsible for isovaleric acid anosmia has been localized to a region of 0.27 cM in chromosome 4 representing about 300,000 bp of DNA (Reed, 1996).

Gene-targeted mice and transgenic mice have provided important information on olfactory transduction and olfactory coding. Mice deficient in either subunit 1 of the cyclic nucleotide-gated channel (Brunet et al., 1996) or in the G-protein G_{olf} (Belluscio et al., 1998) are deficient in their response to all odors tested to date in an EOG assay. These experiments suggest that odor transduction is mediated by a single pathway involving cAMP (see Chapter 7, Ache and Restrepo). In addition, the most compelling evidence indicating that olfactory receptor neurons expressing the same olfactory receptor type project their axons to two glomeruli in each ipsilateral olfactory bulb comes from work in which gene-targeted mice express the axonal marker tau-lac Z (Mombaerts et al., 1996). These experiments provide evidence for a role of receptors in axon targeting (see Chapter 8, McClintock).

A full treatment of the genetics of olfactory recognition in mice is beyond the scope of this chapter. However, it is important to state that mouse genetics has been essential in understanding chemosensory recognition of self through chemosensory (olfactory and vomeronasal) cues. For example, congenic mice differing at only one amino acid in the class I MHC protein can still recognize each other through chemosensory cues (Yamazaki et al., 1994; see Chapter 5, Johnston).

5. SUMMARY

Genetic models can be immensely useful in identifying the in vivo functions of components of the chemosensory transduction pathways. The ability to trace a behavioral defect to a gene allows a great deal of insight into the function of the gene product.

Although much has been learned about the molecular mechanisms of chemosensation in C. *elegans*, much also remains to be examined. The completion of the genome sequence of C. *elegans* also opens up several new avenues of research. Reverse genetic techniques such as transposon or mutagen-mediated deletions and RNA-mediated interference can be used to determine the functions of genes such as the large family of guanylyl cyclases and olfactory receptors. Genetic techniques should also lead to insight into the mechanisms underlying more complex behaviors such as those arising from integration of information from multiple sensory stimuli.

Drosophila exhibits some striking similarities to vertebrates, both in the molecular genetics and the functional organization of its chemosensory systems. There remain, however, many components of chemosensory signaling that need to be identified. A variety of intriguing mutants of both olfaction and taste invite investigation at the molecular level. Some will likely identify fly homologs of previously defined signaling components; others will likely identify novel components. Molecular analysis of these mutants should be greatly facilitated by the *Drosophila* genome project, which is being carried out at an accelerating pace.

The fly exhibits a remarkably rich behavioral repertoire. Genetic analysis of olfaction and taste has demonstrated the feasibility of analyzing these sophisticated behaviors in *Drosophila*. Further analysis should be useful in gaining new insight into higher-order processing of chemosensory signals.

The mouse is proving to be a powerful model organism for the study of olfaction and taste. A number of variants with interesting chemosensory properties have been studied, and the genes responsible for their phenotypes are likely to be defined in molecular terms in the coming years. At the same time, cloned genes will continue to be incisively investigated by gene-targeting so as to acquire a detailed understanding of their roles in chemosensory function.

In summary, it is evident that exploring the molecular basis of chemosensation in genetic organisms has provided a wealth of information. In several cases, this line of experimentation has confirmed biochemical and/or physiological data obtained in other systems, while in other cases it has allowed the identification of new mechanisms. Further exploration of chemoreception in *C. elegans*, *Drosophila*, and the mouse should help to elucidate the diversity of processes used by organisms to sense their chemical environments.

ACKNOWLEDGMENTS

We thank Cori Bargmann and M. De Bruyne for comments on the manuscript. J. C. is supported by grants from the NIH (DC-02174), the NSF, and the HFSP. P. S. is supported by the NIH (GM-56223), Whitehall Foundation, and the Medical Foundation. P. S. is an Alfred P. Sloan Research Fellow, a Searle Scholar, and a Packard Foundation Fellow. We apologize to those investigators whose work we were unable to cite for lack of space. We thank the editors for contributing the section on mouse chemosensory genetics.

REFERENCES

Aceves-Piña, E. and W. Quinn (1979). Learning in normal and mutant *Drosophila* larvae. *Science* **206**:93–96.

Alcorta, E. (1991). Characterization of the electroantennogram in *Drosophila melanogaster* and its use for identifying olfactory capture and transduction mutants. *J Neurophysiol* **65**:702–714.

Anand, A., J. Fernandes, M. Arunan, S. Bhosekar, A. Chopra, N. Dedhia, K. Sequiera, G. Hasan, K. Palazzolo, K. VijayRaghavan, and V. Rodrigues (1990). *Drosophila* "enhancer-trap" transposants: gene expression in chemosensory and motor pathways and identification of mutants affected in smell and taste ability. *J Genet* **69**:151–168.

Anholt, R., R. Lyman, and T. Mackay (1996). Effects of single P-element insertions on olfactory behavior in *Drosophila melanogaster. Genetics* **143**:293–301.

Arora, K., V. Rodrigues, S. Joshi, S. Shanbhag, and O. Siddiqi (1987). A gene affecting the specificity of the chemosensory neurons of *Drosophila. Nature* **330**:62–63.

Ayer, R. K. and J. Carlson (1991). *acj6*: a gene affecting olfactory physiology and behavior in *Drosophila. Proc Nat Acad Sci USA* **88**:5467–5471.

Ayer, R. K. and J. Carlson (1992). Olfactory physiology in the *Drosophila* antenna and maxillary palp: *acj6* distinguishes two classes of odorant pathways. *J Neurobiol* **23**:965–982.

Ayyub, C., J. Paranjape, V. Rodrigues and O. Siddiqi (1990). Genetics of olfactory behavior in *Drosophila melanogaster. J Neurogenet* **6**:243–262.

Bachmanov, A. A., D. R. Reed, Y. Ninomya, M. Inoue, M. G. Tordoff, and R. A. Price (1997). Genetic loci affecting peripheral sensory responses. *Mamm Genome* **8**:545–548.

Bachmanov, A. A., M. G. Tordoff, and G. K. Beauchamp. (1999). Voluntary sodium chloride consumption by mice: differences among five inbred strains. *Behav Genet* **28**:117–124.

Bakalyar, H. A. and R. R. Reed (1990). Identification of a specialized adenylyl cyclase that may mediate odorant detection. *Science* **250**:1403–1406.

Balakrishnan, R. and V. Rodrigues (1991). The *Shaker* and *shaking*-B genes specify elements in the processing of gustatory information in *Drosophila melanogaster. J Exp Biol* **157**:161–181.

Bargmann C.I. (1998). Neurobiology of the *Caenorhabditis elegans* genome. *Science* **282**:2028–2033.

Bargmann, C. I. and H. R. Horvitz (1991a). Chemosensory neurons with overlapping functions direct chemotaxis to multiple chemicals in *C. elegans. Neuron* **7**:729–742.

Bargmann, C. I. and H. R. Horvitz (1991b). Control of larval development by chemosensory neurons in *Caenorhabditis elegans. Science* **251**:1243–1246.

Bargmann, C. I., J. H. Thomas, and H. R. Horvitz (1990). Chemosensory cell function in the behavior and development of *Caenorhabditis elegans. Cold Spring Harbor Symp Quant Biol* **LV**:529–538.

Bargmann, C. I., E. Hartwieg, and H. R. Horvitz (1993). Odorant-selective genes and neurons mediate olfaction in *C. elegans. Cell* **74**:515–527.

Baumann, A., S. Frings, M. Godde, R. Seifert, and U. Kaupp (1994). Primary structure and functional expression of a *Drosophila* cyclic nucleotide-gated channel present in eyes and antennae. *EMBO J* **13**:5040–5050.

Beauchamp, G. K., A. A. Bachmanov, and L. J. Stein (1998). Development and genetics of glutamate taste preferences. *Ann NY Acad Sci* **855**:412–416.

Belluscio, L., G. H. Gold, A. Nemes, and R. Axel (1998). Mice deficient for G_{olf} are anosmic. *Neuron* **20**:69–81.

Ben-Arie, N., D. Lancet, C. Taylor, M. Khen, N. Walker, D. Ledbetter, R. Carozzo, K. Patel, D. Sheer, and H. Lehrach et al. (1994). Olfactory receptor gene cluster on human chromosome 17: possible duplication of an ancestral receptor repertoire. *Hum Molec Genet* **3**:229–235.

Berger, A. J., A. C. Hart, and J. M. Kaplan (1998). $G_{\alpha s}$-induced neurodegeneration in C. *elegans. J Neurosci* **18**:2871–2880.

Borst, A. and M. Heisenberg (1982). Osmotropotaxis in *Drosophila melanogaster. J Comp Physiol* **147**:479–484.

Boughter, J. D. and G. Whitney (1998). Behavioral specificity of the bitter taste gene *Soa Physiol Behav* **63**:101–108.

Bradley, J., J. Li, N. Davidson, H. A. Lester, and K. Zinn (1991). Heteromeric olfactory cyclic nucleotide-gated channels: a subunit that confers increased sensitivity to cAMP. *Proc Natl Acad Sci U S A* **91**:8891–8894.

Breer, H., I. Boekhoff, and E. Tareilus (1990). Rapid kinetics of second messenger formation in olfactory transduction. *Nature* **345**:65–68.

Brundage, L., L. Avery, A. Katz, U. J. Kim, J. E. Mendel, P. W. Sternberg, and M. I. Simon (1996). Mutations in a C. *elegans* $G_{q\alpha}$ gene disrupt movement, egg laying, and viability. *Neuron* **16**:999–1009.

Brunet, L. J., G. H. Gold, and J. Ngai (1996). General anosmia caused by a targeted disruption of the mouse olfactory cyclic nucleotide-gated cation channel. *Neuron* **17**:681–693.

Capeless, C. G., G. Whitney, and E. A. Azen (1992). Chromosome mapping of *Soa*, a gene influencing gustatory sensitivity to sucrose octaacetate in mice. *Behav Genet* **22**:655–663.

Carlson, J. (1991). Olfaction in *Drosophila*: genetic and molecular analysis. *Trends Neurosci* **14**:520–524.

Carlson, J. (1996). Olfaction in *Drosophila*: from odor to behavior. *Trends Genet* **12**:175–180.

Caterina, M. J., M. A. Schumacher, M. Tominaga, T. A. Rosen, J. D. Levine, and D. Julius (1997). The capsaicin receptor: a heat-activated ion channel in the pain pathway. *Nature* **389**:816–824.

Clyne, P., A. Grant, R. O'Connell, and J. Carlson (1997). Odorant response of individual sensilla on the *Drosophila* antenna. *Invertebr Neurosci* **3**:127–135.

Cobb, M. (1996). Precise genotypic and phenotypic characterization of the *Drosophila melanogaster* olfactory mutation *indifferent*. *Genetics* **144**:1577–1587.

Cobb, M., S. Bruneau, and J. M. Jallon (1992). Genetic and developmental factors in the olfactory response of *Drosophila melanogaster* larvae to alcohols. *Proc R Soc Lond B* **248**:103–109.

Coburn, C. and C. I. Bargmann (1996). A putative cyclic nucleotide-gated channel is required for sensory development and function in C. elegans. *Neuron* **17**:695–706.

Colbert, H. A., T. L. Smith, and C. I. Bargmann (1997). OSM-9, a novel protein with structural similarity to channels, is required for olfaction, mechanosensation, and olfactory adaptation in *Caenorhabditis elegans*. *J Neurosci* **17**:8259–8269.

Collet, J., C. A. Spike, E. A. Lundquist, J. E. Shaw, and R. K. Herman (1998). Analysis of *osm-6*, a gene that affects sensory cilium structure and sensory neuron function in *Caenorhabditis elegans*. *Genetics* **148**:187–200.

Daga, A., C. Karlovich, K. Dumstrei, and U. Banerjee (1996). Patterning of cells in the *Drosophila* eye by *lozenge*, which shares homologous domains with AML1. *Genes Dev* **10**:1194–1205.

Dainty, R. H., R. A. Edwards, and C. M. Hibbard (1985). Time course of volatile compound formation during refrigerated storage of naturally contaminated beef in air. *J Appl Bacteriol* **59**:303–309.

De Kramer, J. J. and J. Hemberger (1987). The neurobiology of pheromone reception. In: G. Prestwich and G. Blomquist, eds. *Pheromone Biochemistry*. New York: Academic Press, pp 433–472.

Dethier, V. (1976). *The Hungry Fly*. Cambridge: Harvard Press.

Dhallan, R. S., K-W. Yau, K. A. Schrader, and R. R. Reed (1990). Primary structure and functional expression of a cyclic nucleotide-activated channel from olfactory neurons. *Nature* **347**:184–187.

Du, G. and G. Prestwich (1995). Protein structure encodes the ligand-binding specificity in pheromone-binding proteins. *Biochemistry* **34**:8726–8732.

Dubin, A., N. Heald, B. Cleveland, J. Carlson, and G. Harris (1995). The *scutoid* mutation of *Drosophila* specifically decreases olfactory responses to short-chain acetate esters and ketones. *J Neurobiol* **28**:214–233.

Dusenbery, D. B. (1974). Analysis of chemotaxis in the nematode *Caenorhabditis elegans* by countercurrent separation. *J Exp Zool* **188**:41–47.

Dusenbery, D. B. (1975). The avoidance of D-tryptophan by the nematode *Caenorhabditis elegans*. *J Exp Zool* **193**:413–418.

Dwyer, N. D., E. R. Troemel, P. Sengupta, and C. I. Bargmann (1998). Odorant receptor localization to olfactory cilia is mediated by ODR-4, a novel membrane-associated protein. *Cell* **93**:455–466.

Fadool, D. and B. Ache (1992). Plasma membrane inositol 1,4,5-trisphosphate-activated channels mediate signal transduction in lobster olfactory receptor neurons. *Neuron* **9**:907–918.

Falk, R. and J. Atidia (1975). Mutation affecting taste perception in *Drosophila melanogaster*. *Nature* **254**:325–326.

Ferveur, J. (1997). The pheromonal role of cuticular hydrocarbons in *Drosophila melanogaster*. *Bioessays* **19**:353–358.

Ferveur, J., F. Savarit, C. O'Kane, G. Sureau, R. Greenspan, and J. Jallon (1997). Genetic feminization of pheromones and its behavioral consequences in *Drosophila* males. *Science* **276**:1555–1558.

Fino Silva, I. and R. H. Plasterk (1990). Characterization of a G-protein alpha-subunit gene from the nematode *Caenorhabditis elegans*. *J Mol Biol* **215**:483–487.

Flaherty, L. (1998). Generation, identification and recovery of mouse mutations. *Methods* **14**:107–118.

Golden, J. W. and D. L. Riddle (1982). A pheromone influences larval development in the nematode *Caenorhabditis elegans*. *Science* **218**:578–580.

Golden, J. W. and D. L. Riddle (1984). A pheromone-induced developmental switch in *Caenorhabditis elegans*: temperature-sensitive mutants reveal a wild-type temperature-dependent process. *Proc Natl Acad Sci U S A* **81**:819–823.

Goulding, E. H., J. Ngai, R. H. Kramer, S. Colicos, R. Axel, S. A. Siegelbaum, and A. Chess (1992). Molecular cloning and single-channel properties of the cyclic nucleotide-gated channel from catfish olfactory neurons. *Neuron* **8**:45–58.

Green, M. C. and Witham, B. A. (1991). Handbook on Genetically Standardized JAX Mice. Bar Harbor, Maine: The Jackson Laboratory.

Hall J. (1994). The mating of a fly. *Science* **264**:1702–1714.

Harder, D. B., C. G. Capeless, J. C. Maggio, J. D. Boughter, K. S. Gannon, G. Whitney, and E. A. Azen (1992). Intermediate sucrose octa-acetate sensitivity suggests a third allele at mouse bitter taste locus *Soa* and *Soa-Rua* identity. *Chem Senses* **17**:381–401.

Harder, D. B., K. S. Gannon, and G. Whitney (1996). SW.B6-*Soa*[b] nontaster congenic strains completed and a sucrose octaacetate congenic quartet tested with other bitters. *Chem Senses* **21**:507–517.

Harder, D. B. and G. Whitney (1998). A common polygenic basis for quinine and PROP avoidance in mice. *Chem Senses* **23**:327–332.

Hart A., S. Sims, and J. Kaplan (1995). Synaptic code for sensory modalities revealed by C. *elegans* GLR-1 glutamate receptor. *Nature* **378**:82–85.

Hasan, G. (1990). Molecular cloning of an olfactory gene from *Drosophila melanogaster*. *Proc Natl Acad Sci U S A* **87**:9037–9041.

Hekmat-Scafe, D., A. Steinbrecht, and J. Carlson (1996). Co-expression of two odorant-binding protein homologs in *Drosophila*: implications for olfactory coding. *J Neurosci* **17**:1616–1624.

Helfand, S. and J. Carlson (1989). Isolation and characterization of an olfactory mutant in *Drosophila* with a chemically specific defect. *Proc Nat Acad Sci U S A* **86**:2908–2912.

Hildebrand, J. (1995). Analysis of chemical signals by nervous systems. *Proc Natl Acad Sci U S A* **92**:67–74.

Jallon, J. M. (1984). A few chemical words exchanged by *Drosophila* during courtship and mating. *Behav Gen* **14**:441–478.

Jansen, G., K. L. Thijssen, P. Werner, M. van der Horst, E. Hazendonk, and R. H. Plasterk (1999). The complete family of genes encoding G proteins of *Caenorhabditis elegans*. *Nat Genet* **21**:414–419.

Kaplan, J. and H. R. Horvitz (1993). A dual mechanosensory and chemosensory neuron in *Caenorhabditis elegans*. *Proc Natl Acad Sci U S A* **90**:2227–2231.

Komatsu, H., I. Mori, and Y. Ohshima (1996). Mutations in a cyclic nucleotide-gated channel lead to abnormal thermosensation and chemosensation in *C. elegans*. *Neuron* **17**:707–718.

Korswagen, H. C., J. H. Park, Y. Ohshima, and R. H. Plasterk (1997). An activating mutation in a *Caenorhabditis elegans* G$_s$ protein induces neural degeneration. *Genes Dev* **11**:1493–1503.

Lienhard, M. and R. Stocker (1991). The development of the sensory neuron pattern in the antennal disc of wild-type and mutant (lz^3, ss^a) *Drosophila melanogaster*. *Development* **112**:1063–1075.

Lilly, M. and J. Carlson (1990). *smellblind*: a gene required for *Drosophila* olfaction. *Genetics* **124**:293–302.

Lilly, M., R. Kreber, B. Ganetzky, and J. Carlson (1994a). Evidence that the *Drosophila* olfactory mutant *smellblind* defines a novel class of sodium channel mutants. *Genetics* **136**:1087–1096.

Lilly, M., J. Riesgo-Escovar, and J. Carlson (1994b). Developmental analysis of the *smellblind* mutants: evidence for the role of sodium channels in *Drosophila* development. *Dev Biol* **162**:1–8.

Liman, E. R. and L. B. Buck (1994). A second subunit of the olfactory cyclic nucleotide-gated channel confers high sensitivity to cAMP. *Neuron* **13**:611–621.

Liu, D. T., G. R. Tibbs, and S. A. Siegelbaum (1996). Subunit stoichiometry of cyclic nucleotide-gated channels and effects of subunit order on channel function. *Neuron* **16**:983–990.

Lochrie, M. A., J. E. Mendel, P. W. Sternberg, and M. I. Simon (1991). Homologous and unique G protein alpha subunits in the nematode *Caenorhabditis elegans*. *Cell Reg* **2**:135–154.

Lush, I. E. (1981) The genetics of tasting in mice. I. Sucrose octaacetate. *Genet Res* **38**:93–95.

Maricq A. V., E. Peckol, and C. I. Bargmann (1995). Mechanosensory signalling in *C. elegans* mediated by the GLR-1 glutamate receptor. *Nature* **378**:78–81.

McKenna, M., P. Monte, S. Helfand, C. Woodard, and J. Carlson (1989). A simple chemosensory response in *Drosophila* and the isolation of *acj* mutants in which it is affected. *Proc Natl Acad Sci U S A* **86**:8118–8122.

McKenna, M., D. Hekmat-Scafe, P. Gaines, and J. Carlson (1994). Putative *Drosophila* pheromone-binding proteins expressed in a subregion of the olfactory system. *J Biol Chem* **269**:16340–16347.

Mendel, J. E., H. C. Korswagen, K. S. Liu, Y. M. Hajdu-Cronin, M. I. Simon, R. H. Plasterk, and P. W. Sternberg (1995). Participation of the protein G$_o$ in multiple aspects of behavior in *C. elegans*. *Science* **267**:1652–1655.

Miller, I. J. and G. Whitney (1989). Sucrose octaacetate taster mice have more vallate taste buds than non tasters. *Neurosci Lett* **360**:271–275.

Miyakawa, Y. (1981). Bimodal response in a chemotactic behaviour of *Drosophila* larvae to monovalent salts. *J Insect Physiol* **27**:387–392.

Miyakawa, Y., N. Fujishiro, H. Kijima, and H. Morita (1980). Differences in feeding response to sugars between adults and larvae in *Drosophila melanogaster*. *J Insect Physiol* **26**:685–688.

Mombaerts, P., F. Wang, C. Dulac, S. K. Chao, A. Nemes, M. Mendelsohn, J. Edmondson, and R. Axel (1996). Visualizing an olfactory sensory map. *Cell* 87:675–686.

Montell, C. and G. M. Rubin (1989). Molecular characterization of the *Drosophila* trp locus: a putative integral membrane protein required for phototransduction. *Neuron* 2:1313–1323.

Okazaki, Y. and Y. Hayashizaki (1997). High-speed positional cloning based on restriction landmark genome scanning. *Methods* 13:359–377.

Ozaki, M., K. Morisaki, W. Idei, K. Ozaki, and F. Tokunaga (1995). A putative lipophilic stimulant carrier protein commonly found in the taste and olfactory systems. *Eur J Biochem* **230**:298–308.

Park, Y., C. Caldwell, and S. Datta (1997). Mutation of the central nervous system neuroblast proliferation repressor *ana* leads to defects in larval olfactory behavior. *J Neurobiol* **33**:199–211.

Pelosi P. (1994). Odorant-binding proteins. *Crit Rev Biochem Mol Biol* **29**:199–228.

Perkins, L. A., E. M. Hedgecock, J. N. Thomson, and J. G. Culotti (1986). Mutant sensory cilia in the nematode *Caenorhabditis elegans*. *Dev Biol* **117**:456–487.

Phillips, A., A. Bull, and L. Kelly (1992). Identification of a *Drosophila* gene encoding a calmodulin-binding protein with homology to the trp phototransduction gene. *Neuron* **8**:631–642.

Pikielny, C., G. Hasan, F. Rouyer, and M. Rosbash (1994). Members of a family of *Drosophila* putative odorant-binding proteins are expressed in different subsets of olfactory hairs. *Neuron* **12**:35–49.

Ptacek, K., R. George, R. Barchi, R. Griggs, J. Riggs, M. Robertson, and M. Leppert (1992). Mutations in an S4 segment of the adult skeletal muscle sodium channel cause paramyotonia congenita. *Neuron* **8**:891–897.

Raghu, P. and G. Hasan (1995). The inositol 1,4,5-triphosphate receptor expression in *Drosophila* suggests a role for IP3 signalling in muscle development and adult chemosensory functions. *Dev Biol* **171**:564–577.

Ray, K., V. Hartenstein, and V. Rodrigues (1993). Development of the labellar taste bristles in *Drosophila*. *Dev Biol* **167**:426–438.

Ray, K. and V. Rodrigues (1995). Cellular events during development of the olfactory sense organs in *Drosophila melanogaster*. *Dev Biol* 167:426–438.

Reddy, V., B. Gupta, K. Ray, and V. Rodrigues (1997). Development of the *Drosophila* olfactory sense organs utilizes cell-cell interactions as well as lineage. *Development* **124**:703–712.

Reed, R. R. (1996). Genetic approaches to mammalian olfaction. *Cold Spring Harbor Symp Quant Biol* **61**:165–172.

Riesgo-Escovar, J, C. Woodard, P. Gaines, and J. Carlson (1992). Development and organization of the *Drosophila* olfactory system: an analysis using enhancer traps. *J Neurobiol* **23**:947–964.

Riesgo-Escovar, J., C. Woodard, and J. Carlson (1994). Olfactory physiology in the maxillary palp requires the visual system gene *rdgB*. *J Comp Physiol A* **175**:687–693.

Riesgo-Escovar, J., D. Raha, and J. Carlson (1995). Requirement for a phospholipase C in odor response: overlap between olfaction and vision in *Drosophila*. *Proc Natl Acad Sci U S A* **92**:2864–2868.

Riesgo-Escovar, J., B. Piekos, and J. Carlson (1997a). The *Drosophila* antenna: ultrastructural and physiological studies in wild-type and *lozenge* mutants. *J Comp Physiol A* **180**: 151–160.

Riesgo-Escovar, J., B. Piekos, and J. Carlson (1997b). The maxillary palp of *Drosophila*: ultrastructure and physiology depends on the *lozenge* gene. *J Comp Physiol A* **180**:143–150.

Roayaie, K., J. G. Crump, A. Sagasti, and C. I. Bargmann (1998). The G_α protein ODR-3 mediates olfactory and nociceptive function and controls cilium morphogenesis in C. *elegans* olfactory neurons. *Neuron* **20**:55–67.

Robertson, H. M. (1998). Two large families of chemoreceptor genes in the nematodes *Caenorhabditis elegans* and *Caenorhabditis briggsae* reveal extensive gene duplication, diversification, movement and intron loss. *Genome Res* **8**:449–463.

Rodrigues, V. (1980). Olfactory behavior of *Drosophila melanogaster*. In: International Conference on Development and Behavior of *Drosophila melanogaster*. Tata Institute of Fundamental Research, 1979. The Development and Neurobiology of *Drosophila*. New York: Plenum, pp 361–371.

Rodrigues, V. and O. Siddiqi (1978). Genetic analysis of chemosensory pathway. *Proc Ind Acad Sci* **87B**:147–160.

Rodrigues, V., P. Cheah, K. Ray, and W. Chia (1995). *malvolio*, the *Drosophila* homologue of mouse NRAMP-1 (Bcg), is expressed in macrophages and in the nervous system and is required for normal taste behaviour. *EMBO J* **14**:3007–3020.

Ruvkun G. and O. Hobert (1998). The taxonomy of developmental control in *Caenorhabditis elegans*. *Science* **282**:2033–2041.

Segalat, L., D. A. Elkes, and J. M. Kaplan (1995). Modulation of serotonin-controlled behaviors by G_o in *Caenorhabditis elegans*. *Science* **267**:1648–1651.

Sengupta, P., J. H. Chou, and C. I. Bargmann (1996). *odr-10* encodes a seven transmembrane domain olfactory receptor required for responses to the odorant diacetyl. *Cell* **84**: 899–909.

Shakir, M. A., T. Fukushige, H. Yasuda, J. Miwas, and S. S. Siddiqui (1993). *C. elegans osm-3* gene mediating osmotic avoidance behaviour encodes a kinesin-like protein. *Neuroreport* **4**:891–894.

Shingai, T. and L. M. Beidler (1985). Interstrain differences in bitter taste responses in mice. *Chem Senses* **10**:51–55.

Siddiqi, O. (1983). Olfactory neurogenetics of *Drosophila*. In: V. Chopra, B. Joshi, R. Sharma, and H. Bansal, eds. Genetics: New Frontiers, New Delhi, Oxford and IBH. pp 243–261.

Siddiqi, O. (1987). Neurogenetics of olfaction in *Drosophila melanogaster. Trends Genet* **3**: 137–142.

Sonnhammer, E. L. and R. Durbin (1997). Analysis of protein domain families in *Caenorhabditis elegans. Genomics* **46**:200–216.

Stocker, R. F. and N. Gendre (1989). Courtship behavior of *Drosophila* genetically or surgically deprived of basiconic sensilla. *Behav Genet* **19**:371–385.

Stocker, R., N. Gendre, and P. Batterham (1993). Analysis of the antennal phenotype in the *Drosophila* mutant *Lozenge. J Neurogenet* **9**:29–53.

Stocker, R., M. Tissot, and N. Gendre (1995). Morphogenesis and cellular proliferation pattern in the developing antennal lobe of *Drosophila melanogaster. Roux's Arch Dev Biol* **205**: 62–72.

Sullivan, S. L., M. C. Adamson, K. J. Ressler, C. A. Kozak, and L. B. Buck (1996). The chromosomal distribution of mouse odorant receptor genes. *Proc Natl Acad Sci U S A* **93**:884–888.

Sulston, J. E. and H. R. Horvitz (1977). Post-embryonic cell lineages of the nematode, *Caenorhabditis elegans. Dev Biol* **56**:110–156.

Sulston, J. E., E. Schierenberg, J. G. White, and J. N. Thomson (1983). The embryonic cell lineage of the nematode *Caenorhabditis elegans. Dev Biol* **100**:64–119.

Talluri, S., A. Bhatt, and D. Smith (1995). Identification of a *Drosophila* G protein a subunit ($dG_{q\alpha}$-3) expressed in chemosensory cells and central neurons. *Proc Natl Acad Sci U S A* **92**:11475–11479.

Tanimura, T., K. Isono, T. Takamura, and I. Shimada (1982). Genetic dimorphism in the taste sensitivity to trehalose in *Drosophila melanogaster. J Comp Physiol* **147**:433–437.

Tanimura, T., K. Isono, and M. Yamamoto (1988). Taste sensitivity to trehalose and its alteration by gene dosage in *Drosophila melanogaster. Genetics* **119**:399–406.

Tissot, M., N. Gendre, A. Hawken, K. Stortkuhl, and R. Stocker (1997). Larval chemosensory projections and invasion of adult afferents in the antennal lobe of *Drosophila. J Neurobiol* **32**:281–297.

Tompkins, L. (1979). Developmental analysis of two mutations affecting chemotactic behavior in *Drosophila melanogaster. Dev Biol* **73**:174–177.

Tompkins. L., M. Cardosa, F. White, and T. Sanders (1979). Isolation and analysis of chemosensory behavior mutants in *Drosophila melanogaster. Proc Natl Acad Sci U S A* **76**:884–887.

Troemel, E. R., J. H. Chou, N. D. Dwyer, H. A. Colbert, and C. I. Bargmann (1995). Divergent seven transmembrane receptors are candidate chemosensory receptors in *C. elegans. Cell* **83**:207–218.

Troemel, E. R., B. E. Kimmel, and C. I. Bargmann (1997). Reprogramming chemotaxis responses: sensory neurons define olfactory preferences in *C. elegans. Cell* **91**:161–169.

Vihtelic, T., M. Goebl, S. Milligan, J. O'Tousa, and D. Hyde (1993). Localization of *Drosophila retinal degeneration B*, a membrane-associated phosphatidylinositol transfer protein. *J Cell Biol* **122**:1013–1022.

VijayRaghavan, K., J. Kaur, J. Paranjapem, and V. Rodrigues (1992). The *east* gene of *Drosophila melanogaster* is expressed in the developing embryonic nervous system and is required for normal olfactory and gustatory responses of the adult. *Dev Biol* **154**:23–36.

Vogt, R. G. and L. M. Riddiford (1981). Pheromone binding and inactivation by moth antennae. *Nature* **293**:161–163.

Wang, H.-W., C. J. Wysocki, and G. H. Gold (1993). Induction of olfactory receptor sensitivity in mice. *Science* **260**:998–1000.

Ward, S. (1973). Chemotaxis by the nematode *Caenorhabditis elegans*: identification of attractants and analysis of the response by use of mutants. *Proc Natl Acad Sci U S A* **70**:817–821.

Ward, S., N. Thomson, J. G. White, and S. Brenner (1975). Electron microscopical reconstruction of the anterior sensory anatomy of the nematode *Caenorhabditis elegans*. *J Comp Neurol* **160**:313–337.

Ware R. W., D. Clark, K. Crossland, and R. L. Russell (1975). The nerve ring of the nematode *Caenorhabditis elegans*: sensory input and motor output. *J Comp Neur* **162**:71–110.

Warren, R. P. and R. C. Lewis (1970). Taste polymorphism in mice involving a bitter sugar derivative. *Nature* **227**:77–78.

Wellerdieck, C., M. Oles, L. Pott, S. Korsching, G. Gisselman, and H. Hatt (1997). Functional expression of odorant receptors of the zebrafish *Danio rerio* and of the nematode C. *elegans* in HEK293 cells. *Chem Sens* **22**:467–476.

White, J. G., E. Southgate, J. N. Thomson, and S. Brenner (1986). The structure of the nervous system of the nematode *Caenorhabditis elegans*. *Phil Trans R Soc Lond B* **314**: 1–340.

Whitney, G. and D. B. Harder (1986) Single-locus control of sucrose octaacetate tasting among mice. *Behav Genet* **16**:559–574.

Wong, G. T., K. S. Gannon, and R. F. Margolskee (1996). Transduction of bitter and sweet taste by gustducin. *Nature* **381**:796–800.

Woodard, C., T. Huang, H. Sun, S. Helfand, and J. Carlson (1989). Genetic analysis of olfactory behavior in *Drosophila*: a new screen yields the *ota* mutants. *Genetics* **123**:315–326.

Woodard, C., E. Alcorta, and J. Carlson (1992). The *rdgB* gene of *Drosophila*: a link between vision and olfaction. *J Neurogenet* **8**:17–32.

Wysocki, C. J., G. Whitney, and B. Tucker (1977). Specific anosmia in the laboratory mouse. *Behav Genet* **7**:171–187.

Yamazaki, K., G. K. Beauchamp, F.-W. Shen, J. Bard, and E. A. Boyse (1994). Discrimination of odortypes determined by the major histocompatibility complex among outbred mice. *Proc Natl Acad Sci U S A* **91**:3735–3738.

Yu, S., L. Avery, E. Baude, and D. A. Garbers (1997). Guanylyl cyclase expression in specific sensory neurons: a new family of chemosensory receptors. *Proc Natl Acad Sci U S A* **94**:3384–3387.

Zechman, J. M. and J. N. Labows (1985). Volatiles of *Pseudomonas aeruginosa* and related species by automated headspace concentration-gas chromatography. *Can J Microbiol* **31**:232–237.

Zhang, Y., J. H. Chou, J. Bradley, C. I. Bargmann, and K. Zinn (1997). The *Caenorhabditis elegans* seven-transmembrane protein ODR-10 functions as an odorant receptor in mammalian cells. *Proc Natl Acad Sci U S A* **94**:12162–12167.

Zorio, D., N. Cheng, T. Blumenthal, and J. Spieth (1994). Operons as a common form of chromosomal organization in C. *elegans. Nature* **372**:270–272.

Zwaal, R. R., J. E. Mendel, P. W. Sternberg, and R. H. Plasterk (1997). Two neuronal G proteins are involved in chemosensation of the *Caenorhabditis elegans* dauer-inducing pheromone. *Genetics* **145**:715–727.

4

Chemesthesis: The Common Chemical Sense

BRUCE BRYANT
Monell Chemical Senses Center, Philadelphia, PA

WAYNE L. SILVER
Department of Biology, Wake Forest University, Winston-Salem, NC

1. INTRODUCTION

The ability to detect irritating, noxious chemical stimuli is found in many organisms ranging from bacteria to protists, to annelids, to insects, to vertebrates. G. H. Parker coined the term *common chemical sense* to describe the sensory system responsible for detecting chemical irritants (Parker, 1912). He reported that catfish with their senses of smell and taste eliminated still are able to avoid noxious chemicals presented to their flanks. Parker suggested that free nerve endings in the epithelium are responsible for the detection of chemical irritants and that similar free nerve endings are located over the entire outer surface of other aquatic vertebrates as well as in mucosae in terrestrial organisms. Keele (1962) defined the common chemical sense as the sense of irritation aroused by the action of noxious chemicals on exposed or semiexposed mucous membranes. However, the presence of sensory endings sensitive to chemicals can be

The Neurobiology of Taste and Smell, Second Edition, Edited by Thomas E. Finger, Wayne L. Silver, and Diego Restrepo.
ISBN 0-471-25721-4   Copyright © 2000 Wiley-Liss, Inc.

demonstrated elsewhere by applying chemicals to exposed blister bases (e.g., Keele, 1962) or directly on the skin (e.g., Green and Flammer, 1989). Although Parker (1922) noted that these irritant-detecting receptors resembled pain receptors, he concluded that "the common chemical sense is a true sense with an independent set of receptors and a sensation quality entirely its own." Evidence has accumulated, however, suggesting that the free nerve endings that respond to chemicals, including irritants, do not constitute a separate, independent sense but rather are part of the general somatic sensory system. The chemosensitive fibers appear to be a subset of pain- and temperature-sensitive fibers that occur throughout the skin and mucosal membranes of the nose, mouth, respiratory tract, eye, and anal and genital orifices. In keeping with the notion that the somatic sensory system is being stimulated, the term **chemesthesis** has been used in place of "common chemical sense" to describe the sensations elicited by the chemical stimulation of free nerve endings (Green et al., 1990).

Somatosensory nerves contain different size classes of sensory nerve fibers that mediate a variety of sensations. These include large, fast-conducting A-β mechanoreceptive fibers (6–12 μm diam.; 35–75 m/sec), as well as smaller, thinly myelinated A-δ fibers (1–5 μm diam: 5–30 m/sec) and small, unmyelinated, slow-conducting C fibers (0.2–1.5 μm diam.; 0.5–2 m/sec) that are either thermoreceptive (responding to cold or warm stimuli) or nociceptive (responding to painful thermal, mechanical, and/or chemical stimuli). Nociceptive fibers can be further subclassified according to their neuropeptide content (calcitonin gene-related peptide [CGRP] or substance P) or other cellular markers (lectin binding carbohydrates).

In mammals, chemesthesis is perhaps best exemplified by nerve fibers in the trigeminal (fifth cranial) nerve innervating the mucosae and skin of the mouth, nose, and eyes. Free nerve endings originating from other cranial nerves (e.g., the glossopharyngeal and vagus in the oral-pharyngeal region) as well as spinal nerves also respond to chemical stimuli. Nevertheless, much of the research on chemesthesis, especially as it relates to taste and smell, has been on trigeminal chemoreception, and that subject will be the focus of this chapter.

The trigeminal nerve consists of three main branches: ophthalmic, maxillary, and mandibular divisions (Fig. 4.1). The cell bodies of most trigeminal sensory nerve fibers lie in the trigeminal (Gasserian) ganglia. The ophthalmic branch contains sensory fibers that innervate the cornea, tear glands, part of the nasal cavity, and the skin on the forehead and upper eyelid. Nerve fibers in the maxillary division carry sensory information from the upper portions of the mouth, the palate, parts of the nasal cavity, and the upper lip and skin of the face. The mandibular branch contains both sensory and motor fibers. The sensory fibers innervate the scalp behind the ear, all elements of the lower jaw, the lower lip, the tongue, and the cheeks. The motor fibers supply the muscles of chewing and certain muscles on the floor of the mouth.

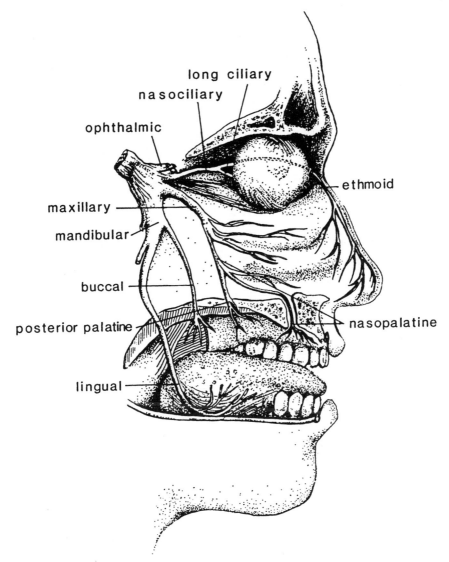

FIG. 4.1. Diagram of the branches of the trigeminal nerve that innervate the nasal and oral cavities. The long ciliary nerve that innervates the cornea (not drawn) is also shown.

2. NASAL CAVITY

2.1. Anatomy

The ethmoid and nasopalatine branches of the trigeminal nerve innervate the nasal cavity (Fig. 4.1). The ethmoid nerve is part of the ophthalmic division while the nasopalatine nerve is part of the maxillary division. Sensory fibers and some sympathetic fibers in the ethmoid nerve are distributed to the anterior nasal mucosa and the external nasal surface (Fig. 4.1). The nasopalatine nerve, which innervates the posterior portion of the nasal cavity (Fig. 1), contains sensory fibers from the maxillary nerve, parasympathetic postganglionic fibers from the sphenopalatine ganglion, and sympathetic postganglionic fibers from the sympathetic ganglia in the neck.

Trigeminal nerve fibers reaching the nasal mucosa ramify repeatedly, terminating in free nerve endings (Cauna et al., 1969). Trigeminal nerve fibers that respond to irritating compounds (see below), contain the neuropeptides substance P and CGRP and extend close to the nasal epithelial surface (Finger et al., 1990b). Electron microscopic serial reconstructions show that the fibers stop at the line of tight junctions only a few micrometers from the surface (Finger et al., 1990b). The proximity of nerve endings so close to the surface suggest that chemical stimuli may simply diffuse from the nasal mucus through the tight junctions, or in the case of lipophilic compounds through the outermost epithelial cells, to interact directly with nasal trigeminal nerve endings.

2.2. Electrophysiology

The earliest recordings from trigeminal nerves were multiunit, integrated responses and primarily examined thresholds to a variety of stimuli (e.g., Ito, 1962; Tucker, 1971; Silver and Moulton, 1982). Thresholds obtained from the rat ethmoid nerve ranged from 3 ppm for octanol to 3020 ppm for methanol (Table 4.1). It appears that the greater the vapor pressure (at 20°C) the lower the threshold, especially for the aliphatic alcohols and acids. Some compounds that

TABLE 4.1 **Thresholds obtained from electrophysiological recordings of the rat ethmoid nerve**

Compound	Threshold (ppm)	Compound	Threshold (ppm)
Octanol	3	Acetic acid	82
Nicotine	5	Diethylamine	108
Heptanol	7	Cyclohexanone	112
Hexanol	8	Benzaldehyde	125
Octanoic acid	9	Benzyl acetate	132
Heptanoic acid	11	Valeric acid	139
Hexanoic acid	16	Formic acid	145
α-Terpineol	19	Butanol	112
l-Carvone	23	Amyl acetate	252
Menthol	27	Ethanol	380
Propionic acid	27	Propanol	457
Linalool	45	Limonene	497
d-Carvone	45	Butyl Acetate	945
Butyric Acid	50	Toluene	2404
Pentanol	59	Methanol	3020

are rated by humans as irritating (e.g., methanol, toluene, and butyl acetate) exhibited high thresholds, while others (e.g., propionic acid and menthol) had relatively low thresholds.

Electrophysiological, whole-nerve studies with capsaicin-treated animals have helped to shed light on the nature of the ethmoid nerve fibers responding to chemical stimuli (Silver et al., 1985, 1991). The administration of capsaicin selectively blocks certain nociceptive neurons (both C and A-δ polymodal nociceptors [PMNs]) so that they are incapable of generating or transmitting impulses (Szolcsanyi et al., 1975). Capsaicin treatment also depletes a variety of neuropeptides in the polymodal nociceptors, including substance P, CGRP, cholecystokinin (CCK), neuropeptide K, somatostatin, and vasoactive intestinal peptide (VIP) (see Holzer, 1988, for review). Since capsaicin treatment depletes neuropeptides, particularly substance P and CGRP, from trigeminal nerve fibers in the nasal cavity (Silver et al., 1991) and eliminates ethmoid nerve responses to chemical stimuli (Silver et al., 1985, 1991), it appears that the chemoreceptive elements in the ethmoid nerve are capsaicin-sensitive, polymodal nociceptors.

A few single unit studies have examined the chemical specificity of nasal trigeminal nerve fiber responses (e.g., Cooper, 1970; Lucier and Egizii, 1989; Anton et al., 1991; Sekizawa and Tsubone, 1994). Many of the trigeminal chemoreceptive units were found to respond to mechanical as well as chemical stimulation. In addition, not all fibers responded to all compounds tested, implying some degree of specificity. Further studies of the ethmoid nerve indicated that thermal (cool) sensitive fibers were stimulated by menthol (Sekizawa et al., 1996).

Slow electrical potentials in response to chemical stimuli have been recorded from the nasal mucosae of rats (Thürauf et al., 1991) and humans (Hummel et al., 1996). These negative mucosal potentials (NMP) appear to be summated receptor potentials from chemical nociceptors, analogous to the electro-olfactogram (EOG) recorded from the olfactory mucosa (see Chapter 7, Ache and Restrepo). The magnitude of the NMP appears to be correlated with stimulus concentration and is eliminated by local anesthetics applied to the nasal cavity. In addition, NMPs were correlated with pain ratings to repetitive stimulation with CO_2 (Hummel et al., 1996). Recently, NMPs were used to determine that the effect of nonsteroidal anti-inflammatory drugs (to decrease tonic pain) is localized in the periphery and is not a central phenomenon (Lötsch et al., 1997a).

Evoked potentials (recorded by electroencephalography) also have been used to examine human nasal trigeminal responses to chemical stimuli. Kobal and Hummel (1988) called potentials evoked by chemesthetic stimuli that were primarily active only on the trigeminal nerve, chemosomatosensory evoked potentials (CSSEP) and those evoked by primarily olfactory stimuli, olfactory evoked potentials (OEP). Maximum CSSEP amplitudes were recorded at parieto-central sites and maximum OEP amplitudes were seen at the vertex (the middle of the top of the head) (Hummel and Kobal, 1992). The separation of trigeminal evoked potentials from olfactory evoked potentials allows for a clinical test of chemosensory function, including such conditions as Alzheimer's disease and Parkinson's disease (Sakuma et al., 1996; Barz et al., 1997).

2.3. Behavior

Several behavioral studies using anosmic animals have examined the ability of the nasal trigeminal system to detect chemical stimuli. After bilateral olfactory nerve section, pigeons can still detect amyl acetate, but at concentrations at least 2.6 log units higher than in intact birds (Walker et al., 1979). In addition, these birds, unlike intact animals, could not discriminate between amyl and butyl acetate. Using a different behavioral paradigm, Walker et al. (1986) found a 2 to 4 log unit difference in thresholds for amyl acetate, butanol, butyl acetate, and benzaldehyde in olfactory nerve-sectioned and intact pigeons. Olfactory nerve-sectioned starlings retained learned odor aversions to phenethyl alcohol, which extinguished after approximately 5 days (Mason and Silver, 1983). The aversion could be re-established subsequently, suggesting that trigeminal chemoreceptors can be used to establish aversions. In rats, the unconditioned bradycardic response to nasally applied 50% CO_2 is mediated by the trigeminal nerve (Yavari et al., 1996.)

Detection thresholds were also elevated in olfactory nerve-sectioned tiger salamanders. Lesioned salamanders were still able to detect amyl acetate, cyclohexanone, butanol, and limonene, but at thresholds 11 to 18 times higher than in intact animals (Silver et al., 1988). (Trigeminal electrophysiological thresholds were 5 to 45 times higher than olfactory electrophysiological thresholds.) These lesioned animals were, however, not able to discriminate between different compounds, when the concentrations are matched for equal response level. The trigeminal nerve is capable of discriminating stimulus concentrations, since these salamanders are capable of avoiding the relatively more intense stimulus.

2.4. Human Psychophysics

Many of the early human psychophysical studies of nasal chemesthesis separated chemesthetic from olfactory stimulation by asking subjects to rate the "irritation" separate from the "odor" of the stimulus (e.g., Moncrieff, 1955; Cain, 1976; Cain and Murphy, 1980). More recent studies use carbon dioxide to stimulate trigeminal receptors and produce irritation (e.g., Cometto-Muñiz and Cain, 1982; Stevens and Cain, 1986). Since carbon dioxide appears to stimulate only trigeminal chemoreceptors and not olfactory receptors, this should provide a means for studying nasal chemesthesis in the absence of olfactory stimulation. In addition, a number of studies have examined nasal trigeminal psychophysics in human anosmics lacking a functional olfactory system (e.g., Doty et al., 1978; Cometto-Muñiz et al., 1997; Laska et al., 1997; Kendal-Reed et al., 1998). A recent report suggests that olfactory stimulation may contribute to nasal irritation in normal subjects (Kendal-Reed et al., 1998). An excellent review of chemesthetic psychophysics can be found in Green and Lawless (1991).

A number of important findings have come from the human psychophysical studies. Among them are: (1) the latency of nasal irritation is long relative to olfactory latency (e.g., 600 msec vs. 1000 msec for high concentrations of n-butanol) (Cain, 1976); (2) repeated stimulation of the nasal cavity with irritants increases the sensation of irritation (Cometto-Muñiz and Cain, 1984); (3) adaptation to nasal chemical irritation is relatively slow (Cain, 1974); (4) the nasal cavity can be desensitized to irritants by capsaicin (Rinder et al., 1994); (5) the nasal trigeminal system can localize stimulation, that is, it can determine to which of the two nostrils the stimulus is delivered, while the olfactory system cannot (Kobal et al., 1989); (6) irritation thresholds are generally higher than odor thresholds for many compounds (Cometto-Muñiz and Cain, 1990); (7) the nasal trigeminal system can recognize and discriminate between different chemicals based on the sensations they elicit (Laska et al., 1997); (8) irritation thresholds, unlike olfactory thresholds, do not follow a circadian rhythm (Lötsch, et al., 1997b).

3. ORAL CAVITY

3.1. Anatomy

Oral chemesthesis is served by nerves coming from the mandibular and maxillary divisions of the trigeminal nerve (Fig. 4.1) as well as by general cutaneous (mucosal) sensory fibers of the glossopharyngeal (ninth cranial) and vagal (tenth cranial) nerves. The sensory terminals of the trigeminal nerves include both nonspecialized free nerve endings as well as the specialized (coiled) nerve endings that serve tactile or thermal sensation. In the tongue, the free nerve endings are found abundantly in the perigemmal epithelium surrounding the taste buds in the fungiform papillae with the greatest density below the stratum corneum near the taste pore (Finger, 1986; Silverman and Kruger, 1989; Suemune et al., 1992). So rich is the trigeminal innervation of the tongue, that it is estimated that 75% of the fibers innervating fungiform papillae are trigeminal in origin (Farbman and Hellekant, 1978).The disposition of nerve terminals in the fungiform papillae places the highest density of terminals where the diffusion barrier of the cornified epithelium is the thinnest (for discussion of chemical penetration of the oral epithelium, see also Chapter 12, Finger and Simon). Filiform papillae are also innervated by trigeminal fibers, which, when present, terminate in the connective tissue core of these papillae and are therefore probably not very accessible to chemical stimuli (Nosrat et al., 1996).

3.2. Electrophysiology

Lingual trigeminal nerves respond to all classes of compounds that produce pain, irritation, or thermal sensations in the mouth. Pungent spices and the active compounds therein (capsaicin from hot red pepper, piperine from black pepper, zingerol and shogaol from ginger, and isobutyl isothiocyanate from horseradish) stimulate some mechanonociceptive fibers (Okuni, 1978). However, the presence of well-characterized nociceptors in the lingual nerve has been poorly documented. Nevertheless, capsaicin-sensitive neurons are clearly present in the oral cavity of mammals as evidenced by the aversiveness of oral capsaicin (see below) and the fact that capsaicin induces extravasation (plasma leakage) in lingual epithelium (Bryant and Moore, 1995). Lingual trigeminal nerves respond to high concentrations of electrolytes applied to the tongue, typically in the molar range, as well as a wide range of nonelectrolytes. Kawamura (1968) reported that nerves that were sensitive to aversively high concentrations of monovalent cations were not sensitive to divalent cations. Sostman and Simon (1991) confirmed that divalent cations were not excitatory, and further determined that the monovalent cation-sensitive neurons were insensitive to noxious mechanical and thermal stimuli. Compounds that are irritating stimulate other types of fibers in addition to polymodal nociceptors. CO_2, which produces the tingling sensation of carbonation, fatty acids, and citric acid all stimulate cold nociceptors as well as cool/acid sensitive lingual nerve fibers (Bryant and Moore, 1995). Ethanol stimulates fast conducting fibers that are sensitive to cooling and tactile stimulation (Hellekant, 1965). Moreover, menthol stimulates cool sensitive fibers (Schaffer et al., 1986; Kosar and Schwartz, 1990). These studies all indicate that there is considerable overlap of chemical sensitivity with pain and thermal modalities and some overlap with the tactile modality.

3.3. Behavior

Surprisingly few studies have examined animal behavioral responses toward prototypical oral irritants. Rozin et al. (1979) determined that rats were averse to food laced with chilipepper with an estimated concentration of 270 ppm capsaicin. Interestingly, despite extensive training with positive reinforcement, rats could not reverse their aversion to capsaicin. Rats also find acids aversive. In unpublished studies (B. P. Bryant, P. A. Moore, and Y. O'Bannon), using a homologous series of fatty acids presented in a drinking assay, aversion to drinking increased as the length of the carbon chain of the acids increased (e.g., the acids become more lipophilic, resulting in greater permeation of the epithelium as well as being more stimulatory of PMNs) (Fig. 4.2). Rats that were desensitized with neonatal capsaicin treatment were less averse to drinking pentanoic acid (30 mM) than control animals, suggesting that capsaicin-sensitive afferents partly mediate the aversion to acid. This interpretation is further supported by the finding that capsaicin desensitization had no effect on taste in rats (Silver et al., 1985) Finally, the relationship between increased rejection of fatty acid solutions and carbon chain length closely parallels the relationship between the ability of fatty acids to stimulate the lingual trigeminal nerve and carbon chain length.

The trigeminal systems of mammals and birds serve much the same function for each group, however, the specific compounds to which these taxa respond are

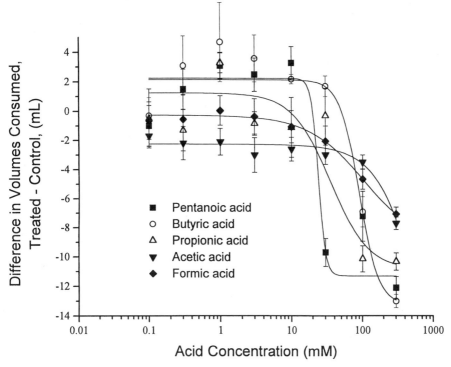

FIG. 4.2. Drinking aversion to fatty acid solutions. Water deprived (2 hrs) rats were tested in a two-choice assay. To reduce the effect of test solution volatiles on nasal trigeminal endings, the spouts of both the control and test solution bottles were wrapped with filter paper soaked in the highest concentration of the acid being tested.

different. Capsaicin, which is typically avoided at 100 ppm by mammals, is essentially without effect in birds (Mason and Clark, 2000). Birds, on the other hand, are very sensitive to compounds such as anthranilates and acetophenones, responding to 500 ppm when presented in drinking water (Mason et al., 1991; Clark et al., 1991).

3.4. Human Psychophysics

Psychophysical studies have revealed some of the sensory processes that characterize the oral nociceptor system. Repeated stimulation of the tongue with capsaicin initially produces increasing amounts of warmth, burning, and pain. This phenomena, termed *sensitization* (Green, 1991), may be due to the continued increase in concentration of irritant at the nerve terminals or to more central processes such as temporal summation. Following a period of 10–15 minutes of no stimulation, subsequent stimulation with capsaicin is frequently without effect. With renewed stimulation and/or higher concentrations of capsaicin, this desensitized state can be overcome and full response levels regained (Green, 1996). Longer-term desensitization, lasting up to a week, occurs following high concentrations of capsaicin (Karrer and Bartoshuk, 1991; McBurney et al., 1997). Consumers of highly pungent cuisines might be expected to show extensive desensitization to capsaicin. However, human subjects who consumed little or no chili pepper showed little or no difference in sensitivity to capsaicin compared with frequent users (Rozin and Shiller, 1980; Prescott and Stevenson, 1995a).

Local differences in the intensity of irritation across the oral mucosa have been reported (Lawless and Stevens, 1988) but it is not clear if this relates to local differences in receptor density or the effectiveness of the permeability barrier of the epithelium. An indication that intensity may depend on receptor density is the finding that people who have more taste papillae on their tongue, the so-called supertasters, give higher pungency ratings to solutions of capsaicin than normal tasters (Bartoshuk et al., 1996). Presumably, with more fungiform taste papillae, there are more lingual nerve endings, with PMNs near the mucosal surface and therefore more neural input from supertasters' tongues.

Psychophysical studies of the interactions of trigeminal stimuli indicate that some irritating stimuli may be mediated by the same sensory pathway. Because irritation due to NaCl and citric acid is cross-desensitized by capsaicin (Gilmore and Green, 1993), these irritants appear to be mediated by a common nociceptive mechanism or neural pathway. That capsaicin is also active via thermal pathways is indicated by interactions between capsaicin and thermal stimuli; capsaicin enhances warming sensations (Green, 1989). The thermal senses also play a role in mediating irritation due to CO_2; pungency due to CO_2 is enhanced by cooling and cooling sensations are enhanced by CO_2 (Green, 1992). That CO_2 irritation is in part mediated by cooling sensitive fibers is suggested by finding that CO_2 stimulates cool sensitive lingual fibers in rats (Komai and Bryant, 1993).

4. CORNEA

4.1. Anatomy

The ocular region is innervated by the ophthalmic branch of the trigeminal nerve (Fig. 4.1), which innervates the cornea, conjunctiva, lacrimal glands, and eyelids. Because the cornea appears to be innervated only by fine caliber axons that have ready access to chemical stimuli, it has served as a model nerve ending preparation. In mammals, an estimated 100–1200 neurons innervate the cornea, depending upon the species. These are primarily small diameter C and A-delta fibers. Nerve endings in the cornea are distributed within the layers of the epithelium. In the rabbit cornea, two sets of endings are located in the epithelium. One set, innervated by unmyelinated C fibers, is oriented vertically, reaching to within 10–20 μm of the surface. The other set of endings, innervated by thinly myelinated fibers that lose their sheaths after entering the cornea, is oriented parallel to the strata of the epithelium (MacIver and Tanelian, 1993a).

Nerve endings in the cornea contain the vasoactive neuropeptides, substance P and CGRP (see Section 7, below). These endings are eliminated by neonatal capsaicin treatment (Gamse et al., 1981). However, their disappearance is followed, in some species, by hyperinnervation by CGRP-containing endings (Marfurt et al., 1993). In vascularized tissue such as skin or mucosa, these neuropeptides play a defensive role by affecting the local microvasculature (see below). In the avascular cornea, where the epithelial turnover rate is quite high, trigeminal endings appear to have a trophic function. Coculture of corneocytes with trigeminal neurons or with substance P promotes mitotic activity and differentiation (Garcia-Hirschfeld et al., 1994). Conversely, the synthesis of substance P in bird trigeminal neurons parallels the wound-induced growth and sprouting of corneal nerve endings, suggesting the influence on nerve endings of an as yet unidentified epithelial derived nerve growth factor (Bee et al., 1988).

4.2. Electrophysiology

Trigeminal endings in the cornea are responsive to irritant compounds that have been tested in the oral and nasal cavities (Dawson, 1962). Similar to findings with cutaneous and oral nociceptors, corneal nerve endings in the cat respond to capsaicin, protons (pH 4.5 or CO_2) and hypertonic NaCl (Gallar et al., 1993; Belmonte et al., 1991). Chemical sensitivity in the cornea of the cat is primarily mediated by polymodal nociceptors. The molecular mechanisms conferring sensitivity to protons and capsaicin are present on the same type of fiber but are not identical; capsazepine, a capsaicin antagonist, inhibits responses to capsaicin but not low pH (Chen et al., 1995). In the rabbit, chemical sensitivity was found only in a subset of C fibers that were insensitive to mechanical or thermal stimulation. These fibers responded to acetylcholine, bradykinin, L-glutamate, and prostaglandin E1 but not histamine or lactic acid (MacIver and Tanelian, 1993b). None of the cooling sensitive fibers nor other mechano- or heat-sensitive fibers were sensitive to these compounds. While this may represent differences between the distribution of mechanisms between species, this may be evidence for a putative chemonociceptor, which is silent and is initially insensitive to mechanical and thermal stimulation (Meyer et al., 1991) but becomes chemically responsive when activated by endogenous inflammatory mediators.

4.3. Human Psychophysics

Psychophysical studies of corneal chemoreception are not extensive. Although it may be common wisdom that only pain is registered from stimulation of the cornea, subjects can discriminate between the sensations elicited by warming and cooling (Kenshalo, 1960). It is apparent from this that different populations of nociceptors are necessarily present on the cornea: PMNs, mechanoheat- and cold nociceptors. In humans, the sensory irritation threshold to capsaicin has been determined to be 6×10^{-8} M (Dupuy et al., 1988). Hypertonic saline also elicits pain in humans (Mandahl, 1993). The irritation threshold to CO_2 has also been determined by applying a humidified stream of CO_2 to the cornea. The threshold is 40% with the intensity of irritation increasing as CO_2 concentration is further increased. Because a strong correlation between the response of cat corneal PMNs to CO_2 and human responses to corneal CO_2 exists, it is likely that PMNs mediate ocular sensations of CO_2 (Chen et al., 1995). Corneal sensitivity to volatile organic compounds (alcohols, ketones, and alkylbenzenes) has been determined (Cometto-Muniz and Cain, 1995) and found to be remarkably similar to nasal pungency thresholds.

5. TRIGEMINAL STIMULI

One classification of trigeminal stimuli is based on the site at which they act (i.e., eyes, nose, mouth, respiratory tract) and the sensations and physiological responses that they elicit (Moncrieff, 1951). One of the best studied of these classes is pungent spices, found in a wide variety of plants (see Govindarajan, 1979; Fujinari, 1997). Pungent vegetables and spices contain such irritating compounds as allyl isothiocyanate (horseradish), eugenol (cloves), allyl isothiocyanate (mustard), sanshool (Szechuan pepper), gingerol and zingerone (ginger), piperine (black pepper), and capsaicin (chili pepper) (Prescott and Stevenson, 1995b; Fujinari, 1997). The last three spices have been particularly well studied and their active components each contain an aromatic ring connected to an alkyl side chain through an amide linkage.

Lacrimators are chemical irritants that elicit tears. They occur in some vegetables and have been developed specifically as chemical warfare agents. A well-known lacrimator is 1-propenyl sulfenic acid, found in onions. Many of these compounds, such as chloracetone, chloracetophenone ("tear gas"), and o-chlorobenzylidene malononitrile have in common a halogen or cyanide group.

Menthol is a chemical that elicits irritating as well as cooling sensations. Because menthol cross-desensitizes capsaicin irritation (Cliff and Green, 1996), it is reasonable to assume that menthol acts on polymodal nociceptors. The cooling sensation, on the other hand, is mediated by low-threshold thermal receptors (Schafer et al., 1986). Compounds that elicit a cooling sensation have a hydrogen bonding group, a compact hydrocarbon skeleton, a correct hydrophilic/hydrophobic balance, and a molecular weight between 150 and 350 (Watson et al., 1978).

Trigeminal nerve endings lie in and below the epithelia of the mouth, nose, and eyes (Suemune et al., 1992; Finger et al., 1990b; Tanelian and MacIver, 1990). Stimuli must pass through the lipid phase of epithelial cell membranes or through the aqueous phase of tight junctions to arrive at the free nerve endings (See Chapter 12, Finger and Simon). $LaCl_3$, which reduces the permeability of epithelial tight junctions, decreases trigeminal nerve responses to ionic stimuli

(Sostman and Simon 1991; Bryant and Moore, 1995) but not to fatty acid stimuli, suggesting that some stimuli must pass through tight junction complexes to arrive at free nerve endings, while others do not (see Chapter 12, Finger and Simon).

That the lipid phase is the primary route of access for hydrophobic irritants is supported by the fact that lipid solubility plays an important role in determining the efficacy of trigeminal stimuli in the mouth, nose, and eyes. Studies with homologous series of chemicals demonstrate that an increase in trigeminal stimulus efficacy occurs with increasing carbon chain length (*electrophysiology*– nose: Silver et al., 1986; mouth: Bryant and Moore, 1995; *psychophysics*–nose and eye: Cometto-Muñiz and Cain, 1995; *reflexively induced respiratory changes*– nose: Hansen and Nielsen, 1994). Therefore, at least some stimuli may gain access to trigeminal nerve endings through the lipid phase of the epithelium. QSAR studies confirm that trigeminal stimuli must be hydrophobic enough to partition into biological membranes (Abraham et al., 1996). For a given compound, irritancy is not solely an issue of the sensitivity of nerve endings to that compound but also access of the compound to the endings.

6. CELLULAR MECHANISMS OF TRIGEMINAL CHEMORECEPTION

6.1. Transduction Mechanisms

Chemical irritants comprise a broad range of compounds that are in themselves either toxic, damaging, or are associated with pathological conditions. Responsiveness to such a broad range of chemical structures necessitates diverse receptive mechanisms (Nielsen, 1991). Because depolarization of sensory nerve endings is the initial step that is common to pain and irritation, any mechanism that depolarizes the appropriate endings will transduce irritants. Transduction of chemical signals by trigeminal neurons is best understood in the case of capsaicin activation of PMNs. The underlying mechanisms have long been understood through studies of the dorsal root ganglia (DRGs), of which trigeminal ganglia are cranial analogs. Evidence for a capsaicin receptor was first seen in the molecular structural requirements for neural activation (Szolcsanyi and Jansco-Gabor, 1975). Electrophysiological studies have determined that capsaicin activates nonspecific cation channels that allow influx of calcium and sodium (Winter, 1987; Liu and Simon, 1996). This initial depolarization caused by the activation of ligand-gated channels is subsequently sustained by voltage-sensitive calcium channels in the endings and transmitted centrally. The anatomical distribution of putative capsaicin receptors was studied using binding of [^{3}H] resiniferatoxin (RTX), an agonist of capsaicin-sensitive neurons. Specific RTX-binding activity was found in the brain stem, sensory ganglia, and spinal cord and was eliminated by systemic capsaicin treatment (Szallasi et al., 1995), consistent with RTX binding sites being identical to capsaicin receptors. Recently, a capsaicin-gated channel has been cloned that is found in trigeminal and dorsal root ganglia and that is selectively expressed in small- to medium-size sensory neurons (Caterina et al., 1997). Because ligands that activate the channel, capsaicin and resiniferatoxin, contain the vanilloid moiety, it is known as a vanilloid receptor, VR1. In addition to being sensitive to vanilloids, VR1 is also activated by noxious heat, and low pH in a manner that is consistent with neurophysiological and psychophysical observations. For instance, capsaicin desensitizes VR1 in a calcium-dependent manner. Moreover, responses of VR1 to capsaicin, noxious heat, and low pH are all antagonized by capsazepine. Finally, low pH potentiates the response of VR1 to both heat and capsaicin, consistent with its role as a polymodal receptive mechanism, which monitors the physiological conditions of peripheral tissue (Tominaga et al., 1998).

Not all neural sensitivity to low pH in peripheral tissues is explained by the pH-sensitivity of VR1. Another mechanism that has been proposed to mediate responses to protons is an acid-sensing ion channel (ASIC), the mRNA of which has been found in the trigeminal ganglion (Lingueglia et al., 1997).

The mechanism by which menthol activates cool-sensitive sensory neurons appears to be through a decrease in calcium conductance (Swandulla et al., 1986). This was demonstrated by studies in which removal of extracellular calcium increased the discharge rate and increases in extracellular calcium decreased the discharge frequency of cooling sensitive receptors (Schaffer et al., 1986).

The mechanisms underlying other forms of pain and irritation caused by endogenous compounds such as histamine, 5HT, and acetylcholine are not fully understood in trigeminal neurons but specific subtypes of known receptors are strongly implicated. Wood and Docherty (1997) have reviewed the role of endogenous mediators and their respective receptors in pain and irritation. The ethmoid nerve responses to nicotine (Allimohammadi and Silver, 2000) as well as the burning sensation of nicotine on the tongue (Dessirrier et al., 1998) appear to be mediated by mecamylamine-sensitive nicotinic receptors. Bradykinin, a potent inflammatory mediator and endogenous algesic peptide, is the ligand for specific receptors in peripheral sensory neurons (McGuirk and Dolphin, 1992).

Many compounds that are irritants are good lipid solvents (e.g., methylene chloride in paint stripper). Irritants may either damage the lipid bilayer of nerve endings or form discrete ion channels in the membrane of a nerve ending, both causing depolarization by allowing ion flux through the disrupted lipid phase. At the high concentrations of capsaicin to which human mouths are exposed in the most pungent cuisines, nonspecific conductance pathways are induced in planar lipid bilayers (Feigin et al., 1995). Moreover, the ability of a series of pungent compounds to induce discrete conductances in artificial lipid bilayers is closely parallel to their threshold to induce Ca^{2+} uptake in capsaicin-sensitive dorsal root neurons. This suggests that, in addition to stimulating receptor-activated ion channels, the pain of higher concentrations of capsaicin may be due to depolarizing currents caused by the formation of nonspecific conductances in the membranes of polymodal nociceptors.

Although depletion of neuropeptides such as substance P has long been cited as a likely cause for desensitization, other mechanisms, such as calcium-activated proteases are also likely (Chard et al., 1995). If extracellular calcium is removed during stimulation, or the influx of calcium through capsaicin-gated channels is blocked by ruthenium red, capsaicin-induced desensitization is blocked. If the availability of extracellular calcium is limited, desensitization is reversible. In addition, if exposure to capsaicin is sustained, involves high concentrations of capsaicin, or occurs during the perinatal period, capsaicin is neurotoxic. In this case, neuronal death is thought to be caused by toxic influx of Na^+ and Ca^{2+} (Holzer, 1991).

6.2. Peri-receptor Modulation of Stimuli

Once a potential irritant has penetrated the outer layer of epithelium, it can act directly or indirectly on the membranes of nerve endings (see Chapter 12, Finger and Simon). In some cases, peripheral metabolism produces the active compound. For example, CO_2 is a relatively inert compound. In water, slow, uncatalyzed dissociation produces protons and bicarbonate ion. Carbonic anhydrase, an enzyme that is present in the oral and nasal epithelia as well as corneal endothelium, can increase the rate of this reaction by up to 20,000-fold. A variety of reports indicate that carbonic anhydrase activity is necessary for trigeminal neural responses to carbonated solutions (Komai and Bryant, 1993) and CO_2 gas (Ericksen and Silver, 1994) and as well as perception of CO_2 pungency (Graber and Kelleher, 1988). Pungent esters, such as ethyl acetate, may derive some of their chemesthetic properties through the enzymatic liberation of the irritants, protons, and alcohols.

7. REFLEXES/EFFECTOR FUNCTION

In addition to their role as sensory afferents giving rise to sensation and appropriate behavioral responses, capsaicin-sensitive, peptidergic sensory neurons may serve as sensory effectors (for reviews see Holzer, 1988; Szolcsányi, 1996a). In a process known as axon reflex, stimulation of trigeminal sensory neurons with noxious or irritating stimuli induces neuropeptide release in the stimulated endings directly and by retrograde transmission in other branches of the same axons (Fig. 4.3). The neuropeptides that are released, substance P and CGRP cause, among other responses, local vasodilation and plasma leakage (extravasation). These tissue defensive actions have lead these neurons to be called noceffectors (Kruger, 1988).

Stimulation of the nasal trigeminal nerve with chemicals may lead to a decrease in respiratory rate, decrease in nasal patency, increase in nasal secretion, and an increase in mucociliary activity, vasodilation, extravasation, and sweating. Peptidergic, capsaicin-sensitive trigeminal fibers have effector functions in the oral cavity as well. These fibers mediate extravasation (Bryant and Moore, 1995) and vasodilation in oral tissues (Izumi and Karita, 1994). Stimulation of lingual nerve fibers (as well as stimulation of chorda tympani and glossopharyngeal nerve

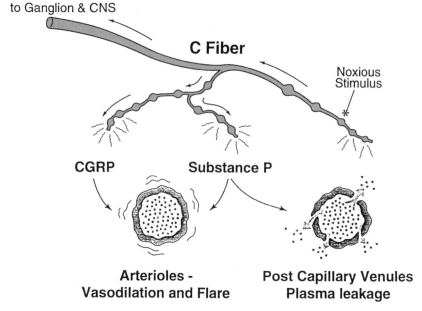

FIG. 4.3. Axon reflex. In addition to causing a signal to be sent toward the central nervous system, noxious stimulation causes retrograde excitation of nerve endings resulting in the release of the vasoactive neuropeptides, substance P and calcitonin gene-related peptide (CGRP).

fibers) also elicits increased salivary flow (Takahashi et al., 1995). In the eye, capsaicin-sensitive sensory effectors in the trigeminal nerve appear to play a role in tearing, contraction of the pupil, vasodilation, increased intraocular pressure, a disruption of the blood-aqueous barrier, ocular inflammation, and corneal wound healing (Butler et al., 1979; Micieli et al., 1992; Krootila et al., 1988; Araki et al., 1994).

While some of these reflexes discussed above may be elicited by trigeminal activation of autonomic fibers, others appear to be mediated by axon reflex (Finger et al., 1990a). Thus, upon stimulation of capsaicin-sensitive neurons, neuropeptides such as substance P and CGRP may be released from the site of stimulation, peripheral terminals, collateral terminals, or central terminals (Kruger, 1988; Szolcsányi, 1996b).

An interesting recent finding is that electrical stimulation of the lingual trigeminal nerve decreases chorda tympani (taste) nerve responses to salt stimuli (Wang et al., 1995). This effect is thought to be due to the effects of neuropeptides released from stimulated trigeminal nerve fibers in the vicinity of the taste buds. Capsaicin stimulation of the tongue also affects the responses of chorda tympani nerve to NaCl (Osada et al., 1997). Earlier studies indicated that some neuropeptides found in trigeminal nerves (e.g. vasopressin and substance P) can affect the activity of taste receptor cells or chorda tympani nerve responses

(Esakov and Serova, 1988). These results suggest that oral trigeminal stimulation may modify, at the peripheral level, the detection of at least some taste stimuli.

Trigeminal stimulation decreases responses from olfactory receptor neurons (Bouvet et al., 1987), presumably due to the local release of neuropeptides, since substance P also inhibits olfactory receptor neurons (Bouvet et al., 1988). In addition to these direct effects on olfactory receptors, trigeminal stimulation appears to modulate the activity of cells in the olfactory bulb (Stone, 1969). While this effect may be due to centrifugal modulation of the olfactory bulb from the central nervous system, a more direct, purely trigeminal route may also be posited. Substance P-/CGRP-containing trigeminal nerve fibers enter the bulb, extending into the glomerular layer (Finger and Böttger, 1993); some of these fibers are collateral branches of fibers innervating the nasal mucosa (Finger and Silver, 1993). Thus, activation of trigeminal chemoreceptors in the nasal cavity could influence olfactory bulb processing directly without the necessity of relay through the brain stem trigeminal sensory nuclei.

Trigeminal nociceptive fibers containing SP and CGRP also innervate the vasculature within the cranium (Saito et al. 1987; Tsai et al. 1988), forming the "trigeminovascular" system. Stimulation of the trigeminovascular system leads to an increase in cortical blood flow, which appears to be responsible for the pain associated with vascular headaches (Macfarlane, 1993; Silberstein, 1992). Electrical stimulation of the nasociliary branch of the trigeminal nerve increases cerebral blood flow (Suzuki et al., 1990), as does nasal stimulation with high concentrations of volatile organic compounds (Major and Silver, 1999), presumably through trigeminal stimulation. This suggests that odor-induced headache may involve the stimulation of trigeminal fibers innervating the nasal mucosa resulting in activation of cerebrovascular nociceptors.

8. SUMMARY

A role of trigeminal chemesthesis is to signal skin and mucosal conditions that are actually or potentially harmful. The stimulus may be either external chemical insult or internal chemical conditions such as inflammation or ischemia. Adaptive behavioral and physiological responses include noxious sensations with associated aversive behavior and local changes in vascular and trophic factor physiology, all of which are defensive in nature. The mechanisms that render trigeminal nerve endings sensitive to chemical stimuli range from specific receptors for endogenous compounds (i.e., neuropeptides, transmitters, and protons) that are released during tissue trauma to ion channels that confer thermal sensitivity. In addition, it is likely that nonspecific perturbation of neural lipid membranes also plays a role in activating chemosensitive nerve endings. Either by directly modulating ion channel activity or through the actions of second messengers, these mechanisms ultimately cause the depolarization of nerve endings with a signal being sent centrally as well as to other peripheral endings. This latter process, axon reflex, is responsible for inducing cutaneous and mucosal secretion of neuropeptides that may modulate activity in nearby taste or olfactory receptor cells as well as having vasoactive and trophic effects in the epithelium in general.

The lipophilicity of potential trigeminal stimuli and effects on membrane permeability are important factors determining potency in irritation. Even if a nerve ending is sensitive to a particular compound, only after the compound has gained access through the integument is it an irritant. Comparative studies of interspecies differences in sensitivity to trigeminal stimuli should shed some light on whether these differences are due to intrinsic sensitivities of trigeminal receptors or simply differences in relative epithelial permeability.

Several aspects of trigeminal chemesthesis warrant further study. Most trigeminally induced sensations involve some degree of outright pain or irritation. This is a clear indication of the functional overlap of chemesthesis with the pain sense and associated behavioral and physiological responses. However, it should also be clear that other low threshold somatosensory modalities, namely tactile and cooling sensations, also are chemically sensitive and are important in everyday sensory experience. In addition, there are indications that these chemically sensitive somatosensory modalities may interact with the senses of olfaction and gustation. Whether these potential interactions are due to actions in the peripheral tissues or higher order of convergence may be addressed by studies ranging from the molecular biological through the cellular physiological to the psychophysical.

REFERENCES

Abraham, M. H., J. Andonian-Haftvan, J. E. Cometto-Muñiz, and W. Cain (1996). An analysis of nasal irritation thresholds using a new solvation equation. *Fundam Appl Toxicol* **31**: 71–76.

Allimohammadi, H. and W. L. Silver (2000). Evidence for nicotinic acetylcholine receptors on nasal trigeminal nerve endings of the rat. *Chem Senses* **25**:61–66.

Anton, F., P. Peppel, I. Euchner, and H. O. Handwerker (1991). Controlled noxious chemical stimulation: responses of rat trigeminal brainstem neurones to CO_2 pulses applied to the nasal mucosa. *Neurosci Lett* **123**:208–211.

Araki, K., Y. Ohashi, S. Kinoshita, K. Hayashi, Y. Kuwayama, and Y. Tano (1994). Epithelial wound healing in the denervated cornea. *Curr Eye Res* **13**:203–211.

Bartoshuk, L. M., V. B Duffy, D. Reed, and A. Williams (1996). Supertasting, earaches and head injury: genetics and pathology alter our taste worlds. *Neurosci Biobehav Rev* **20**: 79–87.

Barz, S., T. Hummel, E. Pauli, M. Majer, C. J. Lang, and G. Kobal (1997). Chemosensory event-related potentials in response to trigeminal and olfactory stimulation in idiopathic Parkinson's disease. *Neurology* **49**:1424–1431.

Bee, J. A., U. Kuhl, D. Edgar, and K. von der Mark (1988). Avian corneal nerves: co-distribution with collagen type IV and acquisition of substance P immunoreactivity. *Invest Ophthalmol Vis Sci* **29**:101–107.

Belmonte, C., J. Gallar, M. A. Pozo, and I. Rebollo (1991). Excitation by irritant chemical substances of sensory afferent units in the cat's cornea. *J Physiol (Lond)* **437**:709–725.

Bouvet J-F., J-C. Delaleu, and A. Holley (1987). Olfactory receptor cell function is affected by trigeminal nerve activity. *Neurosci Lett* **77**:181–186.

Bouvet J-F., J-C. Delaleu, and A. Holley (1988). The activity of olfactory receptor cells is affected by acetylcholine and substance P. *Neurosci Res* **5**:214–223.

Bryant, B. P. and P. A. Moore (1995). Factors affecting the sensitivity of the lingual trigeminal nerve to acids. *Am J Physiol* **268**:R58–R65.

Butler, J. M., W. G. Unger, and B. R. Hammond (1979). Sensory mediation of the ocular response to neutral formaldehyde. *Exp Eye Res* **28**:577–589.

Cain, W. S. (1974). Contribution of the trigeminal nerve to perceived odor magnitude. *Ann NY Acad Sci* **237**:28–34.

Cain, W. S. (1976). Olfaction and the common chemical sense: some psychophysical contrasts. *Sensory Proc* **1**:57–67.

Cain, W. S. and C. L. Murphy (1980). Interaction between chemoreceptive modalities of odour and irritation. *Nature* **284**:255–257.

Caterina, M. J., M. A. Schumacher, M. Tominaga, T. A. Rosen, J. D. Levine, and D. Julius (1997). The capsaicin receptor: a heat-activated ion channel in the pain pathway. *Nature* **389**:816–824.

Cauna, N. K., K. Hinderer, and R. T. Wentges (1969). Sensory receptor organs of the human nasal respiratory mucosa. *Am J Anat* **124**:187–210.

Chard, P. S., D. Bleakman, J. R. Savidge, and R. J. Miller (1995). Capsaicin-induced neurotoxicity in cultured dorsal root ganglion neurons: involvement of calcium-activated proteases. *Neurosci* **65**:1099–1108.

Chen, X. J., J. Gallar, M. A. Pozo, M. Baeza, and C. Belmonte (1995). CO_2 stimulation of the cornea: comparison between human sensation and nerve activity in polymodal nociceptive afferents of the cat. *Eur J Neurosci* **7**:1154–1163.

Clark, L., P. Shah, and J. R Mason (1991). Chemical repellency in birds: relationship between chemical structure and avoidance response. *J Exp Zool* **260**:310–322.

Cliff, M. A. and B. G. Green (1996). Sensitization and desensitization to capsaicin and menthol in the oral cavity: interactions and individual differences. *Physiol Behav* **59**: 487–494.

Cometto-Muñiz, J. E. and W. S. Cain (1982). Perception of nasal pungency in smokers and nonsmokers. *Physiol Behav* **29**:727–731.

Cometto-Muñiz J. E. and W. S. Cain (1984). Temporal integration of pungency. *Chem Senses* **8**:315–327.

Cometto-Muñiz, J. E. and W. S. Cain (1990) Thresholds for odor and nasal pungency. *Physiol Behav* **48**:719–725.

Cometto-Muñiz J. E. and W. S. Cain (1995). Relative sensitivity of the ocular trigeminal, nasal trigeminal and olfactory systems to airborne chemicals. *Chem Senses* **20**:191–198.

Cometto-Muñiz J. E., W. S. Cain, and H. K. Hudnell (1997). Agonistic sensory effects of airborne chemicals in mixtures: odor, nasal pungency and eye irritation. *Percept Psychophys* **59**:665–674.

Cooper, G. P. (1970). Responses of rat ethmoidal units to cutaneous and olfactory stimuli. *Physiologist* **131**:171.

Dawson, W. W. (1962) Chemical stimulation of the peripheral trigeminal nerve. *Nature* **196**:341–345.

Dessirier J. M., M. O'Mahony, J. M. Sieffermann, and E. Carstens (1998). Mecamylamine inhibits nicotine but not capsaicin irritation on the tongue: psychophysical evidence that nicotine and capsaicin activate separate molecular receptors. *Neurosci Lett* **240**:65–68.

Doty, R. L., W. E. Brugger, P. C. Jurs, M. A. Orndorff, P. J. Snyder, and L. D. Lowry (1978). Intranasal trigeminal stimulation from odorous volatiles: psychometric responses from anosmic and normal humans. *Physiol Behav* **20**:175–185.

Dupuy, B., H. Thompson, R. W. Beuerman (1988). Capsaicin: a psychophysical tool to explore corneal sensitivity. *Invest Ophthalmol Vis Sci* **29**(suppl):454.

Ericksen, J. L. and W. L. Silver (1994). Acetazolamide inhibits ethmoid nerve responses to carbon dioxide but not other irritants. *Chem Senses* **19**:466–467.

Esakov, A. E. and Serova, O. N. (1988). Influence of substance P on taste receptor organ and salt intake in rats. *Neurosciences* **14**:321–327.

Farbman, A. I. and G. Hellekant (1978). Quantitative analyses of the fiber population in rat chorda tympani nerves and fungiform papillae. *Am J Anat* **153**:509–522.

Feigin, A. M., E. V. Aronov, B. P. Bryant, J. H. Teeter, and J. H. Brand (1995). Capsaicin and its analogues elicit ion channels in planar lipid bilayers. *Neuroreport* **6**:2134–2136.

Finger, T. E. (1986). Peptide immunocytochemistry demonstrates multiple classes of perigemmal nerve fibers in the circumvallate papilla of the rat. *Chem Senses* **11**:135–144.

Finger, T. E. and B. Böttger (1993). Peripheral peptidergic fibers of the trigeminal nerve in the olfactory bulb of the rat. *J Comp Neurol* **334**:117–124.

Finger, T. E. and W. L. Silver (1993). Single trigeminal ganglion cells have collateral branches in the nasal epithelium and the olfactory bulb. *Chem Senses* **18**:556.

Finger, T. E., M. L. Getchell, T. V. Getchell, and J. C. Kinnamon (1990a). Affector and effector functions of peptidergic innervation of the nasal cavity. In: B. G. Green, J. R. Mason, and M. R. Kare (eds). *Chemical Senses, Vol. 2: Irritation*. New York: Marcel Dekker, pp 1–17.

Finger, T. E., V. St. Jeor, J. C. Kinnamon, and W. L. Silver (1990b). Ultrastructure of substance P- and CGRP-immunoreactive nerve fibers in the nasal epithelium of rodents. *J Comp Neurol* **294**:293–302.

Fujinari, E. M. (1997). Pungent flavor profiles and components of spices by chromatography and chemiluminescent nitrogen detection. In: S. J. Risch and C.-T. Ho (eds). *Spices: Flavor Chemistry and Antioxidant Properties*. Washington, DC: American Chemical Society, ACS Symposium Series 660, pp 98–112.

Gallar, J., M. A. Pozo, R. P. Tuckett, and C. Belmonte (1993). Response of sensory units with unmyelinated fibres to mechanical, thermal and chemical stimulation of the cat's cornea. *J Physiol (Lond)* **468**:609–622.

Gamse, R., S. Leeman, P. Holzman, and F. Lembeck (1981). Differential effects of capsaicin on the content of somatostatin, substance P and neurotensin in the nervous system of the rat. *Naunyn-Schmiedberg's Arch Pharmacol* **317**:140–148.

Garcia-Hirschfeld, J., L. G. Lopez-Briones, and C. Belmonte (1994). Neurotrophic influences on corneal epithelial cells. *Exp Eye Res* **59**:597–605.

Gilmore, M. M. and B. G. Green (1993). Sensory irritation and taste produced by NaCl and citric acid: effects of capsaicin desensitization. *Chem Senses* **18**:257–272.

Govindarajan, V. S. (1979). Pungency: the stimuli and their evaluation. In: J. C. Boudreau (ed). *Food Taste Chemistry.* Washington, DC: American Chemical Society, ACS Symposium Series 115, pp 53–92.

Graber, M. and S. Kelleher (1988). Side effects of acetazolamide: the champagne blues. *Am J Med* **84**:979–980.

Green, B. G. (1989). Sensory interactions between capsaicin and temperature. *Chem Senses* **11**:371–382.

Green, B. G. (1991). Temporal characteristics of capsaicin sensitization and desensitization on the tongue. *Physiol Behav* **49**:501–505.

Green, B. G. (1992). The effects of temperature and concentration on the perceived intensity and quality of carbonation. *Chem Senses* **17**:435–450.

Green, B. G. (1996). Rapid recovery from capsaicin desensitization during recurrent stimulation. *Pain* **68**:245–253.

Green, B. G. and L. J. Flammer (1989). Localization of chemical stimulation: capsaicin on hairy skin. *Somatosensory Motor Res* **6**:553–566.

Green, B. G. and H. T. Lawless (1991). The psychophysics of somatosensory chemoreception in the nose and mouth. In: T. V. Getchell, R. L. Doty, L.M. Bartoshuk, and J. B. Snow Jr. (eds). *Smell and Taste in Health and Disease.* New York: Raven Press, pp 235–253.

Green, B. G., J. R. Mason, and M. R. Kare (1990). *Chemical Senses, Vol. 2: Irritation.* New York: Marcel Dekker, 361 pp.

Hansen, L. F. and G. D. Nielsen (1994). Sensory irritation, pulmonary irritation and structure-activity relationships of alcohols. *Toxicology* **88**:81–99.

Hellekant, G. (1965). The effect of ethyl alcohol on non-gustatory receptors of the tongue of the cat. *Acta Physiol Scand* **65**:243–250.

Holzer, P. (1988). Local effector functions of capsaicin-sensitive sensory nerve endings: involvement of tachykinins, calcitonin gene-related peptide and other neuropeptides. *Neuroscience* **24**:739–768.

Holzer, P. (1991). Capsaicin—cellular targets, mechanisms of action, and selectivity for thin sensory neurons. *Pharmacol Rev* **43**:143–201.

Hummel, T. and G. Kobal (1992). Differences in human evoked potentials related to olfactory or trigeminal chemosensory activation. *Electroencephalogr Clin Neurophysiol* **84**:84–89.

Hummel, T., C. Schiessl, J. Wendler, and G. Kobal (1996). Peripheral electrophysiological responses decrease in response to painful stimulation of the human nasal mucosa. *Neurosci Lett* **212**:37–40.

Ito, K. (1962). Electrophysiological studies on the peripheral olfactory nervous system in mammalia. *Kitahanto J Med Sci* **18**:405–417.

Izumi, H. and K. Karita (1994). The parasympathetic vasodilator fibers in the trigeminal portion of the distal lingual nerve in the cat tongue. *Am J Physiol* **266**:R1517–1522.

Karrer, T. and L. Bartoshuk (1991). Capsaicin desensitization and recovery on the human tongue. *Physiol Behav* **49**:757–764.

Kawamura, Y., J. Okamoto, and M. Funakoshi (1968). A role of oral afferents in aversion to taste solutions. *Physiol Behav* **3**:537–542.

Keele, C. A. (1962). The common chemical sense and its receptors. *Arch Int Pharmacodyn* **139**:547–557.

Kendal-Reed, M., J. C. Walker, W. T. Morgan, M. LaMacchio, and R. W. Lutz (1998). Human responses to propionic acid. I. Quantification of within- and between-participant variation in perception by normosmics and anosmics. *Chem Senses* **23**:71–82.

Kenshalo, D. R. (1960). Comparison of thermal sensitivity of the forehead, lip, conjunctiva, and cornea. *J Appl Physiol* **15**:987–991.

Kobal, G. and C. Hummel (1988). Cerebral chemosensory evoked potentials elicited by chemical stimulation of the human olfactory and respiratory nasal mucosa. *Electroencephalogr Clin Neurophysiol* **71**:241–250.

Kobal, G., S. Van Toller, and T. Hummel (1989). Is there directional smelling? *Experientia* **45**:130–132.

Komai, M. and B. P. Bryant (1993). Acetazolamide specifically inhibits lingual trigeminal nerve responses to carbon dioxide. *Brain Res* **612**:122–129.

Kosar, E. and G. J. Schwartz (1990). Effects of menthol on peripheral nerve and cortical unit responses to thermal stimulation of the oral cavity in the rat. *Brain Res* **513**: 202–211.

Krootila, K., H. Uusitalo, and A. Palkama (1988). Effect of neurogenic irritation and calcitonin gene-related peptide (CGRP) on ocular blood flow in the rabbit. *Curr Eye Res* **7**:695–703.

Kruger, L. (1988). *Morphological features of thin sensory afferent fibers: a new interpretation of 'nociceptor' function.* In: W. Hamman and A. Iggo (eds). *Progress in Brain Research, Vol. 74.* Amsterdam: Elsevier, pp 253–257.

Laska, M., H. Distel, and R. Hudson (1997). Trigeminal perception of odorant quality in congenitally anosmic subjects. *Chem Senses* **22**:447–456.

Lawless, H. T. and D. A. Stevens (1988). Responses by humans top oral chemical irritants as a function of locus of stimulation. *Percep Psychophys* **43**:72–78.

Lingueglia, E., J. R. de Weille, F. Bassilana, C. Heurteaux, H. Sakai, R. Waldmann, and M. Lazdunski (1997). A modulatory subunit of acid sensing ion channels in brain and dorsal root ganglion cells. *J Biol Chem* **272**(47):29778–29783.

Liu, L. and S. A. Simon (1996). Similarities and differences in the currents activated by capsaicin, piperine, and zingerone in rat trigeminal ganglion cells. *J Neurophysiol* **76**:1858–1869.

Lötsch, J., T. Hummel, T. Kraetsch, and G. Kobal (1997a). The negative mucosal potential: separating central and peripheral effects of NSAIDs in man. *Eur J Clin Pharmacol* **52**:359–364.

Lötsch, J., S. Nordin, T. Hummel, C. Murphy, and G. Kobal (1997b). Chronobiology of nasal chemosensitivity: do odor or trigeminal pain thresholds follow a circadian rhythm. *Chem Senses* **22**:593–598.

Lucier, G. E. and R. Egizii (1989). Characterization of cat nasal input responsible for respiratory protective reflexes. *Exp Neurol* **103**:83–89.

Macfarlane, R. (1993). New concepts of vascular headache. *Ann R Coll Surg Engl* **75**:225–228.

MacIver, M. B. and D. L. Tanelian (1993a). Structural and functional specialization of A delta- and C-fiber free nerve endings innervating rabbit corneal epithelium. *J Neurosci* **13**: 4511–4524.

MacIver, M. B. and D. L. Tanelian (1993b). Free nerve ending terminal morphology is fiber type specific for A-delta and C-fibers innervating rabbit corneal epithelium. *J Neurophysiol* **69**:1779–1783.

Major, D. A. and W. L. Silver (1999). Odorants presented to the rat nasal cavity increase cortical blood flow. *Chem Senses* **24**:665–669.

Mandahl, A. (1993). Hypertonic saline test for ophthalmic nerve impairment. *Acta Ophthalmol* **71**:556–559.

Marfurt, C. F., L.C. Ellis, and M. A. Jones (1993). Sensory and sympathetic nerve sprouting in the rat cornea following neonatal administration of capsaicin. *Somatosensory Motor Res* **10**:377–398.

Martenson, M. E., S. L. Ingram, and T. K. Baumann (1994). Potentiation of rabbit trigeminal responses to capsaicin in a low pH environment. *Brain Res* **651**:143–147.

Mason, J. R. and W. L. Silver (1983). Trigeminally mediated odor aversions in starlings. *Brain Res* **269**:196–199.

Mason, J. R., N. J. Bean, P. S. Shah, and L. Clark (1991). Taxon-specific differences in responsiveness to capsaicin and several analogues: correlates between chemical structure and behavioral aversiveness. *J Chem Ecol* **17**:2539–2550.

Mason, J. R. and L. Clark. Chemical senses in birds. (2000) In: G. C.Whitlow (ed). *Avian Physiology*, New York: Academic Press, pp. 39–56.

McBurney, D. H., C. D. Balaban, D. E. Christopher, and C. Harvey (1997). Adaptation to capsaicin within and across days. *Physiol Behav* **61**:181–190.

McGuirk, S. M. and A. C. Dolphin (1992). G-protein mediation in nociceptive signal transduction—an investigation into the excitatory action of bradykinin in a subpopulation of cultured rat sensory neurons. *Neurosciences* **49**:117–128.

Meyer, R. A., K. D. Davis, R. H. Cohen, R.-D. Treede, and J. N. Campbell (1991). Mechanically insensitive afferents (MIAs) in cutaneous nerves of monkey. *Brain Res* **561**:252–261.

Micieli, G., C. Tassorelli, G. Sandrini, F. Antonaci and G. Nappi (1992). The trigemino-pupillary reflex: a model of sensory-vegetative integration. *J Autonom Nerv Syst* **41**: 179–185.

Moncrieff, R. W. (1951). *The Chemical Senses*. London: Leonard Hill, pp. 760.

Moncrieff, R. W. (1955). A technique for comaping the threshold concentrations for olfactory, trigeminal, and ocular irritation. *Q J Exp Psychol* **7**:128–132.

Nielsen, G. D. (1991). Mechanisms of activation of the sensory irritant receptor by airborne chemicals. *Crit Rev Toxicol* **21**:183–208.

Nosrat, C. A., T. Ebendal, and L. Olson (1996). Differential expression of brain-derived neurotrophic factor and neurotrophin 3 mRNA in lingual papillae and taste buds indicates roles in gustatory and somatosensory innervation. *J Comp Neurol* **376(4)**:587–602.

Okuni, Y. (1978). Response of lingual nerve fibers of the rat to pungent spices and irritants in pungent spices. *Shika Gakuho* **78**:135–149.

Osada, K., M. Komai, B. P. Bryant, H. Suzuki, A. Goto, and K. Tsunoda (1997). Capsaicin modifies responses of rat chorda tympani nerve fibers to NaCl. *Chem Senses* **22**: 249–255.

Parker, G. H. (1912). The relations of smell, taste and the common chemical sense in vertebrates. *J Acad Nat Sci Phila* **14**:221–234.

Parker, G. H. (1922). *Smell Taste and Allied Senses in the Vertebrates*. Philadelphia: J.B. Lippincott Co., 192 pp.

Prescott, J. and R. J. Stevenson (1995a). Effects of oral chemical irritation on tastes and flavors in frequent and infrequent users of chili. *Physiol Behav* **58**:1117–1127.

Prescott, J. and R. J. Stevenson (1995b). Pungency in food perception and preference. *Food Rev Int* **11**:665–698.

Rinder. J., P. Stjarne, and J. M. Lundberg (1994). Capsaicin de-sensitization of the human nasal mucosa reduces pain and vascular effects of lactic acid and hypertonic saline. *Rhinology* **32**:173–178.

Rozin, P. and D. Schiller (1980). The nature and acquisition of a preference for chili pepper in humans. *Motiva Emotion* **4**:77–101.

Rozin, P., L. Gruss., and G. Berk (1979). Reversal of innate aversions: attempts to induce a preference for chili peppers in rats. *J Comp Physiol Psychol* **93**:1001–1014.

Saito, K., L-Y. Liu-Chen, and M. A. Moskowitz (1987). Substance P-like immunoreactivity in rat forebrain leptomeninges and cerebral vessels originates from the trigeminal but not sympathetic ganglia. *Brain Res* **403**:66–71.

Sakuma, K., K. Nakashima, and K. Takahashi (1996). Olfactory evoked potentials in Parkinson's disease, Alzheimer's disease and anosmic patients. *Psychiatry Clin Neurosci* **50**:35–40.

Schaffer, K., H. A. Braun, and C. Isenberg (1986). Effect of menthol on cold receptor activity. *J Gen Physiol* **88**:757–776.

Sekizawa, S-I. and H. Tsubone (1994). Nasal receptors responding to noxious chemical irritants. *Respir Physiol* **96**:37–48.

Sekizawa, S-I., H. Tsubone, M. Kuwahara, and S. Sugano (1996). Nasal receptors responding to cold and l-menthol airflow in the guinea pig. *Respir Physiol* **103**:211–219.

Silberstein, S. D. (1992). Advances in understanding the pathophysiology of headache. *Neurology* **42**(suppl 2):6–10.

Silver, W. L. and D. G. Moulton (1982). Chemosensitivity of rat nasal trigeminal receptors. *Physiol Behav* **28**:927–931.

Silver W. L., J. R. Mason, D. A. Marshall, and J. A. Maruniak (1985). Rat trigeminal, olfactory, and taste responses after capsaicin desensitization. *Brain Res* **333**:45–54.

Silver, W. L., J. R. Mason, M. A. Adams, and C. Smeraski (1986). Trigeminal chemoreception in the nasal cavity: responses to aliphatic alcohols. *Brain Res* **376**:221–229.

Silver, W. L., A. H. Arzt, and J. R. Mason (1988). A comparison of the discriminatory ability and sensitivity of the trigeminal and olfactory systems to chemical stimuli in the tiger salamander. *J Comp Physiol A* **164**:55–66.

Silver, W. L., L. G. Farley, and T. E. Finger (1991). The effects of neonatal capsaicin administration on trigeminal nerve chemoreceptors in the rat nasal cavity. *Brain Res* **561**:212–216.

Silverman, J. D. and L. Kruger (1989). Calcitonin gene-related peptide (CGRP) immunoreactive innervation of the rat head with emphasis on specialized sensory structures. *J Comp Neurol* **280**:303–330.

Sostman, A. L. and S. A. Simon (1991). Trigeminal nerve responses in the rat elicited by chemical stimulation of the tongue. *Arch Oral Biol* **36**:95–102.

Stevens, J. C. and W. S. Cain (1986). Aging and the perception of nasal irritation. *Physiol Behav* **37**:323–328.

Stone, H. (1969). Effect of ethmoidal nerve stimulation on olfactory bulbar electrical activity. In: C. Pfaffmann (ed). *Olfaction and Taste III*. New York: Rockefeller University Press, pp 216–220.

Suemune, S., T. Nishimori, M. Hosoi, Y. Suzuki, H. Tsuru, T. Kawata, K. Yamauchi, and N. Maeda (1992). Trigeminal nerve endings of lingual mucosa and musculature of the rat. *Brain Res* **586**:161–165.

Suzuki, N., J. E. Hardebo, J. Kåhrström, and C. Owman (1990). Selective electrical stimulation of postganglionic cerebrovascular parasympathetic nerve fibers originating from the sphenopalatine ganglion enhances cortical blood flow in the rat. *J Cereb Blood Flow Metab* **10**:383–391.

Swandulla, D., K. Schaffer, and H. D. Lux (1986). Calcium channel current inactivation is selectively modulated by menthol. *Neurosci Lett* **68**:23–28.

Szallasi, A., S. Nilsson, T. Farkas-Szallasi, P. M. Blumberg, T. Hokfelt, and J. M. Lundberg (1995). Vanilloid (capsaicin) receptors in the rat: distribution in the brain, regional differences in the spinal cord, axonal transport to the periphery, and depletion by systemic vanilloid treatment. *Brain Res* **703**:175–183.

Szolcsányi, J. (1996a). Capsaicin-sensitive sensory nerve terminals with local and systemic efferent functions: facts and scopes of an unorthodox neuroregulatory mechanism. In: T. Kumazawa, L. Kruger, and K. Mizmura (eds). *Progress in Brain Research, Vol. 113*. Amsterdam: Elsevier Science BV, pp 343–359.

Szolcsányi, J. (1996b). Neurogenic inflammation: reevaluation of an axon reflex theory. In: P. Geppetti and P. Holzer (eds). *Neurogenic Inflammation*. Boca Raton, FL: CRC Press, pp 33–42.

Szolcsanyi, J. and A. Jansco-Gabor (1975). Sensory effects of capsaicin congeners. I. Relationship between chemical structure and pain-producing potency of pungent agents. *Arzneim-Forsch* **25**:1877–1881.

Szolcsanyi, J., A. Jansco-Gabor, and F. Joo (1975). Functional and fine structural characteristics of the sensory neuron blocking effect of capsaicin. *Arch Pharmacol* **287**: 157–163.

Takahashi, H., H. Izumi, and K. Karita (1995). Parasympathetic reflex salivary secretion in the cat parotid gland. *Jpn J Physiol* **45**:475–490.

Tanelian, D. L. and M. B. MacIver (1990). Simultaneous visualization and electrophysiology of corneal A-delta and C fiber afferents. *J Neurosci Methods* **32**:213–222.

Thürauf, N., I. Friedel, C. Hummel, and G. Kobal (1991). The mucosal potential elicited by noxious chemical stimuli with CO_2 in rats: is it a peripheral nociceptive event? *Neurosci Lett* **128**:297–300.

Tominaga, M., M. J. Caterina, A. B. Malmberg, T. A. Rosen, H. Gilbert, K. Skinner, B. E. Raumann, A. I. Basbaum, and D. Julius (1998). The cloned capsaicin receptor integrates multiple pain-producing stimuli. *Neuron* **21**:531–543.

Tsai, S-H., J. M. Tew, J. H. McLean, and M. T. Shipley (1988). Cerebral arterial innervation by nerve fibers containing calcitonin gene-related peptide (CGRP): I. Distribution and origin of CGRP perivascular innervation in the rat. *J Comp Neurol* **271**:435–444.

Tucker, D. (1971). Nonolfactory responses from the nasal cavity: Jacobson's organ and the trigeminal system. In: L. M. Beidler (ed). *Handbook of Sensory Physiology, Vol. IV, Chemical Senses Part 1, Olfaction*. Berlin: Springer-Verlag, pp 151–181.

Walker, J. C., D. Tucker, and J. C. Smith (1979). Odor sensitivity mediated by the trigeminal nerve in the pigeon. *Chem Senses Flav* **4**:107–116.

Walker, J. C., D. B. Walker, C. R. Tambiah, and K. S. Gilmore (1986). Olfactory and nonolfactory odor detection in pigeons: elucidation by a cardiac acceleration paradigm. *Physiol Behav* **38**:575–580.

Wang, Y., R. P. Erickson, and S. A. Simon (1995). Modulation of rat chorda tympani nerve activity by lingual nerve stimulation. *J Neurophysiol* **73**:1468–1483.

Watson, H. R., R. Hems, D. G. Rowsell, and D. J. Spring (1978). New compounds with the menthol cooling effect. *J Soc Cosmet Chem* **29**:185–200.

Winter, J. (1987). Characterization of capsaicin-sensitive neurones in adult rat dorsal root ganglion cultures. *Neurosci Lett* **80**:134–140.

Wood, J. N. and R. Docherty (1997). Chemical activators of sensory neurons. *Annu Rev Physiol* **59**:457–482.

Yavari, P., P. F. McCulloch, and W. M. Panneton (1996). Trigeminally-mediated alteration of cardiorespiratory rhythms during nasal application of carbon dioxide in the rat. *J Auton Nerv Syst* **61**:195–200.

APPENDIX CALCULATION OF STIMULUS CONCENTRATIONS USING THE CLAUSIUS-CLAPEYRON EQUATION

The temperature dependence of the equilibrium vapor concentration (concentration at vapor saturation) is predicted by the Clausius-Clapeyron equation:

$$\text{Log}\,p = -(\Delta H / 2.3 RT) + A \qquad (1)$$

where

   p = vapor pressure (mm Hg)
   R = gas constant (cal/mol)
   T = absolute temperature (degrees Kelvin)
   ΔH = heat of sublimation or vaporization in cal/mol
   A = constant

Equation (1) can be rewritten for vapor concentration, C_{eq}, expressed in ppb using:

$$C_{eq} = [(p)(10^9)]/760 \qquad (2)$$

Substituting equation (2) into equation (1) results in:

$$\text{Log}\,C_{eq} = (a_1)(1/T) + a_0 \qquad (3)$$

where T is the absolute temperature of the vapor and:

$$a_1 = -\Delta H/2.3R \qquad (4)$$

and

$$a_0 = A + 6.119 \qquad (5)$$

A plot of log C_{eq} vs $1/T$ yields a straight line with a slope of a_1 and a y-intercept of a_0.

Therefore, from equation (3), for a given stimulus the equilibrium or saturated vapor concentration can be calculated for any absolute temperature T. This is demonstrated in the following example for amyl acetate.

Amyl acetate has the following vapor pressures (from the *Handbook of Chemistry and Physics*).

Appendix Table 1

Raw Data

Temp °C	Temp °K	1/T °K	Vapor Pressure (mm)	C_{eq}	log C_{eq}
0.0	273.13	3.66E−03	1	1.31E+06	6.12
35.2	308.33	3.24E−03	10	1.31E+07	7.12
62.1	335.23	2.98E−03	40	5.26E+07	7.72
83.2	356.33	2.81E−03	100	1.31E+08	8.12
121.5	394.63	2.53E−03	400	5.26E+08	8.72
142.0	415.13	2.41E−03	760	1.00E+09	9.00

Calculated from
equation 3

From plotting log C_{eq} against $1/T$ and performing a least squares and linear curve fit on the data plotted in the curve the constants can be determined:

$$A_0 = 14.57$$
$$A_1 = 2,298.33$$

Therefore the equation to predict the concentration of the stimulus at vapor saturation is

$$\text{Log}C_{eq} = -2,298.33(1/T) + 14.57$$

To calculate the concentration at vapor saturation at the delivery temperature of 20°C:

$$LogC_{eq} = -2,298.33(1/293.13) + 14.57$$
$$LogC_{eq} = 6.72$$
$$C_{eq} = 5.19 \times 10^6 \, ppb(5,192 \, ppm)$$

From the concentration at vapor saturation the concentration (in ppm) at any fraction of vapor saturation can be calculated. For example

$$10^{-1} = 5192(.1) = 519.2 \, ppm$$
$$10^{-2} = 5192(.01) = 51.92 \, ppm$$

5

Chemical Communication and Pheromones: The Types of Chemical Signals and the Role of the Vomeronasal System

ROBERT E. JOHNSTON

Department of Psychology, Cornell University, Ithaca, NY

1. INTRODUCTION

This chapter reviews the terminology and concepts used in communication by chemical signals, proposes a categorization of the types of signals used (from single chemical compounds to complex mixtures), and discusses the literature on the vomeronasal organ and accessory olfactory system. The goal is to provide a review of chemical communication that deals with the complexity of the processes and avoids oversimplification. In insects there appear to be specialized neural circuits for processing sex pheromones (see Chapter 9, Christensen and White), and some authors have suggested that the vomeronasal system in vertebrates serves similar functions. This hypothesis is reviewed and evaluated in the latter part of the chapter, in which the functions of the vomeronasal organ and accessory olfactory system in tetrapod vertebrates are examined and the controversy about the existence of this system in humans is discussed.

The Neurobiology of Taste and Smell, Second Edition, Edited by Thomas E. Finger, Wayne L. Silver, and Diego Restrepo.
ISBN 0-471-25721-4 Copyright © 2000 Wiley-Liss, Inc.

2. THE NATURE OF CHEMICAL SIGNALS

To understand communication in any sensory channel, one important approach is to study the signals involved and to attempt to determine the specific aspects of a signal that cause a response or provide a particular type of information. This approach has been taken in every system that has been studied, from the dance "language" of honey bees to human language. Scientists working in all sensory domains have developed terminology to refer to communication signals, but the term *pheromone* is unique in that, at least for some authors, the term implies not only properties of the signals but also characteristics at several other levels of analysis, including peripheral sensory mechanisms, central processing and response mechanisms, and the genetic control of these mechanisms. Because the definition and usage of this term has been and continues to be controversial, it is valuable to understand the history of this term and the multiple meanings that it has.

2.1. The Term Pheromone: History and Usage

The term pheromone was coined by Karlson and Lüscher (1959) to refer to "substances which are secreted to the outside by an individual and received by a second individual of the same species, in which they release a specific reaction" (see also Karlson and Butenandt, 1959; Karlson, 1960; Wilson and Bossert, 1963). The term itself comes from the Greek *pherein*, to carry or transfer, and *hormon*, to excite, and was coined to replace a previous term "ectohormone". The basic idea was that pheromones are analogous to hormones, but they served to communicate between organisms rather than between organs within the body. There were several features of this concept. A pheromone was thought to be a substance secreted by special glands, like hormones, and minute amounts of such substances were sufficient to cause a substantial change in behavior or physiology. Although not explicitly part of the definition, early research was based on the assumption that a single chemical compound constituted the signal (Wilson and Bossert, 1963; Linn and Roelofs, 1989). The definition quoted above states that pheromones elicit specific reactions in the receiver; pheromones eliciting behavioral responses were explicitly identified with the ethological concept of a "sign stimulus" or "releasing stimulus" and the responses were seen as an example of "released reactions" to such stimuli. These ethological concepts included the notion that the mechanisms underlying such communication were innately determined and were "hard wired" in the nervous system (Karlson and Lüscher, 1959; Karlson, 1960; Wilson and Bossert, 1963). Also implied was a high degree of determinism in the effects of a stimulus on behavioral or other responses; that is, given the proper circumstances, the specific response was released in a high percentage of cases. Finally, it was expected that there would be specializations of the olfactory system for the detection of specific pheromones, much like specific receptors for hormones within the body.

The coinage of this term and the success of researchers working with insects in identifying single compounds that had predictable effects on behavior or physiology of receivers had a tremendous effect on research on chemical communication in insects and in other taxonomic groups (Whitten, 1966; Johnston et al., 1970; Eisenberg and Kleiman, 1972; Birch, 1974). Along with this explosion of research came the realization that this strict definition of the term pheromone was not sufficient to characterize the myriad of chemical signals that were discovered and the types of effects that they had (Beauchamp et al., 1976; Müller-Schwarze, 1974, 1977; Linn and Roelofs, 1989). Questions have been raised about all aspects of the original concept. In a parallel manner our understanding of hormones has changed significantly; for example, we now realize that there is usually not just one hormone but a set of related hormones, each with a family of receptors with different characteristics, and that the effects observed are often the result of the interaction between the effects of several hormones (Norris, 1996).

At present, different scientists use the term pheromone in very different ways. Many still use the term in a more or less classical way, meaning a specialized signal with either one or several chemical compounds, that causes a specific reaction by means of innate, hard-wired sensory and central response mechanisms. At the other end of the spectrum are those who use the term in a very general way to refer to any chemical signal that has a function in communication. This diversity of usage can lead to misinterpretation of experiments or even faulty logic by researchers themselves; for example, the assumption that because a signal is a "pheromone" there must be specialized receptors or other sensory mechanisms (see also Johnston, 1998). This situation can be confusing to those from other fields who are increasingly being attracted to the study of olfaction.

2.2. Types of Chemical Signals

In order to reduce ambiguity, I propose to simplify the concepts involved by explicitly separating the definitions of the terms for the types of chemical signals (a single chemical compound, a blend or mixture of compounds, etc.) from the underlying mechanisms for response to these signals (i.e., signal detection, sensory coding, response mechanisms, and degree of genetic control of these mechanisms). First, I propose that the term "chemical signal" be used as the generic, all-inclusive term for chemical compounds or mixtures that are released into the environment and, when detected, have a communicative function; that is, they have an effect (behavioral, physiological, or both) on another individual of the same species. Second, "pheromone" should be used for chemical signals that consist of a single chemical compound. Third, "pheromone blend" should be used to refer to a mixture of a small number of compounds that are effective or maximally effective only when they occur in relatively precise ratios (Linn and Roelofs, 1989). Fourth, I suggest the term "mosaic signal" or "odor mosaic" be used to refer to mixtures of a large number of compounds in which many components are important for the full effect. Proportions of at least some components may be important for some functions. Note that these categories are not necessarily mutually exclusive; for example, (a) a component of a pheromone blend that elicits responses on its own could be called a pheromone, and (b) within the complex mixture of a mosaic signal there could be one or more components that contribute to a signal such as an individual signature, but such components also could have an independent effect on a specific behavior (such as attraction or copulation). This scheme has the advantage of preserving the traditional use of the terms "pheromone" and "pheromone blend," while adding a third important category to encompass signals that have previously been given little attention or relegated to a second-class status. For discussion of what is and is not a signal, cross-species communication, and the relevance of evolutionary specialization see Burghardt (1970), Smith (1977), Bradbury and Vehrencamp (1998), Müller-Schwarze (1999), and Sorensen and Stacy (1999).

Some might argue with this definition of the term pheromone because it takes away some of the properties that make this class of signals special (e.g., innate neural mechanisms underlying detection and response). A term that simply refers to the type of chemical signal is preferable, however, because it is less ambiguous and confusing. Issues about the underlying mechanisms and their development are separate, empirical questions that need to be investigated in each case. It should not be assumed, for example, that because a single chemical compound has some strong effect on behavior that the sensory and response mechanisms are genetically controlled, hard-wired processes. Furthermore, a descriptive terminology will make the language of this field more directly parallel to that used in studies of communication in other sensory domains.

2.2.1. A Pheromone: A Signal Consisting of a Single Chemical Compound

Although insect chemical signals were originally believed to be single compounds, most have turned out to be more complex than this. In fact, it is difficult to claim with certainty that any species has a single-compound signal without extensive research to look for other effective compounds. Nonetheless, some cases seem to fit this model. In the gypsy moth, *Lymantria dispar*, one optical isomer, (+)-cis-7,8 epoxy-2-methyloctadecane, is the sex attractant (Roelofs, 1995). Other examples of pheromones are the sex attractants in some species of cockroaches. In the American cockroach (*Periplaneta americana*) males are attracted from a distance by periplanone B; in the brown banded cockroach (*Supella longipalpa*), the attractant is supellapyrone (Roelofs, 1995). Both of these cockroach attractants are unusual, complex compounds that have not been found in other cockroaches, or in any other species of animal or plant, and thus they provide a highly specific signal that is unlikely to be confused in a "noisy" chemical environment.

Several examples of single-compound signals also have been characterized in vertebrates. In the Japanese newt (*Cynops pyrrhogaster*) males attract females from close range by releasing substances from their abdominal gland into the water and directing these substances toward a female by fanning their tails. Kikuyama and colleagues have shown that a decapeptide, given the name sodefrin, is the active compound (Kikuyama et al., 1999). Sodefrin is attractive to reproductively active females, but not to males or to females that are reproductively quiescent. This peptide was sequenced and synthesized, and the synthesized compound has the same level of activity as the native compound. When the nostrils of females are blocked, sodefrin has no effect, indicating that it acts via the olfactory system. It is especially interesting that males of a closely related species, the sword-tailed newt (*Cynops ensicauda*) also produce a decapeptide that attracts conspecific females; this peptide differs from sodefrin by two amino acids. The attractiveness of these two molecules is species-specific, suggesting the possibility of species-specific receptor molecules as well. Proteins may be a common type of signal in amphibians; in two genera of salamanders (*Triturus* and *Plethodon*) glycoproteins play an important role in courtship (Feldhoff et al., 1999; Kikuyama et al., 1999).

Why is it that single molecules are so rare as sex attractants? It is important for individuals to attract and mate with only members of their own species, because mating with individuals of another species will waste time and energy and may result in fewer or no offspring. Only a small percentage of insect species have attractants that consist of a single compound, suggesting that it is either difficult to evolve the molecular machinery for the synthesis of unique molecules or that it is energetically costly, or both. An alternative to species-specific compounds is species-specific ratios of several compounds. The advantage of a system in which specificity is attained by using a ratio of components is that a large number of distinguishable signals can be produced with a small number of chemical compounds, combined in different proportions or different combinations. It is probably easier to evolve a means of producing different relative amounts of a set of compounds that are already present (or the immediate precursors are) than to evolve mechanisms to create new, unique molecules. Likewise, it should be more difficult to evolve a new receptor or other new mechanism for responding to a new molecule than to change the mechanisms involved in evaluating new ratios of the same mixture of compounds. Thus, when new species evolve they should usually modify the mixture rather than evolve new molecules. This hypothesis is supported by the finding that of the approximately 4500 species of Lepidoptera in central Europe only about 150 different chemical compounds are used in sex attractant signals, and most of these are structurally related (Kaisling, 1996).

2.2.2. A Pheromone Blend: Precise Ratios of Components

In many species of insects, the secretion that provides the sex-attractant signal contains a mixture of a small number of structurally related molecules, and furthermore at least two of these compounds must be in a specific proportion in order to obtain maximal responses. Previously, this type of signal has been called a pheromone blend (Linn and Roelofs, 1989), and I maintain this usage. It is worth noting that researchers initially assumed that one of the major components of such mixtures was "the signal." When it became apparent that other components also might be important, the initial assumption was that the role of these minor components was to influence different aspects of a behavioral sequence, such as orientation toward a sex attractant, upwind flight, or cessation of flight in the vicinity of the source. That is, in line with the original notion of a pheromone, it was thought that each component had its own specific effect. More recent research, however, has shown that in general this is not the case. Rather, the signal is the blend (Linn and Roelofs, 1989). In some cases single components may have an effect by themselves, but rarely the full effect. Thus, the sensory system must encode the relative concentrations of several components. An intuitive way to conceptualize this is that the specific mixture creates a unified odor quality (like the smell of an orange or a banana), and that the mixture only "smells right" when it has the correct compounds in the correct proportions.

Several examples should make this clear. In tortricid moths, for example, the response of males to the complete mixture of substances secreted by females is greater than to either the major component alone or a mixture of the two most important components (e.g., in the red banded leaf roller, the oriental fruit moth, and the cabbage looper; for review see Linn and Roelofs, 1989). Furthermore, the specific ratio of some of the components is crucial. In the oriental fruit moth, the female releases four compounds from a gland near her ovipositor when sexually advertising; all contain a 12-carbon chain with one or no double bonds, two are alcohols and two are acetates (Ac). The two acetates and one alcohol act together to attract males, whereas the second alcohol does not seem to be effective unless the first one is present in very low amounts. About 6% of E8-12:Ac relative to Z8-12:Ac is the optimal ratio[1]; greater or lesser amounts of the E8 acetate reduce male responses. The optimal percentage of the third compound is 3% relative to Z8-12:Ac, and again too little or too much of this compound reduces the responses of males. The ratios that elicit optimal response are essentially the same as those emitted by females (Linn and Roelofs, 1989).

As suggested earlier, the importance of these blends is that ratios of compounds provide a simple means of producing a signal that is distinctive compared to that of closely related or sympatric species. Although this discussion of blends has dealt entirely with signals used for sexual attraction of males by females, there is considerable evidence that similar principles may apply in sexual signals of male insects, trail-marking secretions, and recognition of nest mates (Traniello and Robson, 1995; Smith and Breed, 1995).

Among vertebrates, many secretions containing chemical signals are mixtures, but I am not aware of any case in which a specific pheromone blend has been demonstrated (but see Section 2.2.4).

---

[1]The "Z" and "E" refer to whether the two carbons on either side of a carbon–carbon double bond are on the same side (Z, or cis) or the opposite side (E, or trans) of the molecule. "Z8" indicates that at the number eight carbon in the chain, the molecule has a same-side structure.

2.2.3. Mosaic Signals: Multicomponent Signals

I propose that a third type of chemical signal, a mosaic signal, is one that consists of a relatively large number of chemical compounds in which the individual compounds usually do not have an effect by themselves. Such a mixture is presumably perceived as a unified odor quality, much like humans experience the odors of objects such as sweat, coffee, an orange, various types of wine, etc.

One example that has been described in some detail is that of the castoreum of beaver (*Castor canadensis*), which probably has several functions but was assayed in the context of recognition of alien scent marks inside the territory of the home group (Müller-Schwarze, 1992). Beavers mark with this secretion by building a mound of dirt and debris, usually near the bank of their pond, and depositing scent from the castor gland on the top of the mound. When a beaver detects the scent mark of an unfamiliar individual, it engages in a sequence of exploratory behaviors (sniffing from the water, approach by swimming in a zigzag pattern, approach on land, and more sniffing) and then paws at the mound (sometimes obliterating it), straddles the mound, and scent marks over the intruder's mark or on the top of the disturbed mound. Over 60 compounds have been identified from castoreum, but this secretion also contains many compounds that have not yet been identified (Müller-Schwarze, 1992). The composition varies considerably from individual to individual and depends to some extent on diet. A mixture of nine compounds was found to be active in eliciting investigation, pawing, and scent marking, but when each compound was tested alone, none had a significant effect. As more compounds were added to the mixture that was tested, the level of response increased, but none of the mixtures yielded responses as great as to castoreum, suggesting that other, unidentified components are also necessary to achieve the full effect (Müller-Schwarze, 1992).

There are numerous other examples in mammals in which it is likely that multiple components are necessary, even though in many cases not all of the relevant components have been identified or tested. For example, for ewes, extracts of wool or of a scent gland from rams are sufficient to cause increased secretion of luteinizing hormone (LH) and ovulation (Cohen-Tannoudji et al., 1994). Separation of the whole extract into acid and neutral fractions resulted in loss of activity, but the recombination of these fractions was again effective. Two specific components isolated from the neutral fraction were not active, but when combined with the whole acid fraction, the mixture was effective (Cohen-Tannoudji et al., 1994).

Two other examples concern garter snakes and goldfish. In garter snakes, the odor on the dorsal surface of females elicits courtship behavior by males, namely investigation (tongue flicks) and chin rubbing (Mason, 1993). The dorsal surface of female snakes contains a mixture of saturated and unsaturated methyl ketones and additional lipids. Although three individual ketones can elicit some courtship behavior, a mixture of 13 of these ketones was much more effective. Furthermore, addition of a mixture of other skin lipids to this mixture of 13 was even more effective. Mason suggests that the additional lipids may increase the level of responses by making the ketones more available to males (Mason, 1993), but it could also be that these lipids are part of the signal. In courtship among goldfish, two metabolites of prostaglandin F2α released by females stimulate courtship by males, including nudging (inspection of the urogenital opening), chasing, and high levels of activity (Sorensen et al., 1989). Female goldfish release a complex mixture of substances into the water, however, and it now appears that some of these as yet unidentified compounds influence the tendency of males to respond to prostaglandin (Kihslinger and Sorensen, 1999). When male goldfish were exposed to the body odor of immature females of one of five different species of fish, together with prostaglandin F2α, increases in sexual behavior were only seen in those groups exposed to the odor of goldfish (or carp, that hybridize with goldfish); sexual behavior did not increase in response to body odors of the other three species plus prostaglandin. These experiments suggest that prostaglandin may be a key component of a complex mixture (Kihslinger and Sorensen, 1999). Alternatively, this could be an example in which there are two separate signals, both of which influence the probability and vigor of the response, one a sex pheromone and the other a mixture of other compounds that provide contextual information, such as species-identity cues.

The examples above concentrate on signals that have been relatively well characterized chemically and that concern sexual attraction and courtship. Mosaic signals are likely to be involved in a much broader array of communicative functions than this because most of the secretions and excretions produced by vertebrates and at least some of those produced by invertebrates are complex mixtures. In addition, such mixtures are ideally suited for providing information about differences between different classes of individuals, for example, species, sex, social groups or colonies, kin, etc. (Gorman, 1976; Bagneres et al., 1991; Gamboa et al., 1996; Singer et al., 1997). A complex mixture may include information about many different characteristics of the sender. Although one subset of components may be primarily responsible for one particular type of information (e.g., sex or infection by parasites), the response to this odor may depend not only on this information but also on other information in the signal, such as that identifying species, sex, colony, or individual. One way of thinking about mosaic signals is that they are composed of a multidimensional chemical space. They probably are classified (perceived) in a multidimensional perceptual space, and they are relevant to a multidimensional response space. This contrasts sharply with the simpler chemistry of pheromones or pheromone blends, the simpler mechanisms of perception of such signals (Chapter 9, Christensen and White), and the relatively limited number of responses usually associated with such signals.

Some examples of mosaic signals for these types of functions will make these generalizations more concrete. In the north temperate paper wasp (*Polistes fuscatus*), individuals from different colonies (and therefore kin groups) differ in the relative amounts of different cuticular hydrocarbons, and the relative proportions of 10 of these hydrocarbons from different individuals were sufficient to predict acceptance or rejection of an individual by colony members (Gamboa et al., 1996). Differences in cuticular hydrocarbon profiles appear to be a common means of social discrimination in insects, for example, in discrimination of colonies, kin groups, caste, species or subspecies, etc. (e.g., Bagneres et al, 1991; Howard, 1993). Similar results have been found in mammals. Individual Indian mongooses (*Herpestes auropunctatus*), like many mammalian species, recognize different individuals by odor. The concentration of short-chain carboxilic acids in the anal pocket secretion differs across individuals, and mongooses easily discriminate between these secretions from different individuals and between synthetic mixtures of six of these acids that mimic the patterns shown in the anal secretions (Gorman, 1976). In house mice, individuals can discriminate between the urine odors of two mice that are genetically identical except for one locus in the major histocompatibility complex (MHC; Yamazaki et al., 1991). They can even discriminate between the odors of individuals that differ by only a single point mutation in this area of the genome. These discriminations presumably have fitness benefits, because mice prefer to mate with other mice that differ in their MHC type (Yamazaki, 1991; Penn and Potts, 1998). Recent experiments show that differences in the proportions of carboxilic acids in mouse urine are associated with different MHC types, and these differences may be a major part of the signal used to make discriminations between individuals that differ in MHC type (Singer et al., 1997). It is interesting that humans may also base mate choices, in part, on MHC type (Ober, 1999), and odor preferences for individuals differing in MHC type suggest that a similar mechanism may be operating in both humans and mice (Wedekind and Füri, 1997).

2.2.4. Discussion of Types of Signals: Categories or Continua or What?

It is unlikely that all chemical signals fall into the three categories listed above, but at the present time we do not know enough to know if a more detailed taxonomy of signal types would be useful. An example will illustrate some of the difficulties involved. Work with goldfish has provided one of best-understood cases of a complex signalling systems among vertebrates. Several molecules have been identified that are sufficient to elicit behavioral and/or endocrine responses in the context of spawning; these substances act as between-individual signals and also act as within-individual hormones, which has led to their designation as "hormonal pheromones" (Sorensen and Stacy, 1999). Female goldfish secrete two classes of chemically identified signals, steroids and prostaglandins (fatty acids), and in both classes there are several molecules that have overlapping functions during courtship and mating.

In preovulatory female goldfish, a surge in gonadotropin from the pituitary causes an increase in circulating steroids (Sorensen and Stacy, 1999). One of these is 17,20βP (4-pregnen-17α,20β-diol-3-one), which functions within the female to stimulate maturation of the oocyte. This compound is released (along with many other compounds) in a tonic fashion from the gills into the water, and males exposed to 17,20βP show both behavioral and endocrine responses that are mediated through olfaction. Males decrease feeding, increase interactions with females, and have an increase in circulating gonadotropin; this hormonal change leads to an increased volume of milt (sperm and seminal fluid) within 4 hours. Two other steroids that are released by females at the same time (but in urine) also have effects on males, namely androstenedione and the sulfated derivative of 17,20βP, 17,20βP-20S. This sulfated steroid stimulates gonadotropin and milt production in males, whereas androstenedione seems to influence the response to 17,20βP, depending on the ratio of the two compounds. If males are exposed to proportionally large amounts of androstenedione they do not show a gonadotropin response to 17,20βP; it is thought that this is because nonovulatory females secrete a relatively large amount of androstenedione compared to the amount of 17,20βP, and it would be energetically costly for males to respond inappropriately to such females (Sorensen and Stacy, 1999).

A little later, about the time of ovulation, the levels of steroids in the circulation of female goldfish decline greatly, as do the amounts of steroids released into the water, and females become sexually active and attractive to males. These changes in female behavior and the change in their attractiveness to males are caused by an increase in circulating prostaglandins, including prostaglandin F2α (PGF2α) and 15-keto-PGF2α. These two substances have similar effects on male behavior when presented alone, but both together are required to stimulate a gonadotropin surge in males (Sorensen et al., 1989; Sorensen and Stacy, 1999).

Thus, there are two classes of substances released by female goldfish that influence males, and in each class there are several molecules that have independent effects. There appears to be redundancy in the effects of some single-compound signals but interactions between other single-compound signals. As mentioned above (2.2.3) other substances released by females also influence the responses of males in the context of mating behavior (Sorensen et al., 1989; Kihslinger and Sorensen, 1999). In summary, a detailed taxonomy of the types of signals or signal interactions seems premature. The important point to remembr is that there are a variety of ways that individual chemical compounds or mixtures of compounds can interact to influence receivers.

2.2.5. Proteins as the Signal or as Accessory Compounds

There are significant amounts of protein in a number of secretions or excretions that are sources of chemical signals (e.g., mouse urine and hamster vaginal secretions). In a number of these cases the effective signal has been associated with the high-molecular weight fraction, suggesting that the signal might be a protein. The copulatory behavior of male hamsters, for example, is stimulated by a high molecular weight fraction of vaginal secretions. A single protein of the lipocalin type that was purified from this secretion is also effective in stimulating copulation, suggesting that the protein is a sex pheromone ("aphrodisin"; Singer et al., 1987). Recent work, however, suggests that in many cases proteins are not the signal but are pheromone carriers. In the case of male hamster sexual behavior, the gene for aphrodisin was cloned, inserted into bacteria, and the bacteria produced the protein. This protein had little if any effect on male hamsters, suggesting that the sexually arousing signal was one or more ligands bound to the protein despite attempts to strip the protein of such ligands (Singer and Macrides, 1990,). Both the native protein and the bacteria-produced protein were shown to activate cells in the accessory olfactory bulb, as assessed by measurement of the c-fos protein, but the pattern of activated cells was different for the two proteins (Jang et al., 1995, 2000). The fact that cells in the accessory olfactory bulb were activated by both the native and the bacterially produced proteins suggests that the proteins may stimulate receptor cells, but the differing pattern of activation suggests that in vaginal secretion the protein acts as a carrier for other compounds that are the real signal and that these compounds cause a different pattern of activation in the bulb.

Another example in which the role of proteins has been investigated is in house mice. The urine of house mice contains a diverse family of proteins of the lipocalin type, known as major urinary proteins (MUPS). Some authors have concluded that the proteins themselves are the signal (e.g., acceleration of puberty in females by male urine; Mucignat-Caretta et al., 1995). MUPS bind a number of different compounds, however, including those that, by themselves, induce estrus and accelerate puberty (see Section 3), and the primary function of the proteins may be to bind these active components and release them over a long period of time (Beynon et al., 1999; Novotny et al., 1999). It is also conceivable that the proteins may have direct functions in some contexts but be carriers for other molecules that are effective in other contexts (Mucignat-Caretta and Caretta, 1999).

It seems likely that protein–ligand complexes are important for many chemical signals in vertebrates. There may be several advantages of such an arrangement. First, proteins can provide a means of prolonging the half-life of relatively volatile molecules by releasing them at a slow, constant rate once they are deposited in the environment (Beynon et al., 1999). Second, proteins may provide a means for nonpolar molecules to be used in an aqueous mixture, such as urine. Third, the carrier proteins characterized so far are of the same general type (lipocalins) as odorant-binding proteins in the olfactory mucosa, suggesting that they may facilitate interaction of the ligand with receptors in the membranes of receptor cells. Finally, the use of binding proteins may provide a means of conserving the molecules used for a signal across populations, races, or species in order to adapt to different environments by alterations in the proteins used as binders. That is, when adapting to a new environment it may be evolutionarily easier to change the structure of a carrier protein by a few amino acids (and thus changing its binding and release characteristics) than it is to synthesize a new molecule for a signal or to develop new sensory-response mechanisms to compounds that are already present but not used as a signal.

3. MAMMALIAN PHEROMONES AND CHEMICAL SIGNALS

Pheromones and other chemical signals have been particularly difficult to identify in mammals, and because of this it is informative to review and evaluate what is known and to discuss some of the difficulties. Two of the important difficulties in the characterization of chemical signals in mammals are: (a) the secretions produced by mammals tend to be complex mixtures of compounds and are therefore difficult to analyze, and (b) the development of reliable assays is more difficult because the behavior of mammals is more likely to be influenced by a larger number of factors than in some other taxonomic groups—thus the importance of any one cue is likely to be less. Nonetheless a number of chemical signals have been characterized; in some cases a single compound elicits one or more specific responses.

One of the best-known mammalian pheromones is androstenone, which facilitates sexual receptivity in domestic pigs, Sus scrofa. Sows indicate their willingness to mate by "standing" in an immobile posture in response to pressure on the back and other cues, including the sound, sight, and odor of boars. When boars are sexually excited, they "champ" their jaws and secrete large amounts of saliva. The odor of a boar's saliva facilitates the standing response to touch, and androstenone by itself also has this facilitatory effect (Booth, 1980). It is interesting that although attention has focused on androstenone as "the" male pheromone, several other steroids found in saliva (3α-hydroxy-5α-androst-16-ene, androst-4, 16-dien-3-one, and 5α-androst-16-en-3-one) are equally effective (Reed et al., 1974; Booth, 1980). In addition, both preputial gland secretion and urine are effective in facilitating standing by females (Reed et al., 1974; Booth, 1980), suggesting that additional chemical compounds may be effective. This evidence for multiple sources of effective signals and multiple, effective steroids contrasts strongly with the usual impression of androstenone as "the" pheromone in domestic pigs, and suggests that this case should be investigated further. Whether any of the identified signals has a genetically specified effect, in the absence of experience of sows with boars, is yet to be investigated; the available evidence is equally consistent with two interpretations: (1) that several steroids act independently to facilitate standing, and (2) that sows learn what boars smell like and respond to this characteristic odor (of which the steroids are an important part).

Moving from large mammals to yet larger ones, in Asian elephants (Elephas maximus) one substance that is involved in sexual advertisement has been identified from female urine. The urine of female elephants is attractive to males and is especially attractive as ovulation approaches; female urine may also facilitate male sexual behavior. One of the most reliable responses to urine is a behavior pattern known as flehmen, which serves to facilitate access of chemical stimuli to the vomeronasal organ in elephants as well as other mammals, including ungulates and felines. In elephants, flehmen involves placing the tip of the trunk on or just over a source of odor and then placing the tip of the trunk over the vomeronasal duct in the mouth (Rasmussen and Schulte, 1999). Dododecenyl acetate (Z-7-dodencen-1-yl acetate) is a minor component of female urine but its concentration is maximal just prior to ovulation (e.g., maximum levels average 33 μg/ml of urine). This compound elicits a significant increase in the rate of flehmen and may also induce signs of sexual arousal

(Rasmussen et al., 1997). The rate of response to dodecenyl acetate is not as great as to estrous female urine, however, suggesting the involvement of other compounds.

The use of an investigative behavior, such as flehmen or sniffing, as an assay for the effectiveness of a signal has many potential pitfalls. Perhaps the most obvious problem is that an animal can investigate a stimulus for many different reasons, and it often requires multiple controls or a number of different experiments to demonstrate a particular function. In the research on elephants the main difference between the response to dodecenyl acetate and other stimuli was the persistence of the response over repeated trials (Rasmussen et al., 1997; Rasmussen and Schulte, 1999).

Research on the sexual attractant of the golden hamster also illustrates how one must be careful with investigation as a bioassay. The vaginal secretion of female hamsters has many effects on males, including attraction from a distance, stimulation of sexual behavior, release of testosterone into the circulation, and reduction of aggressive behavior (Johnston, 1990). Early studies indicated that one component of the secretion, dimethyl disulfide, was particularly attractive (Singer et al., 1976; for review see Johnston, 1990). The assay used in these studies was one in which an odor was introduced into a male's home cage through a port in the floor, and the time sniffing and digging in the bedding over this port was measured. It was subsequently shown, however, that hamsters show increased investigation in this assay to a wide variety of odors, probably because the stimuli are novel cues in an otherwise familiar and unchanging environment (Johnston, 1981). Other methods of testing for attractiveness of odors are more effective. When male hamsters were tested using one of these methods, in which odors were placed outside a screened door of an arena, dimethyl disulfide was no more attractive to males than control odors. Furthermore, the response of males was no different than the response of females to this odor, whereas males investigated vaginal secretions about five times more than females did. Finally, the response of males to dimethyl disulfide was not dependent on testosterone, as the response to vaginal secretion was (Petrulis and Johnston, 1995). Thus, it appears that dimethyl disulfide, previously identified as a sex attractant pheromone, does not have these functions by itself. The nature of the attractive signal is still not known.

A particularly systematic approach to chemical signals in house mice has been taken by Novotny and his colleagues (*Mus musculus*; Novotny et al., 1999). This group has identified nine compounds that have effects on behavior, physiology, or both. Several of them are male signals that influence female reproductive functions. For example, brevicomin (from male urine) and thiazole (from male preputial glands) promote regular estrous cycles in females and induce estrous cycles in females whose cycles have been suppressed by housing them together in groups. Furthermore, these two compounds, as well as two farnesenes (E,E-α-farnesene and E-β-farnesene) and hydroxymethyl heptanone each individually promote puberty in young females, as assessed by uterine growth. All of these substances bind to the MUPS, discussed above (Section 2.2.5). In addition, brevicomin and thiazole have effects on behaviors such as male–male aggression and attraction of females, apparently signaling the presence of an intact male. The alpha and beta farnesenes may be involved in signaling a dominant status (Novotny et al., 1999). Odors of female house mice also can influence the reproductive physiology of other females. Odors from females can delay puberty in young females and odors of group-housed mice can suppress the estrous cycles of females that are housed in isolation. One compound that causes both of these effects is 2,5-dimethyl pyrazine. This substance occurs in the urine of group-housed females (but not females housed alone) and is dependent on the presence of the adrenal gland (Novotny et al., 1999).

Given the progress in identifying genes for putative olfactory receptors in mice and rats (Chapter 8, McClintock) and in understanding the genome of mice, it should prove highly productive to further characterize chemical signals in mice and to further examine the functions of the substances already identified.

4. THE EVOLUTION AND FUNCTION OF THE VOMERONASAL SYSTEM IN VERTEBRATES

An important issue in trying to understand the mechanisms underlying communication is the degree to which specializations of sensory systems have evolved for the detection and analysis of specific signals. The high degree of specificity of some insect receptor cells and of olfactory glomeruli within the antennal lobe indicates this kind of specialization (Chapter 9, Christensen and White). Indeed, it suggests that there may be receptor molecules specific for pheromones, but no receptor proteins or genes for such proteins have yet been discovered. This section evaluates the hypothesis that the vertebrate vomeronasal organ (VNO) and accessory olfactory bulb (AOB) serve as a pheromone receptor system (for anatomy of the VNO and its projections, see also Price, 1987; Wysocki and Meredith, 1987).

4.1. Evolution of the Vomeronasal Organ

All taxonomic groups of tetrapod vertebrates have at least two major divisions of the nasal chemosensory organs, the main olfactory system and the vomeronasal system (Wysocki, 1979; Eisthen, 1997). In addition, there are branches of the trigeminal nerve that innervate the nasal cavity (see Chapter 4, Bryant and Silver) as well as a variety of other, apparently specialized regions of receptors in the nasal cavity in some vertebrates, including those of the septal organ of Masera and the terminal nerve (Halasz, 1990). The phylogenetic origins of the VNO are obscure, but it is likely that this organ evolved in early tetrapod lineages, since it is present in amphibians and reptiles, suggesting an origin in the last common ancestor of the two groups (Eisthen, 1992, 1997). Consistent with an early origin are observations of the neural projections of the vomeronasal system and the main olfactory system; in frogs, snakes, lizards, and mammals

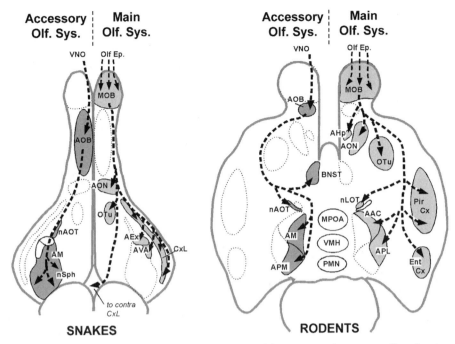

FIG. 5.1. Semischematic diagrams of the projections of the accessory (vomeronasal) and main olfactory systems in snakes and rodents, based on work on garter snakes (redrawn from Lohman and Smeets [1993] based on the work of Lanuza and Halpern [1998]) and on laboratory rats and golden hamsters (redrawn from Johnston, 1985). The projections of the receptor cells of the vomeronasal organ go to the accessory olfactory bulb (AOB) while those of the olfactory epithelium (Olf. Ep.) go to the main olfactory bulb (MOB). These two parts of the bulb project to separate areas of in the CNS. The **AOB** projects primarily to the **nucleus of the accessory olfactory tract** (nAOT), **medial amygdala** (medial nucleus [AM] and nucleus sphericus [nSph] in snakes; medial nucleus [AM] and posteromedial nucleus [APM] in rodents) and to the **bed nucleus of the stria terminalis** (BNST). The **MOB** projects to the **lateral amygdala** (external [AEx] and ventral anterior [AVA] nuclei of the amygdala in snakes; anterior central [AAC] and posterolateral [APL] nucleus in rodents) and to the ventrolateral surface of the brain including: the **anterior olfactory nucleus** (AON), the **olfactory tubercle** (OTu), the **olfactory cortex** (lateral cortex [CxL] in snakes; piriform cortex [PirCx] and entorhinal cortex (EntCx) in rodents) and in rodents, a minor projection to the anterior hippocampus. The figure of the rodent brain also indicates areas of the hypothalamus (medial preoptic area-anterior hypothalamus [MPOA], ventromedial hypothalamus [VMH], and premammilary nucleus [PMN]) receiving higher-order olfactory and accessory olfactory inputs.

these projections display a generally similar pattern (Fig. 5.1; Eisthen, 1992; Lohman and Smeets, 1993; Lanuza and Halpern, 1998). According to this view, the vomeronasal system has been lost secondarily in some taxonomic groups, such as birds, crocodiles, some lizards, some aquatic mammals, and some primates. Adopting either an aerial or an aquatic way of life may explain the loss of this system in birds, crocodiles and aquatic mammals, but it does not explain why the system has been lost in old-world monkeys and apes or in some lizards.

One way that the vomeronasal system differs from the main olfactory sytem in all taxonomic groups is in the morphological types of the receptor cells. Those in the main olfactory system are ciliated while those in the vomeronasal system are microvillar (Eisthen, 1992), suggesting a difference in function. In fish, both types of receptor cells exist, but they are distributed throughout the olfactory epithelium, rather than being sequestered into different areas of the nasal cavity. These two types of cells may represent precursors of the distinct vomeronasal and olfactory organs seen in other, later-evolving taxa of tetrapods.

In the following sections I briefly review the functions of vomeronasal system in taxa that have been investigated.

4.2. Reptiles: Functions of the Vomeronasal System

The vomeronasal organ in reptiles is quite well developed, and often the accessory olfactory bulb is larger than the main olfactory bulb. In many species, stimuli appear to be delivered to the openings of the vomeronasal sac by tongue-flicking actions (Halpern and Kubie, 1984; Schwenk, 1995). As in other tetrapod vertebrates, all reptile vomeronasal organs that have been analyzed histologically have only microvillar receptor cells (Eisthen, 1992).

The functions of the vomeronasal organ have been most thoroughly studied in snakes, in which it mediates both social and nonsocial behaviors. In many species the VNO is essential for recognition of prey by chemical cues, and thus for striking at prey and/or trailing of prey (Burghhart, 1970). For example, garter snakes trail and identify many prey, such as earthworms, by chemical cues. Both trailing and striking at prey are eliminated by cutting the vomeronasal nerve; no effect on these behaviors is observed after cutting the olfactory nerve (Halpern and Kubie 1984). A protein from the skin of earthworms that elicits trailing and striking by snakes has been purified and the transduction of neural responses to this protein has been characterized (Wang et al., 1997). Even in species in which visual or infrared cues guide or elicit strikes, such as in vipers and rattlesnakes, trailing of prey and prey identification after a strike is dependent on chemical senses, probably the VNO (Halpern, 1987), and in rattlesnakes the VNO may mediate striking (Halpern and Kubie, 1984).

The vomeronasal system of snakes is also important for many aspects of social behavior. During the first stage of courtship, a male garter snake presses his chin, lips and, snout against the dorsal surface of the female's back, which has the effect of straightening out the female. The male then aligns his body with that of the female and copulation usually follows. Input from the vomeronasal organ is necessary for chin pressing and copulation, as shown by the effects of nerve cuts; similar cuts of the olfactory nerve have no effect (Halpern and Kubie, 1984). The vomeronasal organ of garter snakes is also necessary for male–male aggressive behavior, trailing of conspecifics, aggregation, and shelter selection, at least as studied in the laboratory (Halpern and Kubie, 1984).

Although studied in less detail, the vomeronasal system of many lizards is also well developed and serves a similarly wide range of functions, including prey identification, prey consumption, selection of other types of foods, and social behavior (Graves and Halpern, 1990; Schwenk, 1995).

4.3. Mammals: Functions of the Vomeronasal System

The most consistent finding across studies in mammals is that the vomeronasal organ is involved in endocrine responses to odors. For example, it has been implicated in testosterone responses of males to odors of females in mice and hamsters, puberty acceleration in young female mice in response to odors of adult males, increases in luteinizing hormone (LH) in female rats to odors from males, pregnancy block in female mice in response to an unfamiliar male, and reproductive activation in female voles to odors of males (for reviews see Johnston, 1985; Halpern 1987; Wysocki and Meredith, 1987; Wysocki, 1989; Meredith and Fernandez-Fewel, 1994; Keverne, 1999). It is important, however, not to overgeneralize from these results (Johnston, 1998). First, not all cases of endocrine responses to odors are mediated by the vomeronasal system. For example, the release of LH in ewes that occurs in response to male odors continues to occur after either lesion of the VNO or section of the vomeronasal nerves (Cohen-Tannoudji et al., 1989). Second, the main olfactory system may also have a role in mediating endocrine responses. For example, although the vomeronasal organ is essential for increases in androgen levels in male hamsters in response to vaginal secretions, lesions of both the vomeronasal system and the main olfactory system are necessary to eliminate such responses in sexually naïve males to estrous females themselves (Pfeiffer and Johnston, 1994). Thus, the vomeronasal system is often important in endocrine responses to social odors, but it does not have this function in all species and it may share responsibility for this function with the main olfactory system in some species.

The vomeronasal system in mammals is also often important in social behavior, especially reproductive behaviors (for reviews see Johnston, 1985, 1998; Wysocki and Meredith, 1987; Wysocki, 1989). Elimination of the VNO or its input to the olfactory bulbs causes deficits in copulatory behavior of male house mice, golden hamsters, and, to a lesser extent, guinea pigs; effects on female mating behavior have been observed in female rats and hamsters (Mackay-Sim and Rose, 1986; Saito and Moltz, 1986). Maternal behavior is affected by damage to the vomeronasal system in laboratory rats and golden hamsters (Fleming et al., 1979; Numan and Sheehan, 1997; Brouette-Lahlou et al., 1999). VNO input is also involved in the control of some odor-stimulated communicative behaviors, such as scent marking in house mice (Wysocki, 1989) but not in hamsters (Johnston, 1992) and ultrasonic calling in both house mice and golden hamsters (Wysocki, 1989; Johnston, 1992). Aggressive behavior also can be influenced by vomeronasal lesions, as shown in house mice (Wysocki and Meredith, 1987; Wysocki, 1989).

Once again, however, it is important to keep this general trend in perspective and to be careful about overgeneralization of this pattern of results (Johnston, 1998). First, in most of the cases listed above no specific chemical compounds have been identified as constituting the signal, so it is not clear whether these responses are being influenced by a pheromone, a pheromone blend, an odor mosaic, or a combination of these types of signals. Second, although it is clear that the vomeronasal system is important for many aspects of social behavior in mammals, its role in other classes of behavior, such as food identification and selection, is not known. The functions of this system have simply not been tested in nonsocial behaviors in mammals. Third, the vomeronasal system is not always involved in mediating responses to social odors. A good example is that of nursing in rabbits. Location of and attachment to the nipple by pups is dependent on as yet unidentified odor cues. Vomeronasal organ removal has no effect on these behaviors, whereas damage to the main olfactory epithelium (by application of $ZnSO_4$) eliminates the response (Hudson and Distel, 1986). Another example is that of the facilitation of the standing response in pigs to pressure on the back. Blocking of the vomeronasal duct has no influence on this reaction or on the attraction of sows to this odor (Dorries et al., 1997). Finally, it is often the case that both the main olfactory and vomeronasal systems are involved in mediating a specific behavior; examples of this sort include male hamster copulatory behavior, ultrasonic calling by female hamsters, and maternal behavior in rats (Johnston, 1998).

Those authors who promote the vomeronasal system as a pheromone receptor system generally equate this view with hard-wired responses that do not involve learning or experiential effects of any kind. There are several examples, however, in which it is clear that the VNO seems to be involved in learning of odor cues. Most thoroughly studied is the case of the pregnancy block effect in house mice; the basic finding is that if female mice are exposed to an unfamiliar male, or the odor of such a male, within four days after mating, most of them do not maintain their pregnancy due to a failure of the fertilized egg to implant in the uterus. Exposure to the mate or to odors of the mate do not have this effect—in fact, they protect against it. This effect thus involves both recognition of the non-mate as an unfamiliar male and an endocrine response to this recognition. In a series of experiments on this phenomenon, Keverne and colleagues have shown that the accessory olfactory bulb is important in the establishment of the memory for the mate and the recognition of the non-mate, and his group and others have characterized mechanisms that might be involved (Okere et al., 1996; Keverne, 1998).

The roles of the vomeronasal system in discrimination and recognition of individuals may vary across species, sexes, sources of scent, or contexts. In golden hamsters, a recent report suggests that the vomeronasal system may be essential for discrimination by male hamsters of individual odors from three different sources in a habituation paradigm (flank glands, vaginal secretions, and feces) but not of a fourth odor (urine; Johnston, 1998, and unpublished results). In female hamsters, however, lesions of the VNO do not eliminate this ability, suggesting a sex difference in vomeronasal organ function in hamsters. The main olfactory system mediates discrimination of individual odors in females (Petrulis et al., 1999). In spiny mice, *Acomys cahrinus*, irrigation of the main olfactory mucosa with zinc sulfate eliminated their ability to recognize familiar nest mates, again indicating the importance of the main olfactory system (Matochik, 1988).

In summary, the vomeronasal system in some species does mediate evolved mechanisms of social communication that may not require previous experience with the signals involved, and the projections of this system to the central nervous system may constitute a pathway largely dedicated to mediating a variety of social interactions. The characterization of this system as a "pheromone receptor" organ is clearly an inaccurate oversimplification, however, because this system detects other types of chemical signals and mediates nonsocial behaviors in many species. Furthermore, the vomeronasal system is not the only chemosensory system that mediates responses to pheromones. The main olfactory system also does this, and it is possible that the sense of taste, the terminal nerve, or the septal organ of Masera could also mediate responses to pheromones in some species (Johnston, 1998; Preti and Wysocki, 1999).

4.4. Do Humans Have a Vomeronasal System?

One of the most controversial questions about the vomeronasal organ in recent years is whether humans possess this sensory organ (Monti-Bloch et al., 1998; Preti and Wysocki, 1999). The lack of convincing evidence for a functional vomeronasal system in old-world primates and apes immediately suggests that one should be cautious in claiming its existence in humans (Wysocki, 1979). There is, however, relatively little published information on old-world monkeys and apes, and consequently there is great need for more research. Several research groups have demonstrated the existence, in the human fetus, of an anatomical structure that has the appearance of the vomeronasal organ in other mammals, including neural connections to the olfactory bulb (Boehm and Gasser, 1993; Kjaer and Fischer-Hansen, 1996). These structures stain positively for LHRH, and this is consistent with other data indicating that neurons that produce LHRH have their origins in fetal tissue associated with the olfactory placode, which also gives rise to the nervus terminalis. At this stage in development the nervous terminalis is closely associated with the vomeronasal organ. According to these authors, however, the fetal VNO and its neural connection to the olfactory bulb disappears at about 19 weeks of age; there is, however, little detailed documentation of this degeneration. If degeneration does occur, it seems extremely unlikely that the vomeronasal organ would reform to appear in the adult (Preti and Wysocki, 1999). Research is needed to more thoroughly describe the developmental history of this organ in humans.

What does seem to be well established for adult humans is that there is usually a depression or pocket in the medial lining of the nasal cavity near the external opening of the nose about where one would expect a vomeronasal organ to be if it did exist (Hummel et al., 1999; Smith et al., 1999). Such structures are not always located, and sometimes this pocket exists on only one side of the nasal cavity. Although this pocket has generally been called the vomeronasal organ by the authors of relevant papers, there is considerable controversy about the function, if any, of this area of epithelium. Proponents of the existence of a human vomeronasal organ have shown that this area does have cells that are somewhat different from those seen in respiratory epithelium (Moran, 1991), but they have not demonstrated the existence of cells with either microvilli or cilia, as found in vomeronasal or olfactory epithelium, and there is no evidence for sensory receptor cells with axons, no less ones with projections to the olfactory bulb or other areas of the brain.

Those who claim that a human vomeronasal organ exists have published data demonstrating a number of effects that are consistent with this hypothesis. After application of certain steroids to the epithelium, this group has observed: (a) changes in electrical potential recorded from this region of the nasal epithelium, (b) changes in affect, (c) changes in measures of autonomic nervous system function, and (d) changes in hormone levels (Monti-Bloch et al., 1998). Critics point out, however, that the mechanisms underlying these responses are not known and that in every case plausible, alternative explanations exist (Preti and Wysocki, 1999). These alternative explanations include: (a) local electrical changes could be due to activation of free nerve endings of the trigeminal nerve or of smooth muscles surrounding the abundant vasculature in the epithelium, or even physical chemical changes ocurring as odorants flow over and interact with a wet surface (a physical junction potential; Mozell, 1962), (b) the substances used are steroids and closely resemble estrogens, androgens, and progestins, and the effects observed could occur by direct action on hormone-sensitive tissues after uptake of the steroids into the circulation, either acting directly or by conversion to other steroids, and (c) the stimuli could act via the olfactory system—although the olfactory epithelium does not appear to respond to putative pheromones with a summated electrical response (e.g., Monti-Bloch & Grosser, 1991; Berliner, et al., 1996) the possible involvement of this system in endocrine responses to such substances was not investigated (Monti-Bloch, et al., 1998). In addition, many of these experiments lack appropriate control conditions. For example, in the experiments with steroids, the authors have tested substances that they believe to be human pheromones, but they have not used other steroids as control stimuli (Preti and Wysocki, 1999).

In sum, proponents of the existence of a human vomeronasal organ and of human pheromones acting through this organ have stimulated much debate and raised interesting issues, but do not at present have convincing evidence for the existence of pheromones in humans, functional vomeronasal organs, or an accessory olfactory system in the brain.

5. CONCLUSION AND SUMMARY

Conceptual difficulties have plagued the literature on olfactory communication and it might help if the concepts, both explicit and implicit, were simplified. I propose that the term "pheromone" be purely descriptive, referring to a chemical signal consisting of a single molecule. Excluded are any characteristics associated with the kinds of sensory processing, response mechanisms, or the degree of genetic control over any part of the detection and response process. In this scheme "pheromone blend" retains its meaning as a mixture of a small number of compounds in which the specific ratios of components are important for communicative function. "Mosaic signals," a new category of signals proposed here, are those in which a relatively large number of compounds are necessary for communicative function. Designation of this category is important because such signals have often been ignored in the past, yet they are important in all taxonomic groups that have been studied in enough detail. These three categories are neither exhaustive nor mutually exclusive, but they are conceptually distinct types of signals that encompass much of the diversity that has been described. Many examples were described of each type of signal; other cases that did not clearly fit into one category were described to emphasize the potential complexities involved.

The latter part of the chapter summarizes what is known about the evolution and functions of the vomeronasal organ and accessory olfactory system in vertebrates and evaluates the hypothesis that this system is specialized for mediating responses to pheromones (that is, the vomeronasal organ as a "pheromone receptor organ"). This system does indeed mediate many responses to chemical signals, but not all of these signals are pheromones. The VNO is also implicated in learned responses to odors and, in some taxonomic groups, is necessary for nonsocial behaviors such as feeding. In mammals, the role of this system in nonsocial behaviors has not been examined. Furthermore, the main olfactory system may be solely responsible for responses to some pheromones, or the main and accessory systems may both be involved. Thus, it is not appropriate to view the vomeronasal organ as either a pheromone receptor organ or as entirely specialized for mediating responses to social signals. In humans, current evidence supports the existence of a pocket in the nasal cavity which is in a position where one would expect to see a vomeronasal organ if there were one. Anatomical evidence, however, does not show convincingly that this pocket has any sensory cells with axons, nor that any cells in this region make connection with the olfactory bulb or other part of the brain. Other evidence that has been advanced to support the existence of a human vomeronasal organ does not rule out alternative explanations for the results. Thus, at the present time there is no convincing evidence for the presence of a functional vomeronasal organ or accessory olfactory system in adult humans.

REFERENCES

Bagneres, A-G., A. Killian, J-L. Clement, and C. Lange, (1991). Interspecific recognition among termites of the genus *Reticulitermes*: evidence for a role for the cuticular hydrocarbons. *J Chem Ecol* **17**:2397–2420.

Beauchamp, G. K., R. L. Doty, D. G. Moulton, and R. A. Mugford (1976). The pheromone concept in mammalian chemical communication: a critique. In: *Mammalian Olfaction,*

Reproductive Processes and Behavior R. L. Doty (ed). New York: Academic Press, pp 143–160.

Berliner, D. L., L. Monti-Bloch, C. Jennings-White and V. Diaz-Sanchez (1996). The functionality of the human vomeronasal organ (VNO): evidence for steroid receptors. *J Steroid Biochem Molec Biol* **58**:259–265.

Beynon, R. J., D. H. L. Robertson, and S. J. Hubbard (1999). The role of protein binding in chemical communication: major urinary proteins in the house mouse. In: *Advances in Chemical Communication in Vertebrates*, R. E. Johnston, D. Müller-Schwarze, and P. W. Sorensen (eds). New York: Plenum, pp 137–148.

Birch, M. C. (ed). (1974). *Pheromones*. Amsterdam: North-Holland.

Boehm, N. and B. Gasser (1993). Sensory receptor-like cells in the human foetal vomeronasal organ. *Neuro Report* **4**:867–870.

Booth, W. D. (1980). Endocrine and exocrine factors in the reproductive behavior of the pig. *Symp Zool Soc Lond* **45**:289–311.

Bradbury, J. W. and S. L. Vehrencamp (1998). *Principles of Animal Communication*. Sunderland, MA: Sinauer.

Brouette-Lahlou, I., F. Godinot, and E. Vernet-Maury (1999). The mother rat's vomeronasal organ is involved in detection of dodecyl propionate, the pup's preputial gland pheromone. *Physiol Behav* **66**:427–436.

Burghardt, G. M. (1970). Defining "communication." In: J. W. Johnston, D. G. Moulton, and A. Turk (eds). *Advances in Chemoreception, Vol. 1: Communication by Chemical Signals*. New York: Appleton-Century-Crofts, pp 5–18.

Cohen-Tannoudji, J., C. Lavenet, A. Locatelli, and J. P. Signoret (1989). Non-involvement of the accessory olfactory system in the LH response of anoestrous ewes to male odour. *J Reprod Fertil* **86**:135–144.

Cohen-Tannoudji, J., J. Einhorn, and J. P. Signoret (1994). Ram sexual pheromone: first approach of chemical identification. *Physiol Behav* **56**:955–961.

Dorries, K. M., E. Adkins-Regan, and B. P. Halpern (1997). Sensitivity and behavioral responses to the pheromone androstenone are not mediated by the vomeronasal organ in domestic pigs. *Brain Behav Evol* **89**:53–62.

Eisenberg, J. F. and D. Kleiman (1972). Olfactory communication in mammals. *Ann Rev Ecol Systematics* **3**:1–31.

Eisthen, H. L. (1992). Phylogeny of the vomeronasal system and of receptor cell types in the olfactory and vomeronasal epithelia of vertebrates. *Microsc Res Techn* **23**:1–21.

Eisthen, H. L. (1997). Evolution of vertebrate olfactory systems. *Brain Behav Evol* **50**: 222–233.

Feldhoff, R. C., S. M. Rollmann, and L. D. Houck (1999). Chemical analysis of courtship pheromones in a plethodontid salamander. In: R. E. Johnston, D. Müller-Schwarze, and P. W. Sorensen (eds). *Advances in Chemical Signals in Vertebrates*. New York: Plenum, pp 117–125.

Fleming, A., F. Vaccarino, L. Tambosso, and P. Chee (1979). Vomeronasal and olfactory system modulation of maternal behavior in the rat. *Science* **203**:372–374.

Gamboa, G. J., T. A. Grudzien, K. E. Espelie, and E. A. Bura (1996). Kin recognition pheromones in social wasps: combining chemical and behavioural evidence. *Anim Behav* **51**:625–629.

Gorman, M. L. (1976). A mechanism for individual recognition by odour in *Herpestes auropunctatus* (Carnivora: Viverridae). *Anim Behav* 24:141–145.

Graves, B. M. and M. Halpern (1990). Roles of vomeronasal organ chemoreception in tongue flicking, exploratory and feeding behaviour of the lizard, *Chalcides ocellatus. Anim Behav* 39:692–698.

Halasz, N. (1990). *The Vertebrate Olfactory System*. Budapest: Akademiai Kiado.

Halpern, M. (1987). The organization and function of the vomeronasal system. *Annu Rev Neurosci* 10:325–362.

Halpern, M. and J. L. Kubie (1984). The role of the ophidian vomeronasal system in species-typical behavior. *Trends Neurosci* 7:472–477.

Howard, R. W. (1993). Cuticular hydrocarbons and chemical communication. In: D. W. Stanley-Samuelson and D. R. Nelson (eds). *Insect Lipids: Chemistry, Biochemistry and Biology*. Lincoln, NE: University of Nebraska Press, pp 179–226.

Hudson, R. and H. Distel (1986). Pheromonal release of suckling in rabbits does not depend on the vomeronasal organ. *Physiol Behav* 37:123–128.

Hummel, T., D. Kühnau, M. Knecht, N. Abolmaali, and K. B. Hüttenbrink (1999). The anatomy of the vomeronasal organ: characterization by means of nasal endoscopy and magnetic resonance imaging. *Chem Senses* 24:622.

Jang, T., A. G. Singer, and F. Macrides (1995). Induction of c-fos-gene product in the male hamster accessory olfactory bulbs by natural and bacterially cloned aphrodisin. *Chem Senses* 20:712–713.

Jang, T., A. G. Singer, and R. J. O'Connell (2000). Induction of c-fos protein in male hamster accessory olfactory bulbs by exposure to cloned and authentic aphrodisin. Submitted for publication.

Johnston, J. W., D. G. Moulton, and A. Turk (eds). (1970). *Advances in Chemoreception, Vol. 1, Communication by Chemical Signals*. New York: Appleton-Century-Crofts.

Johnston, R. E. (1981). Attraction to odors in hamsters: an evaluation of methods. *J Comp Physiol Psychol* 95:951–960.

Johnston, R. E. (1985). Olfactory and vomeronasal mechanisms of communication. In: D. W. Pfaff (ed). *Taste, Olfaction and the Central Nervous System*. New York: Rockefeller University Press, pp 322–346.

Johnston, R. E. (1990). Chemical communication in golden hamsters: from behavior to molecules and neural mechanisms. In: D. A. Dewsbury (ed). *Contemporary Issues in Comparative Psychology*. Sunderland, MA: Sinauer, pp 381–409.

Johnston, R. E. (1992). Vomeronasal and/or olfactory mediation of ultrasonic calling and scent marking by female golden hamsters. *Physiol Behav* 51:1–12.

Johnston, R. E. (1998). Pheromones, the vomeronasal system, and communication. *Ann N Y Acad Sci* 855:333–348.

Kaisling, K.-E. (1996). Peripheral mechanisms of pheromone reception in moths. *Chem Senses* 21:257–268.

Karlson, P. (1960). Pheromones. *Ergebnisse der Biologie* 12:212–225.

Karlson, P. and A. Butenandt (1959). Pheromones (ectohormones) in insects. *Annu Rev Entomol* 4:39–58.

Karlson, P. and M. Lüscher (1959). "Pheromones": a new term for a class of biologically active substances. *Nature* 183:55–56.

Keverne, E. B. (1998). Vomeronasal/accessory system and pheromonal recognition. *Chem Senses* **23**:491–494.

Keverne, E. B. (1999). The vomeronasal organ. *Science* **286**:716–720.

Kihslinger, R. L. and P. W. Sorensen (1999). Pheromonal cues in the goldfish are perceived within the context of the body odor within which they occur. *Chem Senses* **24**:616.

Kikuyama, S., F. Toyodda, T. Iwata, N. Takahashi, K. Yamamoto, H. Hayashi, S. Miura, and S. Tanaka (1999). Female-attracting peptide pheromone in newt cloacal glands. In: R. E. Johnston, D. Müller-Schwarze, and P. W. Sorensen (eds). *Advances in Chemical Signals in Vertebrates*. New York: Plenum, pp 127–136.

Kjaer, I. and B. Fischer-Hansen (1996). The human vomeronasal organ: prenatal develop-mental stages and distribution of luteinizing hormone-releasing hormone. *Eur J Oral Sci* **104**:34–40.

Lanuza, E. and M. Halpern (1998). Efferents and centrifugal afferents of the main and accessory olfactory bulbs in the snake, *Thamnophis sirtalis*. *Brain Behav Evol* **51**:1–22.

Linn, C. E. and W. L. Roelofs (1989). Response specificity of male moths to multicomponent pheromones. *Chem Senses* **14**:421–437.

Lohman, A. H. M. and W. J. A. J. Smeets (1993). Overview of the main and accessory olfactory bulb projections in reptiles. *Brain Behav Evol* **41**:147–155.

Mackay-Sim, A. and J. D. Rose (1986). Removal of the vomeronasal organ impairs lordosis in female hamsters: effect is reversed by luteinising hormone-releasing hormone. *Neuroen-docrinology* **42**:489–493.

Mason, R. T. (1993). Chemical ecology of the red-sided garter snake, *Thamnophis sirtalis parietalis*. *Brain Behav Evol* **41**:261–268.

Matochik, J. A. (1998). Role of the main olfactory system in recognition between individual spiny mice. *Physiol Behav* **42**:217–222.

Meredith, M. and G. Fernandez-Fewel (1994). Vomeronasal system, LHRH, and sex behavior. *Psychoneuroendocrinolgy* **19**:657–672.

Monti-Bloch, L., V. Diaz-Sanchez, C. Jennings-White, and D. L. Berliner (1998). Modulation of serum testosterone and autonomic function through stimulation of the male human vomeronasal organ (VNO) with pregna-4,20-diene-3,6-dione. *J. Steroid Biochem Molec Biol* **65**:237–242.

Monti-Bloch, L. and B. I. Grosser (1991). Effect of putative pheromones on the electrical activity of the human vomeronasal organ and olfactory epithelium. *J Steroid Biochem Molec Biol* **39**:573–582.

Monti-Bloch, L., C. Jennings-White, and D. Berliner (1998). The human vomeronasal system. *Ann N Y Acad Sci* **855**:373–389.

Moran, D. T. (1991). The vomeronasal (Jacobson's) organ in man: ultrastructure and frequency of occurrence. *J Steroid Biochem* **39**:545–552.

Mozell, M. M. (1962). Olfactory mucosal and neural responses in the frog. *Am J Physiol* **20**:353–358.

Mucignat-Caretta, C. and Caretta, A. (1999). Protein-bound odorants as flags of male mouse presence. In: R. E. Johnston, D. Müller-Schwarze, and P. W. Sorensen(eds). *Advances in Chemical Signals in Vertebrates* New York: Plenum pp 359–364.

Mucignat-Caretta, C., A. Caretta, and A. Cavaggioni (1995). Acceleration of puberty onset in female mice by male urinary proteins. *J Physiol* **486**:517–522.

Müller-Schwarze, D. (1974). Olfactory recognition of species, groups, individuals and physiological states among mammals. In: M. C. Birch (ed). *Pheromones*. Amsterdam: North-Holland, pp 316–326.

Müller-Schwarze, D. (1977). Complex mammalian behavior and pheromone bioassay in the field. In: D. Müller-Schwarze and M. M. Mozell (eds). *Chemical Signals in Vertebrates*. New York: Plenum, pp 413–433.

Müller-Schwarze, D. (1992). Castoreum of beaver (*Castor canadensis*): function, chemistry and biological activity of its components. In: R. L. Doty and D. Müller-Schwarze (eds). *Chemical Signals in Vertebrates VI*. New York: Plenum, pp 457–464.

Müller-Schwarze, D. (1999). Signal specialization and evolution in mammals. In: R. E. Johnston, D. Müller-Schwarze, and P. W. Sorensen (eds). *Advances in Chemical Signals in Vertebrates*. New York: Plenum, pp 1–14.

Norris, D. O. (1996). *Vertebrate Endocrinology, (3rd Ed)*. San Diego: Academic Press.

Novotny, M. V., W. Ma, L. Zidek, and E. Daev (1999). Recent biochemical insights into puberty acceleration, estrus induction and puberty delay in the house mouse. In: R. E. Johnston, D. Müller-Schwarze, and P. W. Sorensen (eds). *Advances in Chemical Signals in Vertebrates*. New York: Plenum, pp 99–116.

Numan, M. and T. G. Sheehan (1997). Neuroanatomical circuitry for mammalian maternal behavior. *Ann N Y Acad Sci* **807**:101–125.

Ober, C. (1999). HLA and mate choice. In: R. E. Johnston, D. Müller-Schwarze, and P. W. Sorensen (eds). *Advances in Chemical Communication in Vertebrates*. New York: Plenum, pp 189–199.

Okere, C. O., H. Kaba, and T. Higuchi (1996). Formation of an olfactory recognition memory in mice: reassessment of the role of nitric oxide. *Neuroscience* **71**:349–354.

Penn, D. and W. K. Potts (1998). Chemical signals and parasite-mediated sexual selection. *TREE* **13**:391–395.

Petrulis, A. and R. E. Johnston (1995). A reevaluation of dimethyl disulfide as a sex attractant in golden hamsters. *Physiol Behav* **57**:779–784.

Petrulis, A., M. Peng, and R. E. Johnston (1999). Effects of vomeronasal organ removal on individual odor discrimination, sex-odor discrimination, sex-odor preference, and scent marking by female hamsters. *Physiol Behav* **66**:73–83.

Pfeiffer, C. A. and R. E. Johnston (1994). Hormonal and behavioral responses of male hamsters to females and female odors: roles of olfaction, the vomeronasal system, and sexual experience. *Physiol Behav* **55**:129–138.

Preti, G. and C. J. Wysocki (1999). Human pheromones: releasers or primers—fact or myth. In: R. E. Johnston, D. Müller-Schwarze, and P. W. Sorensen (eds). *Advances in Chemical Signals in Vertebrates*. New York: Plenum, pp 315–332.

Price, J. L. (1987). The central and accessory olfactory systems. In: T. E. Finger and W. L. Silver (eds). *Neurobiology of Taste and Smell*. New York: John Wiley & Sons, pp 179–203.

Rasmussen, L. E. L. and B. A. Schulte (1999). Ecological and biochemical constraints on pheromonal signaling systems in Asian elephants and their evolutionary implications. In: R. E. Johnston, D. Müller-Schwarze, and P. W. Sorensen (eds). *Advances in Chemical Signals in Vertebrates*. New York: Plenum, pp 49–62.

Rasmussen, L. E. L., T. D. Lee, A. Zhang, W. L. Roeloefs, and G. D. Daves (1997). Purification, identification, concentration and bioactivity of (Z)-7-dodecen-1-yl acetate: sex pheromone of the female Asian elephant. *Elaphas maximus*. *Chem Senses* **22**: 417–437.

Reed, H. C., D. R. Melrose, and R. L. S. Patterson (1974). Androgen steroids as an aid to the detection of oestrus in pig artificial insemination. *Br Vet J* **130**:61–66.

Roelofs, W. L. (1995). The chemistry of sexual attraction. In: T. Eisner and J. Meinwald (eds). *Chemical Ecology: The Chemistry of Biotic Interaction*. Washington: National Academy of Sciences, pp 103–117.

Saito, T. R. and H. Moltz (1986). Sexual behavior in the female rat following removal of the vomeronasal organ. *Physiol Behav* **38**:81–87.

Schwenk, K. (1995). Of tongues and noses: chemoreception in lizards and snakes. *Trends Ecol Evol* **10**:7–12.

Singer, A. G., W. C. Agosta, R. J. O'Connell, C. Pfaffmann, D. V. Bowen, and F. H. Field (1976). Dimethyl disulphide: an attractant pheromone in hamster vaginal secretion. *Science* **191**:948–950.

Singer, A. G., W. C. Agosta, and A. N. Clancy (1987). The chemistry of vomeronasally detected pheromones: characterization of an aphrodisiac protein. *Ann N Y Acad Sci* **519**:287–298.

Singer, A. G., G. K. Beauchamp, and K. Yamazaki (1997). Volatile signals of the major histocompatibility complex in male mouse urine. *Proc Natl Acad Sci U S A* **94**:2210–2214.

Singer, A. G. and F. Macrides (1990). Aphrodsin: Pheromone or transducer. *Chem Senses* **15**:199–203.

Smith, B. H. and M. D. Breed (1995). The chemical basis for nestmate recognition and mate discrimination in social insects. In: R. T. Cardé and W. J. Bell (eds). *Chemical Ecology of Insects 2*. New York: Chapman & Hall, pp 287–317.

Smith, T. D., M. I. Siegel, A. M. Burrows, M. P. Mooney, A. R. Burdi, P. A. Fabrizio, and F. R. Clemente (1999). Histological changes in the fetal human vomeronasal epithelium during volumetric growth of the vomeronasal organ. In: R. E. Johnston, D. Müller-Schwarze, and P. W. Sorensen(eds). *Advances in Chemical Communication in Vertebrates*. New York: Plenum, pp 583–591.

Smith, W. J. (1977). *The Behavior of Communicating*. Cambridge, MA: Harvard University Press.

Sorensen, P. W. and N. E. Stacy (1999). Evolution and specialization of fish hormonal pheromones. In: R. E. Johnston D. Müller -Schwarze and P. W. Sorensen (eds). *Advances in Chemical Communication in Vertebrates*. New York: Plenum, pp 15–47.

Sorensen, P. W., N. E. Stacy, and K. J. Chamberlain (1989). Differing behavioral and endocrinological effects of two female sex pheromones on male goldfish. *Horm Behav* **23**:317–332.

Traniello, F. A. and S. K. Robson (1995). Trail and territorial communication in social insects. In: R. T. Cardé and W. J. Bell (eds). *Chemical Ecology of Insects 2*. New York: Chapman & Hall, pp 241–286.

Wang, D., P. Chen, W. Liu, C.-S. Li, and M. Halpern (1997). Chemosignal transduction in the vomeronasal organ of garter snakes: Ca^{2+}-dependent regulation of adenylate cyclase. *Arch Biochem Biophys* **348**:96–106.

Wedekind, C. and S. Füri (1997). Body odor preferences in men and women: do they aim for specific MHC combinations or simply heterozygosity? *Proc R Soc Lond* **264**:1471–1479.

Whitten, W. K. (1966). Pheromones and mammalian reproduction. *Adv Reprod Physiol* **1**:155–177.

Wilson, E. O. and W. H. Bossert (1963). Chemical communication among animals. *Rec Prog Horm Res* **19**:673–710.

Wysocki, C. J. (1979). Neurobehavioral evidence for the involvement of the vomeronasal system in mammalian reproduction. *Neurosci Biobehav Rev* **3**:301–341.

Wysocki, C. J. (1989). Vomeronasal chemoreception: its role in reproductive fitness and physiology. In: J. M. Lakoski, J. R. Perez-Polo and D. K. Rassin (eds). *Neural Control of Reproductive Function*. New York: Alan R. Liss, pp 545–566.

Wysocki, C. J. and M. Meredith (1987). The vomeronasal system. In: T. E. Finger and W. L. Silver (eds). *Neurobiology of Taste and Smell*. New York: John Wiley & Sons, pp 125–150.

Yamazaki, K, G. K. Beauchamp, J. Bard, E. A. Boyse, and L. Thomas (1991). Chemosensory identity and immune function in mice. In: C.J. Wysocki and M. R. Kare (eds). *Chemical Senses, Vol 3: Genetics of Perception and Communication*. New York: Marcel Dekker Inc., pp 211–225.

OLFACTION

6

Cell Biology of Olfactory Epithelium

ALBERT I. FARBMAN

Department of Neurobiology and Physiology, Northwestern University,
Evanston, IL

1. INTRODUCTION

In diverse animals, ranging from worms to arthropods to vertebrates, olfactory receptor neurons share a basic organizational plan. They are bipolar neurons whose dendrites carry the molecular elements necessary for odor transduction, and whose axons extend into the central nervous system. The apical portion of the dendrite is elaborated into cilia or microvillar extensions that increase the available surface area of the cell. Although the focus of this chapter is on the cell biology of the olfactory epithelium in vertebrates, and especially in mammals, I include first an overview of the anatomy of olfactory organs in three commonly studied groups: a nematode (*Caenorhabditis elegans*), and two arthropod classes–insects and crustaceans.

2. INVERTEBRATE OLFACTORY ORGANS

2.1. The "Nose" of *C. Elegans*

The anterior end of the nematode *C. elegans* is characterized by six sensory papillae surrounding the opening to the buccal (oral) cavity. The lateral face of

The Neurobiology of Taste and Smell, Second Edition, Edited by Thomas E. Finger, Wayne L. Silver, and Diego Restrepo.
ISBN 0-471-25721-4   Copyright © 2000 Wiley-Liss, Inc.

the lateral pair of papillae each have a single pore forming the opening to the amphid sensory organ (Ward et al., 1975; Perkins et al., 1986; Bargmann and Mori, 1997), which serves as the nose of the worm (see Fig. 6.1). A specialized socket cell forms the walls of the amphid opening and abuts the surrounding hypodermal cells. Twelve sensory neurons extend dendritic processes into the amphid channel thereby gaining access to external stimuli. Eleven of the neurons in each amphid are chemosensory (Bargmann and Mori, 1997, see also Chapter 3, Sengputa and Carlson). Eight of these neurons (identified as ADF, ADL, ASE, ASG, ASH, ASI, ASJ, and ASK) have long cilia—either branched or unbranched—directly in contact with the environment. These mostly detect water-soluble substances, for example, Na^+, Cl^-, cAMP, biotin, or dauer pheromone. Three chemosensory neurons of each amphid (AWA, AWB, and AWC) have dendritic processes enveloped by a supporting cell called the "amphid sheath cell" or "wing cell." These neurons detect volatile stimuli, for example, diacetyl, thiazole or benzaldehyde. The wing cell also is thought to regulate the extracellular fluids surrounding the dendrites of the chemosensory neurons. The amphid sensory organ also contains another sensory dendrite, of the AFD neuron, which is deeply embedded in the wing cell and which is thermoceptive (Bargmann and Mori, 1997).

2.2. Arthropod Olfactory Structures

Arthropods represent the largest animal phyla including insects, crustaceans (e.g., lobsters and crabs), arachnids (e.g., spiders and scorpions), horseshoe crabs, and the extinct trilobite. Since olfaction has been studied most in insects and crustaceans, their olfactory organs are described below.

2.2.1. Insects

The sense organs of insects (with the exception of compounds eyes) are found in sensilla made from their chitinous cuticle. Cuticular sense organs subserving olfaction are usually found on the antennae. A male silk moth may have as many as 70,000 olfactory sensilla on its antennae (Meng et al., 1989). There is a wide variety of olfactory sensilla varying in shape depending on the species and the function that they serve. They may be hair-shaped up to 500 µm long or peg-like only 10 µm in length. Members of a given species may have several types of olfactory sensilla on their antennae. Despite their variety, sensilla all possess a small group of modified epithelial cells including a bipolar olfactory receptor neuron (ORN) and various sheath cells. The number of ORNs in an insect olfactory sensillum can range from 1 to more than 50, with 2 to 6 being most common (Sutcliffe, 1994). Each ORN has a dendrite extending into the hollow chitinous sensillum (Fig. 6.1) that may contain a number of pores leading to pore tubules (Fig. 6.1). Each ORN also sends an axon into the central nervous system. Thus, insect ORNs are primary receptors as are vertebrate ORNs.

The sensillum trichodeum, which may be up to 500 µm long, is a common, well-studied olfactory sensillum found on moth antennae (Stengl et al., 1992)

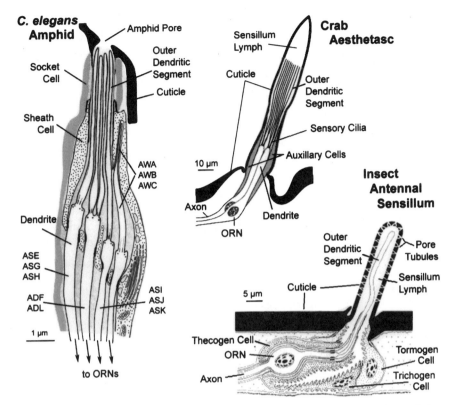

FIG. 6.1. Semischematic diagrams of three invertebrate olfactory organs. Note that scale bars indicate approximate scaling size only. In order to fit the drawings into the space provided, the lengths of the arthropod sensilla shown are very short compared to their width.

Above left: Amphid receptor organ of *C. elegans*. This shows the relationship of the dendrites to the amphid pore; ORNs lie remote from the receptor organ. Three-letter abbreviations (e.g., ASE, ADF, AWA) indicate specific, identified neurons of this organism (see Chapter 3, Sengupta and Carlson for further description). Note that the dendrites of three neurons, AWA, AWB, and AWC, that respond to volatile stimuli, are embedded in the sheath cell and do not have direct access to the pore. The other eight chemosensory amphid neurons respond to nonvolatile stimuli. (Modified with permission from Ward et al. [1975] © John Wiley and Sons.

Below right: Semischematic diagram of a prototypic insect olfactory sensillum with a single receptor neuron (ORN) and three auxiliary cells. The receptor neuron is a bipolar cell, with a dendrite extending into the sensillum, and an axon projecting to the central nervous system. In this diagram only one sensory cell is shown for simplicity. Sensilla may contain two, three, or many more cells. The sensillum is a hollow, fluid-filled tube with a cuticular wall, and there are tiny pores in the cuticle. The fluid, produced by auxiliary cells, protects the dendrite against desiccation and also contains stimulus-binding molecules and enzymes. Access to the sensory cell dendrite from the environment is by way of the cuticular pores. (Reprinted from Farbman, 1992, with the permission of Cambridge University Press).

Above right: Schematic drawing of a typical crustacean aesthetasc. As in the insect sensillum, shown below right, the outer dendritic segments extend into the fluid-filled hollow space formed by the sensillar cuticle. Since there are no pores in the wall of crustacean sensillae, odorant must diffuse through the cuticle wall to gain access to the lymph. (Adapted from Gleeson et al. [1996]).

(Fig. 6.1, lower right). It contains three modified epithelial cells, the trichogen cell, the tormogen cell, and the thecogen cell as well as at least two ORNs. The dendrite on the ORN is separated into inner and outer segments by a short ciliary structure. The ORN is thus divided into three different compartments, each facing different extracellular spaces and fluids. The topmost, the outer dendrite, is naked and is surrounded by the sensillum lymph, which is high in K^+ (Thurm and Kuppers, 1980). The next compartment, the inner dendrite and the soma, are wrapped by the thecogen cell and isolated from the sensillum lymph. The third compartment is the axon, which is tightly covered by a glial cell and is isolated from the other two compartments.

Olfactory stimuli are thought to adsorb on the cuticular surface of the antennal sensillum and then diffuse through the pores and pore tubules into the sensillum lymph (Keil, 1982). The sensillum lymph contains high concentrations of water-soluble proteins that bind odorants (including pheromones) (Pelosi and Maida, 1995). The function of these odorant-binding proteins is not yet known (see also Chapter 7, Ache and Restrepo). However, they have been suggested either to enhance the transport of odorants (pheromones) to the receptors on the dendrites, or to help inactivate odorant in the sensillum lymph, or a combination of the two (Pelosi and Maida, 1995). With the discovery of different classes of binding proteins, a discriminatory role was postulated, that is, only the bound protein/odorant complex would be recognized by membrane-bound receptors.

2.2.2. Crustacean Olfactory Organs

Crustaceans, like insects, have a chitinous exoskeleton giving rise to a variety of sensory sensilla. Also like insects, crustaceans have their olfactory sensilla on their antennae. Crustaceans have two pairs of antennae with the lateral filaments of the first antennae, also called the antennules, usually considered the olfactory organ (Carr et al., 1987). The lateral filament of the antennule contains a tuft of olfactory sensilla or aesthetascs, which are found in all major crustacean groups. In the spiny lobster, an aesthetasc sensillum is about 0.8 mm long and 25 μm in diameter (Grunert and Ache, 1988) while aesthetascs in *Daphnia* are only 40 μm long and 3 μm in diameter (Hallberg et al., 1992). An aesthetasc may contain as many as 300 bipolar ORNs (spiny lobster [Grunert and Ache, 1988]) or as few as 1 (mysid shrimp [Hallberg et al., 1992]). Each ORN gives rise to a dendrite extending into the cuticular aesthetasc, which, unlike insect olfactory sensilla, do not contain pores or pore tubules. Each ORN has an axon that forms the antennular nerve and projects to the central nervous system.

The best studied aesthetascs are found in the decapod crustaceans (lobsters and crabs) (Fig. 6.1, above right). The dendrite of each ORN in the aesthetasc develops two cilia; in some decapods the cilia branch repeatedly. The cilia are bathed in receptor lymph, which fills the space inside the cuticle. The area below the cilia is the inner dendritic segment while that above is called the outer dendritic segment. Auxiliary cells ensheathe the dendrites up to the point where the aesthetasc narrows slightly (constricted region) and the length of dendrites above this point can vary as a function of salinity (Gleeson et al., 1996).

Odor stimuli enter the aesthetasc above the constricted region (Gleeson et al., 1996). Although there are no pores, the aesthetasc cuticle is thin and permeable (Derby et al., 1997). Apparently, the cuticle acts as a molecular sieve allowing small molecules (> 8.5 kDa) to pass through while excluding larger molecules. No olfactory binding proteins have yet been found for crustaceans.

3. OLFACTORY ORGANS OF VERTEBRATES

The olfactory sense organ is a mucous membrane made up of an epithelium and a subepithelial lamina propria of connective tissue, blood vessels, and glands. A thin basement membrane separates the epithelium from the lamina propria (Fig. 6.2). Like mucous membranes elsewhere in the body, olfactory mucous membrane is maintained in a moist condition by glandular secretions that form a thin layer of mucus covering the external epithelial surface.

In vertebrates the olfactory epithelium consists of three major cell types: bipolar sensory neurons (approximately 70–80% of the total number of cells), supporting cells (approximately 15–25%) and two types of basal cells. In a histological section through the olfactory epithelium one sees several layers of cell nuclei. The oval nuclei of supporting cells form a single layer closest to the surface, and the round nuclei of the sensory cells usually make up four to eight or more layers in the lower half to two thirds of the epithelium (Fig. 6.3). Nuclei of the flat (horizontal) or round (globose) basal cells rest on or very near the basement membrane. The total thickness of the olfactory epithelium varies from place to place within the nasal cavity of an individual animal, from as little as 30 μm to as much as 400 μm. Given that the supporting cell nuclei and the basal cell nuclei each form essentially a single layer within the epithelium, the variation in epithelial thickness is directly related to the number of sensory neurons.

In the following I describe the olfactory sensory neuron as a bipolar cell with a dendritic process reaching the epithelial surface and an axonal process that projects to the olfactory bulb. However, it is important to note at the outset that the neuronal population is made up of cells that are at different levels of maturity. Although the structural differences among these neurons may be subtle or not discernible, the neurons do vary in age because new neurons are continuously produced throughout life. In general those neurons with nuclei closer to the basement membrane are younger, and those with nuclei closer to the apical epithelial surface are older (Verhaagen et al., 1989).

3.1. Cytology of the Sensory Neuron

The cell bodies of sensory cells are located in the middle of the epithelium where they are often in direct contact with one another. The cell body is generally

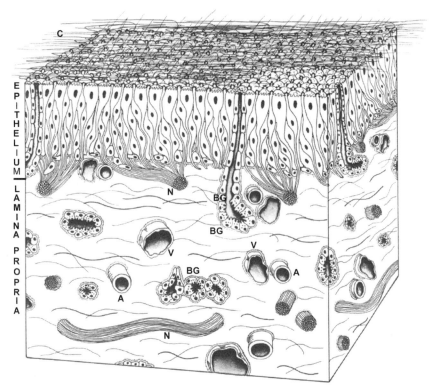

FIG. 6.2. Diagram of the olfactory mucous membrane. The epithelium and lamina propria are shown. The long cilia (C) on the surface are matted in a layer of mucus on the epithelial surface and lie parallel to the surface. Within the lamina propria are Bowman's glands (BG), bundles of olfactory axons (N), and blood vessels, both small arteries and veins (V). Ducts from Bowman's glands open onto the surface. For clarity, the numbers of olfactory nerve bundles and the cell bodies from which they originate are understated in this diagram. (Reprinted from Farbman, 1992, with the permission of Cambridge University Press.)

round to oval and contains a nucleus with clumped chromatin. The proximal pole of the cell body narrows into an axon that joins with other axons to form small nerve bundles or fascicles. These pass out of the epithelium to join other bundles that then project to the olfactory bulb. There they terminate within glomeruli by forming synapses on dendrites of mitral/tufted cells or dendrites of periglomerular cells (see Chapter 9, Christensen and White).

At the apical pole of the cell body is a narrow dendrite that usually ends in an expansion, the dendritic knob, at the epithelial surface. Projecting from the dendritic knob are several cilia, (Fig. 6.4) varying from as few as 4 or 5 to as many as 25 or 30, depending on the species (reviewed in Menco, 1983). In teleost fishes and in neotene amphibians many olfactory sensory neurons carry microvilli as distal appendages. In these animals the ciliated and microvillar cells respond to different categories of odorants.

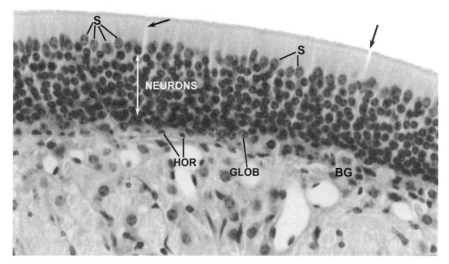

FIG. 6.3. A histological section through rat olfactory epithelium, showing the single layer of pale-staining, oval supporting cell nuclei (s), several layers of dark staining olfactory neuron nuclei (indicated by double-headed white arrow), some horizontal basal cells (HOR), and some globose basal cells (GLOB). A small section of Bowman's gland (BG) is also seen at the bottom of the epithelium, and a few obvious tracks in the upper epithelium representing the ducts of Bowman's gland (black arrows).

The cilia vary in length from about 50 μm in mammals to more than 200 μm in frogs. Menco (1980) estimated that the ciliary membranes increase the receptor surface of the cell by a factor of 25 to 40. Each cilium, as it arises from the dendrite, is about 0.3 μm in diameter. In mammals the cilium becomes narrower about 2–3 μm from its base and tapers toward its tip, where it is about 0.1 μm in diameter. The numerous cilia are packed together as a mat overlying the surface of the olfactory epithelium in a matrix of mucus (Fig. 6.2). From a functional standpoint the cilia are the parts of the sensory neuron first encountered by an odorant in the nasal cavity. However, odorants must first traverse the mucus layer into which the cilia extend (see Chapter 7, Ache and Restrepo). The interaction between odorant and sensory neuron occurs in the ciliary membrane, and there is now good evidence that essentially all components of the olfactory signal transduction apparatus are found in these membranes (cf. reviews by Mori and Yoshihara, 1995; Buck, 1996; see also Chapter 7, Ache and Restrepo). Here we draw attention to evidence for morphological localization of the membrane-associated components of the signal transduction cascade in the membrane of the long, thin distal part of the cilium. Immunocytochemical methods reveal that putative odorant receptors (Menco et al., 1997), G-proteins (Menco et al., 1994), type III adenylyl cyclase (Menco et al., 1992, 1994), and cyclic nucleotide-gated channels (Menco, 1997) all occur in high density on the membrane of the distal cilium where it tapers.

Consequently, the initial transduction events apparently occur in this distal region of the olfactory cilium (see Chapter 7, Ache and Restrepo).

Because olfactory receptor neurons are unique in several ways, there have been efforts to determine whether there is a molecular basis to the uniqueness. Although there is not enough space to discuss this in great detail, we shall discuss one particular molecule briefly, olfactory marker protein (OMP), first discovered in 1972 by Frank Margolis. OMP is a 19 kD protein found in the cytosol of mature receptor neurons from the cilia to the axon terminal. OMP occurs in many classes of vertebrates and appears to be a highly conserved protein localized usually in mature olfactory receptor neurons. Because of this and because it constitutes a significant proportion of the total protein in these cells, nearly 1%, it has been studied intensively to determine its possible function. Unfortunately, the role(s) of this protein is(are) still not completely understood, although two recent studies suggest possible functions. Physiological studies on OMP-null mice have raised the possibility that OMP may play a modulating role in odor detection or signal transduction, because the knockout mice have a longer recovery time between responses to stimuli than wild-type animals (Buiakova et al., 1996). A second possible function has been suggested by the results of studies in which OMP was added to organotypic cultures of rat embryonic olfactory mucosa and induced an increased rate of neurogenesis in a dose-dependent manner (Carr et al., 1998a; Farbman et al., 1998). As will be seen below, however, other growth factors have a similar mitogenic effect in vitro.

3.2. Olfactory Nerve, Ensheathing Cells

The olfactory nerve in vertebrates is made up of the unmyelinated axons originating from the proximal poles of the receptor neuron cell bodies. The structure of the nerve is different from that of other sensory nerves and from other unmyelinated nerves in a few ways. The axons are very fine and have a fairly uniform diameter, about 0.1–0.3 µm. In other unmyelinated nerves individual axons are ensheathed by Schwann cells, one axon to one mesaxon. Olfactory axons are ensheathed several axons to a single mesaxon (Fig. 6.5).

The organization of olfactory axons within bundles suggests the possibility that new axons growing toward the central nervous system from replacement receptor neurons are guided to their targets by cues on preexisting axons. For example, after a receptor neuron dies, the axon of its replacement cell might be guided to the same target by the track formed by adjacent axons in the bundle. However, when massive death is induced in the olfactory epithelium by exposure to methyl bromide, most of the replacement neurons send axonal projections to approximately the same region in the bulb. However, the specific targeting abilities to single glomeruli within the region are compromised, at least within the time frame used in this experiment, 6–8 weeks following exposure to the gas (Carr et al., 1998b; see also Chapter 10, Burd and Tolbert).

Because of their anatomical structure and location, the olfactory receptor neurons provide a point of entry for foreign substances, such as particles and

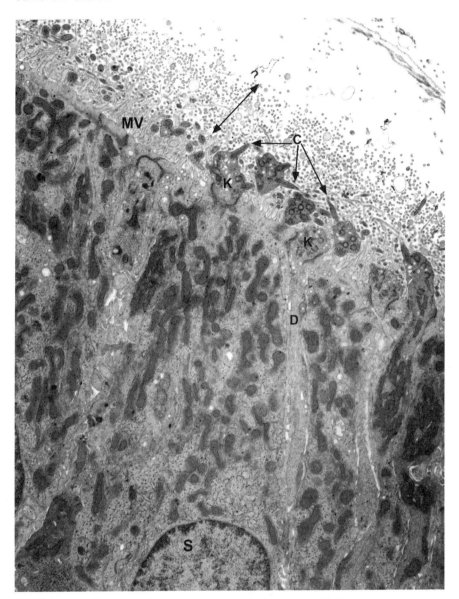

FIG. 6.4. Electron micrograph through the main olfactory epithelium, showing receptor neuron dendrites (D), a dendritic knob (K) and sections through some cilia (C), a supporting cell nucleus (S), and the apical microvilli (MV) of supporting cells. The double-headed arrow indicates the thickness of the ciliary mat, shown here mostly as cross sections of distal ends of cillia (cf. Fig. 6.1.)

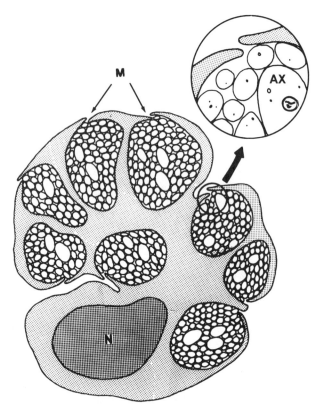

FIG. 6.5. Diagram of the relationship between olfactory axons (AX) and an ensheathing cell; M, mesaxon. N, nucleus of ensheathing cell. (Reprinted from Farbman, 1992, with the permission of Cambridge University Press.)

viruses, to gain entry into the central nervous system via the olfactory bulb. Indeed, once having entered the CNS, viruses and other particles may be transported transneuronally (Shipley, 1985; Baker and Spencer, 1986). Experimental verification for the transport of viruses into the CNS via the olfactory nerve comes from studies using herpes simplex (e.g., Tomlinson and Esiri, 1983; McLean et al., 1989), vesicular stomatitis (Lundh et al., 1987) and other viruses (e.g., Monath et al., 1983; Morales et al., 1988). However, some viruses affect the olfactory neurons minimally (e.g., Sendai virus) but do affect respiratory epithelium (Lundh et al., 1987).

The ensheathing cells of the olfactory nerve differ from typical Schwann cells of peripheral nerves. They originate from the olfactory placodal epithelium (Farbman and Squinto, 1985), whereas Schwann cells originate from the embryonic neural crest. In development of the olfactory nerve there is a progressive increase in the number of axons within each mesaxon (Fraher, 1982) whereas in other peripheral unmyelinated nerves there is a progressive decrease until, in the adult condition, there is only one axon per mesaxon. This means that in the olfactory nerve the axons in a given bundle are essentially in contact with one another and are not individually insulated by ensheathing cell cytoplasm as in other peripheral nerves. A possible consequence of this arrangement is that changes in ionic concentration from activity in one axon (e.g., an increase in the concentration of extracellular potassium associated with an action potential) could reduce the excitability of its neighbors (Daston et al., 1990).

3.3. Neurotransmitter

The responses of olfactory receptor neurons to odorants are communicated to their postsynaptic targets in the bulb by release of a chemical neurotransmitter onto receptors in the target cell membrane. Until very recently the identity of the transmitter had been elusive. Recent improvements in methods of electro-physiological recording from brain slices have permitted investigators to identify the olfactory transmitter as glutamate in turtles (Berkowicz et al., 1994) and in mammals (Ennis et al., 1996; Aroniadou-Anderjaska et al., 1997; Chen and Shepherd, 1997). The glomerular response to olfactory nerve stimulation apparently has two components, suggesting two glutamate receptors. An early component, characterized as a kainate/AMPA receptor-mediated response, is followed by a slow, NMDA receptor-mediated component.

3.4. Supporting Cells

The supporting (sustentacular) cells contain oval nuclei that usually are aligned in a single row, the most superficial of the nuclei in the olfactory epithelium. The apical part of the supporting cell is cylindrical and it reaches the epithelial surface to end in apical microvilli. In amphibians the apical cytoplasm of supporting cells contains secretory granules that are thought to release mucus onto the epithelial surface (Okano and Takagi, 1974) but in mammals there is no evidence for a secretory function. The basal part of the supporting cell is a narrow sliver of cytoplasm that winds between receptor neurons to terminate in a slight expansion usually on the basement membrane, and sometimes on globose basal cells (Goldstein and Schwob, 1996). The expansion on the basement membrane is generally in close relation to acinar cells of Bowman's gland or capillaries in the connective tissue. Because of this proximity, it has been suggested that supporting cells regulate the passage of substances between the underlying connective tissue and the epithelial surface (Rafols and Getchell, 1983). The observation that supporting cells may terminate on the surface of globose basal cells suggests that they communicate with these cells and may participate in regulation of mitosis (Goldstein and Schwob, 1996).

Supporting cells in some species contain an abundance of smooth endoplasmic reticulum in their supranuclear cytoplasm. Abundance of this organelle in other cell types is associated with high levels of ionic flux and supporting cells are indeed thought to participate in regulation of the ionic composition of the mucus layer (Getchell et al., 1984). A prominent smooth endoplasmic reticulum is also associated with detoxification, and an enzyme system associated with this function, the cytochrome P-450 system, is thought to function in the enzymatic degradation of olfactory stimuli and/or other substances in the nasal cavity (e.g., Dahl, 1988; Lazard et al., 1990). In fact the enzymatic activity of the cytochrome P-450 system in olfactory tissue is higher than that in the liver (Reed et al., 1986). Finally, supporting cells also participate in phagocytosis and presumably disposal of dead or dying neurons in the olfactory epithelium, both in the unperturbed animal and in animals in which cell death has been upregulated as a consequence of ablation of the olfactory bulb (Suzuki et al., 1996)

3.5. Basal Cells

Olfactory epithelium contains two different types of basal cells, the flat or horizontal basal cells and the globose basal cells. The two phenotypes can be differentiated by immunological markers and by their morphology as viewed in the electron microscope. The horizontal basal cells lie directly on the basement membrane. They are positive immunocytochemically for certain keratin isotypes, namely keratin 5 and keratin 14 (Suzuki and Takeda, 1991a; Holbrook et al., 1995; Schwob et al., 1995), and also for the epidermal growth factor receptor (Holbrook et al., 1995; Farbman and Buchholz, 1996). With the electron microscope, this type of basal cell is seen to contain bundles of tonofilaments (which accounts for the keratin positive reaction); this cell type also forms a sheath around olfactory axon bundles before they leave the epithelium.

Globose or round basal cells (GBCs), on the other hand, do not express the keratin markers, nor does one see tonofilaments with the electron microscope. These cells, however, do express a specific marker detected by the GBC-1 monoclonal antibody, developed by the Schwob laboratory (Goldstein and Schwob, 1996). GBCs are generally round and are located slightly higher than the layer of flat basal cells, but occasionally are seen inserted within that layer (Holbrook et al., 1995). The GBCs are thought to be the immediate precursors of olfactory neurons, that is, these cells divide and one or both daughter cells differentiate into olfactory neurons (Mackay-Sim and Kittel, 1991a, 1991b; Schwartz Levey et al., 1991; Suzuki and Takeda, 1991b; Caggiano et al., 1994; Holbrook et al., 1995). Cell division in the horizontal basal cell population occurs at a much lower rate than in GBCs (Mackay-Sim and Kittel, 1991a; Holbrook et al., 1995), and their role in epithelial dynamics under unperturbed conditions is less clear. However it has been shown that all cell lineages in olfactory epithelium may be derived from horizontal basal cells. In experiments in which the entire epithelium is destroyed by methyl bromide inhalation the earliest epithelial replacement cells are horizontal basal cells (Schwob et al., 1995). At subsequent time points the entire epithelial cohort of phenotypes differentiates, all presumably from the horizontal basal cells.

4. CELL DYNAMICS IN OLFACTORY EPITHELIUM

4.1. Neurogenesis

Vertebrate olfactory receptor neurons are highly unusual because they apparently have a relatively short life span and are continually replaced throughout life (reviewed by Farbman, 1990, 1992). Cell division of olfactory neuron progenitor cells occurs both in the wound-healing mode, that is, after injury to the olfactory system, and under physiological conditions. In the following I shall consider some aspects of the equilibrium between cell birth and cell death in the olfactory epithelium. More specifically I shall consider (1) how cell division is regulated in the mammalian olfactory epithelium, and (2) how the life span of an olfactory receptor neuron is regulated.

Olfactory cell division is usually assayed by administering a DNA precursor to an animal (or an in vitro test system), allowing some time to pass, and terminating the experiment by removing the tissue and preparing it for the assay. The two commonly used DNA precursors are tritiated thymidine ($[^3H]TdR$) and bromo-deoxyuridine (BrdU). When injected into an animal these precursors are cleared from the circulation in 1–2 h. They are available, then, for a limited period of time to those dividing cells that happen to be in the DNA-synthetic phase (S-phase) of the cell cycle at the time of administration. In culture systems, explants or disaggregated cells have been exposed to the precursor for as little as one hour, to mimic in vivo experiments (Farbman and Buchholz, 1996), or as long as 24 hours (Mahanthappa and Schwarting, 1993). The labeled cells containing the precursor can be revealed histologically by autoradiography (when $[^3H]TdR$ is used as the DNA precursor) or by immunohistochemistry (when BrdU is used). Experiments using these and other methods have shown that globose basal cells in the olfactory epithelium are the neuronal progenitor cells and variable numbers of them incorporate the label, depending on the experimental circumstances (e.g., Graziadei and Monti Graziadei, 1978; Mackay-Sim et al., 1988; Calof and Chikaraishi, 1989; Mackay-Sim and Kittel, 1991a, 1991b; Schwartz Levey et al., 1991; Suzuki and Takeda, 1991b; Caggiano et al., 1994; Weiler and Farbman, 1997). For example, when a naris is occluded in a newborn rat the number of cells incorporating the DNA precursor at postnatal day 30 is reduced ipsilaterally by 40% over a control animal (Farbman et al., 1988). On the other hand, when the olfactory bulb is ablated, the number of cells incorporating the label is doubled, an effect that lasts at least seven weeks postoperatively (Carr and Farbman, 1992).

Whether in a wound-healing situation or under physiological conditions, after several days some labeled cell bodies migrate apically within the epithelium and become mature neurons. However, most others die at various times following incorporation of the precursor (Breipohl et al., 1986; Farbman, 1990; Carr and Farbman, 1992, 1993; Mahalik, 1996); in fact, in control animals most die within 2–4 weeks after incorporation of the DNA marker (Mackay-Sim and Kittel, 1991b). Thus, at any given time, the ages of the surviving olfactory neuronal population are mixed—the younger neurons have their cell bodies in the lower half of the epithelium and the older ones in the upper half (cf. Verhaagen et al., 1989). In this latter study authors used immunocytochemistry to define cells as immature if they were B50/GAP-43 positive, as mature if OMP-positive. These results confirmed in an elegant manner what had been implied nearly a half century earlier (Nagahara, 1940).

In unperturbed adult animals the population of olfactory receptor neurons is in equilibrium once growth is completed, that is, the number of cells that die is approximately equal to the number of new cells produced. In rats this equation is somewhat complicated because the surface area of olfactory epithelium continues to expand, at least until the rat is about 18 months old (Hinds and McNelly, 1981; Meisami, 1989; Apfelbach et al., 1991; Weiler and Farbman, 1997). Consequently, whereas some of the dividing globose basal cells are indeed fated to play a role in replacement, others are incorporated into the expanding epithelial sheet (Weiler and Farbman, 1997) or, as noted above, die before becoming fully mature or soon afterwards.

Several questions arise from these observations. For example, what factors regulate neurogenesis, so that the number of new cells produced will be adequate for replacement and/or growth of the olfactory epithelium? Second, the fact that several cells die before becoming mature raises the question, is a trophic factor necessary for those that survive to maturity?

The specific factors involved in genesis, differentiation, and survival of olfactory receptor neurons are only beginning to be uncovered. For example a number of growth factors have been examined for their possible participation in cell proliferation in both the neuronal and supporting cell lineages in the olfactory epithelium. In organotypic or cell cultures transforming growth factor-α (TGF-α), epidermal growth factor (EGF), amphiregulin, platelet-derived growth factor-AB (PDGF-AB), insulin-like growth factor-1 (IGF-1), TGF-β1, TGF-β2, OMP, and basic fibroblast growth factor (bFGF) were all effective at increasing the number of neuronal progenitor cells that incorporated a DNA-precursor (Mahanthappa and Schwarting, 1993; DeHamer et al., 1994; Farbman and Buchholz, 1996; Carr et al., 1998a). Nerve growth factor had virtually no effect on mitosis.

TGF-α and OMP are at least 100 times more potent than any of the other factors. Both stimulate cell division in vitro in picomolar doses. We have investigated some aspects of TGF-α and members of its receptor family more thoroughly. EGF-receptor is thought to be the receptor that specifically binds and is activated by TGF-α. In situ RT-PCR evidence has shown that the mRNA for EGFR is present in globose basal cells (Rama Krishna et al., 1996) although previous immunohistochemical results had shown that only horizontal basal cells contained the protein (Holbrook et al., 1995; Farbman and Buchholz, 1996). If a receptor is involved directly in initiating the cascade leading to cell division it must be in globose basal cells, the progenitors of the neuron. All four members of the EGF-receptor family, EGF-R, c-ErbB2 (also known as Neu), c-ErbB3 and c-ErbB4 are present in olfactory mucosa, as shown by reverse transcriptase-polymerase chain reaction to assay for mRNAs and in Western blots to assay for the specific proteins (Salehi-Ashtiani and Farbman, 1996; Ezeh and Farbman, 1998). Of these receptors only EGFR, c-ErbB-2, and c-ErbB-4 are localized in the progenitor cells and are therefore in position to respond to growth factors (Salehi-Ashtiani, and Farbman, 1996; Perroteau et al., 1998), and of these only EGFR and c-ErbB-4 have ligand-binding sites. However Neu differentiation factor (neuregulin), a ligand for c-ErbB-3 and c-ErbB-4 is present in the lower layers of epithelium (Salehi-Ashtiani and Farbman, 1996), and could conceivably participate in mitotic regulation.

When TGF-α is injected into the carotid artery of living rats, its receptor in the olfactory epithelium is activated; that is, within 2 minutes following injection EGF-R exhibits tyrosine phosphorylation. When saline is injected into the artery the EGF-R is not phosphorylated (Ezeh and Farbman, 1998). Similar experiments demonstrated that two mitogen-associated-protein-kinases (MAP-kinases—a family of enzymes that act downstream of EGF-R in the mitotic cascade), Erk1 and Erk2, were activated (Ezeh and Farbman, 1998). These results are consistent with the notion that in vivo TGF-α can stimulate mitosis in the olfactory epithelium. However, it is likely that TGF-α is not the only factor regulating olfactory mitosis but that multiple mechanisms are involved in controlling neurogenesis and receptor neuron differentiation in the olfactory sensory epithelium, because our preliminary data show that BrdU uptake is not significantly different between TGF-α-null mutant mice and wild-type mice.

In addition to the molecular complexity and redundancy, it is also clear that the age of the animal is a variable in regulating division of olfactory neuronal progenitor cells and possibly in regulating their average life span. A morphometric analysis showed that the *density* of BrdU-labeled cells, that is, the number of labeled cells per millimeter length of olfactory epithelium, is significantly lower in older rats than in younger ones (Weiler and Farbman, 1997). Interestingly, there was a reciprocal relationship between the increase in total olfactory surface area and the decrease in proliferation density. However, although the *density* of labeled cells declines dramatically with age (at 1 year the density is only 5% of that at birth), the *total number* of dividing progenitors does not change a great deal, certainly by less than a factor of two. These data raised the question of whether the absolute number of progenitor cells in an unperturbed animal is relatively constant with age. If so, it would suggest that any perturbation that upregulates the rate of cell division (such as ablation of the olfactory bulb or olfactory axotomy) does so because at the same time it upregulates cell death. In this scenario, the dying cells may be the source of a trigger that directly or indirectly promotes the upregulation of mitotic rate. Another possibility is that healthy mature cells might provide a signal that downregulates the proliferative rate (Calof et al., 1996).

4.2. Regulation of Life Span—Is There a Trophic Factor?

The earlier works (e.g., Graziadei and Monti Graziadei, 1978) suggested that the average life span of mammalian olfactory neurons was about a month. However, later experiments challenged this notion by showing that olfactory neurons can live much longer, as long as a year in mice raised in a filtered air environment (Hinds et al., 1984; cf. also Mackay-Sim et al., 1988; Mackay-Sim and Kittel, 1991a, 1991b). As noted above, there is morphometric evidence suggesting that olfactory neurons, particularly in older rats, have longer life spans. Moreover, with increasing age there is a greater density of mature neurons, as assessed by counting the number of dendritic knobs in a length of epithelium (Hinds and McNelly, 1981), or estimating the relative proportion of mature neurons (OMP-positive cells) versus immature (B50/GAP43-positive) cells in the epithelium (Verhaagen et al., 1989). In other words the data are consistent with the possibility that, at least in older animals, when the growth rate has slowed, those olfactory neurons that reach maturity do live longer. Other possible explanations for the higher proportion of mature cells and relatively fewer "almost mature" or immature cells include, (1) some postmitotic cells die precociously, for lack of trophic support from either the bulb or elsewhere, and (2) there is an increase in the length of the cell cycle, that is, fewer cells would likely be in the S-phase during the time when a pulse dose of a DNA precursor is available.

There is evidence that the life span of olfactory neurons is influenced by their ability to obtain a putative trophic factor from the olfactory bulb because their life spans are much shorter after bulbectomy (Schwob et al., 1992). After unilateral bulb ablation about 90% of neurons die within 2 weeks after cell division. Indeed, in this experimental model there is a peak in the number of dying cells about 6–7 days after cell division (Carr and Farbman, 1993), the approximate time when growing axons would be expected to reach the olfactory bulb (cf. Miragall and Monti Graziadei, 1982). Moreover, reduction of the number of bulbar mitral cells, one of the cellular targets of olfactory neurons, increases the number of neuronal progenitor cells that take up a DNA precursor, thus suggesting that bulbar mitral cells may be the source of the putative trophic factor (Weiler and Farbman, 1999). In these experiments about 29% of the mitral cells were removed by severing the lateral olfactory tract, which carries the projection of the mitral/tufted cells to higher olfactory centers. The receptor neurons themselves and the periglomerular cells are left intact by this procedure.

The best known trophins that act on the nervous system are the neurotrophins belonging to the nerve growth factor (NGF) family, and some work has been done to investigate the possibility that they might be involved in promoting the survival of olfactory receptor neurons. In addition to NGF the other members of this family are brain-derived neurotrophic factor (BDNF), neurotrophin-3 (NT-3), and neurotrophin-4/5 (NT-4/5). A low-affinity NGF receptor (p75) binds all of the neurotrophins but its activity in promoting neuron survival is unclear. The three high-affinity receptors for this family of neurotrophins are all protein tryrosine kinases. They are known as trk A, the receptor for NGF, trk B, the receptor for both BDNF and NT-4/5, and trk C, the receptor for NT-3 (NT-3 also binds with weaker affinity to both trk A and trk B). When the neurotrophins are located in a target organ they are thought to bind to their receptors at or near the (presynaptic) axon terminal or growth cone. The ligand–receptor complex is taken into the axon by endocytosis, transported retrogradely to the cell body where the neurotrophin exerts its action through activation of different signaling pathways.

Several experiments have been done to examine the effects of various neurotrophins on survival in vitro. In one set of experiments on disaggregated olfactory epithelial cells BDNF in combination with transforming growth factor-β2 (TGF-β2) promoted survival of olfactory neurons, but BDNF alone had no effect (Mahanthappa and Schwarting, 1993). In another experiment BDNF, NT-3, and NT-5 all enhanced survival of neurons, whereas NGF did not (Holcomb et al., 1995). On the other hand NGF was shown to prolong survival of olfactory neurons in tissue culture, but the effects were not evident until after 3 days in vitro (Ronnett et al., 1991). Thus the totality of evidence from in vitro experiments gives us no clear answer concerning the specific effects of the neurotrophins on olfactory neuron survival.

Another approach has been to examine for the presence of the neurotrophin receptors. Trk B and trk C are expressed in mature olfactory receptor neurons when their axons have reached the bulb (Roskams et al., 1996). This suggests that both trks might be involved in differentiation and maintenance or survival of the neurons. Trk A, the high-affinity receptor for NGF, was not present in neurons although there is some evidence that the low-affinity receptor, p75, might be present (Turner and Perez-Polo, 1992). However, more convincing evidence has shown that the low affinity NGF-receptor is localized on ensheathing cells in young animals (Vickland et al., 1991; Ramón-Cueto et al., 1993; Gong et al., 1994) and was not localized on axons or neurons.

If trk B and/or trk C are to be functional on the receptor neurons one might expect to see one or more of their ligands in the target tissue. In situ hybridization studies have shown no evidence of NT-3, the high-affinity ligand for trk C, in the olfactory bulb. Moreover, BDNF, a high-affinity ligand for trk B, was present primarily in the granular layer of the bulb, where it would have no access to trk B receptors on incoming olfactory sensory axons (Guthrie and Gall, 1991). In the latter experiment NT-4, also a high-affinity ligand for trk B, was not investigated. Only NGF was present in abundance in the glomerular layer where receptor neurons terminate, but its high-affinity receptor in the epithelium was localized in a few horizontal basal cells, not in neurons. Thus the evidence in favor of the NGF family of neurotrophins acting as bulb-derived survival factors for the receptor neurons is not compelling. However the possibility that one or more of neurotrophins have paracrine or autocrine effects on the epithelium is not ruled out.

Although the putative bulb-derived trophic factors have not yet been positively identified they probably exist. However, even if one or more trophic factors are present in the bulb and even if they are involved in receptor neuron survival, there are at least two reasons why their existence does not *completely* explain the survival, or failure to survive, of many postmitotic epithelial cells in the neuronal lineage. First, some olfactory neurons, perhaps about 10%, do survive after olfactory bulb ablation; indeed some of these surviving cells endure to maturity and produce considerably more OMP than control cells do (Carr et al., 1998a). In the absence of a bulb these cells must obtain sufficient amounts of survival factor(s) from one or more sources other than the bulb. (Here a survival factor is broadly defined as anything required for the cell's survival, even if acts as an inhibitor of cell death.) Second, many postmitotic cells in the neuronal lineage die before they have had the opportunity to grow an axon all the way to the bulb where they might have access to a putative trophic factor (Breipohl et al., 1986; Carr and Farbman, 1992, 1993; Mahalik, 1996). These precociously dying cells can be identified in one of two ways. First, there are pyknotic bodies near the basal region of the epithelium. Second, pyknotic nuclei labeled with a DNA precursor can be found as early as 24 hours after administration of the label (Carr and Farbman, 1992).

A possible explanation for precocious cell death soon after mitosis is, of course, faulty mitosis. For example, according to the cell cycle/apoptosis hypothesis, withdrawal of trophic support from neurons causes them to attempt uncoordinated re-entry into the cell cycle, which, because it is uncoordinated, leads to apoptotic cell death (Greene et al., 1995). Whereas this may account for the death of some olfactory cells, most cells die a few days to 2 weeks after mitosis. Another possible explanation is that these cells lack a survival-promoting factor, possibly a cytokine or growth factor, that exists in limited amounts *locally*, and they must obtain this local factor early in the differentiation process. This putative cytokine or combination of factors would promote survival and differentiation of those daughter cells in the neuronal lineage that acquire them in sufficient amounts in their early stages of differentiation, that is, before they project an axon to their target organs. In other words, it seems probable that there are at least two sources of survival-promoting factors on which olfactory receptor neurons are dependent. One source is local, that is, within the epithelium, (or very close to the epithelium, perhaps in ensheathing cells or lamina propria) and the other is probably within the bulb. The factors from these two sources act on the receptor neurons at different stages of their development, but these may be overlapping and the two sources may produce more than one factor. Most likely there is redundancy in the system.

4.3. Neuronal Cell Death

It is abundantly clear that neuronal cell death in the olfactory epithelium occurs by apoptosis (Carr and Farbman, 1992; Suzuki et al., 1995, 1996; Mahalik, 1996; Deckner et al., 1997). Moreover, apoptosis in the olfactory epithelium is, in at least two important respects, unlike programmed cell death in other parts of the nervous system. In other systems, cell death involves the deletion of neurons that are no longer "needed" (reviewed by Oppenheim, 1991; Naruse and Keino, 1995). Once deleted these neurons are not replaced. On the contrary, in the olfactory system, dead or dying neurons are replaced. Moreover, in other neuronal systems, programmed cell death occurs at precise times during early development. In the olfactory system cell death occurs throughout adult life and may occur when the cell itself is very young or very old (Breipohl et al., 1986;

Farbman, 1990; Carr and Farbman, 1992; Mahalik, 1996). It seems therefore that cell death in the olfactory system is a more complex phenomenon than in other neuronal systems.

4.4. Genesis of Supporting Cells

Interestingly, there is evidence that the supporting cells in the olfactory epithelium are not replaced to any significant extent (Weiler and Farbman, 1998). Supporting cells do divide at a low rate in rat olfactory epithelium, but a quantitative analysis of their mitotic rate suggests that only enough supporting cells are produced to account for the expansion in olfactory epithelial area. It was also clear from this study that cell proliferation in the supporting cell lineage was regulated by mechanisms different from those that regulate cell division in the neuronal lineage.

5. VOMERONASAL ORGAN

The vomeronasal organ (VNO) in mammals plays a major role in the perception of many non-volatile stimuli including some related to social and/or reproductive behavior (see Chapter 5, Johnston). It is present in most terrestrial mammals, amphibians, and reptiles but is vestigial or absent in fishes, birds, higher primates, and some aquatic amphibians. In mammals, the VNOs are paired, elongated, tube-shaped structures each of which is enclosed in a bony or cartilaginous capsule within the anterior, ventral end of the nasal septum (Fig. 6.6). In mammals the VNO opens via a narrow duct into either the nasal cavity (as in rodents and lagomorphs), or into the nasopalatine canal that connects the nasal and oral cavities (as in carnivores and ungulates). In reptiles, the VNO is completely separated from the nasal cavity by the hard palate and the nasopalatine canal opens directly into the mouth. In amphibians the VNOs remain attached to the main olfactory chambers as diverticula.

By virtue of its location, the VNO is somewhat isolated from the air stream passing into the nasal cavity during respiration or sniffing. Consequently many of the stimuli, whether volatile or nonvolatile, are usually brought into the oral or nasal cavity by licking or nuzzling, as in rodents, or tongue-flicking, as in snakes. The stimuli gain access to the VNO sensory cells by way of the "VN pump" in some animals or by the "flehmen" response in others. The pump operates as follows: There are large, thin-walled blood vessels under autonomic control adjacent to the VNO (Fig. 6.6). In hamsters these vessels become narrowed when their sympathetic innervation is stimulated. Inasmuch as a relatively rigid bony capsule surrounds the vessels and the VNO, narrowing of the vessels allows expansion of the VNO lumen, thus creating a partial vacuum and drawing in volatile compounds and fluids from the oral cavity. Dilation of the vessels, under control of other autonomic nerves, results in reduction of the VNO size and expulsion of fluids and volatile compounds (Meredith and O'Connell, 1979).

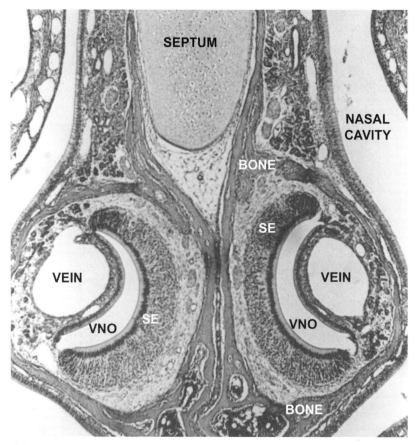

FIG. 6.6. Photomicrograph of a coronal section through a rodent nasal cavity. The crescent-shaped VNOs are near the ventral end of the nasal septum. The thick sensory epithelium (SE) lines the medial side of the VNO lumen in this section—the lateral side contains thin nonsensory epithelium. A thin bony capsule surrounds each VNO. A large vein is located lateral to the VNO.

In several orders of mammals, including ungulates, "flehmen," a distinctive facial grimace involving a curl of the lips accompanied by closure of the external nares, is thought to be associated with the mechanism by which stimuli gain access to the VNO (Melese-d'Hospital and Hart, 1985). Male animals perform flehmen when stimulated to investigate urine from females. Flehmen behavior is apparently a way to determine whether the female is in estrus (see Chapter 5, Johnston).

5.1. Cellular Structure

In cross section, the VNO lumen is crescent-shaped. The sensory epithelium is found only on the concave side of the lumen. The histology is similar to that of

The Neurobiology of Taste and Smell

Editors, Thomas E. Finger, Diego Restrepo, Wayne L. Silver

Copyright © Wiley-Liss, Inc.
ISBN 0-471-25721-4

Cover Credits. TOP: A three-dimensional model of the odorant binding site of olfactory receptor proteins. The colored helices, generated by a computer program, represent membrane-spanning domains based on a predicted structural similarity to the visual receptor protein rhodopsin. Sequence analysis of hundreds of olfactory receptor sequences from various species has identified 17 hyper-variable residues (highlighted), which likely form the odorant recognition pocket. Their enhanced diversity underlies the capacity of the olfactory repertoire to recognize thousands of different odorants. See Chapter 8 for further information. Courtesy of Drs. Tzachi Pilpel and Doron Lancet, both of Crown Human Genome Center, Department of Molecular Genetics, Weizmann Institute of Science, Rehovot 76100, Israel (Doron.Lancet@weizmann.ac.il). LEFT: Photomicrograph of two longitudinally-sectioned taste buds showing cell nuclei in magenta and serotonin-immunoreactive cells in aqua. See Chapter 12 for further information. Courtesy of Dr. Leslie Stone, Colorado State University, Fort Collins, Colorado 80523.

the olfactory epithelium in the main nasal cavity. It is a pseudostratified epithelium containing sensory, supporting, and basal cells. The sensory cell is a bipolar neuron with an apical dendrite reaching the epithelial surface. It differs from neurons in the main nasal cavity in that the apical dendritic appendages are microvilli rather than cilia. On the basal aspect of the cell body the sensory cell cytoplasm narrows into an axon. The individual axons come together to form small bundles beneath the epithelium, and these bundles coalesce into a small number of major bundles that ascend through the connective tissue of the nasal septum, pierce the roof of the nasal cavity, and terminate in the accessory olfactory bulb. Thus the VNO forms the second component of what is, in effect, a dual olfactory system. Although it develops from the embryonic olfactory placode, as does the main olfactory epithelium, it establishes a separate projection pathway to the brain. Moreover, central projections from the accessory olfactory bulb are separate and distinct from those of the main olfactory bulb (see Chapter 9, Chistensen and White).

Recently the receptor neurons in the VNO have been shown to express two different G-proteins in spatially segregated populations (Halpern et al., 1995). Receptor neurons with cell bodies nearer the basal lamina of the epithelium express the α-subunit of G_o whereas those nearer the lumen express the α-subunit of G_i. Because immunoreactivity for both of these G-proteins is present in the axons it has been possible to show that the two populations of neurons project to different regions of the accessory olfactory bulb. The G_o-positive cells project to the posterior half of the accessory bulb whereas the G_i-positive cells project to the anterior half. It has been suggested that these two VNO neuronal populations may code for different types of stimuli.

5.2. Neurogenesis in the VNO

Is the sensory epithelium of the VNO replaced throughout life as is the case in the main olfactory epithelium? As mentioned above, the VNO is isolated from the main respiratory air stream entering the nasal cavity and is therefore less vulnerable to potential damage associated with physical and chemical characteristics of inhaled air (cf. Farbman et al., 1988). In a recent study of the rat VNO at various postnatal ages it was shown that the proliferation density (assayed by the numbers of BrdU-positive cells in a coronal section of the VNO) decreases dramatically until sexual maturity is reached and remains fairly constant thereafter (Weiler et al., 1999). In opossum as well as rat, most proliferating cells are concentrated in the marginal zones, that is, where sensory epithelium is adjacent to nonsensory epithelium (Jia and Halpern, 1998), but a small percentage, less than 20%, is found in the central zones. The rat VNO increases in size continuously throughout most of adult life. The authors argue that the dividing cells in the marginal zones provide a pool for growth of the VNO, whereas those in the central zone provide a pool for neuronal replacement. They suggest further that some of the dividing cells at the margins of the VNO may die before maturation.

Another finding in this study was that the total size of the VNO is related to body size and not to sex, as previously reported (cf. Segovia and Guillamón, 1982). In other words, *the VNO is essentially the same size in males and females of the same body weight.* In the earlier study by Segovia and Guillamón, (1982) VNO sizes of 90-day-old males were compared to those of females at the same age, without taking into account the fact that males are considerably larger than females of the same age (cf. Weiler and Farbman, 1997). This is not to say, however, that there are no differences in responsiveness to sex steroids between VNOs of males and females.

6. SUMMARY

The olfactory receptor cells (ORNs) in both vertebrates and invertebrates are bipolar, primary receptor neurons. The distal process contains all the biochemical elements consonant with its role in transduction of odorants. The receptor neurons extend axons into the central nervous system where they terminate in discrete glomeruli. In arthropods, the olfactory sensory organs include the antennae, antennules, and maxillary palp. These each possess numerous chemosensory hairs, or sensilla, each containing dendrites from several ORNs. Likewise, the amphid receptor organ in *C. elegans* houses the chemosensory dendrites of numerous chemosensory cells. In vertebrates, the main olfactory epithelium resides in the nose or nasal sac and consists of three cell types: supporting cells forming the most superficial layer, receptor constituting the bulk of the epithelium, and two types of basal cells. The globose basal cells are the immediate progenitors of the receptor neurons. Horizontal basal cells, which lie flat against the basal lamina, are multipotent progenitors capable of giving rise to all of the cell types in the epithelium. The olfactory receptor neurons in vertebrates are generated throughout the life of the organism. Newly generated cells either mature to take up residence in the epithelium, or undergo apoptotic cell death. The rate of generation of new cells and the fate of the newly generated cells are regulated by a variety of growth or trophic factors originating from the local epithelium as well as from the olfactory bulb. Many vertebrates possess a specialized part of the olfactory system housed in a distinct vomeronasal organ (VNO). The VNO mostly contains microvillous receptor cells in distinction from the main olfactory epithelium where most, if not all, ORNs are ciliated.

ACKNOWLEDGMENTS

The author thanks the editors for their contributions to this chapter and their critical reading of the manuscript. Some of the work described in this chapter was supported by grants from the National Institutes of Deafness and Other Communicative Disorders, DC 02126, and DC 00347.

REFERENCES

Apfelbach, R., D. Russ, and B. M. Slotnick (1991). Ontogenetic changes in odor sensitivity, olfactory receptor area and olfactory receptor density in the rat. *Chem Senses* **16**:209–218.

Aroniadou-Anderjaska, V., M. Ennis, and M. T. Shipley (1997). Glomerular synaptic responses to olfactory nerve input in rat olfactory bulb slices. *Neuroscience* **79**:425–434.

Baker, H. and R. F. Spencer (1986). Transneuronal transport of peroxidase-conjugated wheat germ agglutinin (WGA-HRP) from the olfactory epithelium to the brain of the adult rat. *Exp Brain Res* **63**:461–473.

Bargmann, C. I. and I. Mori (1997). Chemotaxis and thermotaxis. In: T. B. D. L. Riddle, B. J. Meyer, and J. R. Priess. (eds). *C. elegans II*. Cold Spring Harbor, NY: Cold Spring Harbor Laboratory Press, pp 717–737.

Berkowicz, D. A., P. Q. Trombley, and G. M. Shepherd (1994). Evidence for glutamate as the olfactory receptor cell neurotransmitter. *J Neurophysiol* **71**:2557–2561.

Breipohl, W., A. Mackay-Sim, D. Grandt, B. Rehn, and C. Darrelmann (1986). Neurogenesis in the vertebrate main olfactory epithelium. In: W. Breipohl (ed). *Ontogeny of Olfaction*. Berlin: Springer-Verlag, pp 21–33.

Buck, L. B. (1996). Information coding in the vertebrate olfactory system. *Annu Rev Neurosci* **19**:517–544.

Buiakova, O. I., H. Baker, J. W. Scott, A. Farbman, R. Kream, M. Gillo, L. Franzen, M. Richman, L. M. Davis, S. Abbondanzo, C. L. Stewart, and F. L. Margolis (1996). Olfactory marker protein (OMP) gene deletion causes altered physiological activity of olfactory sensory neurons. *Proc Natl Acad Sci U S A* **93**:9858–9863.

Caggiano, M., J. S. Kauer, and D. D. Hunter (1994). Globose basal cells are neuronal progenitors in the olfactory epithelium: a lineage analysis using a replication-incompetent retrovirus. *Neuron* **13**:339–352.

Calof, A. L. and D. M. Chikaraishi (1989). Analysis of neurogenesis in a mammalian neuroepithelium: proliferation and differentiation of an olfactory neuron precursor in vitro. *Neuron* **3**:115–127.

Calof, A. L., N. Hagiwara, J. D. Holcomb, J. S. Mumm, and J. Shou (1996). Neurogenesis and cell death in olfactory epithelium. *J Neurobiol* **30**:67–81.

Carr, V. M. and A. I. Farbman (1992). Ablation of the olfactory bulb upregulates the rate of neurogenesis and induces precocious cell death in olfactory epithelium. *Exp Neurol* **115**:55–59.

Carr, V. M. and A. I. Farbman (1993). The dynamics of cell death in the olfactory epithelium. *Exp Neurol* **124**:308–314.

Carr, V. M., E. Walters, F. L. Margolis, and A. I. Farbman (1998a). An enhanced olfactory marker protein immunoreactivity in individual olfactory receptor neurons following olfactory bulbectomy may be related to increased neurogenesis. *J Neurobiol* **34**:377–390.

Carr, V. M., G. Ring, S. L. Youngentob, J. E. Schwob, and A. I. Farbman (1998b). HSP70(+) olfactory receptor neuron (ORN) bulbar projections following methyl bromide (MeBr) lesion of the rat olfactory epithelium (OE). *Soc Neurosci Abstr* **24**:1144.

Carr W. E. S., B. W. Ache, and R. A. Gleeson (1987). Chemoreceptors of crustraceans: similiarites to receptors for neuroactive substances in internal tissues. *Environ Health Perspect* **71**:31–46.

Chen, W. R. and G. M. Shepherd (1997). Membrane and synaptic properties of mitral cells in slices of rat olfactory bulb. *Brain Res* **745**:189–196.

Dahl, A. R. (1988). The effect of cytochrome P-450-dependent metabolism and other enzyme activities on olfaction. In: F. L. Margolis and T. V. Getchell (eds). *Molecular Neurobiology of the Olfactory System*. New York: Plenum Press, pp 51–70.

Daston, M. M., G. D. Adamek, and R. C. Gesteland (1990). Ultrastructural organization of receptor cell axons in frog olfactory nerve. *Brain Res* **537**:69–75.

Deckner, M.-L., M. Risling, and J. Frisén (1997). Apoptotic death of olfactory sensory neurons in the adult rat. *Exp Neurol* **143**:132–140.

DeHamer, M. K., J. L. Guevara, K. Hannon, B. B. Olwin, and A. L. Calof (1994). Genesis of olfactory receptor neurons in vitro: regulation of progenitor cell divisions by fibroblast growth factors. *Neuron* **13**:1083–1097.

Derby C. D., H. S. Cate, and L. R. Gentilcore (1997). Perireception in olfaction: molecular mass sieving by aesthetasc sensillar cuticle determines odorant access to receptor sites in the Caribbean spiny lobster *Panulirus argus*. *J Exp Biol* **200**:2073–2081.

Ennis, M., L. A. Zimmer, and M. T. Shipley (1996). Olfactory nerve stimulation activates rat mitral cells via NMDA and non-NMDA receptors in vitro. *NeuroReport* **7**:989–992.

Ezeh, P. I. and A. I. Farbman (1998). Differential activation of ErbB receptors in the rat olfactory mucosa by transforming growth factor-α and epidermal growth factor in vivo. *J Neurobiol* **37**:199–210.

Farbman, A. I. (1990). Olfactory neurogenesis: genetic or environmental controls? *Trends Neurosci* **13**:362–365

Farbman, A. I. (1992). *Cell Biology of Olfaction*. Cambridge: Cambridge University Press, pp 282.

Farbman, A. I. (1994). The cellular basis of olfaction. *Endeavour* **18**:2–8.

Farbman, A. I. and J. A. Buchholz (1996). Transforming growth factor-(α) and other growth factors stimulate cell division in olfactory epithelium in vitro. *J Neurobiol* **30**:267–280.

Farbman, A. I. and L. M. Squinto (1985). Early development of olfactory receptor cell axons. *Devel Brain Res* **19**:205–213.

Farbman, A. I., P. C. Brunjes, L. Rentfro, J. Michas, and S. Ritz (1988). The effect of unilateral naris occlusion on cell dynamics in the developing rat olfactory epithelium. *J Neurosci* **8**:3290–3295.

Farbman, A. I., J. A. Buchholz, E. Walters, and F. L. Margolis (1998). Does olfactory marker protein participate in olfactory neurogenesis? *Ann NY Acad Sci* **855**:248–251.

Fraher, J. P. (1982). The ultrastructure of sheath cells in developing rat vomeronasal nerve. *J Anat* **134**:149–168.

Getchell, T. V., F. L. Margolis, and M. L. Getchell (1984). Perireceptor and receptor events in vertebrate olfaction. *Progr Neurobiol* **23**:317–345.

Gleeson, R., L. McDowell, and H. Aldrich (1996). Structure of the aesthetasc (olfactory) sensilla of the blue crab, *Callinectes sapidus*: transformations as a function of salinity. *Cell Tissue Res* **284**:279–288.

Goldstein, B. J. and J. E. Schwob (1996). Analysis of the globose basal cell compartment in rat olfactory epithelium using GBC-1, a new monoclonal antibody against globose basal cells. *J Neurosci* **16**:4005–4016.

Gong, Q., M. S. Bailey, S. K. Pixley, M. Ennis, W. Liu, and M. T. Shipley (1994). Localization and regulation of low affinity nerve growth receptor expression in the rat olfactory system during development and regeneration. *J Comp Neurol* **344**:336–348.

Graziadei, P. P. C. and G. A. Monti Graziadei (1978). Continuous nerve cell renewal in the olfactory system. In: M. Jacobson (ed). *Handbook of Sensory Physiology, Vol. IX.* Berlin: Springer-Verlag, pp 55–82.

Greene, L. A., S. E. Farinelli, I. Yan, and G. Ferrari (1995). On the mechanisms by which neurotrophic factors regulate neuronal survival and cell death. In: C. F. Ibáñez, T. Hökfelt, L. Olson, K. Fuxe, H. Jörnvall, and D. Ottoson (eds). *Life and Death in the Nervous System.* Oxford: Elsevier Science, pp 55–67.

Grunert, U. and B. W. Ache (1988). Ultrastructure of the aesthetasc (olfactory) sensilla of the spiny lobster, *Panulirus argus. Cell Tissue Res* **251**:95–103.

Guthrie, K. M. and C. M. Gall (1991). Differential expression of mRNAs for the NGF family of neurotrophic factors in the adult rat central olfactory system. *J Comp Neurol* **313**:95–102.

Hallberg, E., K. U. I. Johansson, and R. Elofsson (1992). The aesthetasc concept: structural variations of putative olfactory receptor cell complexes in crustacea. *Micro Res Tech* **22**:325–335.

Halpern, M., L. S. Shapiro, and C. P. Jia (1995). Differential localization of G proteins in the opossum vomeronasal system. *Brain Res* **677**:157–161.

Hinds, J. W. and N. A. McNelly (1981). Aging in the rat olfactory system: correlation of changes in the olfactory epithelium and olfactory bulb. *J Comp Neurol* **203**:441–453.

Hinds, J. W., P. L. Hinds, and N. A. McNelly (1984). An autoradiographic study of the mouse olfactory epithelium: evidence for long-lived receptors. *Anat Rec* **210**:375–383.

Holbrook, E. H., K. E. M. Szumowski, and J. E. Schwob (1995). An immunochemical, ultrastructural, and developmental characterization of the horizontal basal cells of rat olfactory epithelium. *J Comp Neurol* **363**:129–146.

Holcomb, J. D., J. S. Mumm, and A. L. Calof (1995). Apoptosis in the neuronal lineage of the mouse olfactory epithelium: regulation *in vivo* and *in vitro. Dev Biol* **172**:307–323.

Jia C. and M. Halpern (1998). Neurogenesis and migration of receptor neurons in the vomeronasal sensory epithelium in the opossum, *Monodelphis domestica. J Comp Neurol* **400**:287–297.

Keil, T. A. (1982). Contacts of pore tubules and sensory dendrites in antennal chemosensilla of a silkmoth: demonstration of a possible pathway for olfactory molecules. *Tissue Cell* **14**:451–462.

Lazard, D., N. Tal, M. Rubinstein, M. Khen, D. Lancet, and K. Zupko (1990). Identification and biochemical analysis of novel olfactory-specific cytochrome P-450IIA and UDP-glucuronosyl transferase. *Biochemistry* **29**:7433–7440.

Lundh, B., K. Kristensson, and E. Norrby (1987). Selective infections of olfactory and respiratory epithelium by vesicular stomatitis and Sendai viruses. *Neuropath Appl Neurobiol* **13**:111–122.

Mackay-Sim, A., W. Breipohl, and M. Kremer (1988). Cell dynamics in the olfactory epithelium of the tiger salamander: a morphometric analysis. *Exp Brain Res* **71**:189–198.

Mackay-Sim, A. and P. W. Kittel (1991a). On the life span of olfactory receptor neurons. *Eur J Neurosci* **3**:209–215.

Mackay-Sim, A. and P. W. Kittel (1991b). Cell dynamics in the adult mouse olfactory epithelium: a quantitative autoradiographic study. *J Neurosci* **11**:979–984.

Mahalik, T. (1996). Apparent apoptotic cell death in the olfactory epithelium of adult rodents: death occurs at different developmental stages. *J Comp Neurol* **372**:457–464.

Mahanthappa, N. G. and G. A. Schwarting (1993). Peptide growth factor control of olfactory neurogenesis and neuronal survival in vitro: roles of EGF and TGF-β. *Neuron* **10**: 293–305.

Margolis, F. L. (1972). A brain protein unique to the olfactory bulb. *Proc Natl Acad Sci U S A* **69**:1221–1224.

McLean, J. H., M. T. Shipley, and D. I. Bernstein (1989). Golgi-like, transneuronal retrograde labeling with CNS injections of *Herpes simplex* virus type I. *Brain Res Bull* **22**:867–881.

Meisami, E. (1989). A proposed relationship between increases in the number of olfactory receptor neurons, convergence ratio and sensitivity in the developing rat. *Dev Brain Res* **46**:9–19.

Melese-d'Hospital, P. Y. and B. L. Hart (1985). Vomeronasal organ cannulation in male goats: evidence for transport of fluid from the oral cavity to the vomeronasal organ during flehmen. *Physiol Behav* **35**:941–944.

Menco, B. P. M. (1980). Qualitative and quantitative freeze-fracture studies on olfactory and nasal respiratory epithelial surfaces of frog, ox, rat and dog. II. Cell apices, cilia and microvilli. *Cell Tissue Res* **211**:5–30.

Menco, B. P. M. (1983). The ultrastructure of olfactory and nasal respiratory epithelium surfaces. In: G. Reznik and S. F. Stinson (eds). *Nasal Tumors in Animals and Man, Anatomy, Physiology and Epidemiology, Vol. I.* Boca Raton, FL: CRC Press Inc., pp 45–102.

Menco, B. P. M. (1997). Ultrastructural aspects of olfactory signaling. *Chem Senses* **22**:295–311.

Menco, B. P. M., R. C. Bruch, B. Dau, and W. Danho (1992). Ultrastructural localization of olfactory transduction components: the G protein subunit $G_{olf\alpha}$ and type III adenylyl cyclase. *Neuron* **8**:441–453.

Menco, B. P. M., F. D. Tekula, A. I. Farbman, and W. Danho (1994). Developmental expression of G-proteins and adenylyl cyclase in peripheral olfactory systems. Light microscopic and freeze-substitution electron microscopic immunocytochemistry. *J Neurocytol* **23**:708–727.

Menco, B. P. M., A. M. Cunningham, P. Qasba, N. Levy, and R. R. Reed (1997). Putative odour receptors localize in cilia of olfactory receptor cells in rat and mouse: a freeze-substitution ultrastructural study. *J Neurocytol* **26**:691–706.

Meng, L. Z., C. H. Wu, M. Wicklein, K.-E. Kaissling, and H. J. Bestmann (1989). Number and sensitivity of three types of pheromone receptor cells in *Antheraea pernyi* and *A. polyphemus*. *J Comp Physiol A* **165**:139–146.

Meredith, M. and R. J. O'Connell (1979). Efferent control of stimulus access to the hamster vomeronasal organ. *J Physiol* **286**:301–316.

Miragall, F. and G. A. Monti Graziadei (1982). Experimental studies on the olfactory marker protein. II. Appearance of the olfactory marker protein during differentiation of the olfactory sensory neurons of mouse: an immunohistochemical and autoradiographic study. *Brain Res* **329**:245–250.

Monath, T. P., C. B. Cropp, and A. K. Harrison (1983). Mode of entry of a neurotropic arbovirus into the central nervous system. *Lab Invest* **48**:399–410.

Morales, J. A., S. Herzog, C. Kompter, K. Frese, and R. Rott (1988). Axonal transport of Borna disease virus along olfactory pathways in spontaneously and experimentally infected rats. *Med Microbiol Immunol* **177**:51–68.

Mori, K. and Y. Yoshihara (1995). Molecular recognition and olfactory processing in the mammalian olfactory system. *Progr Neurobiol* **45**:585–619.

Nagahara, Y. (1940). Experimentelle Studien über die histologischen Veränderungen des Geruchsorgans nach der Olfactoriusdurchschneidung. Beiträge zur Kenntnis des feineren Baus des Geruchsorgans. *Japan J Med Sci V Pathol* **5**:165–199.

Naruse, I. and H. Keino (1995). Apoptosis in the development of the CNS. *Prog Neurobiol* **47**:135–155.

Okano, M. and S. F. Takagi (1974). Secretion and electrogenesis of the supporting cell in the olfactory epithelium. *J Physiol (Lond.)* **242**:353–370.

Oppenheim, R. W. (1991). Cell death during the development of the nervous system. *Annu Rev Neurosci* **14**:453–501.

Pelosi, P. and R. Maida (1995). Odorant-binding proteins in insects. *Comp Biochem Physiol* **111B**:503–514.

Perkins, L. A., E. M. Hedgecock, J. N. Thomson, and J. G. Culotti (1986). Mutant sensory cilia in the nematode *Caenorhabditis elegans*. *Dev Biol* **117**:456–487.

Perroteau, I., M. Oberto, A. Ieraci, P. Bovolin, and A. Fasolo (1998). ErbB-3 and ErbB-4 expression in the mouse olfactory system. *Ann NY Acad Sci* **855**:255–259.

Rafols, J. A. and T. V. Getchell (1983). Morphological relations between the receptor neurons, sustentacular cells and Schwann cells in the olfactory mucosa of the salamander. *Anat Rec* **206**:87–101.

Rama Krishna, N. S., S. S. Little, and T. V. Getchell (1996). Epidermal growth factor receptor mRNA and protein are expressed in progenitor cells of the olfactory epithelium. *J Comp Neurol* **373**:297–307.

Ramón-Cueto, A., J. Perez, and M. Nieto-Sampedro (1993). *In vitro* enfolding of olfactory neurites by p75 NGF receptor positive ensheathing cells from adult rat olfactory bulb. *Eur J Neurosci* **5**:1172–1180.

Reed, C. J., E. A. Lock, and F. De Matteis (1986). NADPH:cytochrome P-450 reductase in olfactory epithelium. *Biochem J* **240**:585–592.

Ronnett, G. V., L. D. Hester, and S. H. Snyder (1991). Primary culture of neonatal rat olfactory neurons. *J Neurosci* **11**:1243–1255.

Roskams, A. J. I., M. A. Bethel, K. J. Hurt, and G. V. Ronnett (1996). Sequential expression of trks A, B, and C in the regenerating olfactory epithelium. *J Neurosci* **16**: 1294–1307.

Salehi-Ashtiani, K. and A. I. Farbman (1996). Expression of *neu* and neu differentiation factor in the olfactory mucosa of rat. *Int J Devel Neurosci* **14**:801–811.

Schwartz Levy, M. A., D. M. Chikaraishi, and J. S. Kauer (1991). Characterization of potential precursor populations in the mouse olfactory epithelium using immunocytochemistry and autoradiography. *J Neurosci* **11**:3556–3564.

Schwob, J. E., K. E. M. Szumowski, and A. A. Stasky (1992). Olfactory sensory neurons are trophically dependent on the olfactory bulb for survival. *J Neurosci* **12**:3896–3919.

Schwob, J. E., S. L. Youngentob, and R. C. Mezza (1995). The reconstitution of the rat olfactory epithelium after methyl bromide-induced lesion. *J Comp Neurol* **359**:15–37.

Segovia, S. and A. Guillamón (1982). Effects of sex steroids on the development of the vomeronasal organ in the rat. *Dev Brain Res* **5**:209–212.

Shipley, M. T. (1985). Transport of molecules from nose to brain: transneuronal anterograde and retrograde labeling in the rat olfactory system by wheat germ agglutinin-horseradish peroxidase applied to the nasal epithelium. *Brain Res Bull* **15**:129–142.

Stengl, M., H. Hatt, and H. Breer (1992). Peripheral process in insect olfaction. *Annu Rev Physiol* **54**:665–681.

Sutcliffe, J. F. (1994). Sensory bases of attractancy: morphology of mosquito olfactory sensilla—a review. *J Am Mosq Control Assoc* **10**:309–315.

Suzuki, Y. and M. Takeda (1991a). Keratins in the developing olfactory epithelia. *Dev Brain Res* **59**:171–178.

Suzuki, Y. and M. Takeda (1991b). Basal cells in the mouse olfactory epithelium after axotomy: immunohistochemical and electron-microscopic studies. *Cell Tiss Res* **266**:239–245.

Suzuki, Y., J. Schafer, and A. I. Farbman (1995). Phagocytic cells in the rat olfactory epithelium after bulbectomy. *Exp Neurol* **136**:225–233.

Suzuki, Y., M. Takeda, and A. I. Farbman (1996). Supporting cells as phagocytes in the olfactory epithelium after bulbectomy. *J Comp Neurol* **376**:509–517.

Thurn, U. and J. Küppers (1980). Epithelial physiology of insect sensilla. In: *Insect Biology in the Future*. M. Locke and D. S. Smith, (eds) New York: Academic Press, pp 735–763.

Tomlinson, A. H. and M. M. Esiri (1983). Herpes simplex encephalitis. Immunohistological demonstration of spread of virus via olfactory pathways in mice. *J Neurol Sci* **60**:473–484.

Turner, C. P. and J. R. Perez-Polo (1992). Regulation of the low affinity receptor for Nerve Growth Factor, p75^{NGFR}, in the olfactory system of neonatal and adult rat. *Int J Dev Neurosci* **10**:343–359.

Verhaagen, J., A. B. Oestreicher, W. H. Gispen, and F. L. Margolis (1989). The expression of the growth associated protein B50/GAP43 in the olfactory system of neonatal and adult rats. *J Neurosci* **9**:683–691.

Vickland, H., L. E. Westrum, J. N. Kott, S. L. Patterson, and M. A. Bothwell (1991). Nerve growth factor receptor expression in the young and adult rat olfactory system. *Brain Res* **565**:269–279.

Ward, S., N. Thomason, J. G. White, and S. Brenner (1975). Electron microscopical reconstruction of the anterior anatomy of the nematode, *Caenorhabditis elegans*. *J Comp Neurol* **160**:313–338.

Weiler, E. and A. I. Farbman (1997). Proliferation in the rat olfactory epithelium: age-dependent changes. *J Neurosci* **17**:3610–3622.

Weiler, E. and A. I. Farbman (1998). Supporting cell proliferation in the olfactory epithelium decreases postnatally. *Glia* **22**:315–328.

Weiler, E. and A. I. Farbman (1999). Mitral cell loss following lateral olfactory tract transection increases proliferation density in rat olfactory epithelium. *Eur J Neurosci* **11**:3265–3275.

Weiler, E., M. A. McCulloch, and A. I. Farbman (1999). Proliferation in the vomeronasal organ of the rat during postnatal development. *Eur J Neurosci* **11**:700–711.

7

Olfactory Transduction

BARRY W. ACHE
Whitney Laboratory, University of Florida, St. Augustine, FL,
and Departments of Zoology and Neuroscience, University of Florida,
Gainesville, FL

DIEGO RESTREPO
Department of Cellular and Structural Biology, University of Colorado
Health Sciences Center, Denver, CO

1. INTRODUCTION

Odors are detected by primary olfactory receptor neurons in the nose or other olfactory receptor organs such as antenna in insects. These cells number from as few as about ten to millions, depending on the species. In mammals, the cells primarily occur in patches of epithelium that line the superior region of the nasal cavity and constitute the main olfactory organ, although this is not the only chemosensory organ in the nose (see Chapter 6 by Farbman). Bipolar olfactory receptor neurons have a thin unmyelinated axon that reaches the first synaptic relay in the central nervous system, the olfactory bulb, and a single dendrite that extends toward the surface of the epithelium. The distal end of the dendrite supports an arbor of thin hair-like processes, the olfactory cilia, which are bathed in a mucus or fluid layer that overlays the olfactory epithelium.

In invertebrates, olfactory receptor neurons occur in morphologically diverse structures with different specific names, for example, the amphid organ in

The Neurobiology of Taste and Smell, Second Edition, Edited by Thomas E. Finger, Wayne L. Silver, and Diego Restrepo.
ISBN 0-471-25721-4   Copyright © 2000 Wiley-Liss, Inc.

nematodes, the antenna in insects, or the osphradium in some molluscs, that serve as the main olfactory organ of the animal. The olfactory receptor neurons contained in these organs, however, are strikingly similar in general morphology to those found in vertebrates (Ache, 1991; Hildebrand and Shepherd, 1997, see also Chapter 6 by Farbman) (Fig. 7.1). Like their vertebrate counterparts, olfactory receptor neurons from nematodes to squid and insects are primary sensory neurons, bipolar cells with an axon that projects to the central nervous system without synapsing outside the central nervous system. The dendrite of the cells possess a distal arbor of cilia, in many cases more correctly called outer dendritic branches, that extends into a fluid-filled compartment in contact with the environment.

The striking similarity of olfactory receptor cells in different organisms presumably mirrors the common function of the cells. Olfactory receptor cells of all animals serve the same dual function, which is to capture and bind odor molecules, and then convert or transduce the odor signal into an electrical neural signal that informs the brain of the quantity and quality of odorous chemical molecules in the environment. The neural signal has to be able to self-propagate to the central nervous system, which in some animals is more than 10 cm from the peripheral olfactory organ. However, it is important to point out that commonality of function does not necessarily imply similar structure. For the taste system there is a striking dissimilarity in morphology between vertebrates, which use taste bud cells as primary receptors, and invertebrates, which use primary sensory neurons for taste function (see Chapter 12 by Finger and Simon).

Understanding the mechanistic details of olfactory transduction has been, and continues to be, a topic of active investigation in the field, one that integrates information from electrophysiological, biochemical, and molecular approaches. A general picture is beginning to emerge. Olfactory transduction is a multistep process that formally begins with the odor-bound receptor activating a GTP-binding protein that in turn regulates production of an intracellular second messenger. The second messenger directly or indirectly targets an ion channel that generates an initial receptor current. In at least some instances, this current triggers a second ion channel that carries the majority of the receptor current in what is a two-stage amplification cascade. The net receptor current depolarizes the cell, generating a local, graded receptor potential that spreads passively throughout the cell. At some point, it triggers "Hodgkin-Huxley" type, voltage-activated ion channels that are capable of generating action potentials that self-propagate down the axon of the cell to the first olfactory relay in the brain (see Chapter 9 by Christensen and White). This sequence of events results in large molecular amplification of the signal at minimal expense to the signal to noise ratio of the system.

All animals, or for that matter all olfactory receptor cells in the same animal, do not necessarily use the same detailed mechanism to transduce odor signals, although they may well share a limited number of common strategies to transduce odor information. One strategy animals could use would be to have

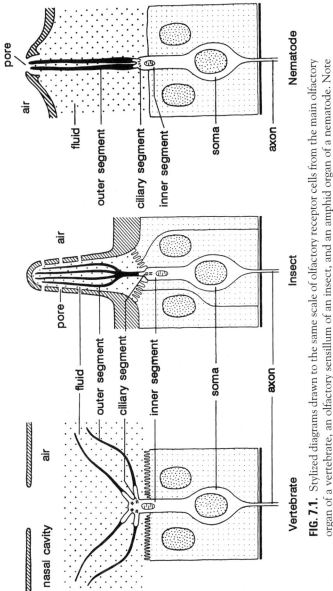

FIG. 7.1. Stylized diagrams drawn to the same scale of olfactory receptor cells from the main olfactory organ of a vertebrate, an olfactory sensillum of an insect, and an amphid organ of a nematode. Note all are primary bipolar receptor neurons that terminate in an arbor of cilia or ciliary-derived processes and send an axon without synapsing to the central nervous system. (After Ache, 1991.) The diagram is not drawn to scale, and the relative distance between the different compartments varies greatly between species. See Chapter 6 by Farbman for more information on structural aspects.

receptor cells that are relatively unsophisticated in terms of their information capacity, that is, cells that serve as throughput filters simply "reporting" the relative efficiency for the odor-binding event. In this case, each cell could produce a relatively coarse output signal that could be refined by the computing power of the central nervous system. The large number of olfactory receptor cells in most animals would favor "downstream" averaging. Cells producing a relatively unrefined signal could operate with less complex intracellular machinery to transduce the odor signal. At the opposite extreme, animals could adapt a more complex strategy in which the receptor cell would use more complex intracellular machinery for olfactory transduction. Such neurons could be capable of fine-tuning, and possibly even integrating, the signal sent to the brain. These two alternatives, of course, are not mutually exclusive. Different cells in the same animal, perhaps in different olfactory organs or in functional subsets of cells the same organ, could adapt different transduction strategies. It is reasonable to expect that the transduction strategy used by a given animal, olfactory organ, or subpopulation of olfactory receptor cells would be shaped by the type of information that needs to be extracted from the odor signal.

Here, we consider the current status of our understanding of the cellular mechanisms implicated in olfactory transduction. The closely related molecular biology of olfactory transduction is considered in the Chapter 8 by McClintock and also should be consulted.

2. EXTRACELLULAR MUCUS MATRIX

While olfactory transduction formally begins with odor binding, understanding olfactory transduction requires understanding the fluid surrounding the olfactory cilia. The olfactory epithelium in vertebrates is covered by a layer of mucus 5–30 µm thick (Menco and Farbman, 1992). This layer, composed of water, electrolytes, and protein, is not homogeneous and can be subdivided into several domains. The mucus layer plays an important functional role in olfactory transduction because it provides the electrolytes necessary for the action of ciliary ion channels, and because odorant molecules, which in many cases are hydrophobic, must penetrate this hydrophilic layer to reach the olfactory receptors on the cilia. Changes in thickness and composition of the mucus layer will affect the response of olfactory neurons to odorants. Mucus is secreted from Bowman's glands and secretion is modulated by adrenergic fibers from the superior cervical ganglion. Fibers immunoreactive for the neuropeptides Y, calcitonin-related peptide, and luteinizing hormone releasing hormone have also been reported around Bowman's glands in the olfactory epithelium (Getchell and Getchell, 1992). A major component of the mucus matrix is olfactomedin, a protein that is secreted from Bowman's glands in a glycosylated form that forms polymers via intermolecular disulfide bonds (Bal and Anholt, 1993). Fully glycosylated olfactomedin is found in the acinar regions of Bowman's glands and in the mucus layer overlaying the olfactory epithelium.

Also present in the mucus are water-soluble olfactory binding proteins (OBP's), which have been isolated, and subsequently cloned from numerous terrestrial species, including insects (Pelosi, 1994). OBP's occur in high concentration and bind a variety of structurally diverse hydrophobic odorants. The function of OBP's is not known, but one possibility is that they increase the effective odorant concentration in the mucus layer that otherwise would be compromised by the partition coefficient of volatile substances in the mucus.

Knowledge of the ionic composition of the fluid bathing the olfactory cilia is especially important in order to understand the effect of opening ciliary ion channels. Ion concentrations in the olfactory mucus of terrestrial vertebrates have been measured using spectrophotometry or ion-sensitive electrodes with varying results that reflect the technical difficulty of obtaining these measurements (Schild and Restrepo, 1998). A very accurate and highly focal technique was used to measure ion concentrations in mammalian olfactory mucus and receptor cells (Reuter, et al., 1998) (Table 7.1). Quite surprising was the relative high value for the K^+ concentration in the mucus, suggesting that the equilibrium potential for K^+ is less negative than the resting potential, and that increasing K^+ conductance would depolarize (excite) the cell, at least up to the equilibrium potential of K^+. A high concentration of extracellular K^+ also surrounds the outer dendrites of insect olfactory receptor cells, and is thought to play a fundamental role in generating the receptor potential (Kaissling, 1995). The Cl^- concentration in the knob of mammalian olfactory receptor cells is also high, close to that in the mucus. The equilibrium potential for currents carried primarily by chloride ions would be very positive, about $+6\,mV$, making Cl^- currents depolarizing (excitatory). Both these ions appear to be actively regulated.

Ion concentrations surrounding olfactory cilia in aquatic species presumably would be set by the ionic composition of the environment, yet these, too, can be regulated. In fully marine species of crabs and lobsters, the ionic composition of the fluid surrounding the olfactory cilia is that of seawater, that is, (mM) Na, 423; K, 9; Mg, 49; Ca, 13 (Gleeson et al., 1993). Marine crabs that migrate into

TABLE 7.1. Ionic Composition of the Olfactory Mucus and the Dendritic Knob of Mammalian Olfactory Receptor Cells Determined by Energy-Dispersive X-ray Microanalysis

Ion	Olfactory Mucus (n = 12)	Dendritic Knob (n = 10)
Cl	55 ± 11	69 ± 19
K	69 ± 10	172 ± 23
Na	55 ± 12	53 ± 31
P	77 ± 13	134 ± 21
S	57 ± 12	174 ± 23

Numbers represent the total number $(X \pm SD)$ of ions (free and bound) in mM.

(From Reuter et al., 1998.)

freshwater, however, appear to maintain at least a sodium concentration around their olfactory cilia equal to that of full seawater by active regulation (Gleeson et al., 1997). Interestingly, the olfactory cilia of the crabs are much shorter in freshwater than they are in saltwater, presumably in response to limitations in the fluid layer that can be ionically regulated.

3. RECEPTOR PROTEINS

Olfactory transduction per se is triggered by the binding of odors to receptor proteins located on the olfactory cilia. Olfactory receptor proteins, all of which so far are members of the G-protein coupled receptor superfamily, are considered in detail in Chapter 8 by McClintock. Of particular relevance to understanding olfactory transduction is whether individual olfactory receptor cells express more than one type of odor receptor protein. Multiple receptors potentially could couple to multiple transduction pathways and confer upon the cell the ability to integrate information transmitted to the central nervous system. As explained in the chapter by McClintock, molecular biological studies suggest that vertebrate receptor cells express only one, or at most a very few, receptor proteins. This finding is not by itself exclusive since cells expressing just two different receptor proteins coupled to different transduction pathways would have markedly increased information capacity. For instance, cells expressing two different receptor proteins could integrate inputs of opposite polarity, a commonly used strategy to increase information capacity in other sensory systems. Indeed, evidence considered below that different odors excite and inhibit individual receptor cells implies that olfactory receptor cells can express more than one type of receptor protein. Molecular evidence confirms that single olfactory receptor cells of the nematode *C. elegans* express multiple receptor proteins (Bargmann and Kaplan, 1998).

4. SECOND MESSENGER SIGNALING

Binding of an odor to the receptor protein stimulates the production of intracellular chemical signals (Fig. 7.2). These "second messengers" target ion channels that gate a flux of ions across the ciliary membrane. The specific enzymes that control second messenger production are regulated by the interaction of the odorant-bound receptor with heterotrimeric GTP-binding regulatory proteins (G-proteins; e.g., Neer, 1995). One G-protein, G_{olf}, was first cloned on the basis of its enriched expression in the olfactory system, and could be localized to the olfactory cilia of vertebrates (Jones and Reed, 1989), suggesting that it has a specific role in olfactory transduction. This protein is a member of the G_s family and mediates stimulation of type III adenylyl cyclase by odor stimuli (Bakalyar and Reed, 1990). Several other G-proteins also occur in the olfactory epithelium of vertebrates. At least one of these, a member of

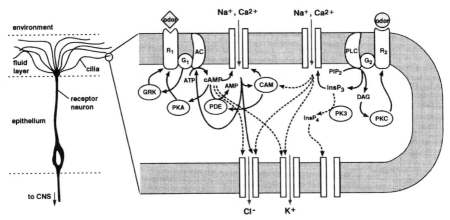

FIG. 7.2. Composite schematic diagram summarizing the intracellular signaling pathways implicated in olfactory transduction. The most completely understood pathway involves a receptor protein (R_1), a GTP-binding protein (G_1, likely G_{olf}), an adenylyl cyclase (AC) that produces adenosine $3',5'$-cyclic monophosphate (cAMP), and a cation channel that is gated directly by cAMP. This is the primary pathway for activating olfactory receptor cells in vertebrates. In some animals, there is evidence, still largely indirect, for a second pathway involving a different receptor protein (R_2), a different GTP-binding protein (G_2), a phospholipase C (PLC) that produces inositol 1,4,5-trisphosphate ($InsP_3$) and diaclglycerol (DAG), and a cation channel that is directly gated by $InsP_3$. These primary transduction pathways can target secondary ion channels that carry some, and possibly most, of the transduction current. Best known is a calcium-activated chloride channel secondarily targeted by the cyclic nucleotide pathway in vertebrate olfactory receptor cells. Other ion channels implicated in olfactory transduction include a calcium-activated potassium channel, a sodium-activated non-selective cation channel, a cyclic nucleotide activated chloride channel, and a channel of unknown selectivity gated by inositol 1,3,4,5-tetrakisphosphate ($InsP_4$). Each pathway also can be modulated by a number of regulatory elements. Regulatory elements implicated in olfactory transduction include a phosphodiesterase (PDE), a G-protein-coupled receptor kinase (GRK), phosphokinases A (PKA) and C (PKC), and calcium-calmodulin (CAM). The diagram shows all known and proposed signaling pathways; all pathways do not occur in every cell, nor in every species. Solid lines represent better established pathways. Dashed lines represent proposed pathways. (After Ache and Zhainazarov, 1995.)

the G_o family, also has been functionally implicated in odor transduction together with G_s (Schander et al., 1998). A member of the G_q family has been functionally implicated in activation of lobster olfactory receptor cells (Fadool et al., 1995).

Two known intracellular signaling molecules or second messengers have been implicated in olfactory transduction: adenosine $3'5'$-cyclic monophosphate (cAMP) and inositol-1,4,5-trisphosphate ($InsP_3$). In rat olfactory ciliary membranes in vitro, odors elevate the concentration of both second messengers rapidly (within about 50 msec) and transiently (Breer and Boekhoff, 1991). One study found that cAMP is produced in response to a wide variety of odors, while

InsP$_3$ is produced in response to primarily (although not exclusively) odors that impart a putrid or otherwise unpleasant sensation in humans (Breer and Boekhoff, 1991). While these findings would suggest that odors are specific to one pathway or the other in rat, they have to be reconciled with other work indicating that odors thought to elicit InsP$_3$ formation also increase cAMP formation in a rat olfactory ciliary membranes in vitro (Jaworsky et al., 1995). The extent to which odors might be specific to one signaling pathway or the other is also uncertain for other animals. While InsP$_3$, but not cAMP, is elevated by pheromones in insect olfactory outer dendritic membranes (Breer et al., 1990) and by amino acids in fish olfactory cilia (Restrepo et al., 1993) in vitro, a wide variety of odors stimulate production of cAMP in amphibian olfactory cilia in vivo (Lowe et al., 1989). Single odors stimulate production of both signal molecules in lobster olfactory cilia in vitro, although with some degree of selectivity (Boekhoff et al., 1994).

Unfortunately, experiments on membrane homogenates do not allow one to discern whether or not both pathways occur in the same receptor cell, where they could potentially interact. Electrophysiological evidence that protein kinase C sensitizes olfactory adenylyl cyclase in vertebrate olfactory receptor cells suggests that crosstalk can occur between the pathways at the level of protein kinases (Frings, 1993). The extent to which the two pathways interact in vivo, and the biochemical mechanisms by which this could occur, however, have yet to be explored in any detail.

4.1. Cyclic Nucleotide-Mediated Signaling

Cyclic nucleotide-mediated signaling, in particular the activation phase of the cascade, is best understood and appears to be the primary signaling pathway in vertebrate olfactory receptor cells. The major components of this pathway have been cloned and sequenced from the olfactory organ (see Chapter 8 by McClintock). Odor-stimulated cAMP targets Ca^{2+}-permeable cAMP-gated nonspecific cation (CNG) channels located in the ciliary membrane (Nakamura and Gold, 1987; Firestein et al., 1991; Frings and Lindemann, 1991; Kurahashi and Kaneko, 1991; Lowe and Gold, 1993a). The CNG channel is permeable to Ca^{2+} ions (Nakamura and Gold, 1987; Kolesnikov et al., 1990; Kurahashi, 1990), and stimulation of olfactory neurons with odorants that are known to induce cAMP formation leads to an increase in intracellular Ca^{2+} (Sato et al., 1991; Restrepo et al., 1993b; Rawson et al., 1997; Leinders-Zufall et al., 1997). Because Ca^{2+} permeates slowly through the CNG channel, other cations are effectively excluded. The main depolarizing effect of cAMP does not appear to be mediated by the CNG channel itself, but rather secondarily by the increase in intracellular calcium $[Ca^{2+}]_i$ that results from opening of the CNG channels (Frings et al., 1995).

Cyclic nucleotide mediated signaling also occurs in invertebrate olfactory receptor cells, although the nature of the conductance(s) targeted by cAMP, and therefore the effect of activating this pathway, is still being explored. As in vertebrates, cAMP appears to be targeting a CNG channel in crustaceans (Hatt and Ache, 1994), insects (Baumann et al., 1994), and nematodes (Coburn and Bargmann, 1996). The main permeant ion in lobster olfactory receptor neurons appears to be K^+, which mediates a current that opposes excitation of the cell (Michel et al., 1991).

4.2. Phosphoinositide-Mediated Signaling

Phosphoinositide-mediated signaling is less well understood, especially in vertebrate olfactory receptor cells (Bruch, 1996). An InsP$_3$ receptor has been localized immunocytochemically to mammalian olfactory cilia (Cunningham et al., 1993), and an InsP$_3$ receptor has been characterized biochemically from a catfish olfactory ciliary membrane preparation (Kalinoski et al., 1992; Restrepo et al., 1992). The biochemically identified receptor appears to be an InsP$_3$-gated nonspecific cation channel (Restrepo et al., 1990). Dialysis of InsP$_3$ into isolated olfactory neurons elicits inward currents at negative holding potentials (Restrepo et al., 1990; Miyamoto et al., 1992a; Okada et al., 1994). Parallel measurements of $[Ca^{2+}]_i$ show that the inward current is accompanied by an increase in $[Ca^{2+}]_i$ (Schild et al., 1995; Kashiwayanagi, 1996). Collectively, these data are consistent with the hypothesis that odor-stimulated InsP$_3$ targets InsP$_3$-gated channels that are either in, or closely associated with, the ciliary membrane, activation of which depolarizes the cell and thereby forms a parallel or alternate excitatory transduction pathway. This hypothesis has been challenged by a report that mice deficient for subunit 1 of the olfactory CNG channel do not respond in an electrophysiological assay (EOG, see below) to any odorants (Brunet et al., 1996). Perhaps, therefore, the InsP$_3$ pathway serves more of a modulatory role in olfactory transduction. This could occur, for instance, if the InsP$_3$ pathway downregulated activity of the olfactory CNG channel, which served as a common element to both input pathways. Clearly, further experiments are necessary to clarify the role of InsP$_3$-mediated signaling in olfactory transduction in vertebrates.

More compelling evidence for InsP$_3$-mediated olfactory signaling comes from invertebrates, particularly the lobster. Antibodies raised against a mammalian InsP$_3$ receptor enhance the odor-evoked receptor current, thereby implicating an InsP$_3$ -mediated pathway in depolarizing lobster olfactory receptor cells (Fadool and Ache, 1992). An InsP$_3$ receptor has been isolated from olfactory organ RNA and localized immunocytochemically to be either in or closely associated with the ciliary membrane (Munger et al., in press). This finding extends earlier functional evidence showing that InsP$_3$-activated unitary currents (Hatt and Ache, 1994) and the enzymatic machinery for InsP$_3$ production (Boekhoff et al., 1994) occur in lobster olfactory cilia. While it is compelling to envision the InsP$_3$ receptor as the final effector in this pathway, this has not been established with certainty. InsP$_3$ does not appear to directly activate or modulate a pheromone-activated ion channel in insect olfactory cilia. Rather, activation of the pheromone channel appears to involve protein kinase C (Zufall and Hatt, 1991). Evidence that a transient $InsP_3$-dependent Ca^{2+} current precedes the cation current associated with pheromone activation in insect olfactory receptor neurons (Stengl, 1994) suggests that $InsP_3$ may act indirectly in insect olfactory receptor cells by inducing a rise in intracellular Ca^{2+} that promotes kinase C-mediated phosphorylation of the channel.

5. ELECTROPHYSIOLOGICAL RESPONSES

As noted earlier, olfactory receptor cells communicate to the central nervous system by generating self-propagating action potentials. Action potentials are generated when the cell membrane is brought to threshold by a depolarizing receptor potential generated in the cilia. The frequency of discharge is proportional to the rate of rise and magnitude of the receptor potential, which in turn is set by the kinetics of the transduction cascade and by capacitive shaping. The number and/or rate of odor evoked action potentials can be counted and used to quantify the output of the cell. Given the large numbers and relatively small size of most olfactory receptor cells, it is easier, and sometimes preferred, to quantify olfactory output using a more global measure of electrical activity called the electro-olfactogram (EOG) in vertebrates, or the electro-antennagram (EAG) in arthropods. EOGs/EAGs are slow, DC potentials recorded across the receptor organ with large, nonfocal electrodes. These potentials are generally assumed to reflect the summed receptor currents flowing along the surface of the organ in response to many co-activated receptor cells. More strongly responding cells and large numbers of similarly responding cells contribute disproportionately to signal, however, potentially biasing the ability of the EOG to measure of the true response profile of the receptor cell population.

5.1. Excitatory Responses

Odorant-induced "cation" currents were first reported by Trotier and MacLeod (1983) in frog and by Anderson and Ache (1985) in lobster olfactory receptor cells. This type of response has been subsequently recorded in olfactory receptor cells from all species studied (review: Hildebrand and Shepherd, 1997).

In vertebrates, in the absence of extracellular calcium and with Ringer solution in the bath, the odor-induced response reverses potential at approximately 0 mV due to activation of the CNG nonspecific conductance. Because of this, the response was originally labeled a "cation" response. In the presence of physiological concentrations of extracellular calcium, however, the response becomes primarily chloride-dependent. The increase in intracellular Ca^{2+} inside the olfactory cilia opens Ca^{2+}-activated Cl^- channels (Kleene and Gesteland, 1991; Kurahashi and Yau, 1993; Lowe and Gold, 1993b). The result of opening Cl^- channels depends on the equilibrium potential for Cl^- in these cells, which as noted earlier would lead to depolarization of the cells in mammals. This holds for amphibian olfactory neurons as well (Kurahashi and Yau, 1993,

Ache and Zhainazarov, 1995). A Cl^- conductance would tend to stabilize the receptor potential in the face of changes in the salt content of the mucus, but data suggest that this channel carries a substantial part of the receptor current. About 40% of the receptor current can be ascribed to the secondary Cl^- current in amphibian (Kurahashi and Yau, 1994) and about 85% in rat (Lowe and Gold, 1993b). Activating the cells via two stages of amplification—cAMP targeting an CNG channel (step 1), then Ca^{2+} influx through the CNG channel targeting a Ca^{2+}-activated Cl^- channel (step 2)—provides a possible mechanism for regulating the total amplification of the system. Changing resting $[Ca^{2+}]_i$ and/or the Ca^{2+} sensitivity of the Cl^- current would regulate the total amplification without appreciably changing the signal-to-noise ratio, which mainly is set by the noise characteristics of the first cascade in two-step amplification cascades.

The odorant-induced "cation" current in arthropods appears to be mediated by phosphoinositide signaling, as discussed earlier, but here too, a secondary channel has been implicated in carrying a major portion of the receptor current in the lobster (Zhainazarov et al., 1998). In the case of the lobster, however, the secondary channel is nonselective for cations (Zhainazarov and Ache, 1997) and is a member of a novel family of ion channels that are activated or modulated by intracellular sodium (Zhainazarov and Ache, 1995a). How the sodium-activated channel is triggered in vivo is still being explored. One possibility is that it is activated by sodium that enters the cell through activation of the primary transduction cascade. Once triggered, activation of the channel presumably would be regenerative since sodium entering the channel would further increase the open probability of the channel and allow more sodium to enter the cell. This is an excellent example of how general organizational principles can be sustained across phylogenetically divergent species, even though the mechanistic details are species dependent.

5.2. Inhibitory Responses

Odors can also suppress as well as enhance the basal or evoked discharge of olfactory receptor neurons in a number of different species. Inhibitory responses were once regarded with caution due to the characteristically low spontaneous activity of olfactory receptor cells and since the responses were recorded in a manner that could have shifted extracellular ionic concentrations (e.g., Getchell, 1986). With more advanced recording techniques, however, inhibition could be associated with specific currents. In the squid, the odor betaine activates a Cl^- conductance that, due to the low internal concentration of Cl^- in these cells, hyperpolarizes the cell (Danaceau and Lucero, 1998). Various odorants hyperpolarize lobster olfactory receptor cells by activating an outward current thought to be mediated by a cAMP-gated K^+ channel (Michel et al., 1991; Michel and Ache, 1992). Odors also inhibit the discharge of olfactory receptor cells in catfish (Kang and Caprio, 1995), mudpuppy (Dubin and Dionne, 1994), and toad (Morales et al., 1996). Inhibitory receptor potentials in toad olfactory receptor cells appear to be mediated by opening of Ca^{2+}-activated K^+ channels of the BK type (Morales et al., 1996). Other types of K^+ (Ca^{2+}) channels occur in olfactory receptor neurons in other species (Maue and Dionne, 1987; Schild, 1989; Miyamoto et al., 1992b, Lucero and Chen, 1997) and could potentially be involved in mediating inhibitory responses, but this has yet to be confirmed experimentally. As noted earlier, the equilibrium potential of K^+ in mammalian olfactory receptor cells is positive of the resting potential of the cells and presumably a K^+ conductance would not mediate an inhibitory current in these cells. Interestingly, dialyzing $InsP_3$ into rat olfactory neurons opens a K^+ (Ca^{2+}) conductance, but the total current (the sum of the K^+ and nonspecific cation components) reverses at $-35\,mV$, a potential more positive than the resting potential for these cells (Okada et al., 1994).

Clearly more work is required to understand the ubiquity of odor-evoked inhibition and the cellular mechanisms involved. Nonetheless, the existence of hyperpolarizing as well as depolarizing receptor potentials adds a new dimension of complexity to ultimately understanding olfactory transduction.

6. ADAPTATION TO PROLONGED STIMULATION

Prolonged exposure to odors decreases the perceived odor intensity, a phenomenon called adaptation. Adaptation is attributable to both central and peripheral processes. The molecular mechanisms that underlie peripheral olfactory adaptation are not completely understood, but are best known from work in vertebrates. At least two processes contribute to peripheral olfactory adaptation. As discussed earlier, odorants increase intracellular Ca^{2+}. Increased intracellular Ca^{2+} is thought to play an important role in olfactory adaptation at the level of the receptor cell because low levels of intracellular Ca^{2+} inhibit adaptation (Kurahashi and Shibuya, 1990). The exact manner in which Ca^{2+} affects adaptation is not known, but probably involves inhibition of the cAMP-gated channel (Kurahashi and Menini, 1997) through both calmodulin-mediated (Liu et al., 1994) and calmodulin-independent (Balasubramanian et al., 1996) pathways. Ca^{2+} also acts by phosphorylating the olfactory receptor protein by receptor-specific kinases and by second messenger-activated kinases through a change in affinity of the olfactory receptors (Dawson et al., 1993; Schleicher et al., 1993). If and how the effects of receptor phosphorylation and Ca^{2+} levels on adaptation are interrelated is not currently known. Finally, prolonged exposure to membrane permeable cGMP analogues (e.g., 8-Br-cGMP) decreases the sensitivity of tiger salamander receptor cells to odors (Leinders-Zufall et al., 1996). Decreased odorant sensitivity also can be elicited by prolonged exposure of the cells to CO. The effect of CO is dependent on the presence of extracellular Ca^{2+}, suggesting that it is mediated by the changes in $[Ca^{2+}]_i$ elicited by opening of CNG-gated channels. These experiments and the effect of inhibitors of the CO/cGMP pathway on olfactory adaptation suggest that CO/cGMP may mediate longer-term adaptation in the cell (Zufall and Leinders-Zufall, 1997). In summary, adaptation is itself a complex process, one that is likely to involve multiple cellular mechanisms.

7. TRANSDUCTION IN VOMERONASAL NEURONS

As mentioned, animals can have more than one olfactory organ. One such structure is the vomeronasal organ or VNO, which occurs along with the olfactory epithelium in the nose of many vertebrates (see Chapter 5 by Johnston and Chapter 6 by Farbman). Like their counterparts in the olfactory epithelium, VNO receptor cells are electrically compact bipolar neurons capable of firing repetitive action potentials (Liman and Corey, 1996; Trotier et al., 1993). VNO neurons possess a Ca^{2+}-sensitive adenylyl cyclase (Wang et al., 1997), but they do not express the other elements of the cyclic nucleotide signaling pathway found in receptor neurons in the olfactory epithelium (i.e., G_{olf}, AC type III, nor the CNG channel; Berghard et al., 1996). Consistent with these observations, manipulation of cAMP levels does not induce a change in conductance in VNO neurons (Liman and Corey, 1996; Trotier et al., 1994). These data indicate that the CNG pathway does not play a primary role in signal transduction in the VNO. The precise mechanism by which odors activate VNO neurons is still unresolved, but may involve InsP₃. Purified aphrodisin from hamster vaginal secretion, a hamster pheromone, as well as volatile and proteinaceous components of urine in rat, stimulate $InsP_3$ production in VNO tissue in the respective animals (Kroner et al., 1996; Krieger et al., 1999; Sasaki et al., 1999). The increase in $InsP_3$ is GTP-dependent, and antibodies against G_o and G_{i2} inhibit different components of the response (Krieger et al., 1999). The partial effect of these G-protein specific antibodies is consistent with the fact that one population of VNO neurons expresses G_o while the rest of the neurons express G_{i2} (Berghard et al., 1996; Jia and Halpern, 1996). The target of $InsP_3$ is not known, although $InsP_3$ elicits an increase in membrane conductance when dialyzed into VNO neurons, suggesting it directly or indirectly targets an ionic conductance (Taniguchi et al., 1995). Recently, a homolog of the trp (transient receptor potential) family of ion channels that include channels functionally implicated in *Drosophila* visual transduction has been identified in VNO neurons (Liman et al., 1999). As suggested earlier, functionally different subsets of olfactory receptor cells in the same animal may use different transduction mechanisms.

8. SUMMARY

Olfactory receptor cells are primary sensory neurons responsible not only for detecting but also for transducing the odor signal. Olfactory transduction occurs in olfactory cilia, which extend from the distal tip of the receptor cells into a mucus or fluid-filled space in direct contact with the odor environment.

Transduction involves a complex, multistage cascade of events triggered by the binding of odor molecules to one or more receptor proteins expressed by the receptor cell. Odor binding, in turn, triggers a G-protein-mediated increase in second messenger (cAMP and/or InsP$_3$) production. Second-messenger activated ion channels, in turn, gate the passage of ions across the ciliary (plasma) membrane. Activation itself appears to be a two-stage process involving secondary activation of an additional ion channel that carries the majority of the receptor current. In some animals, a parallel transduction pathway provides input of opposite polarity into the cell. Numerous steps in the cascade are subject to downregulation by intracellular calcium, and possibly other regulatory factors. These reduce the response of the cell to prolonged stimulation. The complexity of the transduction machinery allows the receptor cell to transmit finely regulated, and possibly even integrated, information about the animal's odor world to the central nervous system.

ACKNOWLEDGMENTS

The work presented here by the authors was supported by NSF grant IBN 9515307 and NIH grant DC 01655 to BWA and NIH grants DC 00566 and DC 00244 to DR. We thank Ms. Lynn Milstead for assistance with the illustrations.

REFERENCES

Ache, B. W. (1991). Phylogeny of smell and taste. In: T. V. Getchell, R. L. Doty, L. M. Bartoshuk, and J. B. Snow (eds). *Smell and Taste in Health and Disease.* New York: Raven Press, pp 3–18.

Ache, B. W. and A. B. Zhainazarov (1995). Dual second-messenger pathways in olfactory transduction. *Curr Opin Neurobiol* **5**:461–466.

Anderson, P. A. V. and B. W. Ache (1985). Voltage- and current-clamp recordings of the receptor potential in olfactory receptor cells *in situ. Brain Res* **338**:273–280.

Bakalyar, H. A. and R. R. Reed (1990). Identification of a specialized adenylyl cyclase that may mediate odorant detection. *Science* **250**:1403–1406.

Bal, R.S. and Anholt, R.R. (1993). Formation of the extracellular mucous matrix of olfactory neuroepithelium: identification of partially glycosylated and nonglycosylated precursors of olfactomedin. *Biochemistry* **32**:1047–1053.

Balasubramanian, S., J. W. Lynch, and P. H. Barry (1996). Calcium-dependent modulation of the agonist affinity of the mammalian olfactory cyclic nucleotide-gated channel by calmodulin and a novel endogenous factor. *J Membr Biol* **152**:13–23.

Bargmann, C.I. and J. M. Kaplan (1998). Signal transduction in the *Caenorhabditis elegans* nervous system. *Annu Rev Neurosci* **21**:279–308.

Baumann, A., S. Frings, M. Godde, R. Seifert, and U. B. Kaupp (1994). Primary structure and functional expression of a *Drosophila* cyclic nucleotide-gated channel present in eyes and antennae. *EMBO J* **13**:5040–5050.

Berghard, A., L. B. Buck, and E. R. Liman (1996). Evidence for distinct signaling mechanisms in two mammalian olfactory sense organs. *Proc Natl Acad Sci U S A* **93**:2365–2369.

Boekhoff, I., W. C. Michel, H. Breer, and B. W. Ache (1994). Single odors differentially stimulate dual second messenger pathways in lobster olfactory receptor cells. *J Neurosci* **14**:3304–3309.

Breer, H. and I. Boekhoff (1991). Odorants of the same odor class activate different second messenger pathways. *Chem Senses* **16**:19–29.

Breer, H., I. Boekhoff, and E. Tareilus (1990). Rapid kinetics of second messenger formation in olfactory transduction. *Nature* **345**:65–68.

Bruch, R. C. (1996). Phosphoinositide second messengers in olfaction. *Comp Biochem Physiol* **113B**:451–459.

Brunet, L. J., G. H. Gold, and J. Ngai (1996). General anosmia caused by a targeted disruption of the mouse olfactory cyclic nucleotide-gated cation channel. *Neuron* **17**:681–693.

Coburn, C. M. and C. I. Bargmann (1996). A putative cyclic nucleotide-gated channel is required for sensory development and function in C. elegans. *Neuron* **17**:695–706.

Cunningham, A. M., D. K. Ryugo, A. H. Sharp, R. R. Reed, S. H. Snyder, and G. V. Ronnett (1993). Neuronal inositol 1,4,5-trisphosphate receptor localized to the plasma membrane of olfactory cilia. *Neuroscience* **57**:339–352.

Danaceau, J. P. and M. T. Lucero (1998). Betaine activates a hyperpolarizing chloride conductance in squid olfactory receptor neurons. *J Comp Physiol A* **183**:225–235.

Dawson, T. M., J. L. Arriza, D. E. Jaworsky, F. Borisy, H. Attramadal, R. J. Lefkowitz, and G. V. Ronnett (1993). Beta-adrenergic receptor kinase-2 and beta-arrestin-2 as mediators of odorant-induced desensitization. *Science* **259**:825–829.

Dubin, A. E. and V. E. Dionne (1994). Action potentials and chemosensitive conductance in the dendrites of olfactory neurons suggest new features for odor transduction. *J Gen Physiol* **103**:181–201.

Fadool, D. A. and B. W. Ache (1992). Plasma membrane inositol 1,4,5-trisphosphate-activated channels mediate signal transduction in lobster olfactory receptor neurons. *Neuron* **9**:907–918.

Fadool, D. A., S. J. Estey, and B. W. Ache (1995). Evidence that a G_q-protein mediates excitatory odor transduction in lobster olfactory receptor neurons. *Chem Senses* **20**:489–498.

Firestein, S., F. Zufall, and G. M. Shepherd (1991). Single odor-sensitive channels in olfactory receptor neurons are also gated by cyclic nucleotides. *J Neurosci* **11**:3565–3572.

Frings, S. (1993). Protein kinase C sensitizes olfactory adenylate cyclase. *J Gen Physiol* **101**:183–205.

Frings, S. and B. Lindemann (1991). Properties of cyclic nucleotide-gated channels mediating olfactory transduction: sidedness of voltage-dependent blockage by Ca^{2+} ions, amiloride, D-600 and diltiazem. *J Gen Physiol* **98**:17a.

Frings, S., R. Seifert, M. Godde, and U. B. Kaupp (1995). Profoundly different calcium permeation and blockage determine the specific function of distinct cyclic-nucleotide-gated channels. *Neuron* **15**:169–179.

Getchell, T. V. (1986). Functional properties of vertebrate olfactory receptor neurons. *Physiol Rev* **66**:772–818.

Getchell, M.L. and T. V. Getchell (1992). Fine structural aspects of secretion and extrinsic innervation in the olfactory mucosa. *Microscop Res Tech* **23**:111–127.

Gleeson, R. A., H. C. Aldrich, J. F. White, and H. G. Trapido-Rosenthal (1993). Ionic and elemental analysis of olfactory sensillar lymph in the spiny lobster, *Panulirus argus*. *Comp Biochem Physiol* **105**:29–34.

Gleeson, R. A., M. G. Wheatly, and C. L. Reiber (1997). Perireceptor mechanisms sustaining olfaction at low salinities: insight from the euryhaline blue crab *Callinectes sapidus*. *J Exp Biol* **200**:445–456.

Hatt, H. and B. W. Ache (1994). Cyclic nucleotide- and inositol phosphate-gated ion channels in lobster olfactory receptor neurons. *Proc Natl Acad Sci U S A* **91**:6264–6268.

Hildebrand, J. G. and G. M. Shepherd (1997). Mechanisms of olfactory discrimination: converging evidence for common principles across phyla. *Annu Rev Neurosci* **20**:595–631.

Jaworsky, D. E., O. Matsuzaki, F. F. Borisy, and G. V. Ronnett (1995). Calcium modulates the rapid kinetics of the odorant-induced cyclic AMP signal in rat olfactory cilia. *J Neurosci* **15**:310–318.

Jia, C. P. and M. Halpern (1996). Subclasses of vomeronasal receptor neurons: differential expression of G proteins (G_{ia2} and G_{oa}) and segregated projections to the accessory olfactory bulb. *Brain Res* **719**:117–128.

Jones, D. T. and R. R. Reed (1989). G_{olf}: an olfactory neuron specific-G protein involved in odorant signal transduction. *Science* **244**:790–795.

Kaissling, K.-E. (1995). Single unit and electroantennogram recordings in insect olfactory organs. In: A. I. Spielman and J. G. Brand (eds). *Experimental Cell Biology of Taste and Olfaction: Current Techniques and Protocols*. Boca Raton FL: CRC Press, pp 361–377.

Kalinoski, D. L., S. B. Aldinger, A. G. Boyle, T. Huque, J. F. Marecek, G. D. Prestwich, and D. Restrepo (1992). Characterization of a novel inositol 1,4,5-trisphosphate receptor in isolated olfactory cilia. *Biochem J* **281**:449–456.

Kang, J. and J. Caprio (1995). In vivo responses of single olfactory receptor neurons in the channel catfish, *Ictalurus punctatus*. *J Neurophysiol* **73**:172–177.

Kashiwayanagi, M. (1996). Dialysis of inositol 1,4,5-trisphosphate induces inward currents and Ca^{2+} uptake in frog olfactory receptor cells. *Biochem Biophys Res Comm* **255**:666–671.

Kleene, S. J. and R. C. Gesteland (1991). Calcium-activated chloride conductance in frog olfactory cilia. *J Neurosci* **11**:3624–3629.

Kolesnikov, S. S., A. B. Zhainazarov, and A. V. Kosolapov (1990). Cyclic nucleotide-activated channels in the frog olfactory receptor plasma membrane. *FEBS Lett* **266**:96–98.

Krieger, J., A. Schmitt, D. Löbell, T. Gudermann, G. Schultz, H. Breer, and I. Boekhoff (1999). Selective activation of G protein subtypes in the vomeronasal organ upon stimulation with urine-derived compounds. *J Biol Chem* **274**:4655–4662.

Kroner, C., H. Breer, A. G. Singer, and R. J. O'Connell (1996). Pheromone-induced second messenger signaling in the hamster vomeronasal organ. *Neuroreport* **7**:2989–2992.

Kurahashi, T. (1990). The response induced by intracellular cyclic AMP in isolated olfactory receptor cells of the newt. *J Physiol* **430**:355–371.

Kurahashi, T. and A. Kaneko (1991). High density cAMP-gated channels at the ciliary membrane in the olfactory receptor cell. *Neuroreport* **2**:5–8.

Kurahashi, T. and A. Menini (1997). Mechanisms of odorant adaptation in the olfactory receptor cell. *Nature* **385**:725–729.

Kurahashi, T. and T. Shibuya (1990). Ca^{2+}-dependent adaptive properties in the solitary olfactory receptor cell of the newt. *Brain Res* **515**:261–268.

Kurahashi, T. and K.-Y. Yau (1993). Co-existence of cationic and chloride components in odorant- induced current of vertebrate olfactory receptor cells. *Nature* **363**:71–74.

Kurahashi, T. and K.-Y. Yau (1994). Olfactory transduction: tale of an unusual chloride current. *Curr Biol* **4**:256–258.

Leinders-Zufall, T., M. N. Rand, G. M. Shepherd, C. A. Greer, and F. Zufall (1997). Calcium entry through cyclic nucleotide-gated channels in individual cilia of olfactory receptor cells: spatiotemporal dynamics. *J Neurosci* **17**:4136–4148.

Leinders-Zufall, T., G. M. Shepherd, and F. Zufall (1996). Modulation by cyclic GMP of the odour sensitivity of vertebrate olfactory receptor cells. *Proc R Soc Lond (Biol)* **263**:803–811.

Liman, E. R. and D. P. Corey (1996). Electrophysiological characterization of chemosensory neurons from the mouse vomeronasal organ. *J Neurosci* **16**:4625–4637.

Liman, E. R., D. P. Corey, and C. Dulac (1999). TRP2: a candidate transduction channel for mammalian pheromone sensory signaling. *Proc Natl Acad Sci U S A* **96**:5791–5796.

Liu, M., T.-Y. Chen, B. Ahamed, J. Li, and K.-Y. Yau (1994). Calcium-calmodulin modulation of the olfactory cyclic nucleotide-gated cation channel. *Science* **266**:1348–1354.

Lowe, G. and G. H. Gold (1993a). Contribution of the ciliary cyclic nucleotide-gated conductance to olfactory transduction in the salamander. *J Physiol (Lond)* **462**:175–196.

Lowe, G. and G. H. Gold (1993b). Nonlinear amplification by calcium-dependent chloride channels in olfactory receptor cells. *Nature* **366**:283–286.

Lowe, G., T. Nakamura, and G. H. Gold (1989). Adenylate cyclase mediates olfactory transduction for a wide variety of odorants. *Proc Natl Acad Sci U S A* **86**:5641–5645.

Lucero, M. T. and N. Chen (1997). Characterization of voltage and Ca^{2+}-activated K^+ channels in squid olfactory receptor cells. *J Exp Biol* **200**:1571–1586.

Maue, R. A. and V. E. Dionne (1987). Patch-clamp studies of isolated mouse olfactory receptor neurons. *J Gen Physiol* **90**:95–125.

Menco, B. P. M. and A. I. Farbman (1992). Ultrastructural evidence for multiple mucous domains in frog olfactory epithelium. *Cell Tissue Res* **270**:47–56.

Michel, W. C. and B. W. Ache (1992). Cyclic nucleotides mediate an odor-evoked potassium conductance in lobster olfactory receptor cells. *J Neurosci* **12**:3979–3984.

Michel, W. C., T. S. McClintock, and B. W. Ache (1991). Inhibition of lobster olfactory receptor cells by an odor-activated potassium conductance. *J Neurophysiol* **65**:446–453.

Miyamoto, T., D. Restrepo, E. J. Cragoe, Jr., and J. H. Teeter (1992a). IP3- and cAMP-induced responses in isolated olfactory receptor neurons from the channel catfish. *J Membr Biol* **127**:173–183.

Miyamoto, T., D. Restrepo, and J. H. Teeter (1992b). Voltage-dependent and odorant-regulated currents in isolated olfactory receptor neurons of the channel catfish. *J Gen Physiol* **99**:505–529.

Morales, B., R. Madrid, and J. Bacigalupo (1996). Calcium influx followed by an elevation of intracellular Ca^{2+} mediate the activation of the inhibitory current induced by odorants in vertebrate olfactory receptor neurons. *FEBS Lett* **402**:259–264.

Munger, S. D., R. A. Gleeson, H. C. Aldrich, N. C. Rust, B. W. Ache, and R. M. Greenberg. Molecular evidence for phosphoinositide-mediated signaling in lobster olfactory receptor neurons *J. Biol Chem* (in press).

Nakamura, T. and G. H. Gold (1987). A cyclic nucleotide-gated conductance in olfactory receptor cilia. *Nature* **325**:442–444.

Neer, E. J. (1995). Heterotrimeric G proteins: organizers of transmembrane signals. *Cell* **80**:249–257.

Okada, Y., J. H. Teeter, and D. Restrepo (1994). Inositol 1,4,5-trisphosphate-gated conductance in isolated rat olfactory neurons. *J Neurophysiol* **71**:595–602.

Pelosi, P. (1994). Odorant-binding proteins. *Crit Rev Biochem Mol Biol* **29**:199–228.

Rawson, N. E., G. Gomez, B. Cowart, J. G. Brand, L. D. Lowry, E. A. Pribitkin, and D. Restrepo (1997). Selectivity and response characteristics of human olfactory neurons. *J Neurophysiol* **77**:1606–1613.

Restrepo, D., I. Boekhoff, and H. Breer (1993). Rapid kinetic measurements of second messenger formation in olfactory cilia from channel catfish. *Am J Physiol Cell Physiol* **264**:C906–C911.

Restrepo, D., T. Miyamoto, B. P. Bryant, and J. H. Teeter (1990). Odor stimuli trigger influx of calcium into olfactory neurons of the channel catfish. *Science* **249**:1166–1168.

Restrepo, D., J. H. Teeter, E. Honda, A. G. Boyle, J. F. Marecek, G. D. Prestwich, and D. L. Kalinoski (1992). Evidence for an $InsP_3$-gated channel protein in isolated rat olfactory cilia. *Am J Physiol* **263**:C667–C673.

Restrepo, D., Y. Okada, and J. H. Teeter (1993). Odorant-regulated Ca^{2+} gradients in rat olfactory neurons. *J Gen Physiol* **102**:907–924.

Reuter, D., K. Zierold, W. Schroeder, and S. Frings (1998). A depolarizing chloride current contributes to chemoelectrical transduction in olfactory sensory neurons *in situ*. *J Neurosci* **18**:6623–6630.

Sasaki, K., K. Okamoto, K. Inamura, Y. Tokumitsu, and M. Kashiwayanagi (1999). Inositol-1,4,5-trisphosphate accumulation induced by urinary pheromones in female rat vomeronasal epithelium. *Brain Res* **823**:161–168.

Sato, T., J. Hirono, M. Tonoike, and M. Takebayashi (1991). Two types of increases in free Ca^{2+} evoked by odor in isolated frog olfactory receptor neurons. *Neuroreport* **2**: 229–232.

Schandar, M., K. L. Laugwitz, I. Boekhoff, C. Kroner, T. Gudermann, G. Schultz, and H. Breer (1998). Odorants selectively activate distinct G protein subtypes in olfactory cilia. *J Biol Chem* **273**:16669–16677.

Schild, D. (1989). Whole-cell currents in olfactory receptor cells of *Xenopus laevis*. *Exp Brain Res* **78**:223–232.

Schild, D., F. W. Lischka, and D. Restrepo (1995). $InsP_3$ causes an increase in apical $[Ca^{2+}]_i$ by activating two distinct components in vertebrate olfactory receptor cells. *J Neurophysiol* **73**:862–866.

Schild, D. and D. Restrepo (1998). Transduction mechanisms in vertebrate olfactory receptor cells. *Physiol Rev* **78**:429–466.

Schleicher, S., I. Boekhoff, J. Arriza, R. J. Lefkowitz, and H. Breer (1993). A-adrenergic receptor kinase-like enzyme is involved in olfactory signal termination. *Proc Natl Acad Sci U S A* **90**:1420–1424.

Stengl, M. (1994). Inositol-trisphosphate-dependent calcium currents precede cation currents in insect olfactory receptor neurons in vitro. *J Comp Physiol A* **174**:187–194.

Taniguchi, M., M. Kashiwayanagi, and K. Kurihara (1995). Intracellular injection of inositol 1,4,5-trisphosphate increases a conductance in membranes of turtle vomeronasal receptor neurons in the slice preparation. *Neurosci Lett* **188**:5–8.

Trotier, D. and P. MacLeod (1983). Intracellular recordings from salamander olfactory receptor neurons. *Brain Res* **268**:225–237.

Trotier, D., K. B. Doving, and J.-F. Rosin (1993). Voltage-dependent currents in microvillar receptor cells of the frog vomeronasal organ. *Eur J Neurosci* **5**:995–1002.

Trotier, D., K. B. Doving, and J.-F. Rosin (1994). Functional properties of frog vomeronasal receptor cells. In: K. Kurihara, N. Suzuki, and H. Ogawa (eds). *Olfaction and Taste XI*. Tokyo: Springer-Verlag, pp 188–191.

Wang, D., P. Chen, W. Liu, C. S. Li, and M. Halpern (1997). Chemosignal transduction in the vomeronasal organ of garter snakes: $Ca(2+)$-dependent regulation of adenylate cyclase. *Arch Biochem Biophys* **348**:96–106.

Zhainazarov, A. B. and B. W. Ache (1995a). Na^+-activated nonselective cation channels in primary olfactory neurons. *J Neurophysiol* **73**:1774–1781.

Zhainazarov, A. B. and B. W. Ache (1995b). Odor-induced currents in *Xenopus* olfactory receptor cells measured with perforated-patch recording. *J Neurophysiol* **74**:479–483.

Zhainazarov, A. B. and B. W. Ache (1997). Gating and conduction properties of a sodium-activated cation channel from lobster olfactory receptor neurons. *J Memb Biol* **156**:173–190.

Zhainazarov, A. B., R. E. Doolin, and B. W. Ache (1998). Sodium-gated cation channel implicated in the activation of lobster olfactory receptor neurons. *J Neurophysiol* **79**:1349–1359.

Zufall, F. and H. Hatt (1991). Dual activation of a sex pheromone-dependent ion channel from insect olfactory dendrites by protein kinase C activators and cyclic GMP. *Proc Natl Acad Sci U S A* **88**:8520–8524.

Zufall, F. and T. Leinders-Zufall (1997). Identification of a long-lasting form of odor adaptation that depends on the carbon monoxide cGMP second-messenger system. *J Neurosci* **17**:2703–2712.

8

Molecular Biology of Olfaction

TIMOTHY S. McCLINTOCK
Department of Physiology, University of Kentucky, Lexington, KY

1. INTRODUCTION

Molecular biology is a thoroughly reductionist approach that seeks to understand the whole by completely understanding the component parts. The increasingly rapid growth in the cloning and sequencing of genes and complementary DNAs (cDNAs) continues to expand opportunities for understanding the function of individual proteins. In addition, the advent of certain recombinant DNA techniques (e.g., heterologous expression, targeted deletion of genes) has made it possible to study the role of single gene products in complex functions in a cell, tissue, or animal in which one or more genes or proteins have been modified. Molecular cloning followed by functional studies involving recombinant DNA manipulations has become a powerful strategy for the study of the olfactory system.

This chapter discusses the cloning of olfactory-related genes, advances made by applying recombinant DNA techniques to the study of olfaction, and the regulation of gene transcription as it applies to the olfactory system. A focus upon processes that are integral and specific to the olfactory system results in an emphasis upon the molecular biology of the peripheral olfactory system. Data from both vertebrate and invertebrate model systems are included because molecular mechanisms are often conserved across phylogenetic groups, or are adapted in informative ways.

The Neurobiology of Taste and Smell, Second Edition, Edited by Thomas E. Finger, Wayne L. Silver, and Diego Restrepo.
ISBN 0-471-25721-4   Copyright © 2000 Wiley-Liss, Inc.

2. MOLECULAR CLONING AND FUNCTIONAL ANALYSIS OF OLFACTORY-RELATED GENES

A cDNA clone is a powerful tool. It can be used to investigate both molecular functional domains within a protein and integrated processes such as biochemical pathways, the behavior of assemblages of cells, and even the behavior of the whole organism. Proving the function of the protein encoded by a cDNA clone requires tests of sufficiency and necessity. Tests of sufficiency typically involve three steps: (1) identifying sites of mRNA expression, most often by in situ hybridization, (2) identifying sites of protein expression using specific antibodies or antisera, and (3) direct testing of the activity of the encoded protein, usually by heterologous expression (i.e., expression in a cell that lacks the protein). Tests of necessity typically involve the production of genetically altered animals, dominant-negative proteins, specific antibodies, or specific pharmacological agents for use in functional experiments. Together, these approaches are used to test the involvement of a gene or protein in olfaction. By extension, they also test the involvement in olfaction of pathways or processes that use the gene or protein.

Table 8.1 is a partial list of genes whose role in the olfactory system is under investigation. The majority of clones on the list are involved in odor detection

TABLE 8.1. A partial list of known olfactory-related genes

Gene	Proposed Role	Sufficiency Location	Sufficiency Function	Necessity	Ref.
		Vertebrates			
Olfactory receptors	Odorant detection	+	+	?	1
$G_{\alpha olf}$	Transduction	+	+	+	2
Adenylyl cyclase III	Transduction	+	+	?	3
Cyclic nucleotide gated channel subunits	Transduction	+	+	+	4
$G_{\alpha q}$	Transduction	+	+	?	5
Inositol 1,4,5-trisphosphate receptor	Transduction	+	+	?	6
$G_{\beta 1}$ and $G_{\beta 2}$	Transduction	+	?	?	7
$G_{\gamma 2}$ and $G_{\gamma 3}$	Transduction	+	?	?	8
Regulator of G-protein signaling (RGS3)	G-protein regulation	+	?	?	9
G-protein coupled receptor kinase (GRK3)	Desensitization	+	+	+	10
Olfactory cell adhesion molecule (OCAM)	Axon fasciculation	+	+	?	11
Olfactory marker protein (OMP)	Growth factor	+	?	?	12
Odorant binding protein (OBP)	Odorant binding	+	+	?	13
Guanylyl cyclase D	Odorant detection	+	+	?	14
Olf/EBF1-3 transcription factors	Gene transcription	+	+	?	15
Roaz	Inhibitor of Olf/EBF	+	+	?	16

(*Continued*)

TABLE 8.1. *(Continued)*

Gene	Proposed Role	Sufficiency Location	Sufficiency Function	Necessity	Ref.
mASH-1 transcription factor	OE development	+	?	+	17
Pax-6 transcription factor	OE development	+	?	?	18
Cytochrome P-450 (CytP450olf1)	Odorant clearance	+	+	?	19
UDP glucuronosyl transferase (UGTolf)	Odorant clearance	+	+	?	20
Olfactomedin	Extracellular matrix	+	+	?	21
	Arthropods				
NorpA (a phospholipase C-β)	Transduction	+	+	+	22
RdgB (a calcium transporter)	Response recovery	+	?	+	23
OS9	Transcription factor	+	?	?	24
Odorant binding proteins (OBP)	Odorant solubility	+	?	?	25
Pheromone binding protein (PBP)	Pheromone solubility	+	?	?	26
snmp-1	Unknown	+	?	?	27
$G_{\alpha q}$	Transduction	+	?	?	28
$G_{\alpha s}$	Transduction	+	?	?	29
Phospholipase C-β	Transduction	+	?	?	30
Acj6, POU domain transcription factor	Odorant specificity	+	?	?	31
	Caenorhabditis				
odr-10, an odorant receptor	Odorant detection	+	+	+	32
tax-2, a CNG channel	Transduction	+	?	+	33
tax-4, a CNG channel	Transduction	+	?	+	34
odr-3, a G-protein	Transduction	+	+/?	+	35
gpa-2, a G-protein	Transduction	+	?	?	36
gpa-3, a G-protein	Transduction	+	?	?	37
osm-9, an ion channel	Transduction	+	?	+	38
daf-11, a guanylyl cyclase	Transduction	+	?	+	39
odr-4	Receptor trafficking	+	?	+	40

Differing nomenclatures for cyclic nucleotide gated channels exist in the literature (Biel et al., 1996). CNG2 was formerly oCNC1, an α subunit. Other cyclic nucleotide gated channel subunits expressed in olfactory receptor neurons are CNG5, the original β subunit, and CNG4.3, an alternatively spliced version of another β subunit gene.

Tests of sufficiency for olfactory function include the location of mRNA or protein in the appropriate cell and demonstration of function by expression of the encoded protein. Tests of necessity include interruption of in vivo olfactory function by eliminating expression of the protein (e.g., targeted deletion of the gene) or by specific block of its activity. +, positive test; ?, not yet demonstrated.

A single recent reference for each gene follows: 1, Buck, 1996; 2, Belluscio et al., 1998; 3, Bakalyar and Reed, 1990; 4, Sautter et al., 1998; 5, Abogadie et al., 1995; 6, Cunningham et al., 1993; 7-8, Bruch et al., 1997; 9, Bruch and Medler, 1996; 10, Peppel et al., 1997; 11, Yoshihara et al., 1997; 12, Carr et al., 1998; 13, 25-26, Pelosi, 1996; 14, Fülle et al., 1995; 15, Wang et al., 1997; 16, Tsai and Reed, 1997; 17, Calof et al., 1996; 18, Davis and Reed, 1996; 19, Nef et al., 1989; 20, Lazard et al., 1991; 21, Yokoe and Anholt, 1993; 22, Riesgo-Escovar et al., 1995; 23, Woodard et al., 1992; 24, Raha and Carlson, 1994; 27, Rogers et al., 1997; 28-30, Xu and McClintock, 1999; 31, Clyne et al., 1999a; 32, Zhang et al., 1997; 33-34, Coburn and Bargmann, 1996; 35, 39, Roayaie et al., 1998; 36-37, Zwaal et al., 1997; 38, Colbert et al., 1997; 40, Dwyer et al., 1998.

and transduction. Genes involved in transcriptional regulation, development, differentiation, structure, and intercellular communication within the olfactory system are as yet under-represented. The isolation of the clones listed was accomplished by various strategies: screening for homologs of known sequences, protein purification, differential subtraction, screening for functional properties, and molecular genetics. A significant impetus for the cloning of olfactory-related genes was the discovery that odorants stimulate the production of cAMP in olfactory cilia (Pace et al., 1985). This led to the use of homology cloning strategies to isolate cDNAs for the putative components of this biochemical pathway: G_{olf}, adenylyl cyclase type III (ACIII), and a cyclic nucleotide gated channel subunit (CNG2). (Jones and Reed, 1989; Bakalyar and Reed, 1990; Dhallan et al., 1990). A trend in subsequent studies has been to identify genes or proteins associated with this first generation of olfactory-related genes. These include G-protein coupled receptor kinase 3 (GRK3), phosducin, CNG4.3, CNG5, and β and γ subunits of G-proteins (Schleicher et al., 1993; Bradley et al., 1994; Liman and Buck, 1994; Bruch et al., 1997; Sautter et al., 1998).

An important criterion for many pioneering efforts to clone olfactory-related genes was specific expression in the olfactory epithelium, because this virtually ensured an important role in olfaction (Margolis, 1988). This approach contributed to the isolation of cDNA clones for olfactory marker protein (OMP), odorant binding proteins (OBPs), G_{olf}, and ACIII, for example (Lee et al., 1987; Rogers et al., 1987; Jones and Reed, 1989; Bakalyar and Reed, 1990). However, we now know that many such genes are expressed in a limited number of cell types in other tissues such as brain or testes. These relatively olfactory-specific proteins highlight critical functions of the olfactory system, but the evolution of olfactory-specific genes is only one mechanism controlling the specificity of function in the olfactory system. Also important are the expression patterns of more commonly expressed genes, the subcellular localization of proteins, and the access of potentially interacting proteins to each other.

An immediate reward from the isolation of a cDNA is its sequence. Analysis of sequences can provide predictions of structure and function. OBP homology to serum lipocalins predicted their ability to bind odorants (Pelosi, 1996). The presence of pleckstrin homology domains in both phosducin and GRK3 predicted that these proteins would compete for binding to $G_{\beta\gamma}$ in olfactory cilia homogenates (Boekhoff et al., 1997). Hydropathy profiles of mammalian olfactory receptors (Buck and Axel, 1991) predicted the identity of the cytoplasmic G-protein coupling domains (McClintock et al., 1997). The sequences of olfactory receptors are especially diverse within transmembrane domains 3–5, a region that participates in ligand binding in some other G-protein coupled receptors, and this region may comprise part of the odorant binding pocket (Fig. 8.1B) (Buck, 1996; Krautwurst et al., 1998). As nucleic acid and protein sequence databases grow and as more functional domains of protein sequences are identified, sequence analysis of olfactory-related genes will be increasingly able to predict function.

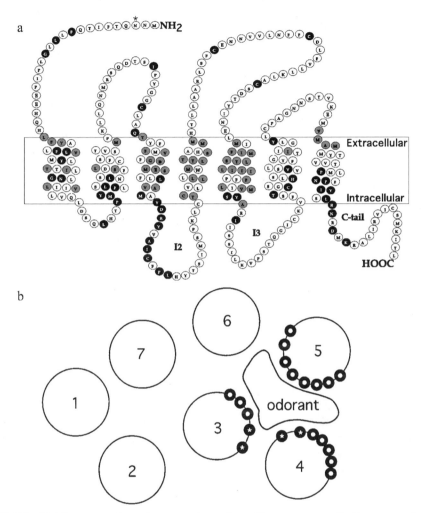

FIG. 8.1. (a) The proposed membrane topology of rat olfactory receptor I3. Positions within the transmembrane domains that are highly variable among the original rat olfactory receptor clones are shown in gray in order to highlight the cluster of highly variable positions in transmembrane domains 3–5 (Buck and Axel, 1991). Residues conserved in more than 90% of 109 rodent olfactory receptor sequences are shown in black (A. A. Gimelbrant, unpublished results). The cytoplasmic loops, I2, I3, and C-tail, that function as G-protein coupling domains are labeled (McClintock et al., 1997). *, the single consensus N-linked glycosylation site that is present in many olfactory receptors. See cover for another molecular model.

(b) A diagram of a proposed organization of the transmembrane domains (1–7) of olfactory receptors, viewed from the extracellular side of the plasma membrane. The putative location of highly variable positions that may constitute the odorant binding pocket are marked with circles. White circles, positions predominantly occupied by hydrophobic amino acids; circles enclosing asterisks, positions occupied predominantly by hydrophilic amino acids. (From Sharon et al. [1998], with permission by New York Academy of Sciences.)

2.1. Olfactory Receptors

Mammalian olfactory receptor cDNAs were first identified in rats by predicted homology to conserved regions of other G-protein coupled receptors (Buck and Axel, 1991). Subsequent studies have identified putative olfactory receptors in lamprey, catfish, zebrafish, mudpuppy, frog, chicken, opossum, mouse, pig, dog, horse, and humans (Mombaerts, 1999). Genomic southern blots estimate the number of mammalian olfactory receptor genes at 500–1000 (Buck, 1996). In general, these genes lack introns and occur in clusters that are spread across most, perhaps all, chromosomes. The olfactory receptor proteins share a predicted topology, and at 310–340 amino acids, are among the shortest G-protein coupled receptors (Fig. 8.1A). They possess relatively few highly conserved amino acids. Two other families of G-protein coupled chemoreceptors, the V1R and V2R families, have been identified in the vomeronasal organ (Dulac and Axel, 1995; Herrada and Dulac; 1997; Matsunami and Buck, 1997; Ryba and Tirindelli, 1997). Receptors similar to the V2R family are also expressed in the fish olfactory epithelium, probably in the microvillous type of olfactory receptor neuron (Cao et al., 1998; Speca et al., 1999). Together, these represent three nonhomologous families of chemosensory receptors in vertebrates (Fig. 8.2). In addition to G-protein coupled receptors, olfactory receptors may also include guanylyl cyclase D. This receptor guanylyl cyclase is expressed in the cilia of olfactory receptor neurons that project axons to a specific subset of glomeruli in the olfactory bulb (Fülle et al., 1995; Juilfs et al., 1997). Even relatively primitive animals also appear to have multiple types of chemosensory receptor proteins. Mining genomic sequence databases identified six distinct families of G-protein coupled receptors that are putative chemosensory receptors in *Caenorhabditis elegans* (Troemel et al., 1995) and a highly diverse multigene family of putative olfactory receptors in *Drosophila* (Clyne et al., 1999b; Vosshall et al., 1999). An interesting parallel with mammals is that the cilia of many of the chemosensory neurons in *C. elegans* also contain receptor guanylyl cyclases that are putative receptor proteins (Yu et al., 1997).

In the rodent olfactory epithelium, expression of olfactory receptor mRNA is restricted to the olfactory receptor neurons, and receptor protein is present in the olfactory cilia (Buck and Axel, 1991; Menco, 1997). Each olfactory receptor is expressed within a defined portion, or zone, of the olfactory epithelium (Buck, 1996). In mammals, there are four major zones, (Mori et al., 1999) (Fig. 8.3). Expression of a receptor within a zone is random. However, there may also be other zones that overlap the original four, such as the small areas of endoturbinate II and ectoturbinate 3 that contain expression of the OR37 receptor subfamily (Kubick et al., 1997). The frequency of neurons expressing a given receptor is usually a few tenths of a percent. This suggests that each ORN expresses at most a few, and more likely just one, olfactory receptor gene. This hypothesis is supported by the observation that axon terminals in the glomeruli of the olfactory bulb contain olfactory receptor mRNA and that labeling for any one receptor mRNA occurs only in a pair of relatively invariant glomeruli (Ressler

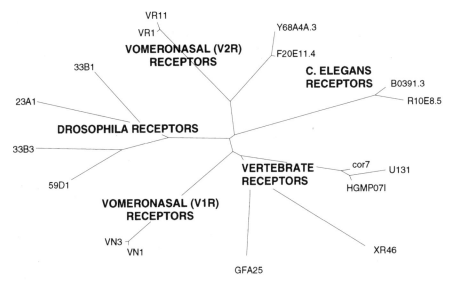

FIG. 8.2. A phylogram depicting the sequence diversity encompassed by the known families of olfactory and vomeronasal receptors. Predicted protein sequences were obtained from the Olfactory Receptor Database (Skoufos et al., 1999), aligned using CLUSTAL W, and used to produce the phylogenetic tree through direct input to PHYLIP via the ClustalW World Wide Web Service (European Bioinformatics Institute) (Higgins et al., 1994). The distances along the branches of the tree approximate the degree of relatedness in the predicted amino acid sequences. The species and Genbank accession numbers for the receptors shown are: rat U131, AAB66333; chicken cor7, CAB01848; human HGMP07I, P30953; frog XR46, CAA69631; goldfish GFA25, AAC64073; *Drosophila* 33B1 and 33B3, AC006240; 23A1 AC005558; 59D1, AC005672; *C. elegans* R10E8.5, CAB04642; B0391.3, CAB03803; F20E11.4, CAB04143; Y68A4A.3, CAA16421; rat VN1, U36785; VN3, U36895; mouse VR1, AF011411; VR11, AF011421. (Prepared by E. Skoufos, Yale University School of Medicine.)

et al., 1994; Vassar et al., 1994). If all receptors participate in the targeting of receptor neuron axons to glomeruli, then these data are difficult to reconcile with any conclusion except that one receptor gene is expressed per receptor neuron. The one neuron–one receptor hypothesis is compelling in the elegantly simple organization of information coding it implies. Some evidence, especially electrophysiological and pharmacological studies from invertebrates and poikilothermic vertebrates, is not consistent with the one neuron–one receptor hypothesis, however (see Chapter 7, Ache and Restrepo).

Demonstrating that olfactory receptors are activated by odorants has proven surprisingly difficult. While at least one fish amino acid odorant receptor of the V2R type expresses well in heterologous systems (Speca et al., 1999), mammalian olfactory receptors become trapped in the endoplasmic reticulum when expressed in most heterologous cells (Gimelbrant et al., 1999). The aspects of the cell biology of the olfactory receptors that prevent their heterologous expression are not yet understood, but in C. *elegans* trafficking of a subset of

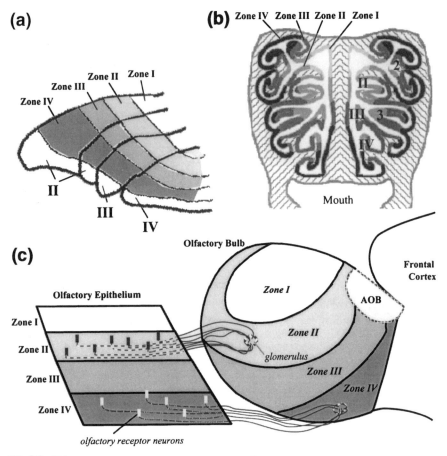

FIG. 8.3. Odorant receptor expression zones in a rodent.

(a) Semischematic medial view of the turbinates of a rat showing the location of the expression zones. Anterior is to the right, similar to the orientation of the schematic drawing in panel C.

(b) Charting of a transverse section through the nasal cavity of a rat rostral to the olfactory bulb. Hatched areas are bone; receptor zones are indicated by four levels of gray similar to those in other panels of this figure. The endoturbinates and ectoturbinates are indicated by Roman and Arabic numerals, respectively.

(c) Semischematic diagram of the projection of the different zones of the olfactory epithelium onto different areas of the olfactory bulb. For Zones II and IV, several olfactory receptor neurons and their axons are shown converging onto single glomeruli within their respective zones. Compare this image to that in Fig. 8.4c. Panels A and B adapted and redrawn from Scott and Brierly, 1999. Panel C adapted from Mori 1999.

olfactory receptors from the receptor neuron cell body to the cilia depends upon a novel chemosensory-specific gene called *odr-4* (Dwyer et al., 1998). Imaginative means of circumventing the expression problem are beginning to appear. Mouse olfactory receptor neurons that were infected in vivo with an adenovirus construct containing olfactory receptor I7 respond to straight chain saturated aldehydes with lengths of 6 to 10 carbons (Zhao et al., 1998). Extending the N-termini of olfactory receptors increased their expression in a cultured cell line and allowed the identification of single odorant agonists for three receptors (Krautwurst et al., 1998). Similarly, the C. *elegans* diacetyl receptor (Odr-10) expressed in HEK293 cells is responsive to a narrow range of odorants, including only diacetyl, pyruvate, and citrate (Wellerdieck et al., 1997; Zhang et al., 1997). While some olfactory receptors appear to be activated by a limited number of closely related chemical structures, consistent with the large number of olfactory receptor genes in mammals, the paucity of information does not yet allow conclusions to be drawn about the breadth of molecular receptive ranges of olfactory receptors in general.

Another function of olfactory receptors is targeting of receptor neuron axons to glomeruli in the olfactory bulb. The observation that mRNA of individual receptors was concentrated in pairs of glomeruli raised the possibility that olfactory receptors participated in the targeting of receptor neuron axons to glomeruli. This was confirmed using transgenic mice in which a receptor gene promoter (receptor P2) drove expression of bicistronic constructs consisting of a receptor, an internal ribosome entry site, and the axonal protein tau fused to β-galactosidase (Mombaerts et al., 1996). When the P2 promoter drove expression of the P2 open reading frame, the mice had β-galactosidase staining in olfactory receptor neurons within one zone in the epithelium and their axons converged on two glomeruli. When the P2 promoter drove expression of receptor M12, however, the axons that contained the β-galactosidase activity now projected to a glomerulus different from that normally innervated by axons containing receptor P2 mRNA or receptor M12 mRNA (Fig. 8.3). In mice in which a receptor gene has been replaced by the fusion of tau and β-galactosidase, the axons containing β-galactosidase do not converge upon a glomerulus (Wang et al., 1998). These results indicate that olfactory receptors are one of several factors that determine the connectivity of axons with glomeruli.

2.2. Olfactory Transduction

In mammals, the expression of $G_{\alpha olf}$, ACIII, and CNG2 in olfactory receptor neurons and their cilia (Menco, 1997) provided evidence of involvement in olfactory transduction. From evidence obtained from functional expression in heterologous systems and gene deletion in mice, it is now clear that $G_{\alpha olf}$ and CNG2 are components of the primary olfactory transduction pathway. Mice lacking functional $G_{\alpha olf}$ or CNG2 genes show greatly reduced responses to odorants in multiunit recordings from the olfactory epithelium (electro-olfactograms) and exhibit reduced suckling, a behavior that is dependent upon olfaction (Brunet et al., 1996; Belluscio et al., 1998). These results provide direct evidence that $G_{\alpha olf}$ and CNG2 are necessary for olfactory transduction and that the subunit composition of olfactory cyclic nucleotide gated channels, which remains to be determined, must include CNG2. It is presumed that ACIII provides the link between $G_{\alpha olf}$ and the cyclic nucleotide-gated channel. The hyposmia of mice lacking $G_{\alpha olf}$ or CNG2 indicates that mammalian olfactory transduction depends solely on the cAMP pathway and appears to contradict previous evidence implicating the IP_3 pathway as a second olfactory transduction pathway (see Chapter 7, Ache and Restrepo). Although the gene knockout effects are clear and compelling, the traditional method used to disrupt the $G_{\alpha olf}$ and CNG2 genes can have effects unrelated to the function of the eliminated gene. These effects are due primarily to two causes (Gerlai, 1996). First, the absence of a protein during development can alter the expression and role of

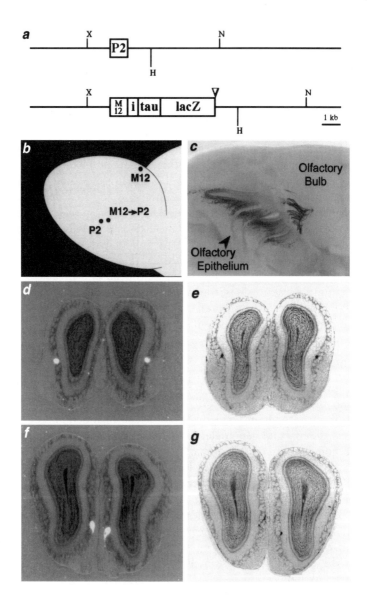

other proteins. Second, the embryonic stem cell and host mice are typically derived from different strains, so genes linked to the recombinant locus by chromosome position can differ in type or sequence and have effects that artifactually appear to result from gene knockout. At this point, it remains possible that olfactory transduction via the IP_3 pathway could have been masked in the $G_{\alpha olf}$- and CNG2-deficient mice.

In invertebrates, the majority of biochemical and physiological evidence indicates that the IP_3 pathway is the predominant mechanism of olfactory transduction (see Chapter 7, Ache and Restrepo). *Drosophila* with mutations in the *norpA* gene, a phospholipase C-β, have olfactory deficits (Riesgo-Escovar et al., 1995). In addition, mutations in the *rdgB* gene, which appears to be a calcium transporter, alters recovery from odor responses (Woodard et al., 1992). Lobsters use both the IP_3 and cAMP pathways for olfactory transduction, consistent with the cloning of $G_{\alpha s}$, $G_{\alpha q}$, G_β and phospholipase C-β cDNAs that are expressed in the olfactory receptor neurons (Xu et al., 1997; Xu and McClintock, 1999). In *C. elegans*, multiple transduction mechanisms also exist and the components of the chemosensory transduction pathways differ between the various types of chemosensory neurons (Sengupta et al., 1996; Colbert et al., 1997; Zhang et al., 1997; Roayaie et al., 1998). In AWA neurons, cloned components include the diacetyl receptor (*odr-10*), a G-protein (*odr-3*) and an ion channel related to the capsaicin receptor family (*osm-9*). In AWC neurons, the cloned components include *odr-3*, another G-protein (*gpa-2*), a guanylyl cyclase (*daf-11*) and two cyclic nucleotide-gated channel subunits (*tax-2, tax-4*). In ASH neurons, the cloned components are *odr-3* and *osm-9*. Studies of the molecular biology of invertebrate olfactory transduction therefore suggest that different taxa employ variations on a common theme involving G-protein mediated pathways to encode odor signals.

◄——

FIG. 8.4. Redirection of olfactory receptor neuron axons in the M12-P2-IRES-tau-lacZ mouse.

(a) The mouse P2 locus (*top*) was altered by homologous recombination to eliminate the P2 receptor open reading frame and place a bicistronic construct under control of the P2 promoter region. The construct contained the M12 receptor, an internal ribosome entry site (IRES), and a fusion of the tau protein with lacZ. Tau is expressed predominantly in axons, so its fusion to lac-Z targets β- galactosidase activity to axons. X, XhoI site; H, HindIII site; N, NcoI site.

(b) Schematic diagram of the right olfactory bulb showing positions of glomeruli where axons expressing P2, M12, or the M12-P2 construct converge.

(c) A whole mount view of a five week old M12-P2-IRES-tau-lacZ homozygous mouse. The convergence of axons containing the β- galactosidase staining upon a single site on the medial aspect of the bulb is apparent.

(d–g) Sections of the olfactory bulbs of a M12-P2-IRES-tau-lacZ heterozygous mouse. In situ hybridization (d, f) shows that the P2 receptor axons converge on two glomeruli, one medial and one lateral. The β-galactosidase staining (f, g) of the axons from receptor neurons expressing the M12-P2-IRES-tau-lacZ construct also converge on a pair of glomeruli. The two pairs of glomeruli are approximately 10 sections apart, placing the M12-P2 glomeruli 200μm posterior to the P2 glomeruli. (From Mombaerts et al. [1996], with permission by Cell Press.)

2.3. Desensitization and Regulation of Olfactory Transduction

Adaptation is a common principle of both sensory systems and intercellular signaling pathways. In systems using G-protein coupled receptor pathways for signaling, part of the adaptation mechanism often involves deactivation of receptors by phosphorylation by specialized kinases, the G-protein coupled receptor kinases (GRKs). In transgenic mice with a null mutation of the GRK3 gene, olfactory transduction appears to be downregulated, a compensation effect that is consistent with a decreased ability to desensitize odorant responses (Peppel et al., 1997). The expression of proteins involved in regulating transduction, such as the phosphodiesterases that hydrolyze cyclic nucleotides, appears to be as finely controlled as the expression of the transduction proteins themselves. The expression of a cGMP-dependent PDE isoform (PDE2) correlates specifically with expression of guanylyl cyclase D while expression of calcium/calmodulin-dependent (PDE1C2) and cAMP-dependent isoforms (PDE4A) correlates specifically with expression of ACIII (Juilfs et al., 1997). Other mechanisms regulating olfactory transduction have yet to be analyzed in detail using recombinant DNA methods.

Desensitization and adaptation in nonmammalian species appear to employ similar mechanisms but molecular analyses have been limited thus far. The cloning of RGS3, a protein that regulates the GTP hydrolyzing activity of certain G-proteins, and evidence of its expression in catfish olfactory receptor neurons, suggests another possible regulatory mechanism for olfactory transduction (Bruch and Medler, 1996). The *osm-9* gene of *C. elegans* that appears to be involved in chemosensory transduction in AWA neurons is limited to a role in adaptation in AWC neurons (Colbert et al., 1997).

2.4. Proteins That Bind and Metabolize Odorants

Major components of the solutions bathing the cilia of olfactory receptor neurons of terrestrial animals are secreted proteins that bind odorants (OBPs) and pheromones (PBPs). Each species appears to express only a few OBPs and even within a species, the OBP sequences are very diverse. Vertebrate OBPs are produced by nasal glands or tubular-acinar glands and secreted into the mucous near the olfactory epithelium. Insect OBPs and PBPs, which are not similar in sequence to mammalian OBPs, are produced by supporting cells of several chemosensory sensilla and secreted into the sensillar lymph (Vogt et al., 1991). Mammalian OBPs bind many, but not all, odorants (Pelosi, 1996). The functional consequences of odorant binding by these proteins remain in doubt, but hypotheses have included concentrating odorants, odorant solubilization, odorant clearance, and presenting odorants to receptors (see also Chapter 7, Ache and Restrepo).

Odorants must be cleared from the vicinity of the olfactory receptors in order for the system to reset itself for subsequent stimulations. Cytochrome P450s and phase II biotransformation enzymes may play a role in this process. Indeed, cDNA clones of a cytochrome P450 (CytP-450olf1), a UDP glucuronosyl transferase (UGTolf), and a phenol sulphotransferase that are unique isoforms expressed abundantly or specifically in the olfactory epithelium have been isolated (Nef et al., 1989; Lazard et al., 1991; Tamura et al., 1998). These biotransformation enzymes are active upon some odorants and presumably contribute to the clearance of odorants.

2.5. Organizers of the Peripheral Olfactory System

The olfactory receptor expression zones of mammalian olfactory epithelia, the regional topographic projection of the epithelium to the bulb, the direct targeting of receptor neuron axons to glomeruli (Dynes and Ngai, 1998) and the convergence of all axons containing the same olfactory receptor mRNA argue for organizing signals in the olfactory system. Some of these signals are likely to be factors that regulate transcription of the olfactory receptors. Others are likely to be signposts that help sort axons to the appropriate glomeruli. As discussed in Section 2.1, the olfactory receptors themselves are one organizing factor. Another is galectin-1, because mice with targeted deletions of mutations of this lectin gene show aberrant targeting of a subset of receptor neuron axons (Puche et al., 1996). Others are likely to be the cell adhesion molecules like OCAM and N-CAM-180 (Treloar et al., 1997; Yoshihara et al., 1997; Kafitz and Greer, 1998). OCAM is expressed only in olfactory receptor expression zones II–IV and the corresponding glomeruli of the ventro-caudal bulb (Yoshihara et al., 1997). OCAM may act to promote homophilic aggregation of axons in the olfactory nerve because heterologous expression of OCAM promotes homophilic aggregation of cells.

3. REGULATION OF GENE TRANSCRIPTION IN THE OLFACTORY SYSTEM

The restricted expression patterns of certain olfactory-related genes, unique aspects of the development of olfactory organs, and the turnover of olfactory receptor neurons all argue that certain aspects of regulation of transcription of some genes are fundamental and specific to the function of the olfactory system. Progress has been made in identifying transcription factors that are part of the mechanism of differentiation of the olfactory receptor neurons. Much less is understood about the molecular biology of embryonic development of the peripheral olfactory system (see Chapter 10, Burd and Tolbert).

3.1. Transcription Factors in Olfactory Receptor Neurons

The discovery of specific binding of proteins in olfactory epithelium homogenates to a DNA motif (TCCC(A/T)NGGAG) in the 5' flanking region of the OMP gene (Kudrycki et al., 1993) led to the isolation of a cDNA clone encoding transcription factor Olf-1 that bound this DNA motif (Wang and Reed, 1993). A differentially spliced form of Olf-1, EBF, was simultaneously identified as a transcription factor for early B lymphocytes (Hagman et al., 1993) and two other members of this Olf-1/EBF (O/E) family were subsequently cloned. The mammalian proteins were named O/E1, O/E2, and O/E3 (Wang et al., 1997). The O/E family members have a highly conserved N-terminal DNA binding domain that can act as a transcriptional activator, a centrally located dimerization domain with a predicted reverse helix-loop-helix (HLH) structure, and a C-terminal domain that can independently activate transcription (Hagman et al., 1995; Wang et al., 1997). All O/E family members bind the Olf-1 consensus site, displaying a similar order of preference for the versions of the O/E sites that flank the $G_{\alpha olf}$, CNG2, and ACIII genes. The O/E proteins can homodimerize and can heterodimerize with each other. During embryonic development, the O/E family members are expressed in several regions of the central nervous system, but as adults the expression of O/E2 and O/E3 is low or nonexistent in most tissues other than the olfactory epithelium. Expression in the olfactory epithelium is in the neural and basal layers and overlaps for all three family members. These results suggest that O/E transcription factors play a broad role in later stages of neural development as well as a role in the differentiation of olfactory receptor neurons. Mice lacking a functional O/E1 gene have a B-cell deficit but no olfactory deficit (Lin and Grosschedl, 1995), suggesting that the O/E family members perform redundant functions in the olfactory epithelium. O/E1 increased transcription of reporter genes under control of the OMP and ACIII promoters, indicating that the O/E family are activators of relatively olfactory-specific genes (Tsai and Reed, 1997). Evidence that the O/E genes are expressed in the basal layer of the epithelium but that expression of genes transcriptionally activated by O/E1-3 cannot be detected there led to the cloning of Roaz, a zinc finger protein (Tsai and Reed, 1997). Roaz suppresses transcriptional activation by O/E1 and is expressed in the basal layer of the olfactory epithelium but not in the mature receptor neurons. Roaz may be a key element in the later stages of differentiation of mature receptor neurons when its disappearance is likely to permit expression of O/E-activated genes.

The regulation of transcription of olfactory receptors is not understood and probably complex. A 6.7 kb upstream region from the mouse olfactory receptor M4 gene was able to recapitulate the zonal expression characteristic of olfactory receptors, suggesting that this upstream region contains some of the regulatory elements of the gene (Qasba and Reed, 1998). There are likely to be sets of transcription factors that are specific to certain subsets of receptor neurons or to receptor expression zones. Retinoic acid-mediated transcription in the dorsolateral olfactory epithelium appears to be one example (Whitesides et al., 1998). In addition to the spatial component of regulation evident in the zonal expression of receptors, temporal regulation also occurs. Fish olfactory receptor genes show consistent times of onset of expression during development (Barth et al., 1997). The expression of receptors is not dependent upon signals emanating from the olfactory bulb (Sullivan et al., 1995). Evidence also exists that one olfactory receptor allele is expressed at the expense of the other, a phenomenon called allelic exclusion (Chess et al., 1994). Although it appears not to correlate with spatial or temporal patterns of olfactory receptor expression, the clustering of olfactory receptor genes (Ben-Arie et al., 1993) may hold clues to the regulation of receptor gene expression. Understanding the regulation of transcription of olfactory receptors is a major unanswered question in the molecular biology of olfaction.

3.2. Development and Differentiation

The molecular biology of olfactory organ development and cell differentiation is intimately related to the regulation of expression of olfactory-related genes and the function of the olfactory system. The O/E transcription factors are likely to be only one branch in a hierarchy of regulatory events during development of the olfactory system. In mammals, homologs of the *Drosophila* proneural gene *achaete-scute* are necessary for development of the olfactory epithelium and are markers for a progenitor cell in the olfactory receptor neuron lineage (Guillemot et al., 1993; Gordon et al., 1995). Another transcription factor, *neurogenin1*, may be a marker for a subsequent progenitor cell in the olfactory receptor neuron lineage (Cau et al., 1997; Calof et al., 1998). The homeodomain protein Pax-6, whose function is necessary for development of the olfactory placode (Hogan et al., 1986), is found in many non-neuronal cells of the olfactory epithelium, suggesting that it marks a cell lineage distinct from the neuronal lineage expressing *achaete-schute* homologs (Davis and Reed, 1996). These discoveries constitute the beginnings of an understanding of the molecular events that regulate olfactory system development and the involvement of other transcriptional regulators presumably await discovery.

4. CONCLUSIONS

In the concluding chapter of the first edition of this text, Lloyd Beidler pointed out that advances in technology often determine progress in neuroscience and predicted that molecular biology techniques would significantly advance our understanding of the function of the chemical senses. This prediction has proved to be accurate. Recombinant DNA techniques have been used to first identify, and then to confirm the roles of many proteins that mediate olfactory transduction. Other areas of olfaction, including the regulation of transcription of olfactory-related genes, olfactory coding, development, and olfactory neurogenesis are also benefiting from recombinant DNA techniques. Many opportunities exist to make further advances using these techniques to investigate the olfactory system.

REFERENCES

Abogadie, F. C., R. C. Bruch, and A. I. Farbman (1995). G-protein subunits expressed in catfish olfactory receptor neurons. *Chem Senses* **20**:199–206.

Bakalyar, H. A. and R. R. Reed (1990). Identification of a specialized adenylyl cyclase that may mediate odorant detection. *Science* **250**:1403–1406.

Barth, A. L., J. C. Dugas, and J. Ngai (1997). Noncoordinate expression of odorant receptor genes tightly linked in the zebrafish genome. *Neuron* **19**:359–369.

Belluscio, L., G. H. Gold, A. Nemes, and R. Axel (1998). Mice deficient in G(olf) are anosmic. *Neuron* **20**:69–81.

Ben-Arie, N., D. Lancet, C. Taylor, M. Khen, N. Walker, D. H. Ledbetter, R. Carrozzo, K. Patel, and D. Sheer (1994). Olfactory receptor gene cluster on human chromosome 17: possible duplication of an ancestral receptor repertoire. *Hum Mol Genet* **3**:229–235.

Biel, M., X. Zong, and F. Hofmann (1996). Cyclic nucleotide-gated ion channels. *Trends Cardivasc Med* **6**:274–280.

Boekhoff, I., K. Touhara, S. Danner, J. Inglese, M. J. Lohse., H. Breer, and R. J. Lefkowitz (1997). Phosducin, potential role in modulation of olfactory signaling. *J Biol Chem* **272**:4606–4612.

Bradley, J., J. Li, N. Davidson, H. A. Lester, and K. Zinn (1994). Heteromeric olfactory cyclic nucleotide-gated channels: a subunit that confers increased sensitivity to cAMP. *Proc Natl Acad Sci U S A* **91**:8890–8894.

Bruch, R. C. and K. F. Medler (1996) A regulator of G-protein signaling in olfactory receptor neurons. *NeuroReport* **7**:2941–2944.

Bruch, R. C., K. F. Medler, H. N. Tran, and J. A. Hamlin (1997). G-protein beta gamma subunit genes expressed in olfactory receptor neurons. *Chem Senses* **22**:587–592.

Brunet, L. J., G. H. Gold, and J. Ngai (1996). General anosmia caused by a targeted disruption of the mouse olfactory cyclic nucleotide-gated channel. *Neuron* **17**:681–693.

Buck, L. B. (1996). Information coding in the vertebrate olfactory system. *Annu Rev Neurosci* **19**:517–544.

Buck, L. and R. Axel (1991). A novel multigene family may encode odorant receptors: a molecular basis for odor recognition. *Cell* **65**:175–187.

Calof, A. L., J. S. Mumm, P. C. Rim, and J. Shou (1996). The neuronal stem cell of the olfactory epithelium. *J Neurobiol* **36**:190–205.

Cao, Y, B. C. Oh, and L. Stryer (1998). Cloning and localization of two multigene receptor families in goldfish olfactory epithelium. *Proc Natl Acad Sci U S A* **95**:11987–11992.

Carr, V. M., E. Walters, F. L. Margolis, and A. I. Farbman (1998). An enhanced olfactory marker protein immunoreactivity in individual olfactory receptor neurons following olfactory bulbectomy may be related to increased neurogenesis. *J Neurobiol* **34**:377–90.

Cau, E., G. Gradwohl, C. Fode, and F. Guillemot (1997). Mash1 activates a cascade of bHLH regulators in olfactory neuron progenitors. *Development* **124**:1611–1621.

Chess, A., I. Simon, H. Cedar, and R. Axel (1994). Allelic inactivation regulates olfactory receptor gene expression. *Cell* **78**:823–834.

Clyne, P. J., S. J. Certel, M. de Bruyne, L. Zaslavsky, W. A. Johnson, and J. R. Carlson (1999a). The odor specificities of a subset of olfactory receptor neurons are governed by Acj6, a POU-domain transcription factor. *Neuron* **22**:339–347.

Clyne, P. J., C. G. Warr, M. R. Freeman, D. Lessing, J. Kim, and J. R. Carlson (1999b). A novel family of divergent seven-transmembrane protein: candidate olfactory receptors in ·Drosophila. *Neuron* **22**:327–338.

Coburn, C. M. and C. I. Bargmann (1996). A putative cyclic nucleotide-gated channel is required for sensory development and function in *C. elegans. Neuron* **17**:695–706.

Colbert, H. A., T. L. Smith, and C. I. Bargmann (1997). OSM-9, a novel protein with structural similarity to channels, is required for olfaction, mechanosensation, and olfactory adaptation in *Caenorhabditis elegans. J Neurosci* **17**:8259–8269.

Cunningham, A. M., D. K. Ryugo, A. H. Sharp, R. R. Reed, S. H. Snyder, and G. V. Ronnett (1993). Neuronal inositol 1,4,5-trisphosphate receptor localized to the plasma membrane of olfactory cilia. *Neuroscience* **57**:339–352.

Davis, J. A. and R. R. Reed (1996). Role of Olf-1 and Pax-6 transcription factors in neurodevelopment. *J Neurosci* **16**:5082–5094.

Dhallan, R. S., K. W. Yau, K. A. Schrader, and R. R. Reed (1990). Primary structure and functional expression of a cyclic nucleotide-activated channel from olfactory neurons. *Nature* **347**:184–187.

Dulac, C. and R. Axel (1995). A novel family of genes encoding putative pheromone receptors in mammals. *Cell* **83**:195–206.

Dwyer, N. D., E. R. Troemel, P. Sengupta, and C. I. Bargmann (1998). Odorant receptor localization to olfactory cilia is mediated by ODR-4, a novel membrane-associated protein. *Cell* **93**:455–466.

Dynes, J. L. and J. Ngai (1998). Pathfinding of olfactory neuron axons to stereotyped glomerular targets revealed by dynamic imaging in living zebrafish embryos. *Neuron* **20**:1081–1091.

Fülle, H. J., R. Vassar, D. C. Foster, R. B. Yang, R. Axel, and D. L. Garbers (1995). A receptor guanylyl cyclase expressed specifically in olfactory sensory neurons. *Proc Natl Acad Sci U S A* **92**:3571–3575.

Gerlai, R. (1996). Gene-targeting studies of mammalian behavior: is it the mutation or the background genotype? *Trends Neurosci* **19**:177–181.

Gimelbrant, A. A., T. D. Stoss, T. M. Landers, and T. S. McClintock (1999). Truncation releases olfactory receptors from the endoplasmic reticulum of heterologous cells. *J Neurochem* **72**:2301–2311.

Gordon, M. K., J. S. Mumm, R. A. Davis, J. D. Holcomb, and A. L. Calof (1995). Dynamics of MASH1 expression in vitro and in vivo suggest a non-stem cell site of MASH1 action in the olfactory receptor neuron lineage. *Mol Cell Neurosci* **6**:363–379.

Guillemot, F., L. C. Lo, J. E. Johnson, A. Auerbach, D. Anderson, and A. L. Joyner (1993). Mammalian achaete-scute homolog 1 is required for the early development of olfactory and autonomic neurons. *Cell* **75**:463–476.

Hagman, J., C. Belanger, A. Travis, C. W. Turck, and R. Grosschedl (1993). Cloning and functional characterization of early B-cell factor, a regulator of lymphocyte-specific gene expression. *Genes Dev* **7**:760–773.

Hagman, J., M. J. Gutch, H. Lin, and R. Grosschedl (1995). EBF contains a novel zinc coordination motif and multiple dimerization and transcriptional activation domains. *EMBO J* **14**:2907–2916.

Herrada, G. and C. Dulac (1997). A novel family of putative pheromone receptors in mammals with a topographically organized and sexually dimorphic distribution. *Cell* **90**:763–773.

Higgins, D., J. Thompson, T. Gibson, J. D. Thompson, D. G. Higgins, and T. J. Gibson (1994). CLUSTAL W: improving the sensitivity of progressive multiple sequence alignment through sequence weighting, position-specific gap penalties, and weight matrix choice. *Nucleic Acids Res* **22**:4673–4680.

Hogan, B. L., G. Horsburgh, J. Cohen, C. M. Hetherington, G. Fisher, and M. F. Lyon (1986). Small eyes (Sey): a homozygous lethal mutation on chromosome 2 which affects the differentiation of both lens and nasal placodes in the mouse. *J Embryol Exp Morphol* **97**:95–110.

Jones, D. T. and R. R. Reed (1989). Golf: An olfactory neuron specific- G protein involved in odorant signal transduction. *Science* **244**:790–794.

Juilfs, D. M., H. J. Fülle, A. Z. Zhao, M. D. Houslay, D. L. Garbers, and J. A. Beavo (1997). A subset of olfactory neurons that selectively express cGMP-stimulated phosphodiesterase (PDE2) and guanylyl cyclase-D define a unique olfactory signal transduction pathway. *Proc Natl Acad Sci U S A* **94**:3388–3395.

Kafitz, K. W. and C. A. Greer (1998) Differential expression of extracellular matrix and cell adhesion molecules in the olfactory nerve and glomerular layers of adult rats. *J Neurobiol* **34**:271–282.

Krautwurst, D., K.-W. Yau, and R. R. Reed (1998). Identification of ligands for olfactory receptors by functional expression of a receptor library. *Cell* **95**:917–926.

Kubick, S., J. Strotmann, I. Andreini, and H. Breer (1997). Subfamily of olfactory receptors characterized by unique structural features and expression patterns. *J Neurochem* **69**:465–475.

Kudrycki, K., C. Stein-Izsak, C. Behn, M. Grillo, R. Akeson, and F. L. Margolis (1993). Olf-1-binding site: characterization of an olfactory neuron-specific promoter motif. *Mol Cell Biol* **13**:3002-3014.

Lazard, D., K. Zupko, Y. Poria, P. Nef, J. Lazarovits, S. Horn, M. Khen, and D. Lancet, (1991). Odorant signal termination by olfactory UDP glucuronosyl transferase. *Nature* **349**:790–793.

Lee, K.-Y., R. G. Wells, and R. R. Reed, (1987). Isolation of an olfactory cDNA: similarity to retinol-binding protein suggests a role in olfaction. *Science* **235**:1053–1056.

Liman, E. R. and L. B. Buck, (1994). A second subunit of the olfactory cyclic nucleotide-gated channel confers high sensitivity to cAMP. *Neuron* **13**:611–621.

Lin, H. and R. Grosschedl (1995). Failure of B-cell differentiation in mice lacking the transcription factor EBF. *Nature* **376**:263–267.

Margolis, F. L. (1988). Molecular cloning of olfactory specific gene products. In: F. L. Margolis and T. V. Getchell (eds). *Molecular Neurobiology of the Olfactory System*. New York: Plenum Press, pp 237–265.

Matsunami, H. and L. B. Buck. A multigene family encoding a diverse array of putative pheromone receptors in mammals. *Cell* **90**:775–784.

McClintock, T. S., T. M. Landers, A. A. Gimelbrant, C. K. Jayawickreme, L. Z. Fuller, B. A. Jackson, and M. R. Lerner (1997). Functional expression of olfactory-adrenergic receptor chimeras and intracellular retention of heterologously expressed olfactory receptors. *Mol Brain Res* **48**:270–278.

Menco, B. P. M. (1997). Ultrastructural aspects of olfactory signaling. *Chem Senses* **22**:295–311.

Mombaerts, P. (1999). Molecular biology of odorant receptors in vertebrates. *Annu Rev Neurosci* **22**:487–509.

Mombaerts, P., F. Wang, C. Dulac, S. K. Chao, A. Nemes, M. Mendelsohn, J. Edmondson, and R. Axel, (1996). Visualizing an olfactory sensory map. *Cell* **87**:675–686.

Mori, K., N. Hiroshi, and Y. Yoshihara (1999). The olfactory Bulb: Coding and Processing of Odor Molecule Information. *Science* **286**:711–715.

Nef, P., J. Heldman, and D. Lazard, T. Margalit, M. Jaye, I. Hanukoglu, and D. Lancet (1989). Olfactory-specific cytochrome P-450. cDNA cloning of a novel neuroepithelial enzyme possibly involved in chemoreception. *J Biol Chem* **264**:6780–6785.

Ngai, J., M. M. Dowling, L. Buck, R. Axel, and A. Chess (1993). The family of genes encoding odorant receptors in the channel catfish. *Cell* **72**:657–666.

Pace, U., E. Hanski, Y. Salomon, and D. Lancet (1985). Odorant-sensitive adenylate cyclase may mediate olfactory reception. *Nature* **316**:255–258.

Pelosi, P. (1996). Perireceptor events in olfaction. *J Neurobiol* **30**:3–19.

Peppel, K., I. Boekhoff, P. McDonald, H. Breer, M. G. Caron, and R. J. Lefkowitz (1997). G protein-coupled receptor kinase 3 (GRK3) gene disruption leads to loss of odorant receptor desensitization. *J Biol Chem* **272**:25425–25428.

Puche, A. C., F. Poirier, M. Hair, P. F. Bartlett, and B. Key (1996). Role of galectin-1 in the developing mouse olfactory system. *Dev Biol* **179**:274–287.

Qasba, P. and R. R. Reed (1998). Tissue and zonal-specific expression of an olfactory receptor transgene. *J Neurosci* **18**:227–236.

Raha, D. and J. Carlson (1994). OS9: a novel olfactory gene of *Drosophila* expressed in two olfactory organs. *J Neurobiol* **25**:169–184.

Ressler, K. J., S. L. Sullivan, and L. B. Buck (1994). Information coding in the olfactory system: evidence for a stereotyped and highly organized epitope map in the olfactory bulb. *Cell* **79**:1245–1255.

Riesgo-Escovar, J., D. Raha, and J. R. Carlson (1995). Requirement for a phospholipase C in odor response: overlap between olfaction and vision in *Drosophila*. *Proc Natl Acad Sci U S A* **92**:2864–2868.

Roayaie, K., J. G. Crump, A. Sagasti, and C. I. Bargmann (1998). The Gα protein ODR-3 mediates olfactory and nociceptive function and controls cilium morphogenesis in C. *elegans* olfactory neurons. *Neuron* **20**:55–67.

Rogers, K. E., P. Dasgupta, U. Gubler, M. Grillo, Y. S. Khew-Goodall, and F. L. Margolis, (1987). Molecular cloning and sequencing of a cDNA for olfactory marker protein. *Proc Natl Acad Sci U S A* **84**:1704–1708.

Rogers, M. E., M. Sun, M. R. Lerner, and R. G. Vogt (1997). Snmp-1, a novel membrane protein of olfactory neurons of the silk moth *Antheraea polyphemus* with homology to the CD36 family of membrane proteins. *J Biol Chem* **272**:14792–14799.

Ryba, N. J., and R. Tirindelli (1997). A new multigene family of putative pheromone receptors. *Neuron* **19**:371–379.

Sautter, A., X. Zong, F. Hofmann, and M. Biel (1998). An isoform of the rod photoreceptor cyclic nucleotide-gated channel β subunit expressed in olfactory neurons. *Proc Natl Acad Sci U S A* **95**:4696–4701.

Schleicher, S., I. Boekhoff, J. Arriza, R. J. Lefkowitz, and H. Breer (1993). A beta-adrenergic receptor kinase-like enzyme is involved in olfactory signal termination. *Proc Natl Acad Sci U S A* **90**:1420–1424.

Scott, J. W. and T. Brierley (1999). A functional map in rat olfactory epithelium. *Chem Senses* **24**:679–690.

Sengupta, P., J. H. Chou, and C. I. Bargmann (1996). *odr-10* encodes a seven transmembrane domain olfactory receptor required for responses to the odorant diacetyl. *Cell* **84**: 899–909.

Sharon, D., G. Glusman, Y. Pilpei, S. Horn-Saban, and D. Lancet (1998). Genome dynamics, evolution, and protein modeling in the olfactory receptor gene superfamily. *Ann N Y Acad Sci* **855**:182–193.

Skoufos, E., M. D. Healy, M. S. Singer, P. M. Nadkarni, P. L. Miller, and G. S. Shepherd (1999). Olfactory receptor database: a database of the largest eukaryotic gene family. *Nucleic Acids Res* **27**:343–345.

Speca, D. J., D. M. Lin, P. W. Sorensen, E. Y. Isacoff, J. Ngai, and A. H. Dittman (1999). Functional identification of a goldfish odorant receptor. *Neuron* **23**:487–498.

Sullivan, S. L., S. Bohm, K. J. Ressler, L. F. Horowitz, and L. B. Buck (1995). Target-independent pattern specification in the olfactory epithelium. *Neuron* **15**:779–789.

Tamura, H., Y. Harada, A. Miyawaki, K. Mikoshiba, and M. Matsui (1998). Molecular cloning and expression of a cDNA encoding an olfactory-specific mouse phenol sulphotransferase. *Biochem J* **331**:953–958.

Treloar, H., H. Tomasiewicz, T. Magnuson, and B. Key (1997). The central pathway of primary olfactory axons is abnormal in mice lacking the N-CAM-180 isoform. *J Neurobiol* **32**: 643–658.

Troemel, E. R., J. H. Chou, N. D. Dwyer, H. A. Colbert, and C. I. Bargmann (1995). Divergent seven transmembrane receptors are candidate chemosensory receptors in C. *elegans*. *Cell* **83**:207–218.

Tsai, R. Y. and R. R. Reed (1997). Cloning and functional characterization of Roaz, a zinc finger protein that interacts with O/E-1 to regulate gene expression: implications for olfactory neuronal development. *J Neurosci* **17**:4159–4169.

Vassar, R., S. K. Chao, R. Sitcheran, J. M. Nunez, L. B. Vosshal, and R. Axel (1994). Topographic organization of sensory projections to the olfactory bulb. *Cell* **79**:981–991.

Vogt, R. G., G. D. Prestwich, and M. R. Lerner (1991). Odorant-binding-protein subfamilies associate with distinct classes of olfactory receptor neurons in insects. *J Neurobiol* **22**: 74–84.

Vosshall, L. B., H. Amrein, P. S. Morozov, A. Rzhetsky, and R. Axel (1999). A spatial map of olfactory receptor expression in the *Drosophila* antenna. *Cell* **96**:725–736.

Wang, F., A. Nemes, M. Mendelsohn, and R. Axel (1998). Odorant receptors govern the formation of a precise topographic map. *Cell* **93**:47–60.

Wang, M. M. and R. R. Reed (1993). Molecular cloning of the olfactory neuronal transcription factor Olf-1 by genetic selection in yeast. *Nature* **364**:121–126.

Wang, S. S., R. Y. Tsai, and R. R. Reed (1997). The characterization of the Olf-1/EBF-like HLH transcription factor family: implications in olfactory gene regulation and neuronal development. *J Neurosci* **17**:4149–4158.

Wellerdieck, C., M. Oles, L. Pott, S. Korshing, G. Gisselman, and H. Hatt (1997). Functional expression of odorant receptors of the zebrafish *Danio rerio* and of the nematode C. *elegans* in HEK293 cells. *Chem Senses* **22**:467–476.

Whitesides, J., M. Hall, R. Anchan, and A.-S. LaMantia (1998). Retinoid signalling distinguishes a subpopulation of olfactory receptor neurons in the developing and adult mouse. *J Comp Neurol* **394**:445–461.

Woodard, C., E. Alcorta, and J. Carlson (1992). The rdgB gene of *Drosophila*: a link between vision and olfaction. *J Neurogenet* **8**:17–31.

Xu, F., B. Hollins, A. M. Gress, T. M. Landers, and T. S. McClintock (1997). Molecular cloning and characterization of a lobster Gα_s-protein expressed in neurons of olfactory organ and brain. *J Neurochem* **69**:1793–1800.

Xu, F. and T. S. McClintock (1999). A lobster phospholipase C-β that associates with G-proteins in response to odorants. *J Neurosci* **19**:4881–4888.

Yokoe, H. and R. R. H. Anholt (1993). Molecular cloning of olfactomedin, an extracellular matrix protein specific to olfactory neuroepithelium. *Proc Natl Acad Sci U S A* **90**: 4655–4659.

Yoshihara, Y., M. Kawasaki, A. Tamada, H. Fujita, H. Hayashi, H. Kagamiyama, and K. Mori (1997). OCAM: A new member of the neural cell adhesion molecule family related to zone-to-zone projection of olfactory and vomeronasal axons. *J Neurosci* **17**:5830–5842.

Yu, S., L. Avery, E. Baude, and D. L. Garbers (1997). Guanylyl cyclase expression in specific sensory neurons: a new family of chemosensory receptors. *Proc Natl Acad Sci U S A* **94**:3384–3387.

Zhang, Y., J. H. Chou, J. Bradley, C. I. Bargmann, and K. Zinn (1997). The *Caenorhabditis elegans* seven-transmembrane protein ODR-10 functions as an odorant receptor in mammalian cells. *Proc Natl Acad Sci U S A* **94**:12162–12167.

Zhao, H., L. Ivic, J. M. Otaki, M. Hashimoto, K. Mikoshiba, and S. Firestein (1998). Functional expression of a mammalian odorant receptor. *Science* **279**:237–242.

Zwaal, R. R., J. E. Mendel, P. W. Sternberg, and R. H. Plasterk (1997). Two neuronal G proteins are involved in chemosensation of the *Caenorhabditis elegans* Dauer-inducing pheromone. *Genetics* **145**:715–727.

9

Representation of Olfactory Information in the Brain

THOMAS A. CHRISTENSEN
Arizona Research Laboratories, Division of Neurobiology,
University of Arizona, Tucson, AZ

JOEL WHITE
Department of Neuroscience, Tufts University School of Medicine, Boston, MA

1. INTRODUCTION

To function efficiently in a complex and changing environment, widely diverse animal species rely on their olfactory sense to communicate with conspecifics and to obtain vital information about their surroundings. It is therefore not surprising that the neural mechanisms governing odor discrimination in the brain have attracted considerable interest for many years. Over the past decade, we have seen unparalleled progress in the ever-growing area of olfactory neuroscience, and thus the time is ripe for new discussion on how an animal's olfactory environment is represented in the brain. This chapter summarizes current models of how olfactory circuits first detect, then recognize and discriminate among an amazing variety of different scents in the environment. We review here the various experimental and computational strategies currently used by researchers to help elucidate the basic principles that underlie the neural processing of odor information in the brain. Throughout this chapter, we draw upon evidence from both vertebrate and invertebrate studies in order to provide a broad base for

The Neurobiology of Taste and Smell, Second Edition, Edited by Thomas E. Finger, Wayne L. Silver, and Diego Restrepo.
ISBN 0-471-25721-4   Copyright © 2000 Wiley-Liss, Inc.

discussion and to focus attention on some of the unique contributions each of these animal groups has made to our present state of knowledge. While the examples focus on selected vertebrates (fish, amphibians, mammals) and invertebrates (insects, crustaceans, mollusks), it is evident that similar principles of organization and function can be found throughout the animal kingdom.

2. WHAT INFORMATION IS ENCODED IN OLFACTORY CIRCUITS?

Before we can discuss how any sensory stimulus is represented in the brain, it is first necessary to identify and classify the different attributes of that stimulus. Only then can we proceed to investigate how each of these attributes is encoded or "extracted" by primary sensory neurons.

2.1. Stimulus Properties

Odorant molecules cannot be classified according to their positions along a single continuous dimension as, for example, wavelengths of light or frequencies of sound. Determining exactly which features of an odorant molecule trigger a neural response remains one of the most daunting challenges to olfactory physiologists today. By analogy to other ligand–receptor interactions, however, it is generally accepted that odor discrimination begins with the binding of an odorant molecule to one or more membrane-bound odorant receptors (Chapter 6, Farbman; Chapter 7, Ache and Restrepo; Chapter 8, McClintock). Shepherd (1987) coined the term *odotope* to help define the different physicochemical characteristics of a molecule that may be important for recognition (molecular weight, functional group, charge, etc.; see also Beets, 1970, and Polak, 1973). Odotope is analogous to the term *epitope* which defines the active determinants of an antigen in the immune system (Section 6.1).

Odorants also occur in different behavioral and environmental *contexts* (Christensen et al., 1998; Alkasab et al., 1999). Olfactory environments are diverse, incorporating a wide range of different types and numbers of odorants. Stimulus context therefore places added constraints on how this information is processed in olfactory circuits. For example, the task of detecting and discriminating among hundreds of environmental odors when searching for food sources is different from the task of detecting the species-specific pheromone signal of a potential mate. In fact, the olfactory system of many insects is clearly divided into separate olfactory subsystems that accomplish the two tasks of pheromonal and nonpheromonal odorant processing (Section 3).

Another basic but frequently overlooked stimulus parameter is *intermittency*, or the natural tendency of odorous molecules in the environment to be patchy and discontinuous. It is now well established that odors possess a distinct spatiotemporal structure, whether delivered in air or water, and recent evidence from vertebrates and invertebrates indicates that the changing time course of an odor stimulus plays an important role in odor perception (reviews: Atema, 1995; Christensen et al., 1996; Murlis, 1997). Thus, information about both the chemical and temporal parameters of an odor stimulus is available to the neural circuits that analyze and discriminate different odorants in the brain. Recent evidence from invertebrates indicates that both types of information are integrated at the first stages of odor processing in the brain, and this is particularly relevant to the perception of odor mixtures (Section 6.4.2).

2.2. Olfactory Receptor Neuron Responses

2.2.1. Perireceptor Events

Prior to any influences on the nervous system, odorants first interact with a number of physical, non-neuronal factors. Most animals actively sample their odorous environment by sniffing (terrestrial vertebrates) or antennal "flicking" (terrestrial and aquatic invertebrates). These behaviors provide a means of controlling odorant application rates and frequency, which likely influence odorant coding. In vertebrates, the nasal cavity can be a complex aerodynamic structure (e.g., Hahn et al., 1993), while in insects and crustacea, viscous boundary layers surrounding the antennae may affect odorant reception (e.g., Atema, 1995). In vertebrates, odorants likely partition into the nasal mucus to differing degrees as they travel through the nasal cavity, producing chromato-graphic effects that may influence coding (Mozell and Jagodowicz, 1973). Once in the mucus (vertebrates) or sensillar fluid (invertebrates), odorants may interact with *odorant-binding proteins* and *odorant-degrading enzymes*, which are thought to be involved in "presenting" odorants to receptor cells and in odorant clearing, respectively (see Chapter 7, Ache and Restrepo). All of these non-neuronal factors are likely to affect subsequent coding of odorant information.

2.2.2. Receptor-Mediated Events

The next step in odor detection involves interactions of the odorant molecule with one or more membrane-bound proteins (receptors) associated with the dendrites of specialized sensory cells, the *olfactory receptor neurons* (ORNs; see Chapter 6, Farbman, and Chapter 8, McClintock). In mammals, ORNs in the olfactory epithelium are situated deep within the nasal cavity (Fig. 9.1A), whereas in insects they are encased within cuticular hairs (or *sensilla*) that are distributed along the main olfactory appendages, the paired antennae (Fig. 9.1B). It is within the ORNs that information about the odor stimulus is first transduced into electrical signals (Chapter 7, Ache and Restrepo) that can then be interpreted by neural circuits in the brain.

2.2.3. ORN Response Tuning

Due to the multidimensional nature of molecular stimuli, it is difficult to make rigorous, quantitative descriptions of ORN sensitivity and selectivity. In

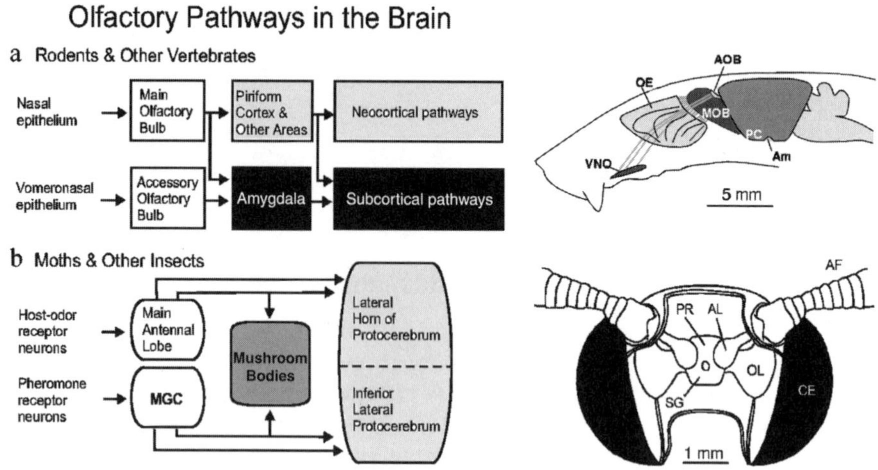

Olfactory Pathways in the Brain

a Rodents & Other Vertebrates

b Moths & Other Insects

FIG. 9.1. Schematic overview and basic anatomy of olfactory pathways in vertebrates and in insects.

(a) In rodents and other vertebrates, the main olfactory bulb (MOB) receives input from ORNs in the olfactory epithelium (OE) and sends projections to higher areas, such as the piriform cortex (PC). The accessory olfactory bulb (AOB) receives input from receptor neurons in the vomeronasal organ (VNO) and sends projections to areas such as the amygdala (Am). The relationship of these structures is shown in a schematic sagittal view of a rat head. The VNO plays a major role in the perception of social chemosensory stimuli in many species, but it has not yet been demonstrated in adult humans (Chapter 5, Johnston, Chapter 6, Farbman).

(b) In moths, the antennal lobe (AL) receives input from ORNs distributed along the length of the antennal flagellum (AF). In many insects, the AL in males is divided into two subsystems, one for processing information about general odorants (main AL), the other devoted to species-specific information about the female sexual pheromone (macroglomerular complex; MGC). Many AL projections converge on the mushroom body in the protocerebrum (PR), whereas some projections bypass the mushroom body completely. Other abbreviations in the schematic frontal view of a moth head: CE, compound eye; OL, optic lobe; SG, subesophageal ganglion.

olfaction, these two properties are inextricably linked—an ORN's selectivity is defined by its sensitivity to all possible odorant molecules. In other sensory systems, data showing sensitivity and selectivity are represented by a *tuning curve*, which is a plot of a cell's response threshold to stimuli along some quality dimension. In studies of vibration sensation in the somatosensory system, for example, tuning curves depict the vibration amplitude that elicits a threshold response as the vibration frequency is systematically varied (Mountcastle et al., 1972). Most ORN studies have recorded responses to a battery of odorants, but at single (often high) concentrations (e.g., Revial et al., 1978). Such studies have led to the description of ORNs as "broadly tuned." These data, however, lack a well-defined *a priori* stimulus dimension and often do not address how changes in concentration affect responsiveness. Studies that measure ORN responses to a

a
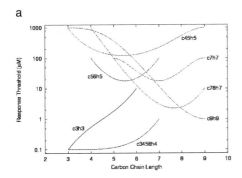

b
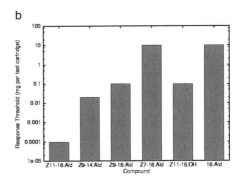

FIG. 9.2. (a) ORN responses from Fig. 7 of Sato et al. (1994), represented as smooth tuning curves fitted to the data. Curves represent the odorant concentration yielding a just noticeable increase in intracellular Ca^{++} to n-fatty acids of increasing carbon chain length. Cell designations are as used by Sato et al. (1994). These "U"-shaped curves are qualitatively similar to tuning curves seen in other sensory systems.

(b) ORN responses to pheromone in insects are narrowly tuned. In the corn earworm moth, the major sex-pheromone component (abbreviated Z11-16:Ald) evokes a dose-dependent electrophysiological response in one type of antennal ORN (adapted from Almaas et al., 1991). Response threshold for Z11-16:Ald is at least 2 log units lower than the stimulus dosage needed to evoke a response to any of five chemically related odorants.

homologous series of straight-chain hydrocarbons come closest to defining a molecular stimulus dimension amenable to investigation.

In one such study, Sato et al. (1994) used homologous series of n-fatty acids and n-aliphatic alcohols to examine rigorously the tuning characteristics of mouse ORNs in vitro. These researchers found that responsive ORNs typically exhibit a lowest threshold to one or a small number of odorants tested. Some cells were narrowly tuned to such molecular features as chain length (e.g., cell c56h5 in Fig. 9.2A) or molecular charge, whereas others were more broadly tuned (cells c45h5, c78h7, and c9h9 in Fig. 9.2A). It is important to emphasize that these studies address tuning characteristics only for the odorants tested—tuning along other quality dimensions using other odorants is undefined. Using a similar approach, but with a stimulus set containing diverse chemical structures, Kafka (1987) described relatively broad yet distinct tuning curves in different ORNs in the silk moth antenna identified by morphological and physiological criteria. A classic example of sensory neurons that are narrowly tuned, often to a single odorant, is found in the subpopulation of ORNs that respond to sex pheromone in insects (Fig. 9.2B; reviewed in Hansson and Christensen, 1999).

Another difficulty in acquiring sufficient data to investigate tuning has been the inability to sample a large number of ORNs having the same response profile (i.e., expressing the same odorant receptor protein(s)). The recent identification of a large gene family that may encode odorant receptor proteins in the vertebrate olfactory epithelium (Buck and Axel, 1991) may provide a means for obtaining such data. For example, it may be possible to express the receptor genes in vitro in a heterologous system, so that individual gene products can be thoroughly investigated (see Chapter 8, McClintock). Combined molecular and physiological approaches provide potentially powerful methods for rigorous investigation of receptor proteins and ORN sensitivity and selectivity.

3. PATHWAYS FOR INFORMATION FLOW

Bundles of axons arising from the ORNs unite to form the *olfactory nerve* in vertebrates, or the *antennal nerve* in invertebrates. These nerves represent the primary route by which odor messages reach the first stage of olfactory processing in the brain. The vertebrate *olfactory bulb*, the insect *antennal lobe*, the crustacean *olfactory lobe*, and the molluscan *procerebral lobe* all represent discrete regions in the brain where olfactory information from the periphery is first processed before being relayed to higher centers for further analysis (Fig. 9.1; reviews: Ache, 1991; Gelperin et al., 1996; Shipley and Ennis, 1996; Hildebrand and Shepherd, 1997). Each of these centers is characterized by a conspicuous arrangement of discrete, rounded modules of synaptic neuropil called *glomeruli* (Fig. 9.3), the site of the

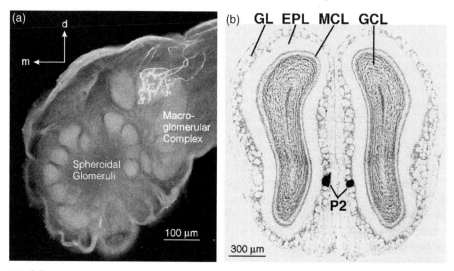

FIG. 9.3. (a) Laser-scanning confocal micrograph of the antennal lobe (AL) of a male sphinx moth, *Manduca sexta*. The AL is anatomically and functionally separated into "main" (spheroidal glomeruli) and "accessory" (macroglomerular complex) centers. A subset of the ORNs that respond selectively to the female sex pheromone were stained with dextran-rhodamine. The stained axons invade only the MGC, revealing a functional topography in the AL glomeruli (see text). d, dorsal; m, medial.

(b) Light micrograph of a coronal section through both mouse MOBs, showing the glomerular layer (GL), the external plexiform layer (EPL), the mitral cell layer (MCL), and the granule cell layer (GCL) (see also Fig. 9.4). A genetic manipulation also stains ORNs expressing the receptor gene sequence P2, showing the convergence of ORN axons onto a single glomerulus in each MOB (another stained glomerulus in each MOB is present in other sections; see also Fig 8.4). (Modified from Mombaerts et al. [1996], with permission.)

first synapse in the olfactory pathway (Section 4.1). Beyond the glomerular level, a comparison of the major levels of organization in the vertebrate olfactory bulb and the insect antennal lobe (Fig. 9.1), as well as higher olfactory centers, reveals a remarkable degree of similarity across phyla (Boeckh and Tolbert, 1993; Christensen et al., 1996; Hildebrand and Shepherd, 1997). In both mammals and insects, the route taken by olfactory information into the brain is anatomically and functionally divided into a *main* and an *accessory* olfactory pathway (Fig. 9.1).

3.1. Main Olfactory Pathway

In mammals, the main olfactory pathway originates in the nasal epithelium and proceeds to the *main olfactory bulb* (Fig. 9.1A). Outputs from the bulb project to higher olfactory centers where they make the second synapse in the pathway. These centers include a number of cortical areas, such as the piriform cortex, the olfactory tubercle, the cortical amygdala, the anterior olfactory nucleus, portions of the hippocampus, the lateral entorhinal cortex, and the insula (Section 5; see also Fig. 5.1 in Chapter 5, Johnston). These areas then project to subcortical and neocortical areas, as well as provide centrifugal connections back to the olfactory bulb (for reviews, see Haberly, 1990, and Shipley and Ennis, 1996). In insects, the main portion of the antennal lobe is comprised of *spheroidal glomeruli* that receive input from nonpheromonal afferent axons (Fig. 9.3A). Projections from the antennal lobe proceed to higher-order sites in the protocerebrum: the calyces of the mushroom body and the lateral horn (Homberg et al., 1988; Section 5). There are extensive interconnections among different areas of the protocerebrum (Rybak and Menzel, 1993; Li and Strausfeld, 1997), with outputs descending from the lateral protocerebrum to motor circuits in the thoracic and abdominal ganglia (Li and Strausfeld, 1997).

3.2. Accessory Olfactory Pathway

In many mammals, reptiles, and amphibians, an accessory olfactory pathway originates in the *vomeronasal organ* (VNO) and projects to an anatomically distinct *accessory olfactory bulb* (AOB), which projects to *subcortical* regions (Fig. 9.1A; see Chapter 5, Johnston, and Chapter 6, Farbman). Behavioral data and the observation that VNO information remains separated from that originating in the nasal epithelium as it proceeds to subcortical centers seems to suggest that the VNO plays a role in the reception of chemical stimuli that communicate a different type of information than the main olfactory system. In most cases, an absolute, exclusive division between accessory and main olfactory function is, however, not always possible (Chapter 5, Johnston). In fish, which lack a VNO and accessory olfactory pathway, there is anatomical and functional evidence for pheromonal and nonpheromonal subdivisions within the olfactory system (Sorensen et al., 1998).

In male moths and other insects, antennal ORNs that are narrowly tuned to components of the female sex pheromone send axons to a specific subset of glomeruli, the *macroglomerular complex* (MGC; Boeckh and Tolbert, 1993; Christensen et al., 1996; Hansson and Christensen, 1999; Fig. 9.3A). The insect MGC is a well-studied example of how groups of functionally and anatomically identifiable glomeruli operate together to extract and discriminate the many different features of an incoming odor signal (Sections 6.1.2 and 6.2). Most of the axons that arise from the male-specific MGC branch in the calyces of the mushroom body, then continue laterally and terminate in the *inferior lateral protocerebrum* (ILP). Other MGC axons bypass the mushroom body and project directly to the ILP (Homberg et al., 1988). The calycal terminals of axons from MGC and spheroidal glomeruli are also anatomically distinct (Homberg et al., 1988; Li and Strausfeld, 1997). Thus pheromonal information remains segregated from nonpheromonal information even at these higher levels of processing in the insect brain.

4. ANATOMY AND PHYSIOLOGY OF THE FIRST PROCESSING STAGE: OLFACTORY BULB AND ANTENNAL LOBE

The first-order processing centers of vertebrates and invertebrates share many anatomical and physiological features. The presence of glomeruli in virtually all olfactory systems studied suggests that these structures have particular significance in olfactory processing.

4.1. The Glomerulus: A Multifunctional Coding Module in the Brain

The first synaptic interactions between ORN axons and their targets in the brain occur within the glomeruli. In vertebrates and in insects, glomeruli are confined to a superficial layer of the bulb or lobe; in insects and certain vertebrates, this geometry produces an array around the circumference of the structure (Fig. 9.3). Numerous workers have used a variety of anatomical, physiological, biochemical, and molecular biological methods to investigate the glomerulus. All of these studies converge on a consensus model of the glomerulus as a fundamental coding unit in olfaction. Analogous to other brain modules that group together neurons having a similar function (e.g., "columns" in visual cortex or "barrels" in somatosensory cortex), the glomerulus and its associated neurons are thought to organize the neural space involved in the early stages of olfactory processing (Leise, 1990; Shepherd, 1992). In the sections below, we outline some of the major organizational features of glomerular circuits, providing background for a conceptual model of how these specialized neuropil structures may contribute to coding the different parameters of an odor stimulus (Section 6.1).

4.1.1. Convergent Connections from Epithelium to Glomeruli

The projection from the olfactory epithelium to the brain appears complex, and the complete rules determining this connectivity have not yet been revealed in any species. In other sensory systems, sensory space is represented so that neighbor relationships in the periphery are maintained at the first level in the CNS, and are often maintained through higher levels. We refer to this property as *parallel topography*. Specific examples include retinotopy in the visual system, ototopy in the auditory system, and somatotopy in the somatosensory system. In general, precise parallel topography does not appear to be present in either vertebrate or invertebrate olfactory systems. Cell counts indicate that in most species, thousands or millions of ORNs converge onto far fewer glomeruli (see Hildebrand and Shepherd, 1997, for review). Furthermore, tracing studies reveal convergent connectivity from widespread regions of the epithelium to discrete areas in the glomerular neuropil and divergent connectivity from localized regions of the olfactory epithelium to widespread regions of the bulb and lobe (Kauer, 1981; Mellon and Munger, 1990; Baier et al., 1994; Schoenfeld et al., 1994; Christensen et al., 1995). Such connectivity patterns are inconsistent with parallel topography.

Lack of parallel topography at this first stage in the olfactory pathway does not mean that the connections are random. Data from a variety of physiological, anatomical, and molecular studies (Kauer and Moulton, 1974; Ressler et al., 1994; Vassar et al., 1994; Cinelli et al., 1995; Mombaerts et al., 1996; Bozza and Kauer, 1998; Hansson and Christensen, 1999) all suggest that in a number of different species, ORNs of a given type (defined as exhibiting the same molecular response profile and/or expressing the same receptor protein(s)) extend axons to only one or a few glomeruli (Fig. 9.3). In situ hybridization studies indicate that cells expressing the same putative receptor mRNA are randomly scattered throughout the olfactory epithelium in catfish (Ngai et al., 1993) or within one of four broad epithelial "zones" in rodents (see Fig. 8.3, Chapter 8, McClintock) (Ressler et al., 1993; Vassar et al., 1993; review: Mori et al. 1999) (but see below). The anatomical, physiological, and molecular biological data therefore suggest that, although a strict parallel topography is lacking in the connection between the epithelium and the glomeruli, an ordered *functional* topography onto the glomerular layer does occur. This functional topography appears to bring together the axons of scattered ORNs of the same type so that they terminate in the same glomerulus. The precise developmental mechanisms that produce this map, as well as the organization of the glomerular map itself, are currently unknown, although the receptor proteins themselves may play a role in its determination (Mombaerts et al., 1996; Wang et al., 1998).

Convergence of widespread ORNs appears to be a general rule in various olfactory systems, but examples of more restricted ORN distributions have been found (insects: Hosl, 1990; Heinbockel et al., 1993; frog: Jiang and Holley, 1992; rodents: Strotmann et al., 1994; Ring et al., 1997). These latter two studies also provide evidence that some ORN types (defined by gene expression or glomerular convergence) are not confined to any one of the four epithelial zones described above. Given that these zones were described using relatively few of the potentially hundreds of putative receptor sequences (Ressler et al., 1993; Vassar et al., 1993), such "exceptions" may become more common as data accumulate. The functional consequences of these various peripheral ORN distributions currently are unknown. Clearly, more studies aimed at characterizing the central projections of ORNs with known ligand specificity (e.g., Bozza and Kauer, 1998; Hansson and Christensen, 1999) are needed to determine the roles of spatial maps and peripheral organization in olfactory coding.

One of the results of ORN-to-glomerulus convergence is amplification of the odorant signal. In insects, the activation threshold for ORNs may be a millionfold higher than for glomerular interneurons responding to the same odorant (Hartlieb et al., 1997). This enormous increase in gain is believed to result from the massive convergence of hundreds of thousands of sensory axons onto many fewer glomerular neurons (van Drongelen et al., 1978; Duchamp-Viret et al., 1989; Hansson and Christensen, 1999). The convergence ratios of ORNs to glomeruli in vertebrates (approximately 25,000:1 for rabbits) are, on average, roughly 10-fold higher than in insects (see discussion in Hildebrand and Shepherd, 1997). The insect ratio climbs to about 50,000:1 when one considers the pheromone-selective ORNs converging onto 2 to 4 MGC glomeruli (Lee and Strausfeld, 1990).

4.2. Intra- and Interglomerular Synaptic Interactions

Synaptic processing in the olfactory bulb and antennal lobe can be best described in terms of the pathways of information flow through the glomerular circuit, and how output from the glomerulus is ultimately determined (Fig. 9.4). The primary connection from the ORN axons to the glomerulus consists of several *feedforward* pathways that initially determine the firing properties of the principal (output) neurons innervating each glomerulus. This diverse population of glomerular output neurons is ultimately responsible for transmitting odor-specific information to higher centers for further processing. Importantly, due to the architecture of these circuits, these processing steps frequently involve interactions with neighboring glomeruli. Glomerular output is carried by *mitral* and *tufted* (M/T) cells in vertebrates and by *projection neurons* in insects and other invertebrates.

Responses of glomerular output cells in both animal groups are often complex (Section 4.3): Odorant stimulation can evoke both *excitatory* and *inhibitory postsynaptic potentials* (EPSPs and IPSPs). The polarity of the response depends on a number of different but specific synaptic interactions within the bulb or lobe. In the glomerulus, sensory axon terminals provide direct excitatory input through monosynaptic contacts onto the apical (primary) dendrites of the output neuron. In vertebrates, this input is thought to be carried by the neurotransmitter glutamate (Chapter 6, Farbman); in insects, acetylcholine is the likely transmitter (Homberg et al., 1995; Christensen et al., 1998). In insects, anatomical and physiological evidence also indicates indirect routes for excitatory input to projection neurons via an intercalated excitatory interneuron (Sun et al., 1997) or through a disinhibitory circuit involving two inhibitory interneurons arranged in series (Christensen et al., 1993; Distler and Boeckh,

MAMMALIAN OLFACTORY BULB   INSECT ANTENNAL LOBE

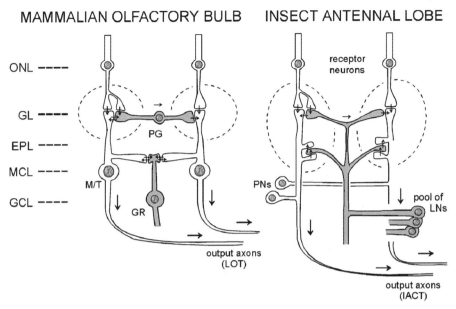

FIG. 9.4. Schematic diagrams of the first-stage processing circuits in the olfactory bulb and antennal lobe. Both circuits are characterized by feedforward excitatory and inhibitory inputs to the apical dendrites of the principal glomerular output elements (bulb: mitral/tufted cell, M/T; lobe: projection neuron, PN), as well as by two levels of lateral interactions mediated by inhibitory circuit elements (bulb: granule cell, GR; periglomerular cell, PG; lobe: local interneurons, LNs). These inhibitory elements are shown shaded, whereas excitatory elements are unshaded. The effect of this inhibition can be local or it can spread laterally (left to right in these diagrams) to shape the output of neighboring glomeruli (dashed outlines). For the sake of clarity, only the major circuit elements and interconnections are shown here. EPL, external plexiform layer; GCL, granule cell layer; GL, glomerular layer; MCL, mitral cell layer; ONL, olfactory nerve layer (see also Fig. 9.3). Olfactory bulb model based on Shepherd (1990); antennal lobe model based on data compiled from moths, cockroaches, and honeybees (reviewed in Boeckh and Tolbert, 1993).

1997a). Polysynaptic pathways for excitatory inputs to M/T cells have also been proposed (review: Shipley and Ennis, 1996).

    Another major synaptic interaction within the glomerulus in both animal groups involves an inhibitory feedforward connection between ORNs and output neurons. Through this path, an intercalated inhibitory interneuron produces an IPSP in the output neuron. Many of these inhibitory circuit elements, *periglomerular* cells in vertebrates and *local interneurons* in invertebrates, are GABAergic (e.g., I_1 in Fig. 9.5A) and serve a critical function in regulating the firing patterns of glomerular output neurons (vertebrates: Duchamp-Viret et al., 1993; insects: Christensen et al., 1993, 1998; crustacea: Wachowiak and Ache, 1998). Inhibitory circuit elements also may target the presynaptic terminals of ORNs, providing another means to regulate input to the glomerulus (Shipley and Ennis, 1996; Distler and Boeckh, 1997b; Wachowiak and Ache, 1998). Thus, the

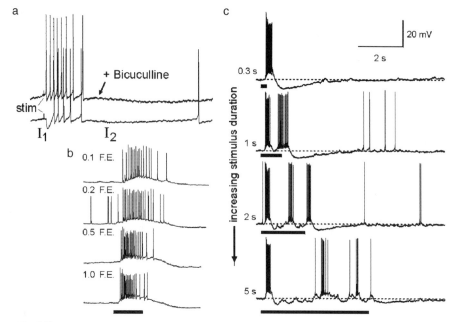

FIG. 9.5. Intracellular recordings from the dendrites of pheromone-selective projection neurons (PNs) in the MGC of male *Manduca sexta*.

(a) Responses to electrical stimulation of the antennal nerve in normal saline (lower trace) and after 5 min in saline containing the GABA$_A$ receptor antagonist bicuculline methiodide (upper trace). PNs respond with a characteristic "triphasic" waveform consisting of an early, bicuculline-sensitive IPSP (I_1), a period of excitation, and a later inhibitory phase (I_2). The latter two response phases are bicuculline-insensitive.

(b) Dose-response relationship using different amounts of the species-specific sex pheromone. Note that the PN uses a rate code to monitor changes in stimulus concentration, and that the burst of action potentials evoked by this 300-msec stimulus copies the stimulus duration. Stimuli shown in fractions of a female-equivalent (F.E.).

(c) The temporal pattern of action potentials in response to a single odor pulse is strongly dependent on pulse duration, and this pattern increases in complexity as the stimulus duration is increased above 300 msec (bar beneath each record). Each successive pulse evokes a greater number of alternating inhibitory and excitatory potentials, and the PN response becomes increasingly oscillatory. These oscillations could be a consequence of extensive synaptic interconnection with GABAergic local neurons in the antennal lobe and suggest the emergence of a different computational mechanism with prolonged stimulation. (Adapted from Christensen et al. [1998].)

complexity of synaptic connections within a glomerulus seems to reflect a remarkable capacity for shaping and adjusting the initial representation of olfactory signals in the brain.

In contrast to the unitary tree-like tuft of processes that innervate one or more glomeruli in the insect antennal lobe, many vertebrate M/T cells possess long *secondary dendrites* that form a second, deeper layer of synaptic connections in the olfactory bulb (called the *external plexiform layer* or EPL; see Fig. 9.3B and Fig. 9.4). In the EPL, the secondary dendrites of M/T cells interact primarily with another inhibitory, GABAergic cell type, the *granule cells*. The distal dendrites of granule cells have spiny processes that make reciprocal dendrodendritic synapses with the secondary dendrites and somata of M/T cells. That is, the dendrites of the granule and M/T cells are both pre- and postsynaptic to each other, thus forming the basis for feedback inhibition of M/T cells. Similar dendrodendritic contacts also are seen between periglomerular cells and the apical dendrites of M/T cells. While quite prevalent in vertebrates, these reciprocal synapses are less common in insects (Distler et al., 1998). Inhibitory effects are instead mediated by local circuits (serial synapses), or they may involve longer intra- or inter-glomerular feedback loops. In insects, furthermore, all synapses are confined to the glomerulus rather than arranged in separate layers, as in the olfactory bulb. While the insect glomerulus is not obviously subdivided, layered organization is a distinguishing characteristic of glomeruli in decapod crustaceans (Schmidt and Ache, 1992). There is some evidence for glomerular subcompartments in vertebrates as well (Halasz and Greer, 1993; Treloar et al., 1996).

The inhibitory elements of the bulb and lobe also provide the substrate for lateral, *inter*glomerular inhibitory interactions (Fig. 9.4). These lateral interactions may mediate a form of "contrast enhancement" in the olfactory system, serving to enhance the molecular contrast between odor messages represented in neighboring glomeruli (Christensen and Hildebrand, 1987; Mori and Shepherd, 1994; Mori et al., 1999; but see Laurent, 1999). This contrast enhancement is disrupted by GABAergic antagonists, which are known to block lateral inhibition (lobe: Christensen et al., 1998; bulb: Yokoi et al., 1995).

4.3. Physiological Responses to Odorants

How do odorant responses of M/T cells and projection neurons compare with those of ORNs? In general, responses of M/T cells are most often described as being more specific, more "narrowly tuned," or more "refined" than ORN responses (Mori and Shepherd, 1994; Yokoi et al., 1995). The nature of such "refinement" is not immediately obvious, because in studies directly comparing ORN and M/T responses in the same species, a higher percentage of M/T cells respond to a given odorant than ORNs, even at low concentrations (Duchamp-Viret et al., 1990; Kang and Caprio, 1995). Descriptions of M/T tuning are difficult to interpret because, in general, cells do not simply increase their firing rate monotonically with increasing odorant concentration (Fig. 9.6). Electro-physiological recordings from M/T cells often reveal different temporal response patterns to the same odorant at different intensities (Kauer, 1974; Meredith, 1986; Hamilton and Kauer, 1985; Cinelli et al., 1995); these patterns are thought to arise from neural interactions within the bulb (previous section). Consistent with this interpretation, computer simulations of neural interactions in olfactory bulb circuits produce responses similar to those seen in physiological recordings (Meredith, 1992; White et al., 1992). As discussed in Section 6.4.1, such

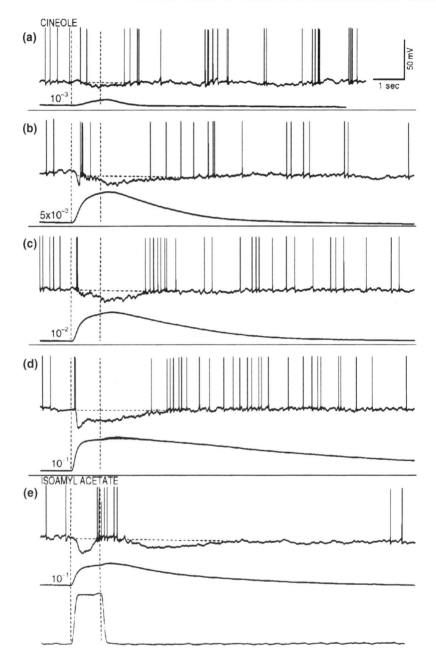

temporal activity patterns are likely to be an important component of odorant information processing.

In insects, antennal lobe projection neurons also can exhibit complex temporal firing properties (see Fig. 9.5C). For example, odorant stimulation can evoke local field potential oscillations in the antennal lobe (20–35 Hz in locusts—Laurent, 1996; up to 50 Hz in moths—Heinbockel et al., 1998). In locusts, projection neurons fire spikes that often are phase-locked to these oscillations. Different odorants elicit different temporal patterns of firing (Laurent, 1996), but the effects of odorant concentration on these patterns has not been described. In moths, there is a strong correlation between pheromone concentration and spike firing rate in many MGC projection neurons (Fig. 9.5B). In these neurons, temporal patterns of spiking often follow closely the temporal properties of the pheromonal stimulus (Christensen and Hildebrand, 1997; Christensen et al., 1996, 1998). MGC projection neurons also exhibit context-dependent responses that become more temporally complex as the duration of odorant simulation increases (Christensen et al., 1998; Fig. 9.5C). We will return to these temporal coding issues in Section 6.4.

5. ANATOMY AND PHYSIOLOGY OF THE SECOND PROCESSING STAGE: PIRIFORM CORTEX AND PROTOCEREBRUM

Similar to the ORN-to-glomerulus connection, projections to secondary olfactory areas are, in general, *distributed*. In mammals (Fig. 9.7A), individual M/T cells project to multiple olfactory areas, as well as to multiple locations within the piriform cortex (Scott, 1981; Ojima et al., 1984). One exception, however, appears to be a topographic connection to the anterior olfactory nucleus (Scott et al., 1985). In insects (Fig. 9.7B), individual projection neurons project to multiple protocerebral sites (Homberg et al., 1988), as well as to distributed sites within the calyces of the mushroom body (Homberg et al., 1988; Li and Strausfeld, 1997). There is also considerable divergence in the projection between the first and second processing stages. In mammals, hundreds of thousands of M/T cells (approximately 250,000 in rat: Meisami and Safari, 1981) project to several million neurons in the piriform cortex (approximately 10^7 in

◄————————————————————————————————————

FIG. 9.6. Intracellular recordings from a salamander M/T cell (upper traces in each panel) showing responses to the odorant cineole over a range of concentrations (a–d) and to a single concentration of isoamyl acetate (e). Lower traces in each panel are recordings of population activity in the olfactory epithelium (electro-olfactogram or EOG). The bottom trace indicates the shape of the stimulus pulse. Vertical dashed lines mark the stimulus onset and termination. Horizontal dashed lines represent the resting membrane potential during periods of hyperpolarization. Note periods of excitation and inhibition in the responses to both odorants, and the change in membrane potential and spike firing over concentrations of cineole. (From Hamilton and Kauer [1985], with permission.)

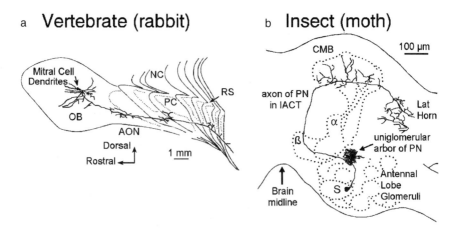

FIG. 9.7. (a) Camera lucida drawing of a dye-filled mitral cell in a rabbit, showing sagittal sections of bulb and telencephalon. AON, anterior olfactory nucleus; NC, neocortex; OB, olfactory bulb; PC, piriform cortex; RS, rhinal sulcus. (Adapted from Ojima et al. [1984], with permission.)

(b) Camera lucida drawing of a dye-filled antennal lobe projection neuron in the sphinx moth, *Manduca sexta*. α, β, alpha and beta lobes of mushroom body; CMB, calyces of mushroom body; IACT, inner antenno-cerebral tract; S, soma. (Adapted from Homberg et al. [1988], with permission.)

opposum; Haberly, 1985). In insects, the number of mushroom body elements can be quite large relative to the number of antennal lobe projection neurons (e.g., 830 projection neurons to 50,000 mushroom body cells in locust; Leitch and Laurent, 1996). Similar to the argument presented for ORN-to-glomerulus connections (Section 4.1.1), distributed and divergent connectivity between the first and second processing stages indicates a lack of parallel topography. While convergence of ORNs of the same functional type appears to be the organizing principle for the ORN-to-glomerulus connection (Section 4.1.1), no such organization has been described for the output connections of the bulb or lobe.

5.1. Synaptic Interactions

In terms of olfactory information processing, more is known about the anatomy and physiology of the piriform cortex and the mushroom body than other higher-order olfactory areas. At first glance, these two areas appear less similar than the circuits of the first processing stage, but do share some common features (Fig. 9.8).

Piriform cortex is composed of three layers (Fig. 9.8A): a superficial fiber/neuropil layer (Layer I), a compact layer of *pyramidal cells* (Layer II), and a deeper cell layer of pyramidal and nonpyramidal cells (Layer III) (see: Haberly, 1985, 1990; Wilson and Bower, 1992; Shipley and Ennis, 1996). Mitral/tufted cell axons enter the piriform cortex via the *lateral olfactory* tract and provide monosynaptic excitatory inputs to pyramidal cells and to a population of feedforward interneurons. In addition to providing the outputs of the cortex,

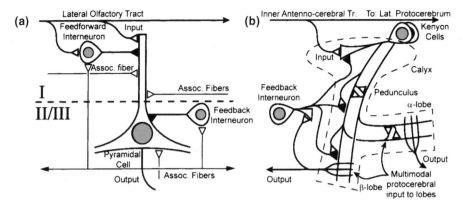

FIG. 9.8. Simplified diagrams of higher-order olfactory target centers: for clarity, only the major connections are illustrated.

(a) Schematic drawing of neural circuits in mammalian piriform cortex. (Adapted from Haberly [1985].)

(b) Schematic drawing of neural circuits in insect mushroom body and protocerebrum. (Adapted from Schürmann, 1987; Leitch and Laurent, 1996; Strausfeld and Li, 1999). Open triangles, excitatory synapses; closed triangles, inhibitory synapses. Note the monosynaptic inputs to the principal processing elements, the prominent role of feedback inhibition, and the presence of excitatory connections between principal output elements in both circuits.

pyramidal cell axon collaterals create an extensive excitatory association-fiber network within the cortex. These collaterals contact other pyramidal cells, the feedforward inhibitory interneurons, and a population of feedback inhibitory interneurons (Fig. 9.8A). A potentially important aspect of this circuit is the conduction velocities of the afferent and association fiber systems. In computer simulations, manipulating these velocities alters the oscillatory dynamics of the piriform cortex (Wilson and Bower, 1992), which may be relevant for temporal information processing (Section 6.4).

The mushroom body is divided into three main areas (Fig. 9.8B): a neuropil region called the *calyx*, a *pedunculus* of axons, and the α and β (plus γ in some species) *lobes* (Schürmann, 1987). The calyces contain the dendrites of "intrinsic" cells of the mushroom body, also known as *Kenyon cells*. The axons of a large population of mostly uniglomerular projection neurons exit the antennal lobe through the *inner antenno-cerebral tract*. These axons terminate in the lateral protocerebrum, with axon collaterals providing distinct, monosynaptic input to Kenyon cells in the calyces (Fig. 9.7B; Schürmann, 1987; Laurent and Naraghi, 1994; Strausfeld and Li, 1999). Kenyon cell axons then project via the pedunculus to the lobes, where multimodal afferents (several classes of "extrinsic" cells) converge from different protocerebral regions. These afferents are presynaptic to Kenyon cell axons, and the Kenyon cell axons are pre- and postsynaptic to each other, as well as presynaptic to efferent neurons. Some of these terminate on GABA-positive extrinsic neurons, suggesting the presence of multiple inhibitory feedback loops in the mushroom body–protocerebral circuits

(Fig. 9.8B). Depending on the species, these GABA-positive cells send processes into the pedunculus or the lobes, and into the calyces (Homberg et al., 1987; Schürmann, 1987; Rybak and Menzel, 1993; Leitch and Laurent, 1996; Brotz et al., 1997; Strausfeld and Li, 1999). Direct synaptic connections between axons in the pedunculus have also been seen, providing a possible means of excitatory interaction between Kenyon cells (Schürmann, 1987; Leitch and Laurent, 1996). A striking feature of mushroom body circuitry is the perpendicular relationship between Kenyon cell axons and extrinsic cell dendrites in the α and β lobes (Schürmann, 1987; Rybak and Menzel, 1993; Li and Strausfeld, 1997). Extrinsic cell dendrites form ordered bands across Kenyon cell axons, an architecture that may hold significance for processing temporal information (Section 6.4).

While numerous data support the classical view of the mushroom bodies as an olfactory information processing center, more recent evidence indicates that the mushroom bodies are involved in a multitude of functions including behavioral conditioning, multimodal integration, associative memory, and control of locomotion (reviewed by Heisenberg, 1998).

5.2. Physiological Responses to Odorants

Relatively few odorant responses in cells of the second processing stage have been described in detail. Recordings in mammals suggest that individual cells in piriform cortex can respond to a number of monomolecular odorants (Haberly, 1969; Tanabe et al., 1975). Patterns of activity revealed using the 2-deoxyglucose method (Cattarelli et al., 1988) and calculations based on single unit recordings (Haberly, 1985) suggest that large numbers of cortical neurons respond to stimulation with monomolecular odorants. Other single unit recordings, however, indicate sparser coding (McCollum et al., 1991).

In locusts, the predominant response of single Kenyon cells to 1-sec pulses of odorant stimuli is a membrane potential oscillation at 20 Hz (Laurent and Naraghi, 1994). The combined oscillations of many Kenyon cells can be recorded extracellularly in the local field potential. Apparently, relatively few Kenyon cells fire action potentials in response to the odorants tested. When present, however, spikes in single cells are phase-locked to the field potential oscillation. Few data are available on odorant selectivity in individual Kenyon cells or extrinsic neurons. Recordings from extrinsic neurons indicate that they are *multimodal*, responding to stimuli from a number of sensory systems (e.g., Li and Strausfeld, 1997). The implication from this and other research is that the mushroom bodies may be a site of association memory in addition to serving a role in olfactory signal processing (see Section 6.3).

6. CODING STRATEGIES AND FUNCTIONAL IMPLICATIONS

6.1. Spatial Codes in the Brain

How does the brain distinguish which of the many possible combinations of ORNs are responsive to a particular odorant? Insight into this question may be gained by considering the multidimensional nature of odorant stimuli, the ORN-to-glomerulus mapping, the functional unit properties of the glomerulus, and the responses of olfactory neurons at various points in the pathway.

6.1.1. Labeled Lines Versus Across-Fiber Patterns

In his classic volume, *The Hungry Fly*, Vincent Dethier discussed two mechanisms by which chemosensory information might be represented in the brain. In the "labeled-line" model, the sensory axons carrying information to the CNS are "absolutely restricted" with respect to selectivity, whereas in an "across-fiber" code, "each stimulus would produce a different and characteristic total response profile" across the entire population of sensory neurons (Dethier, 1976; see also Chapter 1, Finger et al., and Chapter 14, Smith and Davis). While Dethier's discussion focused on the modality of taste, the same principles can and have been applied to olfaction (e.g., Shepherd, 1985). Although these terms appear to represent useful concepts, their application to olfactory coding is not entirely straightforward. On the surface, the ORN-to-glomerulus organization appears to represent a striking example of such a labeled-line system (Section 4.1.1). Numerous investigations, however, indicate that even monomolecular stimuli activate many glomeruli and their associated neurons (Stewart et al., 1979; Kauer, 1988; Duchamp-Viret et al., 1990; Guthrie et al., 1993; Cinelli et al., 1995; Kang and Caprio, 1995; Friedrich and Korsching, 1997). When coupled with the broadly tuned nature of most ORN responses (Section 2.2), these data would seem to support an across-fiber or combinatorial code. Ideas about the nature of such a code have developed over the years (Polak, 1973; Holley and Døving, 1977; Kauer, 1980, 1991; Wright, 1982) and a representation of many of these ideas is shown in Figure 9.9.

In essence, the difficulty in characterizing an olfactory system as either a labeled-line or an across-fiber system lies in how one defines the stimulus. It is necessary to determine whether the "label" is a class of molecules, a single molecule, or only a single feature (odotope) of a molecule. The hypothesis represented in Figure 9.9 suggests that even a monomolecular stimulus may contain multiple odotopes, each detected by a different ORN type. A given odorant would then activate the combination of ORNs that encode the different molecular features (Section 2.2), resulting in an across-fiber response profile that is distributed over a subset of the total population of glomeruli and their associated output neurons.

Another difficulty in applying a strict "labeled-line" or "across-fiber" nomenclature to the olfactory system is that, as generally used, the terms are mutually exclusive and send the implicit message that an olfactory system encodes all odors through one or the other mechanism. As Figure 9.9 suggests, however, olfactory systems may employ elements of both coding strategies; that is, labeled lines may be nested within across-fiber patterns. The MGC system in male moths is one example that illustrates this concept (Section 3.2). Because of the extreme sensitivity and sharply defined responses of both peripheral and central pheromonal pathways to these particular odorants, it has become

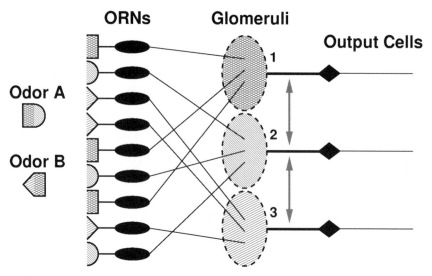

FIG. 9.9. Schematic representation of a hypothesis for coding the molecular features (odotopes) of odors in the olfactory system. This diagram is based on ideas contributed by many olfactory research labs over the past several decades (see text). Odors A and B each contain common features, as well as different features, here represented by geometric shapes and shading. These features are detected by ORNs, with each feature maximally stimulating a different ORN response type. Scattered ORNs of the same response type (i.e., expressing the same receptor protein) project in a convergent manner to individual glomeruli in the olfactory bulb or antennal lobe. Input activity to an individual glomerulus then leads to activity in output cells connected to the glomerulus (only one shown here for each glomerulus), modified by lateral interactions (double-headed arrows). Stimulation with the two odors produces distinct but partially overlapping spatial patterns of activity in the bulb or lobe. In this simple model, Odor A would stimulate output in the form of spike trains from glomeruli 1 and 2, whereas Odor B would evoke output from glomeruli 1 and 3.

commonplace to refer to this system as a labeled line. In fact, several pieces of converging evidence suggest that this strict designation is misleading, and that all available data are consistent with a combinatorial mechanism governing pheromonal coding in the MGC (next section).

6.1.2. A New Concept: Across-Label Coding

Functional and anatomical studies in several moth species have shown that the different molecular features of a pheromone are encoded through interactions within and between the two to four glomeruli that comprise the MGC (reviewed in: Christensen et al., 1996; Vickers et al., 1998; Hansson and Christensen, 1999). At least four functionally distinct classes of output neuron encode information about the pheromone, which in most species is a blend of two chemically related molecules. Two of the four output pathways relay odor-specific information about the two pheromone components, and each arises from a different glomerulus. The other two pathways, however, integrate information about the two components: output neurons in one pathway respond equally well to either component ("pheromone-generalist neurons"), whereas the other pathway responds selectively to the mixture of the two components ("blend neurons"—next section). In several species, a separate MGC glomerulus is the convergence site for ORNs that respond selectively to an odorant (often released by sympatric species) that *inhibits* upwind flight (review: Hansson and Christensen, 1999). These findings therefore show that discrimination of attractive and deterrent odors in these insects is not strictly governed by labeled lines, but depends upon a combinatorial coding process. This process involves the ensemble activity of a diverse array of parallel output pathways from the MGC (Vickers et al., 1998). Thus, given these results as well as evidence for functional specificity of ORNs projecting to the MGC (Section 2.2 and Fig. 9.2B), it appears that this system encodes a specific mixture of odorant molecules rather than a group of odotopes associated with a monomolecular stimulus (preceding section). A hybrid term, *"across-label"* coding, may be more appropriate to describe the processing scheme in this and other olfactory systems characterized by ORNs with defined molecular response profiles. Each glomerulus may serve as the locus for a distinct odorant input, and through parallel distributed processing, different odor stimuli are represented in distinct patterns of activity across partially overlapping subsets of glomeruli.

6.2. Interglomerular Integration—Encoding an Odor Blend

The discussion thus far has emphasized the role of individual glomeruli in olfactory coding. Given the complex circuitry involved (Section 4.2), however, it is not surprising that bulb and lobe output cells do not simply transmit glomerular activity in a one-to-one fashion to the second processing stage. As an example, some projection neurons innervating the moth MGC integrate information about the specific blend of odorants that serve as the species-specific sex pheromone. These so-called *blend* neurons perform a higher level of processing than other neurons that arborize in the same glomerulus, in that their response to the blend is uniquely different from the response to the individual components. In the sphinx moth, *Manduca sexta*, blend neurons may be excited by one pheromone component, inhibited by the other, and display a multiphasic postsynaptic response to the mixture of the two (Christensen and Hildebrand, 1997; Christensen et al., 1998). In some noctuid moths, blend neurons often show no response to the individual blend components, but they respond to the blend with a pronounced burst of action potentials that often far outlasts the stimulation period (review: Hansson and Christensen, 1999). These blend neurons therefore do not simply encode information about the presence of the odor, they respond specifically to the proper ratio of odorants that comprises the sex pheromone in these species.

6.3. Spatial Processing at the Second Stage

The data discussed above suggest that convergence of ORN types as defined by receptor proteins is a main organizing principle in the convergent, nonparallel ORN-to-glomerulus connection (Section 4.1.1). The divergent, but also nonparallel nature of the first stage (bulb or lobe) to second stage (cortex or protocerebrum) connection suggests another processing strategy. From studies on vertebrates, the distributed connections, internal circuitry, and broad odorant response properties have suggested to a number of researchers that the piriform cortex may function as a *content addressable memory* (CAM) (Haberly, 1985; Hasselmo et al., 1990). CAMs are a form of computational neural network that encode memories in a spatially distributed fashion—each element of the network participates in the storage of several memories. Memories are stored through a self-organizing mechanism, whereby connections between network elements are modified based on exposure to input patterns that are to be remembered. Consistent with a memory function, synaptic modification occurs in the piriform cortex (Kanter and Haberly, 1993). As a test of these ideas, a computer simulation of the piriform cortex appears able to function in a way similar to a CAM, storing large numbers of spatial input patterns in a distributed fashion (Hasselmo et al., 1990).

Considerable evidence also indicates a major role for the mushroom body and associated neurons in olfactory memory in insects. The spiking behavior of an identified mushroom body extrinsic neuron changes with odor conditioning (Mauelshagen, 1993) and there is physiological evidence of synaptic potentiation of antennal lobe inputs to the calyces (Oleskevich et al., 1997). In addition, data from a variety of studies in fruitflies (*Drosophila melanogaster*) indicate that biochemical components critical for olfactory learning are located in mushroom body Kenyon cells (Davis, 1996). Whether spatial codes play a role in such a memory function has not been described explicitly.

6.4. Temporal Codes and Encoding Time

In recent years, several studies have provided support for the idea that the brain uses more than spatial activity maps to discriminate one olfactory stimulus from another. It is now quite apparent that olfactory circuits also employ another important parameter in the process of odor representation—time.

6.4.1. Neural Synchronization and Odor Discrimination

As described in Section 4.3, odorant stimulation in locusts induces spike firing in projection neurons that is phase locked to the local field potential oscillations. In the protocerebrum, anatomical and physiological data suggest that the mushroom body may be involved in processing temporal signals such as these (for example, Erber et al., 1987). In particular, the perpendicular orientation of extrinsic cell dendrites across Kenyon cell axons in the lobes suggested to Schürmann (1987) that the mushroom bodies act as "an apparatus for sequential activation and signaling in the millisecond range along time delay lines." Work by Laurent and colleagues (review: Laurent, 1996) has suggested that a given odor may be represented in the antennal lobe by a unique dynamical pattern of synchronized activity across a spatially distributed population of projection neurons. The patterns of projection neuron activity are thought to elicit a subsequent dynamical pattern of spiking in small numbers of mushroom body neurons. The resulting sparse code could thereby increase the memory capacity of the mushroom body (Laurent, 1996). In support of this hypothesis, pharmacological treatments that eliminate projection neuron synchronization reduce the ability of honey bees to discriminate behaviorally between structurally similar odorants, but not between dissimilar odorants (Stopfer et al., 1997). Taken together, the data suggest that the mushroom body is critically important for odorant discrimination and memory. However, some odorant detection ability remains after mushroom body ablation (de Belle and Heisenberg, 1994). It is therefore likely that antennal lobe projections to other protocerebral sites support some form of odorant processing (reviews: Davis, 1996; Kanzaki, 1997), but specific data are lacking.

Oscillatory activity in the olfactory bulb and piriform cortex has been observed in most vertebrates since the earliest recordings from these areas (Adrian, 1942). Data from several studies suggest that oscillations in piriform cortex are primarily due to oscillatory inputs from the olfactory bulb, although intrinsic oscillations have been seen under certain experimental conditions and have been reproduced in computer simulations of piriform cortex (review: Ketchum and Haberly, 1991). While there is little direct experimental evidence regarding the functional significance of these oscillations, researchers have suggested that they serve to "quantize" activity into time intervals matching inherent time constants of the circuit (Wilson and Bower, 1992; Ketchum and Haberly, 1993). This would act to synchronize neural events, which may be important for processing spatio-temporal activity patterns or for altering synaptic efficacy on short time scales (Ketchum and Haberly, 1993).

Although most recent models of piriform cortex focus on spatial patterns of activity (Section 6.3), Haberly (1985) suggests that it is possible that the temporal patterns of activity in mitral cells "are converted to spatial ones as a result of the sequential activation of the cortex and the apparent temporal ordering of feedback through the long association fiber systems." Along these lines, Hopfield (1995) proposed a general mechanism whereby spike latency, rather than firing rate, encodes odorant information. A system of delay lines, as perhaps implemented in the piriform cortex, could then decode this information. In a real-word demonstration of these ideas, White et al. (1998) have developed an artificial chemical detection system that incorporates a computer simulation of the olfactory bulb and a delay line decoding mechanism. The output of these two processing steps represents odorant identity using a spatial code and odorant concentration using a temporal code.

6.4.2. Context-Dependent Odor Processing

The invertebrate, vertebrate, and computational models described above assign a critical function to the temporal patterning of the neural response. They do not, however, take into account that the odor stimulus itself changes (often unpredictably) over time. It is therefore not known if these unique temporal patterns would persist under naturally intermittent stimulus conditions, where stimulus concentrations can change on a millisecond time scale. Another coding model is based on results from several lepidopteran species and asserts that the olfactory system is designed to *encode* time in the process of odor discrimination. More specifically, in many moth projection neurons, especially the *blend neurons* (Section 6.2), the temporal pattern of the response is strongly dependent on the temporal pattern of the odorant stimulus itself (Fig. 9.5C). In these glomerular circuits, the molecular identity of the odor cannot be uniquely represented in the temporal pattern of evoked activity because spike patterns are not tied to a coherent network oscillation, and changes in stimulus timing and concentration evoke central activity patterns that reflect these changes. When combined with behavioral evidence, these data suggest that context-dependent odor responses in the brain provide the animal with vital information about the changing environmental conditions in natural odor plumes, in addition to information about the molecular identity of the stimulus (Christensen et al., 1998). This model of odor representation is thus distinctly different from previous models, which focus mainly on the molecular parameters of an odorant, because it also stresses the importance of stimulus temporal properties in shaping the neural representation of an olfactory stimulus.

6.5. Integrating Space and Time

Although we have discussed spatial and temporal codes separately, olfactory representations undoubtedly combine the two into a single *spatio-temporal* code. In essence, it seems that olfactory information at each level of the pathway is represented not only by *which* cells fire action potentials, but also *when* they fire them. The odorant information contained in such a code would be a combinatorial representation of the multiple chemical features of the odorant stimulus, whether found on an individual odorant molecule or in a blend of odorants. The olfactory code is additionally constrained to function with short stimulus durations found in natural odor plumes or as occur with sniffing. A code that utilizes spike *timing* can accommodate such brief stimulus events (Rieke et al., 1997). A spatiotemporal code thus represents a flexible, robust means of representing the complex multidimensional nature of odorant stimuli as they evolve over time.

7. CONCLUSION

Over the past decade, olfactory neuroscientists have made remarkable advances toward understanding how the brain encodes olfactory signals, but much work remains to be done. A thorough understanding of these processes is still hindered by our general lack of knowledge about the optimal odor ligands that activate olfactory receptors, and the behavioral and social relevance of these numerous and varied odor signals to the life of the animal. In the realm of information processing, we are only just beginning to understand the importance of combining our exploration of single-unit spike activity with that of population codes. Such information will give a complete sense of both the individual elements that make up olfactory circuits and the operation of the circuits as functional ensembles. Using this approach, we are not only beginning to understand how the chemical parameters of an odor stimulus are encoded within and between glomerular circuits, we are learning how the spatiotemporal features of the stimulus environment are processed. Odorant molecular properties, circuit processing, and stimulus timing all affect odor perception.

By comparing results obtained from insects and other invertebrates with data from mammals and other vertebrates, it is clear that certain principles underlying the glomerular organization of the primary olfactory projection into the brain have been conserved across phyla. Equally interesting, however, are the important differences, even between closely related species, and these continue to caution us against generalizing our results before sufficient evidence is at hand. In sum, important strides have been made, but only through the concerted efforts of anatomists, physiologists, molecular biologists, behaviorists, physicists, and computational biologists. Indeed, if we have learned nothing else from studying the olfactory system, we have learned that sniffing out the brain's olfactory code will surely involve a broad, multidisciplinary approach. If our rate of progress over the recent years is any indication, the near future will be profoundly rich and exciting.

8. SUMMARY

In this chapter, current ideas on the neural representation of olfactory information using data drawn from a variety of invertebrate and vertebrate species are reviewed. Beginning with the olfactory receptor neurons in the periphery, the olfactory pathway, including the neuronal circuits of the first and second olfactory processing stages are briefly described. Then the response properties of ORNs and how these responses suggest a distributed olfactory code in the brain are detailed. In the first processing stage of both invertebrates and vertebrates, the glomerulus emerges as an important module for organizing olfactory information. A wealth of data indicates that an individual glomerulus is the convergence point of ORNs that display a single response type, and thus likely express the same receptor protein(s). The anatomy and physiology of both the first and second processing stages are consistent with a distributed olfactory code, and recordings from both areas also indicate the importance of a temporal component. We discuss possible ways that such a spatiotemporal code may support odorant discrimination, as well as enable odorant tracking over time. While there are striking similarities in olfactory systems across species, there are also important differences, suggesting that olfactory systems have evolved multiple coding mechanisms in order to solve a wide variety of different olfactory tasks.

ACKNOWLEDGMENTS

We wish to thank our co-workers, particularly our many friends in the labs of John Hildebrand and John Kauer, who have provided endless hours of stimulating conversation, challenging comments, and helpful ideas. We also thank Nicholas Strausfeld and Mike Meredith for helpful suggestions. Supported by grants from the USDA (97-35302-4868) to T.A.C, from NIH (AI-23253 and DC-02751) to John Hildebrand, and from ONR (N00014-95-1-1340) and DARPA (9636-A-065) to J.W.

REFERENCES

Ache, B. W. (1991). Phylogeny of smell and taste. In: T. V. Getchell, R. L. Doty, L. M. Bartoshuk, and J. B. Snow, Jr. (eds). *Smell and Taste in Health and Disease*. New York: Raven Press, pp 3–18.

Adrian, E. D. (1942). Olfactory reactions in the brain of the hedgehog. *J Physiol* 100:459–473.

Alkasab, T. K., T. C. Bozza, T. A. Cleland, K. M. Dorries, T. C. Pearce, J. White, and J. S. Kauer (1999). Characterizing complex chemosensors: Information theoretic analysis of olfactory systems. *Trends Neurosci* 22:102–108.

Almaas, T. J., T. A. Christensen, and H. Mustaparta (1991). Chemical communication in heliothine moths I. Antennal receptor neurons encode several feature of intra and interspecific odorants in the male corn earworm moth *Helicoverpa zea*. *J Comp Physiol A* 169:249–258.

Atema, J. (1995). Chemical signals in the marine environment: dispersal, detection, and temporal signal analysis. *Proc Natl Acad Sci U S A* 92:62–66.

Baier, H., S. Rotter, and S. Korsching (1994). Connectional topography in the zebrafish olfactory system: random positions but regular spacing of sensory neurons projecting to an individual glomerulus. *Proc Natl Acad Sci U S A* 91:11646–11650.

Beets, M. G. (1970). The molecular parameters of olfactory response. *Pharm Rev* 22:1–34.

Boeckh, J. and L. P. Tolbert (1993). Synaptic organization and development of the antennal lobe in insects. *Microsc Res Tech* 24:260–280.

Bozza, T. C. and J. S. Kauer (1998). Odorant response properties of convergent olfactory receptor neurons. *J Neurosci* 18:4560–4569.

Brotz, T. M., B. Bochenek, K. Aronstein, R. H. French-Constant, and A. Borst (1997). Gamma-aminobutyric acid receptor distribution in the mushroom bodies of a fly (*Calliphora erythrocephala*): a functional subdivision of Kenyon cells. *J Comp Neurol* 383:42–48.

Buck, L. and R. Axel (1991). A novel multigene family may encode odorant receptors: a molecular basis for odor recognition. *Cell* 65:175–187.

Cattarelli, M., L. Astic, and J. S. Kauer (1988). Metabolic mapping of 2-deoxyglucose uptake in the rat piriform cortex using computerized image processing. *Brain Res* 442:180–184.

Christensen, T. A. and J. G. Hildebrand (1987). Functions, organization, and physiology of the olfactory pathways in the Lepidopteran brain. In: A. P. Gupta (ed). *Arthropod Brain: Its Evolution, Development, Structure, and Functions.* New York: John Wiley and Sons, pp 457–484.

Christensen, T. A. and J. G. Hildebrand (1997). Coincident stimulation with pheromone components improves temporal pattern resolution in central olfactory neurons. *J Neurophysiol* **77**:775–781.

Christensen, T. A., B. R. Waldrop, I. D. Harrow, and J. G. Hildebrand (1993). Local interneurons and information processing in the olfactory glomeruli of the moth *Manduca sexta. J Comp Physiol A* **173**:385–399.

Christensen, T. A., I. D. Harrow, C. Cuzzocrea, P. W. Randolph, and J. G. Hildebrand (1995). Distinct projections of two populations of olfactory receptor axons in the antennal lobe of the sphinx moth *Manduca sexta. Chem Senses* **20**:313–323.

Christensen, T. A., T. Heinbockel, and J. G. Hildebrand (1996). Olfactory information processing in the brain: encoding chemical and temporal features of odors. *J Neurobiol* **30**:82–91.

Christensen, T. A., B. R. Waldrop, and J. G. Hildebrand (1998). Multitasking in the olfactory system: context-dependent responses to odors reveal dual GABA-regulated coding mechanisms in single olfactory projection neurons. *J Neurosci* **18**:5999–6008.

Cinelli, A. R., K. A. Hamilton, and J. S. Kauer (1995). Salamander olfactory bulb neuronal activity observed by video rate, voltage-sensitive dye imaging. III. Spatial and temporal properties of responses evoked by odorant stimulation. *J Neurophysiol* **73**:2053–2071.

Davis, R. L. (1996). Physiology and biochemistry of *Drosophila* learning mutants. *Physiol Rev* **76**:299–317.

de Belle, J. S. and M. Heisenberg (1994). Associative odor learning in *Drosophila* abolished by chemical ablation of mushroom bodies. *Science* **263**:692–695.

Dethier, V. G. (1976). *The Hungry Fly.* Cambridge, MA: Harvard University Press.

Distler, P. G. and J. Boeckh (1997a). Synaptic connection between identified neuron types in the antennal lobe glomeruli of the cockroach, *Periplaneta americana.* I. Uniglomerular projection neurons. *J Comp Neurol* **378**:307–319.

Distler, P. G. and J. Boeckh (1997b). Synaptic connection between identified neuron types in the antennal lobe glomeruli of the cockroach *Periplaneta americana.* II. Local multiglomerular interneurons. *J Comp Neurol* **383**:529–540.

Distler, P. G., C. Gruber, and J. Boeckh (1998). Synaptic connection between GABA-immunoreactive neurons and uniglomerular projections neurons within the antennal lobe of the cockroach *Periplaneta americana. Synapse* **29**:1–13.

Duchamp-Viret, P., A. Duchamp, and M. Vigouroux (1989). Amplifying role of convergence in olfactory system a comparative study of receptor cell and second order neuron sensitivities. *J Neurophysiol* **61**:1085–1094.

Duchamp-Viret, P., A. Duchamp, and G. Sicard (1990). Olfactory discrimination over a wide concentration range. Comparison of receptor cell and bulb neuron abilities. *Brain Res* **517**:256–262.

Duchamp-Viret, P., A. Duchamp, and M. Chaput (1993). GABAergic control of odor-induced activity in the frog olfactory bulb: electrophysiological study with picrotoxin and bicuculline. *Neuroscience* **53**:111–120.

Erber, J., U. Homberg, and W. Gronenberg (1987). Functional roles of the mushroom bodies in insects. In: A. P. Gupta (ed). *Arthropod Brain: Its Evolution, Development, Structure, and Functions*. New York: John Wiley and Sons, pp 485–511.

Friedrich, R. W. and S. I. Korsching (1997). Combinatorial and chemotopic odorant coding in the zebrafish olfactory bulb visualized by optical imaging. *Neuron* 18:737–752.

Gelperin, A., D. Kleinfeld, W. Denk, and I. R. Cooke (1996). Oscillations and gaseous oxides in invertebrate olfaction. *J Neurobiol* 30:110–122.

Guthrie, K. M., A. J. Anderson, M. Leon, and C. Gall (1993). Odor-induced increases in c-fos mRNA expression reveal an anatomical "unit" for odor processing in olfactory bulb. *Proc Natl Acad Sci U S A* 90:3329–3333.

Haberly, L. B. (1969). Single unit responses to odor in the prepyriform cortex of the rat. *Brain Res* 12:481–484.

Haberly, L. B. (1985). Neuronal circuitry in olfactory cortex: anatomy and functional implications. *Chem Senses* 10:219–238.

Haberly, L. B. (1990). Comparative aspects of olfactory cortex. In: E. G. Jones and A. Peters (eds). *Cerebral Cortex Vol 8B*. New York: Plenum Publishing, pp 137–166.

Hahn, I., P. W. Scherer, and M. M. Mozell (1993). Velocity profiles measured for airflow through a large-scale model of the human nasal cavity. *J Appl Physiol* 75:2273–2287.

Halasz, N. and C. A. Greer (1993). Terminal arborizations of olfactory nerve fibers in the glomeruli of the olfactory bulb. *J Comp Neurol* 337:307–316.

Hamilton, K. A. and J. S. Kauer (1985). Intracellular potentials of salamander mitral/tufted neurons in response to odor stimulation. *Brain Res* 338:181–185.

Hansson, B. S. and T. A. Christensen (1999). Functional characteristics of the antennal lobe. In: B. S. Hansson (ed). *Insect Olfaction*. Berlin: Springer, pp 125–161.

Hartlieb, E., S. Anton, and B. S. Hansson (1997). Dose-dependent response characteristics of antennal lobe neurons in the male moth *Agrotis segetum* (Lepidoptera: Noctuidae). *J Comp Physiol A* 181:469–476.

Hasselmo, M. E., M. A. Wilson, B. P. Anderson, and J. M. Bower (1990). Associative memory function in piriform (olfactory) cortex: computational modeling and neuropharmacology. *Cold Spring Harb Symp Quant Biol* LV:599–610.

Heinbockel, T., T. A. Christensen, and J. G. Hildebrand (1993). Receptive fields of sex-pheromone responsive neurons in the antennal lobes of *Manduca sexta*. *Soc Neurosci Abstr* 19:126–126.

Heinbockel, T., P. Kloppenburg, and J. G. Hildebrand (1998). Pheromone-evoked potentials and oscillations in the antennal lobes of the sphinx moth *Manduca sexta*. *J Comp Physiol A* 182:703–714.

Heisenberg, M. (1998). What do the mushroom bodies do for the insect brain? An introduction. *Learning Memory* 5:1–10.

Hildebrand, J. G. and G. M. Shepherd (1997). Mechanisms of olfactory discrimination: converging evidence for common principles across phyla. *Annu Rev Neurosci* 20:595–631.

Holley, A. and K. B. Døving (1977). Receptor sensitivity, acceptor distribution, convergence and neural coding in the olfactory system. In: J. LeMagnen and P. MacLeod (eds). *Olfaction and Taste VI*. London: IRL Press, pp 113–123.

Homberg, U., T. G. Kingan, and J. G. Hildebrand (1987). Immunocytochemistry of GABA in the brain and suboesophageal ganglion of *Manduca sexta*. *Cell Tissue Res* 248:1–24.

Homberg, U., R. A. Montague, and J. G. Hildebrand (1988). Anatomy of antenno-cerebral pathways in the brain of the sphinx moth *Manduca sexta*. *Cell Tissue Res* **254**: 255–281.

Homberg, U., S. Hoskins, and J. G. Hildebrand (1995). Distribution of acetylcholinesterase activity in the deutocerebrum of the sphinx moth *Manduca sexta*. *Cell Tissue Res* **279**:249–259.

Hopfield, J. J. (1995). Pattern recognition computation using action potential timing for stimulus representation. *Nature* **376**:33–36.

Hosl, M. (1990). Pheromone-sensitive neurons in the deutocerebrum of *Periplaneta americana*: receptive fields on the antenna. *J Comp Physiol A* **167**:321–327.

Jiang, T. and A. Holley (1992). Some properties of receptive fields of olfactory mitral/tufted cells in the frog. *J Neurophysiol* **68**:726–733.

Kafka, W. A. (1987). Similarity of reaction spectra and odor discrimination: single receptor cell recordings in *Antheraea polyphemus* (Saturniidae). *J Comp Physiol A* **161**:867–880.

Kang, J. and J. Caprio (1995). Electrophysiological responses of single olfactory bulb neurons to amino acids in the channel catfish, *Ictalurus punctatus*. *J Neurophysiol* **74**: 1421–1434.

Kanter, E. D. and L. B. Haberly (1993). Associative long-term potentiation in piriform cortex slices requires GABA$_A$ blockade. *J Neurosci* **13**:2477–2482.

Kanzaki, R. (1997). Pheromone processing in the lateral accessory lobes of the moth brain: flip-flopping signals related to zigzagging upwind walking. In: R. T. Carde and A. K. Minks (eds). *Insect Pheromone Research: New Directions*. New York: Chapman and Hall, pp 291–303.

Kauer, J. S. (1974). Response patterns of amphibian olfactory bulb neurones to odour stimulation. *J Physiol* **243**:695–715.

Kauer, J. S. (1980). Some spatial characteristics of central information processing in the vertebrate olfactory pathway. In: H. van der Starre (ed). *Olfaction and Taste VII*. London: IRL Press, pp 227–236.

Kauer, J. S. (1981). Olfactory receptor cell staining using horseradish peroxidase. *Anat Rec* **200**:331–336.

Kauer, J. S. (1988). Real-time imaging of evoked activity in local circuits of the salamander olfactory bulb. *Nature* **331**:166–168.

Kauer, J. S. (1991). Contributions of topography and parallel processing to odor coding in the vertebrate olfactory pathway. *Trends Neurosci* **14**:79–85.

Kauer, J. S. and D. G. Moulton (1974). Responses of olfactory bulb neurones to odour stimulation of small nasal areas in the salamander. *J Physiol* **243**:717–737.

Ketchum, K. L. and L. B. Haberly (1991). Fast oscillations and dispersive propagation in olfactory cortex and other cortical areas: a functional hypothesis. In: J. L. Davis and H. Eichenbaum (eds). *Olfaction: A Model System for Computational Neuroscience*. Cambridge, MA: The MIT Press, pp 69–100.

Ketchum, K. L. and L. B. Haberly (1993). Synaptic events that generate fast oscillations in piriform cortex. *J Neurosci* **13**:3980–3985.

Laurent, G. (1996). Dynamical representation of odors by oscillating and evolving neural assemblies. *Trends Neurosci* **19**:489–496.

Laurent, G. (1999). A systems perspective on early olfactory coding. *Science* **286**: 723–728.

Laurent, G. and M. Naraghi (1994). Odorant-induced oscillations in the mushroom bodies of the locust. *J Neurosci* **14**:2993–3004.

Lee, J. K. and N. J. Strausfeld (1990). Structure, distribution and number of surface sensilla and their receptor cells on the olfactory appendage of the male moth *Manduca sexta*. *J Neurocytol* **19**:519–538.

Leise, E. M. (1990). Modular construction of nervous systems: a basic principle of design for invertebrates and vertebrates. *Brain Res Rev* **15**:1–23.

Leitch, B. and G. Laurent (1996). GABAergic synapses in the antennal lobe and mushroom body of the locust olfactory system. *J Comp Neurol* **372**:487–514.

Li, Y. and N. J. Strausfeld (1997). Morphology and sensory modality of mushroom body extrinsic neurons in the brain of the cockroach, *Periplaneta americana*. *J Comp Neurol* **387**:631–650.

Mauelshagen, J. (1993). Neural correlates of olfactory learning paradigms in an identified neuron in the honeybee brain. *J Neurophysiol* **69**:609–625.

McCollum, J., J. Larson, T. Otto, F. Schottler, R. Granger, and G. Lynch (1991). Short-latency single unit processing in olfactory cortex. *J Cogn Neurosci* **3**:293–299.

Meisami, E. and L. Safari (1981). A quantitative study of the effects of early unilateral olfactory deprivation on the number and distribution of mitral and tufted cells and of glomeruli in the rat olfactory bulb. *Brain Res* **221**:81–107.

Mellon, D. and S. D. Munger (1990). Nontopographic projection of olfactory sensory neurons in the crayfish brain. *J Comp Neurol* **296**:253–262.

Meredith, M. (1986). Patterned response to odor in mammalian olfactory bulb: the influence of intensity. *J Neurophysiol* **56**:572–597.

Meredith, M. (1992). Neural circuit computation: complex patterns in the olfactory bulb. *Brain Res Bull* **29**:111–117.

Mombaerts, P., F. Wang, C. Dulac, S. K. Chao, A. Nemes, M. Mendelsohn, J. Edmondson, and R. Axel (1996). Visualizing an olfactory sensory map. *Cell* **87**:675–686.

Mori, K. and G. M. Shepherd (1994). Emerging principles of molecular signal processing by mitral/tufted cells in the olfactory bulb. *Semin Cell Biol* **5**:65–74.

Mori, K., N. Hiroshi, and Y. Yoshihara (1999). The olfactory bulb: coding and processing of odor molecule information. *Science* **286**:711–715.

Mountcastle, V. B., R. H. Lamotte, and G. Carli (1972). Detection thresholds for stimuli in humans and monkeys: comparison with threshold events in mechanoreceptive afferent nerve fibers innervating the monkey hand. *J Neurophysiol* **35**:122–136.

Mozell, M. M. and M. Jagodowicz (1973). Chromatographic separation of odorants by the nose: retention times measured across *in vivo* olfactory mucosa. *Science* **181**:1247–1249.

Murlis, J. (1997). Odor plumes and the signal they provide. In: R. T. Carde and A. K. Minks (eds). *Insect Pheromone Research: New Directions*. New York: Chapman and Hall, pp 221–231.

Ngai, J., A. Chess, M. M. Dowling, N. Necles, E. R. Macagno, and R. Axel (1993). Coding of olfactory information: topography of odorant receptor expression in the catfish olfactory epithelium. *Cell* **72**:667–680.

Ojima, H., K. Mori, and K. Kishi (1984). The trajectory of mitral cell axons in the rabbit olfactory cortex revealed by intracellular HRP injection. *J Comp Neurol* **230**:77–87.

Oleskevich, S., J. D. Clements, and M. V. Srinivasan (1997). Long-term synaptic plasticity in the honeybee. *J Neurophysiol* **78**:528–532.

Polak, E. H. (1973). Multiple profile–multiple receptor site model for vertebrate olfaction. *J Theor Biol* **40**:469–484.

Ressler, K. J., S. L. Sullivan, and L. B. Buck (1993). A zonal organization of odorant receptor gene expression in the olfactory epithelium. *Cell* **73**:597–609.

Ressler, K. J., S. L. Sullivan, and L. B. Buck (1994). Information coding in the olfactory system: evidence for a stereotyped and highly organized epitope map in the olfactory bulb. *Cell* **79**:1245–1255.

Revial, M. F., A. Duchamp, and A. Holley (1978). Odour discrimination by frog olfactory receptors: a second study. *Chem Senses Flav* **3**:7–21.

Rieke, F., D. Warland, R. de Ruyter van Steveninck, and W. Bialek (1997). *Spikes: Exploring the Neural Code.* Cambridge, MA: MIT Press.

Ring, G., R. C. Mezza, and J. E. Schwob (1997). Immunohistochemical identification of discrete subsets of rat olfactory neurons and the glomeruli that they innervate. *J Comp Neurol* **388**:415–434.

Rybak, J. and R. Menzel (1993). Anatomy of the mushroom bodies in the honey bee brain: the neuronal connections of the alpha-lobe. *J Comp Neurol* **334**:444–465.

Sato, T., J. Hirono, M. Tonoike, and M. Takebayashi (1994). Tuning specificities to aliphatic odorants in mouse olfactory receptor neurons and their local distribution. *J Neurophysiol* **72**:2980–2989.

Schmidt, M. and B. W. Ache (1992). Antennular projections to the midbrain of the spiny lobster. II. Sensory innervation of the olfactory lobe. *J Comp Neurol* **318**:291–303.

Schoenfeld, T. A., A. N. Clancy, W. B. Forbes, and M. Foteos (1994). The spatial organization of the peripheral olfactory system of the hamster. Part I: Receptor neuron projections to the main olfactory bulb. *Brain Res Bull* **34**:183–210.

Schürmann, F. W. (1987). The architecture of the mushroom bodies and related neuropils in the insect brain. In: A. P. Gupta (ed). *Arthropod Brain: Its Evolution, Development, Structure, and Functions.* New York: John Wiley and Sons, pp 231–264.

Scott, J. W. (1981). Electrophysiological identification of mitral and tufted cells and distributions of their axons in olfactory system of the rat. *J Neurophysiol* **46**:918–931.

Scott, J. W., E. C. Ranier, J. L. Pemberton, E. Orona, and L. E. Mouradian (1985). Pattern of rat olfactory bulb mitral and tufted cell connections to the anterior olfactory nucleus pars externa. *J Comp Neurol* **242**:415–424.

Shepherd, G. M. (1985). Are there labeled lines in the olfactory pathway? In: D. W. Pfaff (ed). *Taste, Olfaction, and the Central Nervous System.* New York: The Rockefeller University Press, pp 307–321.

Shepherd, G. M. (1987). A molecular vocabulary for olfaction. *Ann N Y Acad Sci* **510**: 98–103.

Shepherd, G. M. (1990). Contribution toward a theory of olfaction. In: K. Colbow (ed). *Frank Allison Linville's R.H. Wright Lectures on Olfactory Research.* Burnaby, B C: Simon Frasier University, pp 61–109.

Shepherd, G. M. (1992). Modules for molecules. *Nature* **358**:457–458.

Shipley, M. T. and M. Ennis (1996). Functional organization of the olfactory system. *J Neurobiol* **30**:123–176.

Sorensen, P. W., T. A. Christensen, and N. E. Stacey (1998). Discrimination of pheromonal cues in fishes: emerging parallels with insects. *Curr Opin Neurobiol* **8**:458–467.

Stewart, W. B., J. S. Kauer, and G. M. Shepherd (1979). Functional organization of rat olfactory bulb analysed by the 2-deoxyglucose method. *J Comp Neurol* **185**:715–734.

Stopfer, M., S. Bhagavan, B. H. Smith, and G. Laurent (1997). Impaired odour discrimination on desynchronization of odour-encoding neural assemblies. *Nature* **390**:70–74.

Strausfeld, N. J. and Y. Li (1999). Organization of olfactory and multimodal afferent neurons supplying the calyx and pedunculus of the cockroach mushroom bodies. *J Comp Neurol* **409**:603–625.

Strotmann, J., I. Wanner, T. Helfrich, A. Beck, C. Meinken, S. Kubick, and H. Breer (1994). Olfactory neurones expressing distinct odorant receptor subtypes are spatially segregated in the nasal neuroepithelium. *Cell Tissue Res* **276**:429–438.

Sun, X. J., L. P. Tolbert, and J. G. Hildebrand (1997). Synaptic organization of the uniglomerular projection neurons of the antennal lobe of the moth *Manduca sexta*: a laser scanning confocal and electron microscopic study. *J Comp Neurol* **379**:2–20.

Tanabe, T., M. Iino, and S. F. Takagi (1975). Discrimination of odors in olfactory bulb, pyriform-amygdaloid areas, and orbitofrontal cortex of the monkey. *J Neurophysiol* **38**:1284–1296.

Treloar, H., E. Walters, F. Margolis, and B. Key (1996). Olfactory glomeruli are innervated by more than one distinct subset of primary sensory olfactory neurons in mice. *J Comp Neurol* **367**:550–562.

van Drongelen, W., A. Holley, and K. B. Døving (1978). Convergence in the olfactory system: quantitative aspects of odour sensitivity. *J Theor Biol* **71**:39–48.

Vassar, R., J. Ngai, and R. Axel (1993). Spatial segregation of odorant receptor expression in the mammalian olfactory epithelium. *Cell* **74**:309–318.

Vassar, R., S. K. Chao, R. Sticheran, J. M. Nunez, L. B. Vosshall, and R. Axel (1994). Topographic organization of sensory projections to the olfactory bulb. *Cell* **79**:981–991.

Vickers, N. J., T. A. Christensen, and J. G. Hildebrand (1998). Combinatorial odor discrimination in the brain: attractive and antagonist odor blends are represented in distinct combinations of uniquely identifiable glomeruli. *J Comp Neurol* **400**:35–56.

Wachowiak, M. and B. W. Ache (1998). Multiple inhibitory pathways shape odor-evoked responses in lobster olfactory projection neurons. *J Comp Neurol* **182**:425–434.

Wang, F., A. Nemes, M. Mendelsohn, and R. Axel (1998). Odorant receptors govern the formation of a precise topographic map. *Cell* **93**:47–60.

White, J., T. A. Dickinson, D. R. Walt, and J. S. Kauer (1998). An olfactory neuronal network for vapor recognition in an artificial nose. *Biol Cybern* **78**:245–251.

White, J., K. A. Hamilton, S. R. Neff, and J. S. Kauer (1992). Emergent properties of odor information coding in a representational model of the salamander olfactory bulb. *J Neurosci* **12**:1772–1780.

Wilson, M. and J. M. Bower (1992). Cortical oscillations and temporal interactions in a computer simulation of piriform cortex. *J Neurophysiol* **67**:981–995.

Wright, R. H. (1982). *The Sense of Smell*. Boca Raton, FL: CRC Press.

Yokoi, M., K. Mori, and S. Nakanishi (1995). Refinement of odor molecule tuning by dendrodendritic synaptic inhibition in the olfactory bulb. *Proc Natl Acad Sci U S A* **92**:3371–3375.

10

Development of the Olfactory System

GAIL D. BURD
Department of Molecular and Cellular Biology,
University of Arizona, Tucson, AZ

LESLIE P. TOLBERT
Arizona Research Laboratories, Division of Neurobiology,
University of Arizona, Tucson, AZ

1. INTRODUCTION

The organizational features of olfactory systems pose a number of unique developmental challenges. For instance, olfactory receptor axons typically project onto their target olfactory neuropils in complex topographic patterns, each receptor neuron expresses only one or so of the many similar olfactory receptor genes, and in many species new receptor neurons are constantly generated and incorporated into the circuitry. Such features suggest that development in olfactory systems, while sharing some mechanisms with other sensory systems, may also involve novel mechanisms underlying cell genesis and maintenance, cellular specification, axon guidance, and specificity of developing synapses.

Despite major differences in overall body plan, the wide variety of vertebrates and invertebrates that have been studied have olfactory systems that share many common features of cellular organization. For example, in most systems examined, bipolar receptor neurons in a sheet of sensory epithelium extend axons into the central nervous system (CNS) to synapse with target

The Neurobiology of Taste and Smell, Second Edition, Edited by Thomas E. Finger, Wayne L. Silver, and Diego Restrepo.
ISBN 0-471-25721-4   Copyright © 2000 Wiley-Liss, Inc.

neurons in conspicuous spheroidal glomeruli. Perhaps not surprisingly, the cellular and molecular mechanisms of development of these systems also share common themes. In both vertebrates and invertebrates, development of the olfactory system can be divided into the following major events: induction of the peripheral olfactory organ and specification of receptor neurons, axon outgrowth by these neurons and contact with the CNS, induction and early maturation of the recipient CNS structures, development of synaptic glomeruli, and functional maturation of receptor neurons and their target CNS neurons.

In this chapter, we discuss developmental mechanisms underlying the assembly of olfactory systems, with special focus on three diverse species, the mouse *Mus musculus*, the frog *Xenopus laevis*, and the moth *Manduca sexta*, that have widely differing schedules of development. While we include reference to work in other species, our goal is to use these three species to elucidate mechanisms that underlie development in broad sectors of the animal kingdom. In addition, this review covers the development of the peripheral olfactory organ and the first central olfactory neuropil, but not of higher processing centers, and focuses strictly on developmental mechanisms with only passing reference to the remarkable neural plasticity that occurs in olfactory systems in adults.

2. OVERVIEW OF OLFACTORY SYSTEM DEVELOPMENT

A brief review of the normal sequence of development in three divergent species will reveal the range of scenarios of olfactory development. Intriguingly, the different scenarios lead to strikingly similar cellular structures. Figure 10.1 illustrates the main stages of development.

2.1. Mouse

In mice, many sensory structures of the head derive from specialized sets of cells called placodes. Olfactory placodes on each side of the midline invaginate to form olfactory pits, which give rise to both a primary olfactory epithelium for detection of ordinary odors and a vomeronasal organ for detection of pheromones (Cuschieri and Bannister, 1975). Cell proliferation and differentiation in the olfactory pits take place early in embryonic development (see Table 10.1) (Cuschieri and Bannister, 1975), and very quickly olfactory axons begin to grow out of the developing epithelium to the brain, only microns away (Hinds, 1972b; Hinds and Hinds, 1976). As the axons extend, a set of neurons (see Schwanzel-Fukuda, 1999) and special "ensheathing" glial cells (Valverde et al., 1992) migrates with them. The earliest axons from the primary olfactory epithelium extend deep into the undifferentiated telencephalon and contact the proliferative cells lining the ventricles (Hinds, 1972b; Gong and Shipley, 1995). After axonal contact, the telencephalon forms a discrete bulge, as olfactory bulb neurons proliferate. In the main olfactory bulb, mitral cells are born first, followed

DEVELOPMENT OF THE OLFACTORY SYSTEM

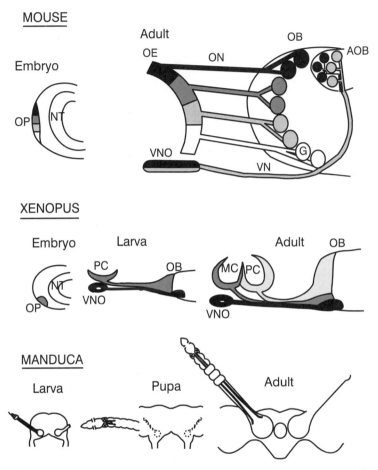

FIG. 10.1. Schematic diagram illustrating development of the olfactory system of the mouse *Mus musculus*, the frog *Xenopus laevis*, and the moth *Manduca sexta*. In the mouse, development occurs in a continuum from embryonic to postnatal stages; in *Xenopus*, embryonic development proceeds into larval stages and many tissues are modified or develop *de novo* during the metamorphic transition from larva to adult; in *Manduca*, most of the adult tissues are born *de novo* during metamorphosis. Different functional zones in the olfactory epithelium and projections to the olfactory bulb are shown in different shades of gray, with four zones in the main olfactory system and two zones in the accessory system of the mouse, three functional zones in *Xenopus*, and at least two functional zones in *Manduca*. A = antenna, AL = antennal lobe, AOB = accessory olfactory bulb, G = glomerulus, GL = glomerular layer, LA = larval antenna, LAC = larval antennal center, NT = neural tube, MC = middle cavity, OB = olfactory bulb, OE = olfactory epithelium, ON = olfactory nerve, OP = olfactory placode, PC = principal cavity, VN = vomeronasal nerve, and VNO = vomeronasal organ. Diagram is not drawn to scale.

TABLE 10.1. Time table for olfactory system development

Feature	Manduca[*]	Xenopus[**]	Mouse
Olfactory periphery forms	Pupal st. 1[13]	St. 23[12]	E10-E18[2]
First ORN axons contact CNS	Late pupal st. 3[13]	St. 29/30[1]	E12[7]
First CNS neurons differentiate	?	St. 32[1]	E14[8]
VNO forms	NA	St. 37/38[12]	E11[2]
Projection neurons/ mitral cells born	Larval stages[5]	St. 11-postmet.[3]	E11–E13[6]
First synapses in OT	Before pupal st. 2[11]	St. 36[1]	E14[9]
Glial cells surround protoglomeruli	Pupal st. 5 and st. 6[10]	?	[†]E18–P0 (rat)[15,16]
First glomeruli defined	Pupal st. 6[14,10]	St. 36[1]	[†]E20-P0 (rat)[15,16]
Interneurons/PG and GR cells born	Larval-pupal st. 3[5]	St. 41–postmet.[3]	E18–P20+[6]
Receptors can respond to odors	Pupal st. 14–16[5]	?	[†]E19 (rat)[4]
End of metamorphosis/birth	Pupal st. 18[13]	St. 66[12]	E19+[6]

[*]Times given for development of adult olfactory system, which arises almost entirely during metamorphosis. The larva also has a rudimentary olfactory system, which may contribute a small amount to the adult system.

[**]A new peripheral olfactory epithelium forms and an existing epithelium is modified during metamorphosis.

[†]Gestational time for rats is 1–2 days longer than for mice, e.g., E19 rat is roughly equivalent to E18 mouse.

CNS = central nervous system, GR = granule cell, ORN = olfactory receptor neuron, OT = olfactory target, PG = periglomerular cells, postmet. = postmetamorphosis, st. = stage, VNO = vomeronasal organ.

References: (1) Byrd and Burd, 1991; (2) Cuschieri and Bannister, 1975; (3) Fritz et al., 1996; (4) Gesteland et al., 1982; (5) Hildebrand et al., 1997; (6) Hinds, 1968; (7) Hinds, 1972a; (8) Hinds, 1972b; (9) Hinds and Hinds, 1976; (10) Oland et al., 1990; (11) Oland and Tobert, 1996; (12) Reiss and Burd, 1997b; (13) Sanes and Hildebrand, 1976; (14) Tolbert et al., 1983; (15) Treloar et al., 1999; (16) Valverde et al., 1992.

by tufted cells and then periglomerular cells and granule cells (Hinds, 1968). The neurons of the accessory olfactory bulb follow a similar developmental pattern, but each class of neuron is born earlier than in the main bulb (Hinds, 1968).

As subsequent olfactory receptor axons grow into the developing olfactory bulb, they terminate in a superficial layer, the presumptive glomerular layer. While an incomplete picture of development exists for mice (Hinds, 1972a), in rats, receptor axons are known to begin to organize into glomerular aggregates early (Valverde et al., 1992), before target mitral dendrites show signs of glomerular tufting; soon afterward, individual mitral cells send dendrites into multiple developing glomeruli and only later retract to leave one glomerular tuft (Treloar et al., 1999). Although synapses form in excess and then are pruned postnatally (Hinds and Hinds, 1976), there is no concomitant pruning of the axonal arbors of receptor neurons: from birth, the terminal arbors are virtually identical to those in older animals (Klenoff and Greer, 1998). Postnatal development, however, may occur, as new glomeruli form (LaMantia et al., 1992).

In mice, the time of onset of odorant responses in receptor cells has not been elucidated, but in the rat, action potentials have been recorded (Gesteland et al., 1982) just days after axons begin to grow out of the developing olfactory epithelium (Farbman and Squinto, 1985) and selective odorant responses begin to be observed two days after olfactory axons form synapses in the olfactory bulb (Gesteland et al., 1982). This suggests that the olfactory bulb may participate in refining the physiological responses of the olfactory receptor neurons, but molecular biology data suggest that the receptor neurons express single odorant receptor genes prior to contact with the olfactory bulb (Sullivan et al., 1995).

2.2. Frog

Early olfactory system development in Xenopus is quite similar to that in the mouse, but various features of the system are modified during metamorphosis. At the neural plate stage, cells fated for the olfactory placodes reside in two anterior cell clusters, and after neural tube closure, these cells begin to form the olfactory placodes (Klein and Graziadei, 1983; Eagleson and Harris, 1990). The earliest olfactory receptor axons extend into the adjacent, rostral neural tube (Byrd and Burd, 1991).

Like mice, larval Xenopus have two regions of olfactory epithelium: a principal-cavity olfactory epithelium and a vomeronasal epithelium (see Reiss and Burd, 1997b). Unlike most other vertebrates, however, Xenopus develops a third area of olfactory epithelium at metamorphosis. Triggered by the secretion of thyroid hormone, the sensory epithelium in the principal cavity of the nasal capsule, previously exposed only to water-borne odorants, is transformed from water-sensing to air-sensing. At the same time, a new area of sensory epithelium develops in the middle cavity to assume the water-sensing role in the adult (Venus et al., 1998). During metamorphosis, the types of olfactory receptor neurons and supporting cells in the principal cavity change to the adult structure and biochemistry, while the cells of the new middle cavity epithelium assume the features of the cells in the larval principal cavity (Frietag et al., 1995; Hansen et al., 1998; Petti et al., 1999; Burd, 1999).

Development of the olfactory bulb in Xenopus is also split into two phases, matching the step-wise development of the olfactory epithelium. The olfactory bulb produced during embryogenesis and early larval stages receives input from the receptor neurons of the principal cavity; axons from the vomeronasal organ innervate the accessory olfactory bulb (Reiss and Burd, 1997a). Later, during metamorphosis, a new, dorsal region of the olfactory bulb is generated (Fritz et al., 1996), and the principal-cavity neurons shift their projection to this new dorsal bulb. At the same time, the axons from the new water-sensing middle cavity project to the region of the bulb previously innervated by the principal cavity afferent axons (Reiss and Burd, 1997a). Neurogenesis continues in both the dorsal and ventral olfactory bulbs at least until the end of metamorphosis (Fritz et al., 1996). The glomeruli emerge as loosely segregated neuropil structures from a single, homogeneous neuropil (Byrd and Burd, 1991), unlike glomeruli in mice and moths. Even in adult frogs, glomeruli are not clearly defined; there are few periglomerular cells and glial cells in the glomerular layer, and primary dendrites of mitral cells can bifurcate and form two glomerular tufts (Scalia et al., 1991). Also unlike mice and moths, the right and left olfactory bulbs of larval *Xenopus* join along the midline to produce a single, fused olfactory bulb (Byrd and Burd, 1991). The accessory olfactory bulbs, however, remain separated. By the onset of metamorphic climax, many olfactory axons from one side of the animal cross the midline and distribute to glomeruli on the contralateral side (Byrd and Burd, 1993).

2.3. Moth

Olfactory development is even more heavily concentrated into a postembryonic metamorphic period in the moth *Manduca sexta*. In species such as moths and flies that undergo a complete metamorphosis, rudimentary antennae that develop during embryogenesis are shed and replaced by large, complex antennae that arise from imaginal disks during the pupal stage of metamorphosis (Sanes and Hildebrand, 1976). In *Manduca*, receptor neurons are located in specialized olfactory sensory hairs, or sensilla, in 2 cm-long antennae (see Chapter 9, Christensen and White). The olfactory receptor neurons of the adult antenna are born early after the imaginal disks evert (Sanes and Hildebrand, 1976), and immediately begin to extend axons down the antenna toward the brain. The first axons reach the brain about a day later; others continue to arrive over many subsequent days. During this period of ingrowth, axons develop spontaneous activity (see Oland and Tolbert, 1996). Since olfactory receptor genes have not yet been identified in *Manduca*, it is not known when olfactory receptor proteins begin to be expressed by receptor neurons in the antenna, but electrophysiological responses to odors are not detectable until just before the adult moth emerges (see Hildebrand, 1985).

Like the olfactory receptor neurons of the antennae, their targets, the antennal lobes of the adult brain, arise essentially *de novo* during metamorphosis in *Manduca* (see Hildebrand, 1985). While small, simple antennal lobes are generated during embryogenesis to serve the rudimentary larval antennae, the adult antennal lobes arise almost completely from a small set of neuroblasts that divide throughout larval life but whose progeny do not differentiate until metamorphosis (see Hildebrand et al., 1997). Unlike in mouse or frog, by the time axons from the developing adult antenna begin to reach the brain, all of the neurons of the antennal lobe have been born (see Oland and Tolbert, 1996). Their cell bodies lie outside a coarse neuropil formed by their branching neurites, and the neuropil is surrounded by glial cells. The first receptor axons pierce the glial border to terminate in a fringe around the coarse neuropil (Oland et al., 1990). As more axons continue to grow in, the terminals segregate into spheroidal "protoglomeruli." The earliest protoglomeruli form in the area of entry of the antennal axons, and mature into sexually dimorphic glomeruli (see Rössler et al., 1999a); the male-specific glomerular complex is specific for processing female sex pheromone (Hansson et al., 1991), so is in some ways similar to the accessory olfactory bulb innervated by the vomeronasal organ in mice and frogs. About a day after protoglomeruli form, they become surrounded by glial cells. The neurites of uniglomerular projection neurons (in many ways analogous to the mitral/tufted cells of vertebrates) reach outward to overlap with the axon terminals almost as soon as protoglomeruli form (Malun et al., 1994), whereas the neurites of local interneurons of the antennal lobe lag behind until after the glia form glomerular borders (Oland et al., 1990). By about two thirds of the way through metamorphosis, the antennal lobe is virtually indistinguishable from that of an adult animal at the light- and electron-microscopic levels (Tolbert et al., 1983; Oland et al., 1990), although electrophysiological maturation continues (Tolbert et al., 1983; see Hildebrand, 1985).

In summary, development of the peripheral and central olfactory systems in the three species of special focus here proceeds along dramatically different timetables. In mice, olfactory development occurs along a continuum, during embryological and early postnatal stages, while in *Xenopus* and *Manduca* it involves two stages of development, during embryonic and early larval stages and during metamorphosis. The metamorphosis in *Xenopus* is incomplete, while in *Manduca* it is almost complete. As we shall discuss below, however, even though timing is different, cellular mechanisms, where they are understood, appear to be quite similar.

3. MECHANISMS UNDERLYING OLFACTORY SYSTEM DEVELOPMENT

3.1. Molecules Involved in Early Patterning of the Olfactory System

Compared to other neural systems, relatively little is known about the molecular mechanisms that control development of olfactory structures. Only a few developmental control genes have been identified that are likely to play roles in induction and early differentiation of the olfactory placodes and olfactory bulbs in vertebrates; even less is known about specific genes controlling development of the olfactory system in invertebrates.

Certain secreted factors are likely to play important roles in the formation of olfactory structures. The gene for cerberus, a factor secreted by the anterior endomesoderm, is one of the earliest genes shown to be involved in induction of head structures including subsequent formation of olfactory placodes (Bouw-meesta et al., 1996). Another secreted factor, retinoic acid, has been implicated in both placode and bulb morphogenesis (Anchan et al., 1997).

Genes for a number of transcription factors, such as *Distal-less*, *Otx2*, *NeuroD*, *fork head*, and *Lim-1*, have been localized to the olfactory placode and/or developing olfactory bulb in *Xenopus*, but roles for most of these factors remain unknown (see Reiss and Burd, 1997b). Gli3, a zinc-finger-type transcription factor, is critical for bulb development in mice (Hui and Joyner, 1993), apparently through a molecular pathway that is involved in attachment of the olfactory nerve to the telencephalon (Naruse et al., 1994). Another transcription factor, Pax-6, is expressed in the olfactory system of *Xenopus* (Hirsch and Harris, 1997) and is critical for normal development of the olfactory bulb and olfactory placode in mouse (Grindley et al., 1995).

3.2. Effects of Circulating Hormones

In both vertebrates and invertebrates, circulating hormones appear to govern particular aspects of olfactory system development. Thyroid hormone stimulates amphibian metamorphosis and has been shown to play a crucial role in vertebrate neural development (Grave, 1977). It is important for normal development of the olfactory epithelium and vomeronasal epithelium of the rat (Segovia et al., 1982; Paternostro and Meisami, 1996) and the olfactory epithelium of *Xenopus* (Burd, 1992). Hypothyroidism in both species leads to the presence of significantly fewer olfactory receptor axons, suggesting potential influences of thyroid hormone on neuronal genesis and/or maturation and survival in the olfactory epithelium. A middle cavity can be induced to form in *Xenopus* by precocious treatment with thyroid hormone (Petti and Burd, 1995), but it is not clear whether thyroid hormone is essential for the induction of this structure. In addition, thyroid hormone appears to be important for certain olfactory behaviors in fish at metamorphosis. For example, thyroid hormone stimulates young migratory salmon to imprint on stream odors, and this odor memory is retained for 10 months without further odor stimuli or hormonal treatment (Hasler and Scholz, 1983). Some part of the odor memory may reside within the olfactory receptor neurons themselves (Nevitt et al., 1994).

Ecdysteroids, a class of steroid hormones, orchestrate tissue development and reorganization during metamorphosis in insects. While the ecdysteroids have been shown to have profound control over neuronal development in thoracic and abdominal portions of the central nervous system of *Manduca*, it has been difficult to show an effect of ecdysteroids on neurons of the antennal lobe. Presumably the hormones influence production of antennal-lobe neurons during larval stages, although this has not been shown directly. The hormones also appear to have a very minor effect on branching patterns of antennal-lobe neurons as assayed in cell culture experiments, but the most striking effect of the hormones in the antennal lobe is their control of glial cell division (see Oland and Tolbert, 1996).

3.3. Chemical Signals Between Cells During Early Development

In vertebrates, the survival of olfactory receptor neurons is dependent upon innervation of the olfactory bulb (see Section 3.5 below). Removal of the olfactory bulb in adult rodents results in rapid turnover of olfactory receptor neurons (Schwob et al., 1992). The receptor neurons develop, project axons toward the telencephalon, and then die. Thus, it appears that the olfactory receptor neurons require a growth or trophic factor from the olfactory bulb for their survival. In *Xenopus*, however, required maturation and survival factors are present in other brain regions as well (Higgs and Burd, 1997).

The olfactory epithelium of rodents expresses trophic factors that are likely to play a role in adult neural plasticity (Farbman and Buchholz, 1996; Miwa et al., 1998). Whether these factors also participate in the development of this system remains to be determined. Insulin-like growth factor-1 (IGF-1) seems to participate in development and neural plasticity in this system (Ye et al., 1997); it appears to influence the rate of neurogenesis of olfactory receptor neurons *in vivo* (Pixley et al., 1998). Furthermore, fibroblast growth factor-1 can stimulate morphological differentiation of ensheathing cells and olfactory receptor neurons during development (Key et al., 1996). In regenerating adult olfactory receptor neurons, the receptors for specific neurotrophins appear sequentially (Roskams et al., 1996), suggesting that a sequence of neurotrophins, some of which may be provided by the olfactory bulb, may influence the development of receptor neurons.

3.4. Development Independent of Interactions Between the Olfactory Periphery and the CNS

3.4.1. Initial Maturation of Olfactory Receptor Cells

Cellular interactions with the central olfactory target are not necessary for the initial differentiation of olfactory receptor neurons. Molecular markers for neurons, and some markers for mature olfactory receptor neurons, are present in the receptor neurons of rodents before the olfactory axons enter the telencephalon. In mice, neural cell adhesion molecule (N-CAM) begins to be expressed in the olfactory placodes, odorant receptor genes are expressed in developing receptor neurons, and growth-associated protein-43 (GAP-43) is expressed in the receptor axons before the axons make contact with the brain (Miragall et al., 1989; Sullivan et al., 1995). Similarly, olfactory marker protein (OMP), a specific marker for olfactory receptor neurons, and other proteins associated with the transduction machinery of the olfactory receptor neurons are expressed before olfactory axons form synapses with olfactory bulb neurons, and contact with the olfactory bulb is not necessary for their expression (Sullivan et al., 1995). In *Manduca*, the olfactory receptor neurons and support cells of the antennal olfactory epithelium complete much of their morphological differentiation before the receptor axons reach the brain, and can do so even if the brain has been removed (Sanes et al., 1976; see Tolbert et al., 1983).

3.4.2. Early Differentiation Without Olfactory Axon Innervation

Several findings indicate that contact by olfactory receptor axons does not induce an olfactory bulb in an otherwise undetermined portion of the telencephalic vesicle; instead the axons must instruct maturational changes of a telencephalic region that already is determined to give rise to olfactory bulb. Even when receptor axons are prevented from contacting the CNS, the part of the telencephalon that normally would form the olfactory bulb continues to exhibit markers of olfactory determination, such as the cell adhesion molecule O-CAM in mouse (Yoshihara et al., 1997) and the KAL protein encoded by the Kallmann syndrome gene in chick (Lutz et al., 1994). Thus, O-CAM and KAL expression indicate that some level of olfactory bulb differentiation occurs even in the absence of olfactory axon ingrowth.

3.5. Effects of Olfactory Receptor Axons on Target Development

3.5.1. Effects of Removal of Olfactory Receptor Axons

Some of the earliest axons to grow into the brain from the olfactory epithelium in mice penetrate deep into a restricted region of the telencephalic vesicle (Hinds, 1972b), where they induce mitotic changes in the telencephalic ventricular zone that produce a conspicuous bulb (Gong and Shipley, 1995). Olfactory receptor axons are implicated also in the maturation of the olfactory bulb. Removal of olfactory placodes during early embryonic development in Xenopus has revealed that innervation of the rostral neural tube by olfactory axons is essential for the development of a conspicuous olfactory bulb (see Reiss and Burd, 1997b) and may even influence development of more caudal telencephalic regions (Graziadei and Monti-Graziadei, 1992). Loss of olfactory innervation during metamorphosis in Xenopus also leads to a reduction in the number of olfactory bulb neurons that mature (Burd, 1999).

In contrast, in the cockroach Periplaneta americana, which undergoes partial metamorphosis, no early interaction between olfactory receptor axons and the brain is required during embryogenesis to produce the primary olfactory centers (Salecker and Boeckh, 1995). Furthermore, in insects such as Manduca that undergo complete metamorphosis, the larval antennal lobes are replaced by large adult antennal lobes during metamorphosis, even if the antennal anlage are removed (Hildebrand et al., 1979).

Even in these species, and in fact in every species so far examined, formation of glomeruli requires the presence of olfactory receptor axons. Olfactory receptor axons have been found to have the special property that they form glomerulus-like "knots" even when they grow into ectopic regions of the brain or form a blind neuroma (mouse and frog: Graziadei and Monti-Graziadei, 1986; moth: Rössler et al., 1999a). Hildebrand et al. (1979) observed that if antennal receptor axons are prevented from innervating the antennal lobe during development in Manduca, an antennal lobe develops, but the resulting lobe is essentially aglomerular. Without receptor-neuron input, development is abnormal from the moment the axons would normally have begun to reach the brain (see Oland and Tolbert, 1996). Instead of undergoing the morphological changes described above for normal development, glial cells remain restricted to a rim surrounding the neuropil, multiglomerular local interneurons of the lobe develop diffuse rather than tufted branching patterns, and uniglomerular projection neurons develop arbors that are restricted in extent but larger than the normal glomerular size (Oland et al., 1990). Partial deafferentation experiments have revealed that axons from as little as 12% of the antenna are sufficient to precipitate the cellular changes that lead to the formation of glomeruli, although innervation from over 25% of the antenna is necessary for the normal number of glomeruli to develop (see Oland and Tolbert, 1996).

In many areas of the nervous system, cell numbers adjust to produce an appropriate numerical match between the number of afferent axons and the size of the target. In *Xenopus* and other vertebrate olfactory systems, the number of receptor axons innervating the olfactory bulb has been related to the number of mitral cells or mitral/tufted cells in adults and during development (Byrd and Burd, 1991; see Reiss and Burd, 1997b).

In summary, manipulation of olfactory receptor axons during development in vertebrates and invertebrates has revealed that these axons influence a number of aspects of target development, including the production and/or maintenance of target neurons, the branching patterns of those neurons, and glial morphology and behavior.

3.5.2. Ectopic Innervation Experiments

Ectopic innervation studies have provided further support for the hypothesis that olfactory axons stimulate CNS neurons to express particular olfactory bulb–neuron characteristics. For example, when olfactory placodes are transplanted to different head regions in *Xenopus*, olfactory axons sometimes enter the diencephalon and induce glomerulus-like structures that may involve dendrites of CNS neurons (e.g., Stout and Graziadei, 1980). When the olfactory bulb is removed in adult rats, olfactory axons project to the rostral forebrain; some forebrain neurons contacted by the olfactory axons express immunoreactivity for tyrosine hydroxylase, common for a class of olfactory bulb neuron, but not for forebrain neurons (Guthrie and Leon, 1989). This almost certainly is related to the finding that tyrosine hydroxylase immunoreactivity is regulated by olfactory axon innervation in the developing and adult mammalian olfactory bulb (McLean and Shipley, 1988). The mechanisms involved in this signaling are unknown but may simply involve activity (Baker, 1990; Liu et al., 1999). Together, these findings suggest that olfactory axons can induce biochemical programs in neurons, whether or not they are the normal post-synaptic target.

Similarly, in *Manduca*, olfactory receptor axons can induce changes in abnormal target neurons. For example, the male-specific receptor axons of genetically male antennae transplanted onto female hosts have the ability to induce a male-like macroglomerular complex in the host antennal lobe (see Hildebrand, 1985), and this complex includes the dendrites of female neurons that normally would not branch in a macroglomerular pattern (Rössler et al.,

1999a). Not only do the male-specific axons terminate in the induced macroglomerular complex, but they induce changes in synaptic connectivity, and the (female) animal now responds behaviorally to female sex pheromone (see Hildebrand et al., 1997). These experiments provide unequivocal evidence for a strong influence of antennal axons on development of target neurons.

3.5.3. Role of Neural Activity

Naris Closure. Is neuronal activity essential for the influence of receptor axons on olfactory target development? In the mammalian olfactory system, unilateral naris closure in the neonatal period deprives the closed nasal cavity of odor stimulation and is sufficient to produce changes in the development of the olfactory epithelium (Stahl et al., 1990) and morphological changes in the olfactory bulb on the deprived side (see Brunjes, 1994). In rats, reductions occur in the numbers of neurons in the bulb, apparently due to cell death and not to a reduction in proliferation of precursor cells, but the number of glomeruli is unaffected (see Brunjes, 1994; Najbauer and Leon, 1995). Naris closure in neonatal rats also results in changes in neurotransmitters in the olfactory bulb (Baker, 1990). It is not clear, however, whether the primary cause of the morphological and biochemical changes following naris occlusion is decreased synaptic activation (Philpot et al., 1997) or reduction in a trophic factor (e.g., NGF: Gómez-Pinilla et al., 1989) supplied by the axons.

Early Olfactory Experience. Early olfactory experience during a critical period also can have a lasting effect on the development of central olfactory targets. For example, Leon and his colleagues determined that, during the first postnatal week of life in rats, a single odor exposure paired with a naturally reinforcing tactile stimulus results in enhanced preference for the trained odor for at least two weeks (Leon, 1992). Coincident with the induced behavioral change, a number of cellular changes occur, including enhanced 2-deoxyglucose (2-DG) uptake in specific glomeruli, and enhanced density of Fos-immunoreactive cells, enlargement of individual glomeruli, and increased numbers of juxtaglomerular cells (see Leon, 1992, 1998). The mechanisms underlying these changes may be related to enhanced levels of norepinephrine in the olfactory bulb during the critical period of training (Rangel and Leon, 1995).

Action Potential Blockade. While the findings outlined above indicate a role for activity in mammals, one experiment in the moth suggests that organization of olfactory neuropil into glomeruli appears to be largely independent of activity. The antenna exhibits no response to puffs of odor until very late in development, just a few days before the adult moth emerges from the pupal case (Schweitzer et al., 1976). Spontaneous activity, however, can be recorded from the antennal nerve during the period when glomeruli are forming in the antennal lobe. When all Na^+-based activity, including spontaneous activity, is blocked with tetrodotoxin, neuronal and glial components of glomeruli develop apparently normally (see Oland and Tolbert, 1996). Whether individual axons are targeted to the correct glomeruli remains a question, although the presence of a normal-appearing macroglomerular complex in treated animals suggested that the positional specificity of innervation was preserved for at least a subset of receptor axons.

3.6. Role of Glial Cells in Development of Glomeruli

In both vertebrates and invertebrates, glial cells play an important role in glomerulus formation. Recent investigations have implicated several types of glial cells in glomerulus development in mammals. Specialized ensheathing cells that migrate into the brain along olfactory axons penetrate into the prospective glomerular layer and form partial envelopes around the developing glomeruli (Valverde et al., 1992). Another type of cell, astrocytes that are decorated with growth-inhibiting tenascin, is hypothesized to form a barrier to axon ingrowth that causes glomeruli to develop at the edge of the astrocyte territory deeper in the bulb (Gonzalez and Silver, 1994). The summed correlative evidence suggests that glial cells might interact with receptor axons and be involved in several ways in glomerulus formation.

This idea has been tested directly in developing *Manduca*, where the number of glial cells can be reduced while maintaining apparently normal numbers of antennal and antennal-lobe neurons (Oland et al., 1988; see Oland and Tolbert, 1996). When glial numbers were reduced to approximately 25% of normal, glomeruli did not form. Olfactory receptor axons did initially form protoglomeruli, but without the normal glial envelope to constrain them, the axons dispersed before target neurons grew outward to form synapses. The constraint of axonal growth to individual glomeruli, and subsequent similar constraint of dendritic growth, may be due to the existence of tenascin-like molecules on the surfaces of the glial cells (Krull et al., 1994).

3.7. Pathfinding and Targeting by Olfactory Receptor Axons

In other sensory systems, the afferent axons form a topographic map of the sensory surface in the CNS. In contrast, in olfactory systems receptor neurons sparsely distributed over wide areas of the olfactory sensory epithelium converge on one or a few glomeruli (LeGros Clark and Warwick, 1946; Mombaerts et al., 1996, see below). In both the main and accessory olfactory systems, only a coarse zonal topography exists; within zones, axons must seek out other axons with similar response properties and glomerular targets.

In *Xenopus*, the first olfactory axons project a distance of only a few microns to the forebrain (Byrd and Burd, 1991), but during metamorphosis axons from transformed olfactory receptor neurons must project much longer distances to a new target and axons from newly generated receptor neurons must find a distant target (Reiss and Burd, 1997a, 1997b). The olfactory axons could be directly targeted to specific bulb regions or competition for target sites could be involved. Competition between principal-cavity and middle-cavity afferent axons for targets in the ventral olfactory bulb, however, is not a factor in targeting the principal cavity afferent axons to their new target. The olfactory axons from the principal cavity are retargeted to the dorsal olfactory bulb even when the ventral bulb is devoid of afferent axons from the middle cavity (Reiss and Burd, 1997a). Since the middle and principal cavities have different functions (sensing odorants in water versus air, respectively; Venus et al., 1998) and different forms of odorant receptors (fish-like versus mammal-like, respectively; Frietag et al., 1995), the segregation of axons from the two cavities to different regions of the olfactory bulb may be based on the differential receptor expression.

In *Manduca*, it is clear that certain receptor neurons send their axons to particular glomeruli. For instance, the male-specific receptor neurons that respond to components of the female pheromone project to a specific macroglomerular complex in the antennal lobe, and axons responsive to different components of the pheromone terminate in different parts of the macroglomerular complex (see Chapter 9, Christensen and White). Experiments involving surgical rerouting of the antennal nerve have revealed that olfactory receptor axons generally steer toward their normal target, the antennal lobe, no matter at what ectopic site they first encounter the brain (Oland et al., 1998). The male-specific receptor axons also find and terminate in the appropriate, dorsolateral, part of the antennal lobe, no matter from what angle they enter the lobe (Rössler et al., 1999a).

Thus, across species, except on a very coarse level, the glomeruli generally are not arrayed in a pattern that reflects the pattern of distribution of the receptor neurons in the epithelial sheet. Therefore, the olfactory system faces an unusual challenge in its development; axons from dispersed sensory neurons must find each other and find targets without a simple topographic matching of pre- and post synaptic cells.

3.7.1. Axon Trajectories and Receptor-Type Specificity Involved in Targeting

How do the receptor axons sort according to their odor specificities, and how do they find their appropriate CNS targets? As reviewed above, in situ hybridization experiments (Sullivan et al., 1995) have shown that receptor neurons express olfactory receptor proteins before they send their axons into the CNS, so the formal possibility that the initial projection is random and that receptor neurons are instructed by their targets to express particular receptor proteins can be eliminated. Receptor axons must instead be destined for particular glomeruli as they enter the bulb. Do they find their targets via broad exploration, or do they hone in directly on their targets? Are the olfactory receptor proteins per se involved in finding their targets?

Recently, individual developing receptor axons and their growth cones have been examined in a number of vertebrate and invertebrate species. Axons in the developing olfactory nerve of the mouse (Whitesides and LaMantia, 1996) are fasciculated and tipped by simple growth cones; as they enter the brain they form complex, lamellate growth cones, suggestive of exploration of their environment. Differences in fasciculation and in growth-cone complexity suggest that developing olfactory receptor axons explore broadly as they make choices where the olfactory epithelium abuts the frontonasal mesenchyme, in the mesenchyme itself, and within the telencephalon. In contrast, most of the receptor-axon growth cones in *Manduca* are relatively simple throughout their trajectory, even at the point where the axons enter the brain and sort into target- (glomerulus-) specific bundles (Oland et al., 1998). These findings lead to the tentative suggestion that in these species, the establishment of relationships between axons that have common targets must be accomplished primarily by short-range molecular interactions (such as would be mediated by cell-adhesion molecules; see below). Even after they start to form protoglomeruli, there is no evidence of exuberant outgrowth of axonal branches followed by pruning in *Manduca*, as is reported in the rat (Santacana et al., 1992).

Experiments of Mombaerts et al. (1996) and Wang et al. (1998) have looked in detail at the role of olfactory receptor proteins in axon targeting in mice. They have shown that switching the particular olfactory receptor protein expressed by receptor neurons alters their choice of target. Intriguingly, in some cases, the new target is neither the correct target for the unaltered receptor neurons nor the correct target for receptor neurons that normally express the receptor protein that is now being expressed. If receptor neurons are prevented from expressing any olfactory receptor protein, the axons do not converge on any particular glomerulus (Wang et al., 1998). Thus the olfactory receptor proteins clearly play some role in targeting, but probably do not account fully for specific choice of target. The most parsimonious hypothesis is that the receptor proteins themselves perform a function in homophilic adhesion, producing fasciculation of growing axons with matching receptor proteins (Singer et al., 1995).

3.7.2. Cell-Surface and Extracellular Matrix Molecules

Several groups have examined the distributions of putative cell adhesion molecules in developing olfactory systems. The embryonic form of N-CAM appears not only in the primordium of the olfactory bulb in the telencephalon, as reviewed above, but also on a subset of receptor axons in the developing and adult mouse olfactory bulb (Miragall et al., 1989; Whitesides and LaMantia, 1996). Other putative cell adhesion molecules, including the KAL protein, which is absent in patients with Kallmann's syndrome (Franco et al., 1991), have been found on olfactory receptor axons (Miragall et al., 1989, Whitesides and LaMantia, 1996) and on the surfaces of the Schwann-like cells that migrate with the growing receptor axons (Miragall et al., 1989). Both N-CAM and KAL (see Schwanzel-Fukuda, 1999) appear to be important for the migration of LHRH-positive neurons from the olfactory placode into the brain.

The recently discovered O-CAM, which has distinct similarities to N-CAM, is present very early in development on receptor neurons of some but not all zones of the main olfactory epithelium and on their axon terminals in the appropriate zones of the main olfactory bulb (Mori et al., 1997). It also is present on a subset of receptor neurons in the vomeronasal organ and their axon terminals in the rostral half of the accessory olfactory bulb (Yoshihara et al., 1997) as well as on the mitral/tufted cells in the caudal half of the accessory olfactory bulb (von Campenhausen et al., 1997). O-CAM can function as a homophilic adhesion molecule, so is thought to play an adhesive role (Yoshihara et al., 1997).

In addition to molecules in the CAM family, the extracellular-matrix component laminin and heparan sulphate proteoglycans are present early in development in the rat, first in the basement membranes of the olfactory epithelium and the telencephalon and then increasingly restricted to the axons of olfactory receptor neurons (Treloar et al., 1996), suggesting possible roles in early axon guidance. In mice, endogenous lectins, which may bind to carbohydrate ligands on the surfaces of olfactory receptor axons and are present on multiple cell types, appear to be involved in sorting and fasciculation of the receptor axons (Puche et al., 1996).

In *Manduca*, a subset of olfactory receptor axons express fasciclin II, a cell adhesion molecule closely related to N-CAM and O-CAM, as they grow into the antennal lobe and form glomeruli; these axons are scattered throughout the antennal nerve until they reach the glia-rich "sorting zone" at the entrance to the antennal lobe, where they fasciculate with other fasciclin II-positive axons targeted for specific glomeruli. If the number of glial cells in the sorting zone is severely reduced, fasciclin-positive axons do not fasciculate with each other, and many axons misroute (Rössler et al., 1999b), suggesting that an interaction with glia is necessary for fasciclin II-based axonal sorting.

Current research adresses the hypothesis that a hierarchy of adhesion molecules, including CAM's, extracellular-matrix molecules, proteoglycans, and the olfactory receptor proteins themselves, guides axons to ever more restricted "addresses" in the olfactory bulb/antennal lobe.

4. SUMMARY AND OUTLOOK FOR FUTURE RESEARCH

Although olfactory systems develop along lines in many ways similar to those governing the development of other sensory systems, olfactory systems have several features that create unique or unusual challenges for development. These features raise specific questions:

- Sensory neurons in the olfactory epithelium are derived from cranial ectodermal placodes. What genes are involved in determination and patterning of the olfactory placode?

- Olfactory axons grow into the rostral forebrain, and, while they do not appear to induce the earliest aspects of olfactory bulb differentiation, they are necessary for the maturation of the olfactory bulb. What are the growth factors or signaling molecules involved in bulb maturation, and what factors from the bulb participate in maintaining mature olfactory receptor neurons?

- Olfactory receptor neurons must express only one or two of up to a thousand potential olfactory receptor proteins. What factors influence which receptor proteins they will express?

- Receptor neurons in the olfactory epithelium are laid out in a two-dimensional array that does not appear to reflect a two-dimensional map of some aspect of the olfactory world. Is there an underlying functional order to the array?

- Axons from dispersed receptor neurons must find targets probably without a simple topographic guidance system. What guides them to the appropriate part of the target neuropil, and how do they establish glomeruli?

- As the olfactory receptor axons grow into the brain, in mammals, they are accompanied by a mass of migrating cells. What is the role of these cells?

- Olfactory systems incorporate a novel type of glial cell at the interface between the peripheral and central nervous system. In the moth, such glial cells are important for axon sorting. In vertebrates, do these glial cells play important roles in guidance and/or sorting of olfactory receptor axons during development and adult plasticity?

- In some species, olfactory receptor cells are replaced throughout life, and new axons must continuously find their way into the CNS and to the appropriate glomeruli. Are the molecular mechanisms for the generation of new receptor neurons and guidance of ingrowing receptor axons in the adult similar to those underlying development?

As reviewed in this chapter, investigators are beginning to elucidate the answers to many of these questions. Studies to date have revealed that there is an intricate communication between olfactory receptor neurons and the neurons and glial cells in their target areas. The studies we have cited and many others omitted in the interest of brevity provide important insights into olfactory development, and thereby augment our overall understanding of neural development.

REFERENCES

Anchan, R. M., D. P. Drake, E. A. Gerwe, C. F. Haines, and A.-S. LaMantia (1997). A failure of retinoid-mediated induction accompanies the loss of the olfactory pathway during mammalian forebrain development. *J Comp Neurol* **379**:1–15.

Baker, H. (1990). Unilateral, neonatal olfactory deprivation alters tyrosine hydroxylase expression but not aromatic amino acid decarboxylase or GABA immunoreactivity. *Neuroscience* **36**:761–771

Bouwmeesta, T., S.-H. Kim, Y. Sasai, B. Lu, and E. De Robertis (1996). Cerberus is a head-inducing secreted factor expressed in the anterior endoderm of Spemann's organizer. *Nature* **382**:595–601.

Brunjes, P. C. (1994). Unilateral naris closure and olfactory system development. *Brain Res Rev* **19**:146–160.

Burd, G. D. (1992). Development of the olfactory nerve in the clawed frog, *Xenopus laevis*. II. Effects of hypothyroidism. *J Comp Neurol* **315**:255–263.

Burd, G. D. (1999). Development of the olfactory system in the African clawed frog, *Xenopus laevis*. In: R. L. Hyson and F. Johnson (eds). *Biology of Early Influences*. New York: Kluwer Academic Plenum Publ, pp. 153–170.

Byrd, C. A. and G. D. Burd (1991). Development of the olfactory bulb in the clawed frog, *Xenopus laevis*: a morphological and quantitative analysis. *J Comp Neurol* **314**: 79–90.

Byrd, C. A. and G. D. Burd (1993). Morphological and quantitative evaluation of the development of the olfactory bulb after olfactory placode transplantation. *J Comp Neurol* **331**:551–563.

Cuschieri, A. and L. H. Bannister (1975). The development of the olfactory mucosa in the mouse: light microscopy. *J Anat* **119**:277–286.

Eagleson, G. W. and W. A. Harris (1990). Mapping of the presumptive brain regions in the neural plate of *Xenopus laevis*. *J Neurobiol* **21**:427–440.

Farbman, A. I. and J. A. Buchholz (1996). Transforming growth factor-α and other growth factors stimulate cell division in olfactory epithelium in vitro. *J Neurobiol* **30**:267–280.

Farbman, A. I. and L. M. Squinto (1985). Early development of olfactory receptor cell axons. *Dev Brain Res* **19**:205–213.

Franco, B., S. Guioli, A. Pragliola, B. Incerti, B. Bardoni, R. Tonlorenzi, R. Carrozzo, E. Maestrini, M. Pieretti, P. Taillon-Miller et al. (1991). A gene deleted in Kallmann's syndrome shares homology with neural cell adhesion and axonal path-finding molecules. *Nature* **353**:529–536.

Frietag, J., J. Krieger, J. Strotmann, and H. Breer (1995). Two classes of olfactory receptors in *Xenopus laevis*. *Neuron* **15**:1383–1392.

Fritz, A., D. Gorlick, and G. D. Burd (1996). Neurogenesis in the olfactory bulb of the frog *Xenopus laevis* shows unique patterns of during embryonic development and metamorphosis. *Int J Dev Neurosci* **14**:931–943.

Gesteland, R. C., R. A. Yancey, and A. I. Farbman (1982). Development of olfactory receptor neuron selectivity in the rat fetus. *Neuroscience* **7**:3127–3136.

Gómez-Pinilla, F., K. M. Guthrie, M. Leon, and M. Nieto-Sampedro (1989). NGF receptor increase in the olfactory bulb of the rat after early unilateral deprivation. *Dev Brain Res* **48**:161–165.

Gong, Q. and M. T. Shipley (1995). Evidence that pioneer olfactory axons regulate telencephalon cell cycle kinetics to induce the formation of the olfactory bulb. *Neuron* **14**: 91–101.

Gonzalez, M. de L. and J. Silver (1994). Axon-glia interactions regulate ECM patterning in the postnatal rat olfactory bulb. *J Neurosci* **14**:6121–6131.

Grave, G. D. (1977). Thyroid hormones and brain development. New York: Raven Press.

Graziadei, P. P. C. and A. G. Monti-Graziadei (1986). Principles of organization of the vertebrate olfactory glomerulus: an hypothesis. *Neuroscience* 19:1025–1035.

Graziadei, P. P. C. and A. G. Monti-Graziadei (1992). The influence of the olfactory placode on the development of the telencephalon in Xenopus laevis. *Neuroscience* 46:617–69.

Grindley, J. C., D. R. Davidson, and R. E. Hill (1995). The role of Pax-6 in eye and nose development. *Development* 121:1433–1442.

Guthrie, K. M. and M. Leon (1989). Induction of tyrosine hydroxylase expression in rat forebrain neurons. *Brain Res* 497:117–131.

Hansen, A., J. O. Reiss, C. L. Gentry, and G. D. Burd (1998). Ultrastructure of the olfactory organ in the clawed frog, Xenopus laevis, during larval development and metamorphosis. *J Comp Neurol* 398:273–288.

Hansson, B. S., T. A. Christensen, and J. G. Hildebrand (1991). Functional organization of the macroglomerular complex in the antennal lobe of the sphinx moth Manduca sexta. *J Comp Neurol* 312:264–278.

Hasler, A. D. and A. T. Scholz (1983). Olfactory imprinting and homing in salmon: investigations into the mechanisms of the imprinting process. New York: Springer-Verlag.

Higgs, D. M. and G. D. Burd (1999). The role of the brain in metamorphosis of the olfactory epithelium in the frog, Xenopus laevis. *Dev Brain Res* 118:185–195.

Hildebrand, J. G. (1985). Metamorphosis of the insect nervous system: influences of the periphery on the postembryonic development of the antennal sensory pathway in the brain of Manduca sexta. In: A. I. Selverston (ed). *Model Neural Networks and Behavior.* Plenum Publishing, pp 129–148.

Hildebrand, J. G., L. M. Hall, and B. C. Osmond (1979). Distribution of binding sites for ^{125}I-labeled alpha-bungarotoxin in normal and deafferented antennal lobes of Manduca sexta. *Proc Natl Acad Sci U S A* 76:499–503.

Hildebrand, J. G., W. Rössler, and L. P. Tolbert (1997). Postembryonic development of the olfactory system in the moth Manduca sexta: primary-afferent control of glomerular development. *Semin Cell Dev Biol* 8:163–170.

Hinds, J. W. (1968). Autoradiographic study of histogenesis in the mouse olfactory bulb. I. Time of origin of neurons and neuroglia. *J Comp Neurol* 134:287–304.

Hinds, J. W. (1972a). Early neuron differentiation in the mouse olfactory bulb. I. Light microscopy. *J Comp Neurol* 146:233–252.

Hinds, J. W. (1972b). Early neuron differentiation in the mouse olfactory bulb. II. Electron microscopy. *J Comp Neurol* 146:253–276.

Hinds, J. W. and P. L. Hinds (1976). Synapse formation in the mouse olfactory bulb: I. Quantitative studies. *J Comp Neurol* 169:15–40.

Hirsch, N. and W. A. Harris (1997). Xenopus Pax-6 and retinal development. *J Neurobiol* 32:45–61.

Hui C. C. and A. L. Joyner (1993). A mouse model of greig cephalopolysyndactyly syndrome: the extra-toes J mutation contains an intragenic deletion of the Gli3 gene. *Nat Genet* 3:241–246.

Key, B., H.B. Treloar, L. Wangerek, M.D. Ford, and V. Nurcombe (1996). Expression and localization of FGF-1 in the developing rat olfactory system. *J Comp Neurol* **366**: 197–206.

Klein, S. L. and P. P. C. Graziadei (1983). The differentiation of the olfactory placode in *Xenopus laevis*: a light and electron microscopic study. *J Comp Neurol* **217**:17–30.

Klenoff, J. R. and C. A. Greer (1998). Postnatal development of olfactory receptor cell axonal arbors. *J Comp Neurol* **390**:256–267.

Krull, C. E., L. A. Oland, A. Faissner, M. Schachner, and L. P. Tolbert (1994). In vitro analyses indicate a potential role for tenascin-like molecules in the development of insect olfactory glomeruli. *J Neurobiol* **25**:989–1004.

LaMantia, A.-S., S. L. Pomeroy, and D. Purves (1992). Vital imaging of glomeruli in the mouse olfactory bulb. *J Neurosci* **12**:976–988.

Le Gros Clark, W. E. and R. T. T. Warwick (1946). The pattern of olfactory innervation. *J Neurol Neurosurg Psychiatry* **9**:101–107.

Leon, M. (1992). The neurobiology of filial learning. *Annu Rev Psychol* **43**:337–398.

Leon, M. (1998). Catecholaminergic contributions to early learning. *Adv Pharm* **42**: 961–964.

Liu N., E. Cigola, C. Tinti, B. K. Jin, B. Conti, B. T. Volpe, and H. Baker (1999). Unique regulation of immediate early gene and tyrosine hydroxylase expression in the odor-deprived mouse olfactory bulb. *J Biol Chem* **274**:3042–3047.

Lutz, B, S. Kuratani, E. I. Rugarli, S. Wawesik, C. Wong, F. R. Bieber, A. Ballabio, and G. Eichele (1994). Expression of the Kallmann syndrom gene in human fetal brain and in the manipulated chick embryo. *Hum Mol Genet* **3**:1717–1723.

Malun, D., L. A. Oland, and L. P. Tolbert (1994). Uniglomerular projection neurons participate in early development of olfactory glomeruli in the moth *Manduca sexta*. *J Comp Neurol* **347**:1–22.

McLean, J. H. and M. T. Shipley (1988). Postmitotic, postmigrational expression of tyrosine hydroxylase in olfactory bulb dopaminergic neurons. *J Neurosci* **8**: 3658–3669.

Miragall, F., G. Kadmon, and M. Schachner (1989). Expression of L1 and NCAM cell adhesion molecules during development of mouse olfactory system. *Dev Biol* **135**: 272–286.

Miwa, T., N. Uramoto, T. Ishimaru, M. Furukawa, K. Shiba, and T. Morjizumi (1998). Retrograde transport of nerve growth factor from olfactory bulb to olfactory epithelium. *Neuroreport* **9**:153–155.

Mombaerts, P., F. Wang, C. Dulac, S. K Chao, A. Nemes, M. Mendelsohn, J. Edmonson, and R. Axel (1996). Visualizing an olfactory sensory map. *Cell* **87**:675–686.

Mori K., A. Tamada, S. Mitsui, Y. Yoshihara, and I. Naruse (1997). Development of primordial olfactory bulb in *Pdn/Pdn* mice that lack olfactory axon innervation. *Soc Neurosci Abstr* **23**:2077.

Najbauer, J. and M. Leon (1995). Olfactory experience modulated apoptosis in the developing olfactory bulb. *Brain Res* **674**:245–251.

Naruse, I., Y. Fukui, H. Keino, and M. Taniguchi (1994). The arrest of luteinizing hormone-releasing hormone neuronal migration in the genetic arhinencephalic mouse embryo (*Pdn/Pdn*). *Dev Brain Res* **81**:178–184.

Nevitt, G. A., A. H. Dittman, T. P. Quinn, and W. J. Moody, Jr. (1994). Evidence for a peripheral olfactory memory in imprinted salmon. *Proc Natl Acad Sci U S A* **91**:4288–4292.

Oland, L. A., and L. P. Tolbert (1996). Multiple factors shape development of olfactory glomeruli: insights from an insect model system. *J Neurobiol* **30**:92–109.

Oland, L. A., L. P. Tolbert, and K. L. Mossman (1988). Radiation-induced reduction of the glial population during development disrupts the formation of olfactory glomeruli in an insect. *J Neurosci* **8**:353–367.

Oland, L. A., G. Orr, and L. P. Tolbert (1990). Construction of a protoglomerular template by olfactory axons initiates the formation of olfactory glomeruli in the insect brain. *J Neurosci* **10**:2096–2112.

Oland, L. A., W. M. Pott, M. R. Higgins, and L. P. Tolbert (1998). Targeted ingrowth and glial relationships of olfactory receptor axons in the primary olfactory pathway of an insect. *J Comp Neurol* **398** 119–138.

Paternostro, M. A. and E. Meisami (1996). Essential role of thyroid hormones in maturation of olfactory receptor neurons: an immunocytochemical study of number and cytoarchitecture of OMP-positive cells in developing rats. *Int J Dev Neurosci* **14**:867–880.

Petti, M. A. and G. D. Burd (1995). Thyroid hormone induces changes in the nasal cavities that parallel those observed at metamorphosis. *Chem Senses* **20**:756.

Petti, M. A., S. F. Matheson, and G. D. Burd (1999). Differential antigen expression during metamorphosis in the tripartite olfactory system of the African clawed frog, *Xenopus laevis*. *Cell Tiss Res* **297**:383–396.

Philpot B. D., T. C. Foster, and P. C. Brunjes (1997). Mitral/tufted cell activity is attenuated and becomes uncoupled from respiration following naris closure. *J Neurobiol* **33**: 374–386.

Pixley, S. K., N. S. Dangoria, K. K. Odoms, and L. Hastings (1998). Effects of insulin-like growth factor 1 on olfactory neurogenesis in vivo and in vitro. *Ann N Y Acad Sci* **855**: 244–247.

Puche, A. C., F. Poirier, M. Hair, P. F. Bartlett, and B. Key (1996). Role of galectin-1 in the developing mouse olfactory system. *Dev Biol* **179**:274–287.

Rangel, S. and M. Leon (1995). Early odor preference training increases olfactory bulb norepinephrine. *Dev Brain Res* **85**:187–191.

Reiss, J. O. and G. D. Burd (1997a). Metamorphic remodeling of the primary olfactory projection in *Xenopus*: developmental independence of projections from olfactory neuron subclasses. *J Neurobiol* **32**:213–222.

Reiss, J. O. and G. D. Burd (1997b). Cellular and molecular interactions in the development of the *Xenopus* olfactory system. *Semin Cell Dev Biol* **8**:171–179.

Roskams, A. J., M. A. Bethel, K. J. Hurt, and G. V. Ronnett (1996). Sequential expression of Trks A, B, and C in the regenerating olfactory neuroepithelium. *J Neurosci* **16**: 1294–1307.

Rössler, W., P. W. Randolph, L. P. Tolbert, and J. G. Hildebrand (1999a). Axons of olfactory receptor cells of transsexually grafted antennae induce development of sexually dimorphic glomeruli in *Manduca sexta*. *J Neurobiol* **38**:521–541.

Rössler, W., L. A. Oland, M. R. Higgins, J. G. Hildebrand, and L. P. Tolbert (1999b). Development of a glia-rich axon sorting zone in the olfactory pathway of the moth *Manduca sexta*. *J Neurosci* **19**:9865–9877.

Salecker, I. and J. Boeckh (1995). Embryonic development of the antennal lobes of a hemimetabolous insect, the cockroach *Periplaneta americana*: light and electron microscopic observations. *J Comp Neurol* **352**:33–54.

Sanes, J. and J. G. Hildebrand (1976). Structure and development of antennae in the moth, *Manduca sexta*. *Dev Biol* **51**:282–299.

Santacana, M., M. Heredia, and F. Valverde (1992). Transient pattern of exuberant projections of olfactory axons during development in the rat. *Dev Brain Res* **70**:213–222.

Scalia, F., G. Gallousis, and S. Roca (1991). A note on the organization of the amphibian olfactory bulb. *J Comp Neurol* **305**:435–442.

Schwanzel-Fukuda, M. (1999). Origin and migration of luteinizing hormone-releasing hormone neurons in mammals. *Microsc Res Tech* **44**:2–10.

Schweitzer, E. S., J. R. Sanes, and J. G. Hildebrand (1976). Ontogeny of electroantennogram responses in the moth *Manduca sexta*. *J Insect Physiol* **2**:955–960.

Schwob, J. E., K. E. Mieleszko Szumowski, and A. A. Stasky (1992). Olfactory sensory neurons are trophically dependent on the olfactory bulb for their prolonged survival. *J Neurosci* **12**:3896–3919.

Segovia, S., M. C. R. Del Cerro, and A. Guillamon (1982). Effects of neonatal thyroidectomy on the development of the vomeronasal organ in the rat. *Dev Brain Res* **5**:206–208.

Singer, M. S., G. M. Shepherd, and C. A. Greer (1995). Olfactory receptors guide axons. *Nature* **377**:19–20.

Stahl, B., R. Hudson, and H. Distel (1990). Effects of reversible nare occlusion on the development of the olfactory epithelium in the rabbit nasal septum. *Cell Tiss Res* **259**:272–281.

Stout, R. P. and P. P. C. Graziadei (1980). Influence of the olfactory placode on the development of the brain in *Xenopus laevis* (Daudin): I. Axonal growth and connections of the transplanted olfactory placode. *Neuroscience* **5**:2175–2186.

Sullivan, S. L., S. Bohm, K. J. Ressler, L. F. Horowitz, and L. B. Buck (1995). Target-independent pattern specification in the olfactory epithelium. *Neuron* **15**:779–789.

Tolbert, L. P., S. G. Matsumoto, and J. G. Hildebrand (1983). The development of synapses in the antennal lobes of the moth *Manduca sexta*. *J Neurosci* **3**:1158–1175.

Treloar, H. B., V. Nurcombe, and B. Key (1996). Expression of extracellular matrix molecules in the embryonic rat olfactory pathway. *J Neurobiol* **31**:41–55.

Treloar, H. B., A. L. Purcell, and C. A. Greer (1999). Glomerular formation in the developing rat olfactory bulb. *J Comp Neurol* **413**:289–304.

Valverde, F., and M. Santacana, M. Heredia (1992). Formation of an olfactory glomerulus: morphological aspects of development and organization. *Neuroscience* **49**:255–275.

Venus B., J. R. Wolff, and G. D. Burd (1998). Functional anatomy of the olfactory system of *Xenopus laevis*. *Soc Neurosci Abstr* **24**:909.

von Campenhausen H., Y. Yoshihara, and K. Mori (1997). OCAM reveals segregated mitral/tufted cell pathways in developing accessory olfactory bulb. *Neuroreport* **8**:2607–2612.

Wang, F., A. Nemes, M. Mendelsohn, and R. Axel (1998). Odorant receptors govern the formation of a precise topographic map. *Cell* **93**:47–60.

Whitesides, J. G. and A.-S. LaMantia (1996). Differential adhesion and the initial assembly of the mammalian olfactory nerve. *J Comp Neurol* **373**:240–254.

Ye, P., Y. Umayahara, D. Ritter, T. Bunting, H. Auman, P. Rotwein, and A. J. D'Ercole (1997). Regulation of insulin-like growth factor I (IGF-I) gene expression in brain of transgenic mice expressing an IGF-I-luciferase fusion gene. *Endocrinology* **138**:5466–5475.

Yoshihara Y., M. Kawasaki, A. Tamada, H. Fujita, H. Hayashi, H. Kagamiyama, and K. Mori (1997). OCAM: a new member of the neural cell adhesion molecule family related to zone-to-zone projection of olfactory and vomeronasal axons. *J Neurosci* **17**:5830–5842.

11

Human Olfaction

NANCY E. RAWSON
Monell Chemical Senses Center, Philadelphia, PA

1. INTRODUCTION

Olfaction has been traditionally considered our least important and least developed sense compared to most other species, yet we are able to detect and, with training, discriminate among thousands of odorants at concentrations below the limit of detection of virtually any instrument (Amoore and Hautal, 1983). This system is essential for survival in many species, and in humans plays an important role in avoidance of danger (e.g., smoke, gas leaks, spoiled food), food selection, and in our social interactions; and deficits or abnormalities in olfactory ability can adversely affect our diets and lifestyles (Mattes and Cowart, 1994). A large body of data has accumulated to provide a detailed picture of the physiological basis for detection of odorants and transduction into electrical signals that travel to the olfactory bulb (see Chapter 7, Ache and Restrepo). The vast majority of these data has been collected in animals other than humans. Recent data indicate that while human olfactory receptor neurons (ORNs) exhibit many biological characteristics similar to those of other species studied, some features related to odorant specificity, response characteristics, and basic neuronal properties are unusual (Restrepo et al., 1993b; Rawson et al., 1997). These results emphasize the importance of investigating human tissue in order to understand human olfactory physiology. This chapter will present current

The Neurobiology of Taste and Smell, Second Edition, Edited by Thomas E. Finger, Wayne L. Silver, and Diego Restrepo.
ISBN 0-471-25721-4   Copyright © 2000 Wiley-Liss, Inc.

knowledge on the function and features of the human olfactory system in health and disease, providing a perspective ranging from the cellular to the behavioral.

2. CHEMICAL AND PERCEPTUAL CHARACTERISTICS OF ODORANTS

To begin to understand olfaction, we must start with the stimulus— an odorant may be any volatile chemical that interacts with the olfactory epithelium to produce an olfactory sensation. The key chemical features that render a chemical "odorous" include its volatility and its solubility in the mucus overlying the olfactory receptor cells. All available evidence indicates that both chemical and molecular features (e.g., molecular weight, molecular mass and shape, polarity, resonance structure, types of bonds, and side groups) are critical for determining the odorous characteristics of a chemical. In order to be odorous, a chemical must be sufficiently volatile (a vapor pressure typically above 0.01 mm Hg at environmental temperatures), and typically are small—with a molecular weight less than 300—and contain a strong hydrophobic region and weaker polar region (Silver and Walker, 1997). These traits determine the molecule's hydrophobicity, which can be measured as the octanol/water partition coefficient. This parameter is among the most reliable for predicting the effectiveness of a chemical as an odorant.

Odor qualities have been classified in a variety of ways, but schemes based on perceptual qualities reflect neither chemical nor physiological referents. Of all sensory modalities, olfaction is the only one for which the perceptual qualities are described by their source—smoky, fruity, etc.—rather than by descriptors that reference or relate to features of the stimulus that are responsible for its quality, such as loud (decibels) or low (frequency). Thus, odors are more intimately related to their origins in our vocabulary than other sensory stimuli. In other words, chemicals that bear little resemblance structurally can smell the same, and chemicals that are nearly identical structurally can exhibit very different perceptual qualities (Fig. 11.1). Thus, no "primary odors" have been confirmed, in the same manner that "primary colors" are commonly described (see also discussion in Chapter 9 by Christensen and White). Part of the reason for this may be that many odor qualities that we encounter and perceive as a "single" odor are actually complex mixtures of volatile chemicals. The aroma of coffee, for instance, contains over 800 different chemicals, although only 20 to 30 of them are primarily responsible for the characteristic quality (Deibler et al., 1998). The technique used to identify these components is "gas chromatograph-olfactometry," in which the mixture is separated using a gas chromatograph (GC) that ejects the separated volatiles out a port; a person sniffs the output and records the time particular qualities are perceived. The time points and descriptive information can then be matched to a particular peak on the GC readout to identify particular chemicals. This technique can help to identify "off-notes" or unpleasant qualities as well as the key volatile flavor components of a mixture. By

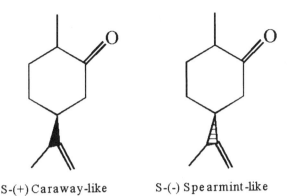

S-(+) Caraway-like   S-(-) Spearmint-like

FIG. 11.1. The stereoisomers of carvone exhibit very different perceptual qualities.

recombining the identified components, the elements needed to recreate the desired quality can be determined.

Structurally, odorants may be classified according to common chemical features such as aldehydes or alcohols, and according to such features as chain length, number and polarity of side groups, etc. These basic chemical traits play an important part in determining the volatility and mucus solubility of the odorant. These features, as well as the three-dimensional structure of the molecule determines what receptor protein a chemical is able to interact with and thus what pattern of olfactory neuron activity it will elicit. Perhaps the first structure-activity scheme proposed was that of Amoore, who suggested that, like a key fitting into a lock, the overall molecular shape was a primary predictor of activity (Amoore, 1967). However, no structure-based classification scheme yet devised is consistently capable of predicting odor quality across more than a limited number of possibilities, so new approaches have been suggested. These include schemes that take into account the physical attributes that result from chemical structure interacting with its environment such as the resonance or vibration resulting from charge interactions (Turin, 1996). Fragrance chemists now use molecular modeling techniques to search for the "olfactophore"—the minimum set of structural features that produce a particular olfactory quality—on the basis of the molecular similarity of sets of chemicals that produce that quality (Frater et al., 1998). These studies, while practically important commercially, may also help us to understand and predict the most likely conformation of the receptors and ligand bindings sites. However, if most odorants interact with multiple receptors as recently proposed (Malnic et al., 1999), the physiological implications may be limited (but see Chapter 9, Christensen and White). Until more is known about how the receptors actually interact with odorous ligands, the role of the odorant binding proteins, the composition of the mucus, and whether existing receptor-mediated mechanisms can account for all of olfactory perception, a truly accurate scheme for classifying odor qualities to develop models that can reliably predict perceptual impact remains a challenge to the field.

3. ANATOMY, PHYSIOLOGY, AND DEVELOPMENT

The human nose contains about 12 million olfactory receptor neurons located within the olfactory epithelium (Silver and Walker, 1997). As in other species, the olfactory epithelium also contains supporting cells, immature neurons and basal cells (Moran et al., 1982; see also Chapter 6, Farbman). This epithelium is overlaid by a layer of mucus, which is continuously in motion due to the action of the motile cilia of the respiratory epithelial cells. The mucus serves as both a passive protective layer and an active player in immunity against infection and in olfaction. Glands within the human nose produce 1–2 l of mucus per day, which is about 96% water and 4% glycoproteins (Lanza and Clerico, 1995). The viscosity and composition of the mucus determine what chemicals are able to penetrate to the receptors below and contributes to removal of the odorants.

The olfactory epithelium is commonly depicted as being limited to the region within the olfactory cleft. This misconception appears to have evolved from an early paper that stated that the olfactory neuroepithelium was restricted to the superior turbinate and nasal septum (von Brunn, 1892). However, figures in this paper indicated extension onto the middle turbinate and septum (Lanza and Clerico, 1995). This has been confirmed anatomically (Feron et al., 1998) and using functional methods to study biopsies from these regions (Restrepo et al., 1993b). While variability exists between individuals, a majority of adults has olfactory epithelium extending down onto the middle turbinate and septum, representing approximately several square centimeters of sensory area (Fig. 11.2).

The human olfactory system is anatomically complete before birth, and perception of odorous chemicals via the amniotic fluid and early in infancy may even play a role in development of later responses and preferences to odorant stimuli in our environment (see Mennella and Beauchamp, 1996, for review and Chapter 10, Burd and Tolbert). Little is known about whether developmental changes occur in humans with respect to the types of odorants that can be detected or discriminated, although one study reported increases in sensitivity to the musky odor androstenone during adolescence (Dorries et al., 1989). Whether the types of receptors expressed or other anatomical changes occur during development is unknown.

The process of odorant detection and signal transduction in human olfactory receptor neurons (ORNs) appears to be very similar to that of other mammalian species, although some differences also have been observed. A large number of olfactory receptor genes have been identified from studies of human DNA, and these genes are present on nearly every chromosome and represent over 1% of the human genome (Rouquier et al., 1998). These genes code for G-protein-linked receptor proteins containing seven potential transmembrane-spanning domains, and exhibit a high degree of sequence homology with those of other mammalian species (see also Chapter 8, McClintock). However, it has recently been found that in humans, many of the sequences identified as olfactory

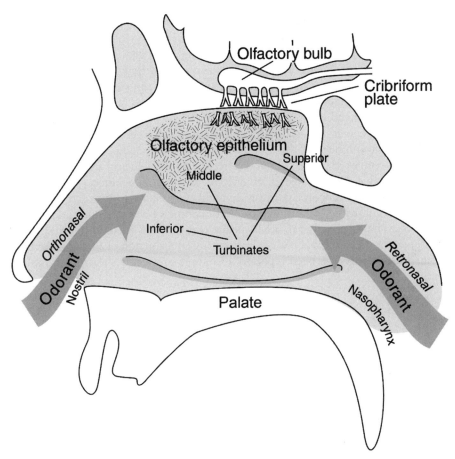

FIG. 11.2. Semischematic diagram depicting the human nasal cavity and extent of olfactory epithelium determined by physiological and histological studies. Volatile chemicals can gain access to the sensory epithelium either via the nares, that is, orthonasally, or via the nasopharynx during chewing and swallowing, that is, retronasally.

receptor-like sequences are nonexpressed pseudogenes (Crowe et al., 1996). Further, a great degree of sequence polymorphism exists between individuals even in the receptor genes that can be expressed (Trask et al., 1998). It is not yet clear whether these genetic variations affect the odorant affinity or specificity of the expressed receptors, but it is possible that these differences significantly affect the perceived qualities of odors between individuals. Such differences might also influence the ability to discriminate among odorants (e.g., see Amoore, 1977).

Studies of transduction pathways in human ORNs have used cells isolated from tissue obtained during nasal sinus surgery and biopsies taken under local anesthesia from the high middle turbinate and apposed septum (Lovell et al., 1982; Lowry and Pribitkin, 1995) (see Fig. 11.2.). These studies employ electrophysiological and imaging methods to examine the responses to odorant stimuli and pharmacological agents administered while the cells are maintained over several hours in a physiological buffer (Thurauf et al., 1996). These investigations suggest that human ORNs function similarly in many respects to those of other species (see Chapter 7, Ache and Restrepo), although direct demonstration of many of the events in the pathways remain to be accomplished. Specifically, stimulation with an odorant elicits an increase in intracellular calcium and a membrane depolarization (Fig. 11.3a). This rise in intracellular calcium is dependent on extracellular calcium and, for certain odorants, can be blocked by an inhibitor of the cyclic nucleotide gated channel (Restrepo et al., 1993b; Rawson et al., 1997). In addition, electrophysiological studies have also demonstrated a cAMP-gated, nonselective cation conductance (Schild et al., 1995). These data support a pathway involving G-protein activation of adenylate cyclase leading to cAMP production and opening of a cyclic nucleotide-gated channel, as has been found in other species (see Chapter 7, Ache and Restrepo). Evidence for an additional pathway mediating transduction of some odorants in human ORNs also has been reported. Odorants previously shown to stimulate

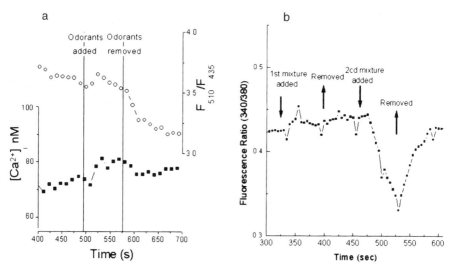

FIG. 11.3. The effect of odorant stimulation on human ORNs isolated from biopsies (as in Rawson et al., 1997).

(a) Membrane potential and $[Ca^{2+}]_i$, measured with the fluorescent indicators di-8-ANEPPS and fura-2, respectively. The upper line shows the increase in membrane potential (depolarization) that occurs concurrently with the increase in $[Ca^{2+}]_i$ (lower trace). The odorant mixture is applied via the superfusion bath and consists of citralva, citronellal, eugenol, geraniol, hedione, menthone and phenylethylalcohol (PEA), 100 mM each.

(b) The response of a human ORN to odorant stimulation in which a decrease in the fluorescence ratio of fura-2 indicating decreasing $[Ca^{2+}]_i$ is observed in response to the second odorant mixture, but not the first. Stimuli were applied via the superfusion bath; second mixture consists of ethyl vanillin, isovaleric acid, lilial, lyral, phenylethylamine, and triethylamine, 100 mM each (first mixture as in A).

production of IP$_3$ in isolated rat olfactory cilia (Breer and Boekhoff, 1991) have been found to trigger an increase in intracellular calcium in human ORNs that is not affected by the cyclic nucleotide-gated channel blocker, but is prevented by neomycin, an inhibitor of phospholipase C (Rawson et al., 1997). IP$_3$-mediated pathways for odorant transduction have been demonstrated clearly in invertebrate species, but their presence and role in mammalian species remains open to discussion. Whether the neomycin-sensitive calcium responses are indeed independent of cAMP is not yet known, but what does seem clear is that multiple pathways are needed to account for odorant-stimulated calcium responses in human ORNs.

Studies with isolated human ORNs also indicate that these cells exhibit some intriguing differences from other species studied. For instance, a significant proportion of human ORNs responds to odorants with a *decrease* in intracellular calcium concentration (Fig. 11.3b; also Rawson et al., 1997). In some species, odorants can either activate or inhibit different receptor cells (e.g., Maue and Dionne, 1987; Morales et al., 1995; see also Chapter 9, Christensen and White). This dual modality may serve to increase the coding and discriminatory capabilities of the system by improving the signal-to-noise ratio (see Chapter 7, Ache and Restrepo). Clearly, further work is needed to determine whether human ORNs can be inhibited by some odorants and the potential role and mechanisms for odorant-stimulated decreases in intracellular calcium levels. Human olfactory receptor cells also appear to be more selective than are those of rat. Under conditions in which cells are isolated using similar procedures, one-third to one-half of rat olfactory neurons respond to multiple stimuli (Restrepo et al., 1993a; Tareilus et al., 1995). In contrast, of 173 olfactory neurons from adult human subjects aged 18 to 45 that were studied with calcium imaging techniques and the same odorant mixtures, none have been observed to respond to more than one mixture (Rawson et al., 1997, and unpublished data). Thus, it may be that odor quality coding occurs in a somewhat different fashion in humans than in rats, with a greater emphasis on identification than sensitivity.

4. THE STUDY OF OLFACTORY FUNCTION

4.1. Psychophysical Measures and What They Mean

Psychophysical testing methods for olfactory function are in general more difficult to carry out and to interpret than the comparable types of tests used to evaluate taste function. This complexity is due to rapid adaptation, which is comparatively minor in the taste system (see Breslin, Chapter 16); the comparably vast number of potential stimuli; the lack of an odor classification scheme that bears any relevance to the physiological processes underlying the perception of odor qualities; and the greater difficulty in controlling the stimulus and its delivery. Olfactory performance is evaluated in human subjects using both threshold and suprathreshold psychophysical testing methods (see Doty, 1991,

for review). The most commonly used tests include determination of the lowest
odorant concentration that an individual can detect, called detection threshold,
and identification of odors that are present at suprathreshold concentrations
using a forced multiple choice questionnaire.

4.1.1. Detection Threshold Measures

Detection thresholds are determined by administering pairs of air streams that
contain either clean, deodorized air as a blank, or an odorant. A variety of
delivery systems are possible, either from squeeze bottles or an instrument that
controls air flow and odorant release called an olfactometer (Prah et al.,
1995) (see also Chapter 4, Bryant and Silver). Subjects report which sample has
an odor, and varying concentrations are administered until the subject is able to
reliably detect the difference between the blank and the odorant-containing
sample. A variety of techniques can be used to carry out the test, all of which are
quite time-consuming and require a considerable amount of subject cooperation
and attention. The method of constant stimuli involves presenting stimuli whose
concentrations are varied randomly. The method of limits presents an alternating
ascending and descending concentration series, while the staircase method,
initially uses only increasing stimuli concentrations presented in a forced-choice
procedure until the subject responds correctly, when a lower concentration is
presented; the point at which several reversals occur around a given concentration is taken as the threshold. Methods based on signal detection theory
present a single concentration at a time, interspersed with blanks and the
proportion of correct responses is used to determine the threshold. These
methods are described in more detail in Doty (1991). Overcoming the problem
posed by adaptation is one of the biggest difficulties in these techniques, and
generally an ascending series is used with the intent that lower concentrations
are less likely to influence perception of higher concentrations than vice-versa.
The "detection threshold" is a relative measure, as the concentration of odorant
actually impacting on the sensory epithelium is unknown, and will vary with the
stimulation device and stimulus in a way that cannot be easily predicted. In spite
of the variety of techniques employed and criteria used to determine "reliable
detection" among laboratories, detection thresholds obtained within a given lab
are fairly stable and reliable upon repeated testing of healthy individuals (Doty
et al., 1995). Because these tests can be time-consuming and labor-intensive to
administer, most clinical studies determine thresholds for only one or two
odorants. Thus it is important to realize that performance for the odorant used
may not be representative of all odorants one might be exposed to. For instance,
recent data obtained in our clinic for two odorants, phenyl ethyl alcohol (PEA)
and lyral, suggest that age-associated differences may be heterogeneous and may
relate to differences in the sensitivity of the pathways mediating cellular responses to these odorants to the effects of aging. Whereas the majority of subjects
over 65 exhibited detection thresholds for PEA that were in the hyposmic range
(below the 50th percentile for adults 18–45), the majority of these same subjects

were able to detect lyral at concentrations within the same range of detection thresholds obtained from younger subjects (Rawson et al., 1998).

Detection threshold measures are difficult to determine with great accuracy due to problems related to contaminants in the chemicals, poor control over the stimulus delivery, and variable attentiveness of the subject. Another difficulty is the need to distinguish between detection of the odor and detection of the irritation that many odorous chemicals can elicit at higher concentrations. A useful technique for distinguishing these sensory modalities is the use of the lateralization threshold—when stimulated uninasally, irritants can be localized, while pure odorants cannot (Wysocki et al., 1992). For instance, the olfactory detection threshold for hexanol is about 10 ppm, while its lateralization threshold is about 1000 ppm (Cometto-Muniz and Cain, 1998). The variability in testing methods, stimulus delivery systems, sources of odorants and analysis and reporting methods make comparison of data among laboratories problematic. The American Society for Testing and Materials published a compilation of reported threshold values for a large number of olfactory and taste stimuli in 1978 (Fazzalair, 1978), but an updated volume is clearly needed. Some laboratories express values in parts per billion (ppb) based on analysis of the volatile components of the air or headspace in the container, while others report the concentration in the vehicle the odorant is dissolved in. While it is difficult to compare among laboratories, some values obtained from the literature or our clinic for commonly employed odorants are presented in Table 11.1.

TABLE 11.1. Representative detection thresholds for some commonly tested odorants for subjects who are not anosmic

Odorant	Population	Range	Reference
Isovaleric acid (sweaty)	$n=24$	$10^{-3.3}-10^{-0.5}$ ppm	Meilgard et al., 1978
Phenylethyl–alcohol (PEA; rose)	$n=197$, 18–59 yrs	$10^{-1.2}-10^{-8.6}$ v/v	Rawson, unpublished data
Amyl acetate (banana)	$n=163$, 20–50 yrs	$10^{-4.5}-10^{-6.5}$ v/v	Deems and Doty, 1987
Lyral (floral/fruity)	$n=28$, 18–39 yrs	0.003%–0.00015% (25th and 75th% ile, respectively)	Rawson et al., 1995
$(+)$ α–ionone (sweet)	$n=13$	0.03–5.90 ppm	Polak et al., 1989
$(-)$ α–ionone (violet)		0.10–21.00 ppm	
$(+)$–carvone (caraway)	$n=8$	$10^{-3}-4$ ppm	Polak et al., 1989
$(-)$–carvone (minty)		$10^{-3.2}-2$ ppm	
m–xylene (organic)	$n=20$ (Japan)	$10^{-1.9}$ ppm	Hoshika et al., 1993
	$n=8$ (Netherlands)	$10^{-0.9}$ ppm (median values)	
H_2S (rotten eggs)	$n=20$ (Japan)	$10^{-3.3}$ ppm	Hoshika et al., 1993
	$n=8$ (Netherlands)	$10^{-3.5}$ ppm (median values)	
Pyridine (rank, unpleasant)	$n=63$	$10^{-0.5}-10^{-3.5}$ ppm	Stevens et al., 1989

4.1.2. Suprathreshold Measures

Suprathreshold measures avoid some of the technical difficulties inherent in threshold determinations, and include odor identification and discrimination tests and intensity estimations.

Odor identification tests in a scratch-and-sniff format are quick and easy to administer and have been used extensively as a 'first pass' approach to evaluate olfactory function in clinical and research settings. The University of Pennsylvania Smell Identification Test (UPSIT) is widely used and is available commercially (Sensonics, Inc., Philadelphia PA). Odorants are presented in encapsulated form on cards and the subject scratches the card to reveal the odor. Whether presented in this format, or from sniff bottles, cardboard sticks ("Sniffin" sticks; Hummel et al., 1997) or other devices, after presentation the subject selects from several descriptors in a multiple-choice format. The descriptors are items associated with the odor, such as 'gasoline' or 'peanuts', rather than odor qualities such as 'fruity' or 'musky'. Thus, correct identification requires not only ability to perceive the odor and familiarity with it, but also with the item selected to represent it. For this reason, these measures incorporate a cognitive component, which in some cases may be a disadvantage, and difficulties can arise if the subject has other cognitive or motor deficits that make this testing protocol difficult to carry out. The descriptors also carry an inherent cultural bias, and cultural differences in familiarity, perceived pleasantness, intensity (Ayabe-Kanamura et al., 1998), and even sensitivity (see Table 11.1) have been reported that should be taken into consideration in designing these studies and interpreting their results.

Odor discrimination tests assess the ability of the subject to perceive that two odors are different. These tests can be carried out as a forced-choice determination of whether two odors are 'the same' or 'different', or can be used to determine the "least detectable difference" in concentrations of the same odor. Alternatively, a triangle test may be used, in which three bottles are given, with two containing the same odor and the third containing a different odor, and the subject is asked to select the one that is different. This kind of measure can potentially provide information about whether two odors are likely to act at the same receptor(s). Such tests are also used extensively in the development of foods, fragrances and fragrance-containing products to compare between potential odorous chemicals or ingredients to be used to match to a particular 'gold standard'.

Measures of odor intensity, or **magnitude estimation**, provide some advantages over threshold detection measures, in that they are generally easier to carry out, require fewer concentrations of stimuli, and are less susceptible to contamination effects that can occur with very low odorant concentrations (Doty, 1991). In these tests, the subject sniffs the odorant from a bottle or other device and rates the intensity using a line scale or a numeric or textural category scale (e.g., $0 =$ no odor, up to 5 for very strong odor). The labeled magnitude scale (LMS) incorporates descriptors (e.g. 'strongest imaginable odor') with a numeric scale, and produces results that differ slightly from pure magnitude estimation scales based on numeric ratings alone (Green et al., 1996). A great deal of discussion has been carried out as to the best methods for devising these scales, analyzing the data generated, and ensuring against bias due to individual or group differences in the way the subjects understand or use the scale (see Doty, 1991 for review).

In the flavor and fragrance industry, a different approach to odor identi-fication is used, in which panelists are trained to recognize the specific 'notes' of odorous compounds and a whole lexicon of descriptors have been developed specifically for the purposes of evaluating flavors and fragrances. While no universally acceptable dictionary has been devised, many common terms are used such as woody, leathery, floral, oriental, fresh, and floral-aldehyde (Rossiter, 1996). Each of these general categories includes a host of more specific descri-ptors, enabling the perfumer or flavorist to precisely describe the qualities of a particular chemical or mixture in a consistent fashion. A skilled perfumer is able to discriminate and categorize thousands of odorant qualities. Specialized training appears not to influence absolute sensitivity, as in a study of professional wine tasters; the ability to discriminate and identify odors was superior to that of controls, but detection thresholds were similar (Bende and Nordin, 1997). It would be intriguing to determine whether this type of training and experience influences the physiology of the olfactory system!

4.2. Evoked Potentials and Functional Imaging

Techniques that aim to both objectify olfactory function measures and provide insight into the central mechanisms governing olfactory perception are also being used to investigate olfactory performance and physiology. Olfactory evoked potentials can be recorded using the same technique as electroencephalograms, in which electrodes are attached to the cranium and electrical fields generated by the activity of large populations of neurons are detected (Kobal and Hummel, 1988). As these evoked potentials are very small, it is necessary to average many recordings and use mathematical filtering techniques to separate signal from noise. Thus, if activity is increasing in some neurons and decreasing in others, the effect would not be detected. However, in a study of two patients with Parkinson's disease (PD) and two controls, Barz et al. (1997) reported a signi-ficant delay in the onset of EEG activity in the PD patients vs. controls in response to exposure to a variety of odors. In contrast, EEG patterns elicited by a trigeminal stimulus (CO_2), were no different between these subjects (see Chapter 5, Bryant and Silver).

Functional magnetic resonance imaging (fMRI) and PET (positron emission tomography) scanning are also beginning to be used in the study of olfaction, particularly in areas where traditional psychophysical methods are difficult, such as in patients with Down's syndrome (Murphy and Jinich, 1996). A recent report using fMRI was able to reveal differences in the brain areas activated in response to sniffing unscented air vs. scented air (Sobel et al., 1998). This study demonstrated that sniffing clean air activated primarily the region of the piriform cortex and the medial and posterior orbito-frontal gyri in the frontal lobe. The addition of odorants to the clean air activated regions in the lateral and anterior orbito-frontal gyri. It is important to realize that these methods are measuring changes in blood flow rather than neural activity directly, and represent changes over fairly broad regions (typically 3 – 12 mm). Nevertheless, as our understanding of the neural basis of olfactory perception grows, these types of studies may aide in characterizing or diagnosing a variety of types of neural dysfunctions as well as extending our understanding of how odor perception is carried out in the central nervous system (see also Chapter 9 by Christiansen and White). With improvements in the technology that are anticipated in the future, it may be possible to generate maps of neural activity within discrete brain regions that correlate with perceptual odor qualities. If sufficient consistency were shown across subjects, it would enable comparison between individuals for clinical diagnostic as well as product development purposes.

5. MODULATION OF OLFACTORY PERCEPTION: PLASTICITY IN THE SYSTEM

Modulation of olfactory function may occur at many levels, and can be rapid (milliseconds—seconds—minutes) or prolonged (hours—days—years).

In addition to naturally occurring adaptation processes, environmental exposures, medications, aging, nutritional status, pregnancy, mental illness and a variety of diseases and disorders can affect olfactory perception.

Physiologically, modulation occurs through a number of cellular processes that take place peripherally in the receptor cell and centrally in the brain. The cellular process of adaptation is responsible for quickly terminating the response to odorant stimuli and allowing the cell to prepare for the next stimulus (see Breer and Boekhoff, 1992 for review). This process is dependent on an influx of intracellular calcium (Kurahashi and Menini, 1997; Kurahashi and Shibuya, 1990) and involves phosphorylation of proteins within the receptor signal transduction cascade (Boekhoff et al., 1992; Zufall et al., 1991 and see also Chapter 7, Ache and Restrepo). While these studies have been carried out in non-human vertebrates, current evidence supports similar cellular adaptation mechanisms in human ORNs. In ORNs from other species, cAMP or diacylglycerol that is produced in response to odorant stimulation activates protein kinases (PK's). While the precise targets of these kinases have not been fully elucidated, inhibition of specific protein kinases (PKA and PKC) prolongs the calcium influx that occurs in response to odorant stimuli in human ORNs (Gomez G. et al., 2000). Thus, phosphorylation-dependent inactivation of the cation channels or intermediate transduction elements is responsible for terminating the receptor cell response to a given odorant. Medications or conditions that alter cellular calcium homeostasis could enhance or reduce this process, altering olfactory perception.

Behaviorally, short-term adaptation (or desensitization) occurs within a minute of odorant exposure and resensitization can take several minutes. This short-term phenomenon is likely to relate to both receptor cell processes and perireceptor processes, such as the time required to clear the odorant from the mucus. Adaptation processes also occur in the central portions of the olfactory system, but are less well understood. It is generally thought that receptor cell adaptation is a short-term phenomenon, while adaptation centrally may last hours, days or weeks. Behaviorally, continuous exposure to an odorant for two weeks at home can increase the detection thresholds for that odorant in a test setting (i.e. reduce sensitivity) even after removal of the odorant from the home (Figure 11.4; Dalton and Wysocki, 1996; Brand et al., 1996). This finding suggests the occurrence of a long-lasting physiological change that affects sensitivity. Cross-adaptation can also occur, in which sensitivity to one odorant is reduced by exposure to a different odorant (Pierce Jr. et al., 1996). This phenomenon can occur between chemicals that differ perceptually or structurally, and little is known about the physiological basis, although structurally similar chemicals are presumably detected via the same receptor(s). A practical implication of this, however, is that it might be exploited as a tool to suppress perception of unpleasant or bothersome odors such as breath or body odor in the same way that sound masking is being used to suppress tinnitus (Kemp and George, 1992). However, more studies are needed to characterize the properties of cross-adapting odors to be able to predict what types of chemicals might be effective for specific types of odors.

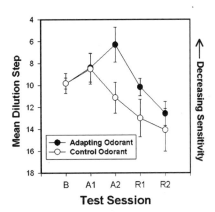

FIG. 11.4. Odor detection thresholds averaged across 10 subjects for two odorants (benzyl acetate and citralva). For half of the subjects benzyl acetate served as the adapting odorant and citralva served as the control; assignment of odorants to conditions were reversed for the remaining subjects. Sensitivity to the adapting odorant (either benzyl acetate or citralva) decreased during the adaptation phase (A1, A2), while subjects were given continuous, at-home exposure to one of the odorants. After two weeks without exposure, threshold sensitivity had recovered (R1, R2). (P. Dalton, unpublished data.)

While the classic phenomenon of sensitization is rare in olfaction, the ability to detect androstenone can be induced by repeated androstenone exposure in subjects that are initially anosmic to it (Wysocki et al., 1989). This effect may be due to reorganization at the peripheral level, since the electrical response recorded across the olfactory epithelium (electro-olfactogram) is induced in mice treated similarly (Wang et al., 1993). Whether repeated stimulation is inducing changes at the level of receptor expression or is affecting other components of the detection/transduction pathway remains unknown.

Hormones and neuropeptides acting at central or peripheral levels may also modulate olfactory sensitivity. Receptors for dopamine, cholecystokinin, serotonin and insulin have been localized within the olfactory bulb of a variety of vertebrate species (see Halasz and Shepherd, 1983 for review), and in some cases, shown to be capable of modulating the responses of the cells on which they reside (e.g. Fadool and Levitan, 1998; Brunig et al., 1999). Estrogen may act directly at receptors in the olfactory bulb (Shughrue et al., 1997) and may modulate the effects of retinoic acid, which is known to be important in the development of the olfactory system (Anchan et al., 1997). This may contribute to the finding of changes in olfactory function throughout the menstrual cycle and of a tendency for females to perform better on some olfactory function tests than males (see Dorries, 1992 for review). Further, the age-associated decline in olfactory function begins later among women than it does in men (Wysocki and Gilbert, 1989). More studies are needed to understand the impact of these neuromodulators on human olfaction.

The implication of these kinds of phenomena is that the olfactory world of any given individual at any given moment is likely to be unique. Such physiological and environmental effects are also likely to contribute to the high individual variability seen in psychophysical data on olfactory threshold tests (see Table 1 and Stevens et al., 1988).

6. OLFACTORY DYSFUNCTION/PATHOLOGIES

Abnormalities in Olfactory Perception: Anosmia, Hyposmia, Dysosmia and Phantosmia

It is estimated that at least 2.7 million Americans experience some form of chronic olfactory dysfunction (Hoffman et al., 1998). Anosmia, or the inability to detect any odor, is rare, and if present from birth, may be due to a genetic condition called Kallmann's syndrome. In this genetic disorder the olfactory bulbs fail to develop due to failure of the olfactory axons to extend from the developing olfactory placode to the telencephalon (see also Chapter 10, Burd and Tolbert). These individuals are also hypogonadotropic, as the proper migration of neurons from the placode to the brain is necessary for the development of the central circuits controlling reproductive development. Interestingly, the olfactory epithelium of these individuals continues to produce functional olfactory neurons, in spite of their lack of connection to functional central circuits (Rawson et al., 1995).

An inability to detect a particular odor, termed a specific anosmia, is thought to be a rare phenomenon, but instances have been reported for a number of chemicals, including isovaleric acid and androstenone (Amoore, 1967; 1977). It is possible that a larger number of specific anosmias exist that have yet to be identified due to the difficulty in screening large numbers of people for a large number of odors. Genetic differences in receptor expression are likely to account for these conditions, as anosmia to musk is inherited as a simple recessive trait (Whissell-Buechy and Amoore 1973). Advances in our understanding of the human genome are likely to identify the specific gene deletions or mutations causing these anosmias in the near future.

Head trauma can also cause anosmia, due to damage to the olfactory nerve from coup-contra-coup forces (i.e. rapid movement in opposite directions) resulting from a sharp blow to the head. In these cases, anosmia often persists even after all anatomical signs of damage are gone. However, in some cases anosmia changes to dysosmia (inappropriate odors) or phantosmia (odors detected in the absence of a volatile chemical). In approximately 30-40% of cases, recovery is partial (hyposmia) and some cases recover completely over months or years (Cowart et al., 1997). Olfactory dysfunction can also occur following viral or bacterial infections, nasal sinus disease and allergic rhinitis (Cowart et al., 1993). In chronic nasal sinus disease, the mucus clearance rate is significantly reduced compared to controls, but this was not correlated with changes in elasticity or viscosity, suggesting changes in ciliary motility or other parameters are likely as well (Atsuta and Majima, 1998). In some of these cases, hyposmia may be secondary to inflammation, and treatment with anti-inflammatory nasal sprays may alleviate the problem. A few reports suggest that vitamin A therapy may aid recovery in these conditions, and in view of the importance of retinoids in development of the olfactory epithelium (Anchan et al., 1997), this treatment warrants further study. However, in most instances, little can be done to cure or treat the condition (Cowart et. al., 1997).

A wide range of medications also has been reported to adversely affect olfaction (see Ackerman and Kasbekar, 1997 for review). These vary from agents acting at cellular pathways known to be involved in olfaction such as calcium channel blockers, to chemicals that may damage the mucosa such as cocaine, methotrexate or cytosine arabinoside. Others, such as diuretics, antihistamines or anticholinergics, may alter mucus composition and therefore impede access of the odorants to the receptors.

7. OLFACTION, AGING, AND NEURODEGENERATIVE DISEASES

Age-associated changes in olfactory function are well established, although the basis for these changes is uncertain. With increasing age, many individuals experience a loss of olfactory function, but the degree of loss and age at which it may become noticeable varies considerably with the individual. Studies have reported age-associated deficits in several parameters of olfactory function, including odor identification and discrimination (Schiffman, 1977; Schiffman et al., 1979; Doty et al., 1984; Wysocki and Gilbert, 1989), detection thresholds (Schemper et al., 1981; Deems and Doty, 1987; Stevens et al., 1988), and intensity perception (Stevens and Cain, 1985; Ennis and Hornung, 1988). Whether these changes are uniform across odors (Stevens et al., 1988) or heterogeneous (Wysocki and Gilbert, 1989) remains unclear. Data also suggest that older subjects adapt more quickly to odors and resensitize more slowly (Stevens et al., 1989). This sensory deficit adversely affects not only quality of life (e.g., enjoyment of foods), but can endanger survival as well—45% of elderly subjects were found to be unable to detect the warning odor in natural gas (ethyl mercaptan) at the safety standard level (Cain and Stevens, 1989)!

The reasons for these age-associated declines in function may relate to prior history of disease and medication usage, exposures to pollutants or irritants, genetic factors, nutritional status, or behavioral factors such as smoking. At present, we cannot say whether the observed association between olfactory loss is due the natural aging process per se (i.e., chronological age), or to an age-associated increase in susceptibility to the effects of potentially detrimental factors, the accumulation of multiple insults, or other health problems (i.e., physiological age). Investigations into the possible causes of olfactory impairment must, therefore, be approached at an individual level and must consider physiological as well as chronological age if we are to gain insight into this sensory deficit.

Anatomical studies into the aging of the olfactory system have found structural changes associated with aging. Histological studies in humans (Paik et al., 1992) and animals (Hirai et al., 1996) have found that with age, the neuroepithelium becomes patchy, with increased infiltration by respiratory epithelium. These studies imply that the ability of the olfactory epithelium to regenerate deteriorates with age, and that a simple loss of mature ORNs may explain the commonly reported age-related loss of olfactory sensitivity. However, in animals only a small portion of olfactory epithelium need be intact for olfactory function to be apparently unimpaired, and a recent study found no age-associated differences in rats in the intensity or patterns of electro-olfactogram responses to five odorants (Loo et al., 1996). Thus, it is likely that factors other than epithelial area contribute to age-related changes in olfactory performance. Work in our laboratory examining olfactory biopsies from subjects ranging in age from 18 to 84 suggests that the degree of reduction in olfactory area in the elderly may be overestimated. We have found that biopsies obtained from subjects over 65 years are just as likely to produce odor-responsive olfactory neurons as are biopsies from subjects 45 years old or younger (Rawson et al., 1998).

Less is known about physiological changes occurring in the olfactory system of humans or animal models that may contribute to impaired olfactory ability with age or degenerative disease. Recent efforts are exploring the use of noninvasive techniques such as olfactory-evoked potentials (Morgan et al.,

1997) to study differences in responses to odorants in subjects with impaired olfaction of various etiologies. These approaches support psychophysical evidence for age-associated changes in olfactory function but do not indicate the precise location or nature of the physiological changes responsible. Recent data describing the functional characteristics of olfactory receptor neurons isolated from olfactory tissue biopsies taken from volunteers ranging in age from 18 to 84 whose olfactory performance is also tested suggest that age-associated changes in olfactory performance are not due simply to a loss of responsive ORNs. These studies indicated that the neurons obtained from older subjects were even more likely to respond to odor stimuli. However, the response frequency to two odorant mixtures designed to activate either cAMP or IP_3-mediated transduction pathways differed from the frequency observed among ORNs from younger subjects. Proportionally fewer cells from older subjects responded to the odorant mixture containing adenylate cyclase activators and more to the PLC-activating mixture compared to ORNs from younger subjects. In addition, ORNs were found that responded to both mixtures, while among younger subjects, ORNs respond only to one odorant or mixture (Rawson et al., 1997). Further, the differential response frequency of the ORNs was reflected in the subject's olfactory sensitivity. In other words, older subjects were better able to detect the odorant lyral than phenylethyl alcohol (PEA), and ORNs obtained from their biopsies were more likely than younger subjects' ORNs to respond to the lyral-containing mixture (IP_3-related) than to the PEA-containing mixture (cAMP mediated). Thus, changes in olfaction with age may not be uniform, and may relate to underlying changes in the physiological mechanisms by which ORNs respond to odor stimuli.

Changes in olfactory function occur in a number of neurodegenerative disorders such as Alzheimer's dementia (Koss et al., 1988), Huntington's disease (Kramer and Siegelbaum, 1992) and Parkinson's disease (Doty et al., 1993). The ability to obtain ORNs via biopsy from living subjects provides a useful tool to investigate the biological basis of these diseases, as well as potentially aiding in diagnosing and evaluating alternative treatments. Alzheimer's disease (AD) represents the most frequent form of dementia, and in spite of extensive research, its causes remain enigmatic and treatments are of limited success (Ginsberg et al., 1997). Olfactory impairment and neuroanatomical changes in the central portions of the olfactory system occur early in the development of this disease, and olfactory testing has been explored as a diagnostic aid and as an indicator of the severity of dysfunction. For instance, in AD, deficits in odor identification ability have been found that are more severe than those seen in age-matched controls and are independent of other cognitive changes (Lehrner et al., 1997). Olfactory neurons and cell cultures also have been used to investigate the cellular processes that are affected by Alzheimer's disease, and to study basic cellular mechanisms that may underlie development of the plaques and tangles that are the pathological hallmarks of the disease (Johnson et al., 1994). While some histological studies have reported AD-specific neuropathology within the olfactory epithelium (Talamo et al., 1989; Yamagishi et al., 1994), others using different markers, have noted similar features in AD, non-AD, and healthy brain tissue (Crino et al., 1995). Although central components of the olfactory system may be more overtly affected, these studies do support the notion that changes occurring in the olfactory neurons may reflect those occurring within the CNS, and that further study of olfactory physiology in these patients may be useful in furthering our understanding of these diseases.

Impaired ability to identify odors with normal or reduced sensitivity has been reported in a number of psychiatric disorders such as schizophrenia and depression (Serby et al., 1990; Saoud et al., 1998), but the extent to which these dysfunctions are disease-related versus medication-related is unclear. As olfactory tissue is both an accessible source of neuronal tissue, and amenable to functional studies of transduction elements potentially involved in the etiology of these disorders, studies using tissue from patients with these disorders hold promise to yield insight into the biochemical or molecular pathways that lead to these diseases. Such an approach may also be useful to help us to understand the molecular action of a whole host of neurally active drugs, and perhaps even as an in vitro tool to prescreen agents for efficacy for a given patient.

8. THE ROLE OF OLFACTION IN OUR LIVES AND WELL-BEING

8.1. Chemosensory Stimulation of Cephalic Phase Responses

While the contribution of odors to our enjoyment of food is clear, the sensory components of food play an even more important physiological role. The orosensory stimuli associated with food ingestion are responsible for triggering what are termed "cephalic phase responses". These responses, studied first by Pavlov in the early 1900s include stimulation of salivary flow, gastric acid, and also preabsorptive release of insulin, c-peptide, and glucagon (see Teff, 1996, for review). These cephalic phase responses prepare the body to receive incoming nutrients and to metabolize food optimally. Olfactory, together with visual and gustatory, stimuli thus become physiologically essential to insure optimal metabolism and nutritional status. When these stimuli are absent, as when patients are fed through nasogastric tubes or intravenously, blood glucose levels remain elevated longer following nutrient administration. Thus, it appears that one role for these effects is to prevent large swings in nutrient availability. The more sensory modalities stimulated, the more robust the cephalic phase response (Feldman and Richardson, 1986). The physiological importance of this response is evident from studies administering a small dose of insulin to imitate the cephalic phase response in obese subjects who exhibited an attenuated response. This treatment significantly improved glucose tolerance in obese subjects even in the face of basal hyperinsulinemia (Teff and Engelman, 1996). The neural and metabolic bases for this phenomenon are currently under active investigation.

8.2. Odor-Associations and Odors as a Cue for Learning

The potency of chemosensory stimuli to evoke memories was noted by Proust, who, in response to memories elicited by a particular type of tea and cake, wrote,

> But when from a long-distant past nothing subsists, after the people are dead, after the things are broken and scattered, taste and smell alone, more fragile but more enduring, more unsubstantial, more persistent, more faithful, remain poised a long time, like souls, remembering, waiting, hoping, amid the ruins of all the rest; and bear unflinchingly, in the tiny and almost impalpable drop of their essence, the vast structure of recollection. (From Marcel Proust, *Remembrance of Things Past. Vol. 1: Swann's Way: Within a Budding Grove.*)

The emotional evocativeness of odor-associated memories is likely due to the close connections between the olfactory system and areas of the brain involved with emotion (amygdala), and learning and memory (hippocampus) (see Herz and Engen, 1996, for review). In a study in which subjects were asked to recall picutres that were shown with or without concurrent odor exposure, recall accuracy was unaffected by odor, but the descriptions provided during recall were more emotional than when the pictures were viewed in the absence of an odor (Herz and Cupchik, 1995). Other studies have shown that odors are more potent at supporting context-dependent memory than other types of cues such as visual or auditory (Cann and Ross, 1989). Thus, the same ambient odor present during learning and testing significantly enhances recall of the learned information, and this is particularly true when the cue is presented in an emotionally charged situation, such as just before an exam (Herz, 1997).

8.3. Olfaction and Cognition

Beyond these physical and experiential considerations, cognitive factors clearly impact on our perceptions about olfactory stimuli, and can affect performance on psychophysical tests. For instance, Dalton reported that subjects advised that a particular odor was potentially hazardous rated the odor as more intense than when the same subjects were told that (the same) odor was a natural extract (Dalton, 1996). In another study, subjects given positive information about the odor they were being exposed to exhibited the most adaptation and the least perceived irritation, compared to those given neutral or negative information about the odor (Dalton et al., 1997). Further, workers chronically exposed to acetone perceived that odor as more intense and more irritating than nonexposed subjects, while perceptions about a different odor (phenylethylalcohol) did not differ between the two groups (Wysocki et al., 1997). Thus, olfaction does not occur "in a vacuum," and is intimately influenced by our experience with and information about the odor itself.

8.4. Aromatherapy—Can Odors Affect Mood?

Clearly, odors can color the perceptions we have about ourselves and the world around us in both positive and negative ways, and can influence our behavior.

These effects are most likely due to learned associations (see Knasko, 1996, for review). Exposure to a pleasant odor triggered recall of more positive personal memories, and odors can alter self-rated mood in positive or negative ways (see Herz and Engen, 1996, for review). In offices, factories, and stores, odors are being explored for the potential to improve performance and influence behavior. Research into these effects, however, suggests that the impact of odors on these parameters is influenced by many factors such as hedonics and congruency with the surroundings, and is inconsistent and difficult to predict. For instance, different studies report that a pleasant odor may improve, impair, or have no effect (Knasko, 1996) on performance of tasks such as arithmetic or word construction. The often advertised ability of certain odors to "improve mood" and lessen various health complaints propounded by aromatherapy advocates is likely to be dependent to a great extent on the individual's associations with the odor and the beliefs and expectations that are linked to those associations. In addition, physiological or pharmacological effects of the odorous chemicals that are likely to be absorbed via the epithelium or nasal mucosa are also possible. For example, within minutes after a massage with a lavender-scented oil, detectable levels of the main olfactory components (linalool and linalyl acetate) could be measured in plasma (Buchbauer, 1993)! Studies using electroencephalographic (EEG) recordings have investigated whether inhaled odors can alter mental alertness. The theta component of the EEG reflects shifts in attention, and reduced theta activity correlates with less anxiety (Lorig and Schwartz, 1989). Results of such studies have been inconsistent, but do generally suggest that some odors can alter theta and alpha wave activity. Martin (1998) reported significantly reduced theta activity in 21 subjects during exposure to spearmint odor, while activity during exposure to almond, cumin, and strawberry was not different from deodorized air. For these subjects, the spearmint odor was also rated as the most pleasant.

8.5. Chemical Communication/Social Interactions

In humans, volatile chemicals that may be odorous or not overtly odorous can influence our perceptions about others, figure prominently in mother–infant interactions, and may even play a role in mate selection. Mothers and infants can recognize each other by scent alone within hours of birth, and infants are more attentive and responsive to their mother's odors than odors from unrelated females (see Schaal, 1988, for review). Volatile chemicals such as those in garlic and vanilla are transmitted into breast milk and can thereby influence the sensory world and behavior of the human infant (for review see Mennella and Beauchamp, 1996). Maternal garlic ingestion, or vanilla addition to formula, increased the duration of the feeding bout and amount of sucking exhibited by infants who had little or no prior experience with these odorous chemicals (Mennella and Beauchamp, 1991, 1996). Undoubtedly, factors in addition to olfactory sensations contribute to these findings, but they do point out the important role that odors can play in early development and behavior. These pheromonal-types of chemical effects involved in social interactions are discussed in more detail in Chapter 5 by Johnston.

9. SUMMARY

Although humans do not depend on their sense of smell for survival in the same way that most other mammals do, our ability to detect odors is superior to any machine yet devised, and with training, it is possible to identify thousands of volatile chemicals. Unlike other sensory stimuli, odor quality descriptors remain unconnected to any common feature referents, and in spite of beginning to understand the molecular receptors responsible for detection, we have yet to discover what those referents that determine odorant quality might be. This may also explain in part the commonly experienced "tip of the nose" phenomenon—the ability to recognize an odor as familiar, without being able to actually identify it. Even when we are unable to name an odor, its perception is often sufficient to trigger emotionally potent associative memories. Further, olfactory stimuli play an important part in triggering reflexes designed to prepare the body for digestion and metabolism, and influence our social interactions in ways that are just beginning to be explored.

Research into the physiological mechanisms underlying human olfaction has progressed tremendously in the last decade. Advances in technology have enabled us to literally watch olfaction occurring at the cellular and organismal level. We can observe individual cells detecting stimuli and converting that information into an electrical signal that is relayed to the central nervous system. We can also observe changes in brain activity in living subjects as the electrical signals are decoded into the perceptions we experience as "odors." Yet much remains to be learned and while some aspects of the system reflect those seen throughout the animal kingdom, some features appear to be unlike those seen in other mammals. We have yet to understand how such a remarkably diverse array of perceptual qualities can be encoded and maintained in the face of constant replacement of the odorant detecting cells.

In spite of its comparatively great regenerative ability, olfaction is susceptible to impairment from a variety of causes. Natural variations in anatomical and physiological characteristics can impact on the relative accessibility of odors emanating from internal or external sources to the olfactory epithelium, their ability to penetrate the mucus and the detection, and transduction mechanisms necessary for perception. This system, designed for great sensitivity, is susceptible to "noise" that could lead to inappropriate perceptions if environmental or physical conditions upset the delicate balance between "signal" and "noise" in any way. Medications, nutritional status, hormonal status, and current and past disease conditions can all impact on the ability of the olfactory system to be maintained and function properly. Genetic differences come into play through determination of the population of receptors that may be expressed, which in some cases might be influenced by prior exposure. Thus, the olfactory worlds of each of us may be as individual as our fingerprints, and as unimportant as we may consider it among our five senses, our olfactory perceptions play an intimate role in how we interact with the world around us.

"Odors, when sweet violets sicken,
Live within the sense they quicken."
From: "Music" by Percy Bysshe Shelley

ACKNOWLEDGMENTS

The editorial assistance of Wayne Silver, and contributions by Pamela Dalton and Marcia Pelchat are gratefully acknowledged in the preparation of this chapter. Work described here was funded in part by NIH DC08276 and DC00214 and a grant from the Van Ameringen Foundation.

REFERENCES

Ackerman, B. H. and N. Kasbekar (1997). Disturbances of taste and smell induced by drugs. *Pharmacotherapy* **17**:482–496.

Amoore, J. E. (1967). Molecular shape and odour: pattern analysis of PAPA. *Nature* **216**:1084–1087.

Amoore, J. E. (1977). Specific anosmia and the concept of primary odors. *Chem Senses Flav* **2**:267–281.

Amoore, J. E. and E. Hautal (1983). Odor as an aid to chemical safety: odor thresholds compared with threshold limit values and volatilities for 214 chemicals in air and water dilution. *J Appl Toxicol* **3**:272–290.

Anchan, R. M., D. P. Drake, C. F. Haines, E. A. Gerwe, and A.-S. LaMantia (1997). Disruption of local retinoid-mediated gene expression accompanies abnormal development in the mammalian olfactory pathway. *J Comp Neurol* **379**:171–184.

Atsuta, S. and Y. Majima (1998). Nasal mucociliary clearance of chronic sinusitis in relation to rheological properties of nasal mucus. *Ann Otol Rhinol Laryngol* **107**:47–51.

Ayabe-Kanamura, S., I. Schicker, M. Laska, R. Hudson, H. Distel, T. Kobayakawa, and S. Saito (1998). Differences in perception of everyday odors: a Japanese-German cross-cultural study. *Chem Senses* **23**:31–38.

Barz, S., T. Hummel, E. Pauli, M. Majer, C. J. Lang, and G. Kobal (1997). Chemosensory event-related potentials in response to trigeminal and olfactory stimulation in idiopathic Parkinson's disease. *Neurology* **49**:1424–1431.

Bende, M. and S. Nordin (1997). Perceptual learning in olfaction: professional wine tasters versus controls. *Physiol Behav* **62**:1065–1070.

Boekhoff, I., S. Schleicher, J. Strotmann, and H. Breer (1992). Odor-induced phosphorylation of olfactory cilia proteins. *Proc Natl Acad Sci U S A* **89**:11983–11987.

Brand, J. G., N. E. Rawson, and D. Restrepo (1996). Signal transduction in olfaction and symptomatology of olfactory dysfunction. *Rev Int Psychopathol* **22**:281–303.

Breer, H. and I. Boekhoff (1991). Odorants of the same odor class activate different second messenger pathways. *Chem Senses* **16**:19–29.

Breer, H. and I. Boekhoff (1992). Termination of second messenger signaling in olfaction. *Proc Natl Acad Sci U S A* **89**:471–474.

Brunig, I., M. Sommer, H. Hatt, and J. Bormann (1999). Dopamine receptor subtypes modulate olfactory bulb gamma-aminobutyric acid type A receptors. *Proc Natl Acad Sci U S A* **96**:2456–2460.

Buchbauer, G. (1993). Biological effects of fragrances and essential oils. *Perfumer and Flavorist* **18**:19–24.

Cain, W. S. and J. C. Stevens (1989). Uniformity of olfactory loss in aging. *Ann N Y Acad Sci* **561**:29–38.

Cann, A. and D. A. Ross (1989). Olfactory stimuli as context cues in human memory. *Am J Psychol* **102**:91–102.

Cometto-Muniz, J. E. and W. S. Cain (1998). Trigeminal and olfactory sensitivity: comparison of modalities and methods of measurement. *Int Arch Occup Environ Health* **71**:105–110.

Cowart, B. J., I. M. Young, R. S. Feldman, and L. D. Lowry (1997). Clinical disorders of smell and taste. In: G. K. Beauchamp and L. M. Bartoshuk (eds). *Handbook of Perception and Cognition: Tasting and Smelling.* San Diego: Academic Press, pp 175–198.

Cowart, B. J., K. Flynn-Rodden, S. J. McGeady, and L. D. Lowry (1993). Hyposmia in allergic rhinitis. *J Allergy Clin Immunol* **91**:747–751.

Crino, P. B., B. Greenberg, J. A. Martin, V. M.-Y. Lee, W. D. Hill, and J. Q. Trojanowski (1995). beta-Amyloid peptide and amyloid precursor proteins in olfactory mucosa of patients with Alzheimer's disease, Parkinson's disease, and Down's syndrome. *Ann Otol Rhinol Laryngol* **104**:655–661.

Crowe, M. L., B. N. Perry, and I. F. Connerton (1996). Olfactory receptor-encoding genes and pseudogenes are expressed in humans. *Gene* **169**:247–249.

Dalton, P. (1996). Odor perception and beliefs about risk. *Chem Senses* **21**:447–458.

Dalton, P. and C. J. Wysocki (1996). The nature and duration of adaptation following long-term odor exposure. *Percep Psychophys* **58**:781–792.

Dalton, P., C. J. Wysocki, M. J. Brody, and H. J. Lawley (1997). The influence of cognitive bias on the perceived odor, irritation and health symptoms from chemical exposure. *Int Arch Occup and Environ Health* **69**:407–417.

Deems, D. A. and R. L. Doty (1987). Age-related changes in the phenyl ethyl alcohol odor detection threshold. *Trans Penn Acad Ophthalmol Otolaryngol* **39**:646–650.

Deibler, K. D., T. E. Acree, and E. H. Lavin (1998). Aroma analysis of coffee brew by gas chromatography-olfactometry. In: E. T. Contis, C.-T. Ho, C. J. Mussinan, T. H. Parliament, F. Shahidi and, A. M. Spanier (eds). *Food Flavors: Formation, Analysis and Packaging Influences.* Amsterdam: Elsevier Science B.V., pp 69–78.

Dorries, K. M. (1992). Sex differences in olfaction in mammals. In: Serby, M. and K. Chobor (eds). *The Science of Olfaction.* New York: Springer-Verlag, pp 245–275.

Dorries, K. M., J. H. Schmidt, G. K. Beauchamp, and C. J. Wysocki (1989). Changes in sensitivity to the odor of androstenone during adolescence. *Dev Psychobiol* **22**:423–435.

Doty, R. L. (1991). Olfactory system. In: T. V. Getchell, R. L. Doty, L. M. Bartoshuk, and J. B. Snow Jr. (eds). *Smell and Taste in Health and Disease.* New York: Raven Press, pp 175–203.

Doty, R. L., P. Shaman, S. L. Applebaum, R. Giberson, L. Siksorski, and L. Rosenberg (1984). Smell identification ability: changes with age. *Science* **226**:1441–1443.

Doty, R. L., L. I. Golbe, D. A. McKeown, M. B. Stern, C. M. Lehrach, and D. Crawford (1993). Olfactory testing differentiates between progressive supranuclear palsy and idiopathic Parkinson's disease. *Neurology* **43**:962–965.

Doty, R. L., D. A. McKeown, W. W. Lee, and P. Shama (1995). A study of the test-retest reliability of ten olfactory tests. *Chem Senses* **20**:645–656.

Ennis, M. P. and D. E. Hornung (1988). Comparisons of the estimates of smell, taste and overall intensity in young and elderly people. *Chem Senses* **13**:131–139.

Fadool, D. A. and I. B. Levitan (1998). Modulation of olfactory bulb neuron potassium current by tyrosine phosphorylation. *J Neurosci* **18**:6126–6137.

Fazzalair, F. A. (1978). *Compilation of Odor and Taste Threshold Values Data.* Baltimore: American Society for Testing and Materials, 497 pp.

Feldman, M. and C. T. Richardson (1986). Role of thought, sight, smell, and taste of food in the cephalic phase of gastric acid secretion in humans. *Gastroenterology* **90**:428–433.

Feron, F., C. Perry, J. J. McGrath, and A. Mackay-Sim (1998). New techniques for biopsy and culture of human olfactory epithelial neurons. *Arch Otolarygol Head Neck Surg* **124**: 861–866.

Frater, G., J. A. Bajgrowicz, and P. Kraft (1998). Fragrance chemistry. *Tetrahedron* **54**:7633–7703.

Ginsberg, S. D., P. B. Crino, V. M.-Y. Lee, J. H. Eberwine, and J. Q. Trojanowski (1997). Sequestration of RNA in Alzheimer's disease neurofibrillary tangles and senile plaques. *Ann Neurol* **41**:200–208.

Gomez G., N. E. Rawson, B. Cowart, L. D. Lowry, E. A. Pribitkin, and D. Restrepo (2000). Modulation of odor-induced increases in $[Ca^{2+}]$; by inhibitors of protein kinases A and C in rat and human olfactory receptor neurons. *J. Neurosci* (In Press).

Green, B. G., P. Dalton, B. J. Cowart, G. Shaffer, K. Rankin, and J. Higgins (1996). Evaluating the "labeled magnitude scale" for measuring sensations of taste and smell. *Chem Senses* **21**:323–334.

Halasz, N. and G. M. Shepherd (1983). Neurochemistry of the vertebrate olfactory bulb. *Neuroscience* **10**:579–619.

Herz, R. (1997). Emotion experienced during encoding enhances odor retrieval cue effectiveness. *Am J Psychol* **110**:489–505.

Herz, R. and G. C. Cupchik (1995). The emotional distinctiveness of odor-evoked memories. *Chem Senses* **20**:517–528.

Herz, R. S. and T. Engen (1996). Odor memory: review and analysis. *Psychol Bull Rev* **3**:300–313.

Hirai, T., S. Kojima, A. Shimada, T. Umemura, M. Skai, and C. Itakura (1996). Age related changes in the olfactory system of dogs. *Neuropathol Appl Neurobiol* **22**:531–539.

Hoffman, H. J., E. K Ishii, and R. H. Macturk (1998). Age-related changes in the prevalence of smell/taste problems among the United States adult population. *Ann N Y Acad Sci* **855**:716–722.

Hoshika, Y., T. Imamura, G. Muto, L. J. Van Gemert, J. A. Don, and J. I. Walpot (1993). International comparison of odor threshold values of several odorants in Japan and in The Netherlands. *Environ Res* **61**:78–83.

Hummel, T., B. Sekinger, S. R. Wolf, E. Pauli, and G. Kobal (1997). "Sniffin' sticks": olfactoy performance assessed by the combined testing of odor identification, odor discrimination and olfactory threshold. *Chem Senses* 22:39–52.

Johnson, G. S., J. Basaric-Keys, H. A. Ghanbari, R. S. Lebovics, K. P. Lesch, C. R. Merril, T. Sunderland, and B. Wolozin (1994). Protein alterations in olfactory neuroblasts from Alzheimer donors. *Neurobiol Aging* 15:675–680.

Kemp, S. and R. N. George (1992). Masking of tinnitus induced by sound. *J Speech Hear Res* 35:1169–1179.

Knasko, S. C. (1996). Human responses to ambient olfactory stimuli. In: R. B. Gammage and B. A. Berven (eds). *Indoor Air and Human Health.* Boca Raton, FL: Lewis Pub., pp 107–123.

Kobal, G. and C. Hummel (1988). Cerebral chemosensory evoked potentials elicited by chemical stimulation of the human olfactory and respiratory nasal mucosa. *Electroencephalogr Clin Neurophysiol* 71:241–250.

Koss, E., J. M. Weiffenbach, J. V. Haxby, and R. P. Friedland (1988). Olfactory detection and identification performance are dissociated in early Alzheimer's disease. *Neurology* 38:1228–1232.

Kramer, R. H. and S. A. Siegelbaum (1992). Intracellular Ca^{2+} regulates the sensitivity of cyclic nucleotide-gated channels in olfactory receptor neurons. *Neuron* 9:897–906.

Kurahashi, T. and A. Menini (1997). Mechanism of odorant adaptation in the olfactory receptor cell. *Nature* 385:725–729.

Kurahashi, T. and T. Shibuya (1990). Ca^{2+}-dependent adaptive properties in the solitary olfactory receptor cell of the newt. *Brain Res* 515:261–268.

Lanza, D. C. and D. M. Clerico (1995). Anatomy of the human nasal passages. In: R. L. Doty (ed). *Handbook of Olfaction and Gustation.* New York: Marcel Dekker, pp 53–73.

Lehrner, J. P., T. Brucke, P. Dal-Bianco, G. Gatterer, and I. Kryspin-Exner (1997). Olfactory functions in Parkinson's disease and Alzheimer's disease. *Chem Senses* 22:105–110.

Loo, A. T., S. L. Youngentob, P. F. Kent, and J. E. Schwob (1996). The aging olfactory epithelium: neurogenesis, response to damage, and odorant-induced activity. *Int J Dev Neurosci* 14:881–900.

Lorig, T. S. and G. E. Schwartz (1989). EEG activity during food and relaxation imagery. *Imagin Cogn Pers* 8:201–208.

Lovell, M. A., B. W. Jafek, D. T. Moran, and C. Rowley III (1982). Biopsy of human olfactory mucosa. *Arch Otolaryngol* 108:247–249.

Lowry, L. D. and E. A. Pribitkin (1995). Collection of human olfactory tissue. In: A. I. Spielman and J. G. Brand (eds). *Experimental Cell Biology of Taste and Olfaction.* Boca Raton, FL: CRC Press, pp 47–48.

Malnic, B., J. Hirono, T. Sato, and L. B. Buck (1999). Combinatorial receptor codes for odors. *Cell* 96:713–723.

Martin, G. N. (1998). Human electroencephalographic (EEG) response to olfactory stimulation: two experiments using the aroma of food. *Int J Psychophysiol* 30:287–302.

Mattes, R. D. and B. J. Cowart (1994). Dietary assessment of patients with chemosensory disorders. *J Am Diet Assoc* 94:187–195.

Maue, R. A. and V. E. Dionne (1987). Patch-clamp studies of isolated mouse olfactory receptor neurons. *J Gen Physiol* 90:95–125.

Meilgaard, M., G. V. Civille, and B.T. Carr (1978). *Sensory Evaluation Techniques.* Boca Raton, FL: CRC Press, pp 113–118.

Mennella, J. A. and G. K. Beauchamp (1991). Maternal diet alters the sensory qualities of human milk and the nursling's behavior. *Pediatrics* **88**:737–744.

Mennella, J. A. and G. K. Beauchamp (1996). The early development of human flavor preferences. In: E. D. Capaldi (ed). *Why We Eat What We Eat.* Washington, DC: American Psychological Association, pp 83–112.

Morales, B., P. Labarca, and J. Bacigalupo (1995). A ciliary K^+ conductance sensitive to charibdotoxin underlies inhibitory responses in toad olfactory receptor neurons. *FEBS Lett* **359**:41–44.

Moran, D. T., J. Carter Rowley III, B. W. Jafek, and M. A. Lovell (1982). The fine structure of the olfactory mucosa in man. *J Neurocytol* **11**:721–746.

Morgan, C. D., J. W. Covington, M. W. Geisler, J. Polich, and C. Murphy (1997). Olfactory event-related potentials: older males demonstrate the greatest deficits. *Electroencephalogr Clin Neurophysiol* **104**:351–358.

Murphy, C. and S. Jinich (1996). Olfactory dysfunction in Down's syndrome. *Neurobiol Aging* **17**:631–637.

Paik, S. I., M. N. Lehman, A. M. Seiden, H. J. Duncan, and D. V. Smith (1992). Human olfactory biopsy: the influence of age and receptor distribution. *Arch Otolaryngol Head Neck Surg* **118**:731–738.

Pierce Jr., J. D., C. H. Wysocki, E. V. Aronov, J. B. Webb, and R. M. Boden (1996). The role of perceptual and structural similarity in cross-adaptation. *Chem Senses* **21**:223–237.

Prah, J. D., S. B. Sears, and J. C. Walker (1995). Modern approaches to air dilution olfactometry. In: R.L. Doty (ed). *Handbook of Olfaction and Gustation.* New York: Marcel Dekker, pp 227–255.

Polak, E. H, A. M Fombon, C. Tilquin, and P. H Punter (1989). Sensory evidence for olfactory receptors with opposite chiral selectivity. *Behav Brain Res* **31**:199–206.

Rawson, N. E., J. G. Brand, B. J. Cowart, L. D. Lowry, E. A. Pribitkin, V. M. Rao, and D. Restrepo (1995). Functionally mature olfactory neurons from two anosmic patients with Kallmann syndrome. *Brain Res* **681**:58–64.

Rawson, N. E., G. Gomez, B. Cowart, J. G. Brand, L. D. Lowry, E. A. Pribitkin, and D. Restrepo (1997). Selectivity and response characteristics of human olfactory neurons. *J Neurophysiol* **22**:1606–1613.

Rawson, N. E., G. Gomez, B. Cowart, and D. Restrepo (1998). The use of olfactory receptor neurons (ORNs) from biopsies to study changes in aging and neurodegenerative diseases. *Ann N Y Acad Sci* **855**:701–707.

Restrepo, D., Y. Okada, and J. H. Teeter (1993a). Odorant-regulated Ca^{2+} gradients in rat olfactory neurons. *J Gen Physiol* **102**:907–924.

Restrepo, D., Y. Okada, J. H. Teeter, L. D. Lowry, and B. Cowart (1993b). Human olfactory neurons respond to odor stimuli with an increase in cytoplasmic Ca^{++}. *Biophys J* **64**:1961–1966.

Rossiter, K. J. (1996). Structure-odor relationships. *Chem Rev* **96**:3201–3240.

Rouquier, S., S. Taviaux, B. J. Trask, V. Brand-Arpon, G. van den Engh, J. Demaille, and D. Giorgi (1998). Distribution of olfactory receptor genes in the human genome. *Nat Genet* **18**:243–250.

Saoud, M., T. Hueber, H. Mandran, J. Dalery, and T. d'Amato (1998). Olfactory identification deficiency and WCST performance in men with schizophrenia. *Psychiatry Res* **81**:251–257.

Schaal, B. (1988). Olfaction in infants and children: developmental and functional perspectives. *Chem Senses* **13**:145–190.

Schemper, T., S. Voss, and W. S. Cain (1981). Odor identification in young and elderly persons: sensory and cognitive limitations. *J Gerontol* **36**:446–452.

Schiffman, S. S. (1977). Food recognition by the elderly. *J Gerontol* **32**:586–592.

Schiffman, S., M. Orlandi, and R. P. Erickson (1979). Changes in taste and smell with age: biological aspects. In: J. M. Ordy and K. Brizzee (eds). *Sensory Systems and Communication in the Elderly.* New York: Raven Press, pp 247–268.

Schild, D., F. W. Lischka, and D. Restrepo (1995). InsP$_3$ causes an increase in apical $[Ca^{2+}]_i$ by activating two distinct current components in vertebrate olfactory receptor cells. *J Neurophysiol* **73**:862–866.

Serby, M., P. Larson, and D. Kalkstein (1990). Olfactory sense in psychoses. *Biol Psychol* **28**:829–830.

Shughrue, P. J., M. V. Lane, and I. Merchenthaler (1997). Comparative distribution of estrogen receptor-alpha and -beta mRNA in the rat central nervous system. *J Comp Neurol* **388**:507–525.

Silver, W. L. and J. C. Walker (1997). Odors. In: J. J. Lagowski (ed). *Macmillan Encyclopedia of Chemistry.* New York: Macmillan Publishing Co., pp 1–8.

Sobel, N., V. Prabhakaran, J. E. Desmond, G. H. Glover, R. L. Goode, E. V. Sullivan, and D. E. Gabrirell (1998). Sniffing and Smelling: separate subsystems in the human olfactory cortex. *Nature* **392**:282–286.

Song, M. R., S. K. Lee, Y. W. Seo, H. S. Choi, J. W. Lee, and M. O. Lee (1998). Differential modulation of transcriptional activity of oestrogen receptors by direct protein-protein interactions with retinoid receptors. *Biochem J* **336**:711–717.

Stevens, J. C. and W. S. Cain (1985). Age-related deficiency in the perceived strength of six odorants. *Chem Senses* **10**:517–529.

Stevens, J. C., W. S. Cain, and R. J. Burke (1988). Variability of olfactory thresholds. *Chem Senses* **13**:643–653.

Stevens, J. C., W. S. Cain, F. T. Schiet, and M. W. Oatley (1989). Olfactory adaptation and recovery in old age. *Perception* **18**:265–276.

Talamo, B. R., R. A. Rudel, K. S. Kosik, V. M.-Y. Lee, S. Neff, L. Adelman, and J. S. Kauer (1989). Pathological changes in olfactory neurons in patients with Alzheimer's disease. *Nature* **337**:736–739.

Tareilus, E., J. Noe, and H. Breer (1995). Calcium signaling in olfactory neurons. *Biochem Biophys Acta* **1269**:129–138.

Teff, K. L. (1996). Physiological effects of flavour perception. *Trends Food Sci Technol* **7**:448–452.

Teff, K. L. and K. Engelman (1996). Oral sensory stimulation improves glucose tolerance: effects on post-prandial glucose, insulin, C-peptide and glucagon. *Am J Physiol* **270**:R1371–R1379..

Thurauf, N., M. Gjuric, G. Kobal, and H. Hatt (1996). Cyclic nucleotide-gated channels in identified human olfactory receptor neurons. *Eur J Neurosci* **8**:2080–2089.

Trask, B. J., C. Friedman, A. Martin-Gallardo, L. Rowen, C. Akinbami, J. Blankenship, C. Collins, D. Giorgi, S. Iadonato, F. Johnson, W. L. Kuo, H. Massa, T. Morrish, S. Naylor, O. T. H. Nguyen, S. Rouquier, T. Smith, D. J. Wong, J. Youngblom, and G. van den Engh (1998). Members of the olfactory receptor gene family are contained in large blocks of DNA duplicated polymorphically near the ends of human chromosomes. *Hum Mol Genet* **7**:13–26.

Turin, A. (1996). Spectroscopic mechanism for primary olfactory reception. *Chem Senses* **21**:773–791.

von Brunn, A. (1892). Beitrage zur mikroskopischen anatomie der menschlichen nasenhohle. *Arch Mikr Anat* **39**:632–651.

Wang, H.-W., C. Wysocki, and G. H. Gold (1993). Induction of olfactory receptor sensitivity in mice. *Science* **260**:998–999.

Whissell-Buechy, D. and J. E. Amoore (1973). Odor-blindness to musk: simple recessive inheritance. *Nature* **242**:271–273.

Wysocki, C. J. and A. N. Gilbert (1989). National Geographic smell survey: effects of age are heterogeneous. In: C. Murphy, W. S. Cain, and D. M. Hegsted (eds). *Nutrition and the Chemical Senses in Aging.* New York: New York Academy of Science, pp 12–28.

Wysocki, C. J., K. M. Dorries, and G. K. Beauchamp (1989). Ability to perceive androstenone can be acquired by ostensibly anosmic people. *Proc Nat Acad Sci U S A* **86**:7976–7978.

Wysocki, C. J., B. G. Green, and T. P. Malia (1992). Monorhinal stimulation as a method for differentiating between thresholds for irritation and odor (abstract). *Chem Senses* **17**: 722–723.

Wysocki, C. J., P. Dalton, M. J. Brody, and J. H. Lawley (1997). Acetone odor and irritation thresholds obtained from acetone-exposed factory workers and from control (occupationally unexposed) subjects. *Am Ind Hyg Assoc J* **58**:704–712.

Yamagishi, M., Y. Ishizuka, and K. Seki (1994). Pathology of olfactory mucosa in patients with Alzheimer's disease. *Ann Otolaryngol Rhinol Laryngol* **103**:421–427.

Zufall, F., G. M. S. Shepherd SH, and S. Firestein (1991). Inhibition of the olfactory cyclic nucleotide-activated ion channel by intracellular calcium. *Proc R Soc Lond B Biol Sci* **246**:225.

GUSTATION

12

Cell Biology of Taste Epithelium

THOMAS E. FINGER
Rocky Mountain Taste and Smell Center, Department of Cellular and Structural Biology, University of Colorado Health Sciences Center, Denver, CO

SIDNEY A. SIMON
Departments of Neurobiology and Anesthesiology, Duke University Medical Center, Durham, NC

1. INTRODUCTION

The sense of taste is a chemosensory system intimately associated with selection and rejection of potential foodstuffs (Dethier, 1993; See also Chapter 13; Glendinning et al.). The receptor cells for taste are situated in the epithelium of body parts involved in manipulation and ingestion of food, for example, lips, oral cavity, tongue, pharynx, and, in invertebrates, various cephalic appendages. For vertebrates, taste is clearly defined, that is, the sense mediated by specialized epithelial cells that are arrayed in discrete sensory endorgans (*taste buds*) and that synapse onto the distal ends of primary sensory neurons of the facial, glosso-pharyngeal, or vagus nerves. In contrast, in invertebrates, the sense of taste is mediated via bipolar sensory neurons that have a distal process reaching the surface of the epithelium and a central process extending directly into the central nervous system. Thus, although the sense of taste can be described in both vertebrates and invertebrates, the systems are not homologous and are structurally dissimilar. Nonetheless, many aspects of stimulus transduction and

The Neurobiology of Taste and Smell, Second Edition, Edited by Thomas E. Finger, Wayne L. Silver, and Diego Restrepo.
ISBN 0-471-25721-4   Copyright © 2000 Wiley-Liss, Inc.

processing appear similar (see Chapter 13 Glendinning et al.; Chapter 14, Smith and Davis).

Complicating this picture is that many aquatic animals—both vertebrate and invertebrate—have innumerable chemoreceptors scattered across their body or concentrated on locomotory appendages. For the purposes of this chapter, we will not consider all of these diverse chemoreceptors to be taste even though in some species they may be utilized to locate food in the environment. For example, most aquatic vertebrates possess a chemosensory system mediated by solitary chemoreceptor cells scattered across the body surface, that are capable of detecting food, and are sometime utilized like a taste system (Silver and Finger, 1984; Whitear, 1992). The solitary chemoreceptor cells are dissimilar from a taste system, however, in terms of peripheral organization, innervation, and, sometimes, in utilization.

This chapter will provide an overview of the structure of invertebrate and vertebrate taste epithelia, with special emphasis on the functional organization of vertebrate taste buds and their surrounding epithelium.

2. SENSORY ENDORGANS FOR TASTE

Virtually all metazoan animals possess chemoreceptor cells associated with the mouth and cephalic end of the gut tube. The general organization of these various chemoreceptors in the different invertebrate phyla is similar, but vary greatly in detail.

2.1. Invertebrates

All invertebrate taste receptor cells, unlike those in vertebrates, are bipolar primary sensory neurons. The dendrites of such cells extend outward, usually ending just below a pore or channel in the cuticle or exoskeleton through which chemicals gain access to the neuronal membrane. The central process of these receptor cells extends into the central nervous system. We describe below the general organization of taste organs in a variety of commonly studied invertebrates.

2.1.1. Nematoda (*Caenorhabditis elegans*)

The nematode *C. elegans* is an important model organism that has been put to good use in the study of the molecular and cellular basis of olfaction (see Chapter 3, Sengupta and Carlson). Far less is known about taste functions in this species. Its mouth is surrounded by three pairs of lips each having an inner labial sensillum at its apex. The dendrites of two sensory neurons occur in each sensillum, one (IL1) being mechanosensory, the other (IL2) chemosensory (Tabish et al., 1995). Little is known about the physiology of these cells, but in non feeding "dauer" larvae the openings in the cuticle at the end of the sensilla are not patent, thus potential food stimuli cannot reach the receptor neuron membrane. Some neurons of the pharynx also may be chemosensory and may be involved in local control of pharyngeal motility. These cells too might be considered to be part of the taste system just as in vertebrates taste buds occur in the pharynx and upper esophagus.

Some investigators (e.g., Bargmann et al., 1993) have considered some chemosensory (ASE) neurons of the amphid sensilla (see Chapter 6, Farbman) also to be part of the taste system because these receptor cells respond to potential foodstuffs, for example, biotin and cAMP. These ASE neurons do not, however, fit our definition of taste neurons because they are not closely associated with mouth parts or pharynx. Likewise, although olfactory receptor neurons in fishes respond to amino acids, they are not part of the taste system (Caprio et al., 1993).

2.1.2. Arthropoda (Insects and Crustaceans)

Aquatic and terrestrial arthropods are replete with chemosensory sensilla on appendages that contact food. Chemosensory receptor cells occur in various sensilla of mouthparts as well as hairs on limbs, for example, tarsi (See Fig. 12.1). Although the maxillary and mandibular chemoreceptors clearly serve taste functions, as we use the term, the designation of tarsal receptor cells as "taste" receptors is less clear-cut. A reasonable argument can be made, however, that the chemoreceptors on the limbs of arthropods are taste receptors since the legs and mouthparts are serially homologous. Therefore, receptor organs on the limbs that are structurally similar to the "taste" receptor organs on the mouthparts should also be considered taste receptors.

Taste receptor organs in arthropods generally are situated in raised sensilla with a specialized apical cuticle bearing a single pore at its apex. These so-called uniporous sensilla (those with a single major opening at their tip) are innervated by neurons with apical dendrites encased in a single dendritic sheath formed by supporting cells of the endorgan (Shields, 1994; Singh, 1997). The apical pore of the sensilla, which opens into the dendritic sheath, is filled with mucoid substances or fibrillar plugs that may affect permeability to potential chemical stimuli (Zacharuk and Shields, 1991). Presumably, chemical stimuli penetrate the apical pore to dissolve in the fluid inside the dendritic sheath. Carrier proteins may play a role in presentation of potential stimuli to the receptors located on the dendritic membrane. The dendritic sheath with its enclosed dendritic distal segments lies within the fluid-filled sensillar sinus formed by the outer sheath, or tormogen, cell. (See Chapter 13, Glendinning et al., for discussion of ion fluxes in transduction.)

2.2. Vertebrates

Taste buds, the sensory endorgans mediating taste in vertebrates, are small collections of specialized, elongate epithelial cells. Taste buds first appear phylogenetically coincident with the vertebrate lineage. Primitive taste buds, also

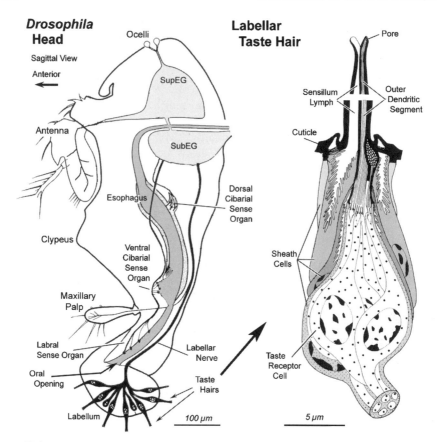

FIG. 12.1. Diagrams of taste receptors on the fly *Drosophila* modified with permission after the original work of R. N. Singh (1997), © J. Wiley and Sons. **Left:** Sagittal section through the head of a fly showing the principal organs bearing taste sensilla: maxillary palp, labellum, labrum, and cibarial organs. The nerves innervating these sturctures are also shown in their course toward the subesophageal ganglion (SubEG). The antenna carries the olfactory organ (See Chapter 6, Farbman) whose neurons project into the antennal lobe of the supraeso-phogeal ganglion (SupEG). The alimentary canal is shaded dark gray. **Right:** Higher magnification view of a taste sensillum from the labellum. Several taste receptor cells situated at the base of the sensillum extend apical dendrites to reach the single pore at the apex. The taste receptor cells are surrounded by several types of sheath cells similar to those surrounding the olfactory sensilla (see also Figure 6.1 of Chapter 6, Farbman).

called terminal buds (Baatrup, 1983), are present in lampreys, but are absent in present-day chordates such as hagfish. Throughout the vertebrate lineage, taste buds share several morphological features. For example, taste buds consist of several types of modified epithelial cells, at least some of which synapse onto the distal processes of ganglion cells of either the facial (CN VII), glossopharyngeal (CN IX), or vagus (CN X) nerves (see also Chapter 14, Smith and Davis). Taste buds occur within the oropharynx, but, in many species, also may be situated on the epiglottis, lips, and, in aquatic species, across the entire body surface. Despite this widespread distribution, taste buds always are innervated by branches of the facial, glossopharyngeal, or vagus nerves. An example of this principle is that the taste buds situated on the tail of a catfish are innervated by a recurrent branch of the facial nerve (Herrick, 1905).

Aquatic anamniote vertebrates as well as hagfish possess dispersed epidermal chemosensory cells that may be related phylogenetically to taste buds (Finger, 1997). The solitary chemosensory cells (SCCs) are similar to the individual cells of taste buds in terms of being elongate epithelial cells that synapse onto the distal processes of remote ganglion cells. However, SCCs differ from taste buds because they are seldom grouped into compact endorgans. Rather, SCCs are scattered as isolated cells across much of the epithelium and may be present even in the olfactory and oral cavities. In hagfish (chordates, but not vertebrates), SCCs can form aggregates (Braun, 1998) reminiscent of vertebrate taste buds, but this is the exception rather than the rule. SCCs, as isolated cells, can be situated anywhere along the body surface and may be innervated by any cutaneous nerve including spinal and trigeminal nerves (Whitear, 1992). Thus because of their nonspecific innervation and utilization in nonfeeding contexts (Peters et al., 1991), SCCs are not considered part of the taste system.

3. TASTE BUDS

In all vertebrates, taste buds contain a variety of cell types. These include elongate cells stretching from basal lamina to the surface of the epithelium, proliferative basal cells lying adjacent to the basal lamina, and "edge" or "peri-gemmal" cells forming a transition to the surrounding nongustatory epithelium (Knapp et al., 1995). The majority of cells in a taste bud are elongate, columnar epithelial-type cells with a tapering apical process that may reach the epithelial surface. In some species, for example, frogs, taste endorgans contain differentiated supporting cells, that is, cells that do not synapse with nerve fibers or with other sensory cells in the bud (Osculati and Sbarbati, 1995). It is a matter of some controversy as to whether supporting cells are present in taste buds of other species (Reutter and Witt, 1993). In mice, however, all morphological types of taste cells are reported to synapse onto afferent nerve fibers (Kinnamon et al., 1985), and thus are assumed to be involved in the transduction or transmission of gustatory information. The apex of a taste bud may be capped with specialized mucus, or may lie in an epithelial invagination called a taste pore. Interestingly, in the posterior but not anterior tongue, the taste pore is filled with an electron dense substance (called the pore substance) of unknown function that is apparently secreted by cells of the taste bud (Reutter and Witt, 1993).

3.1. Taste Bud Structure in Various Taxa

Taste cells exhibit characteristics similar to both neurons and epithelial cells. Like epithelial cells, taste receptor cells occur in epithelia, synthesize cytokeratins, and are continually replaced. Taste receptor cells are neuron-like, in that they have voltage-gated ion channels (Roper, 1983) and synapse onto neurons.

In most species, taste buds have the overall shape of an onion, or a tulip bulb. Depending on the species, a taste bud contains between 20 and 100 cells and has a width ranging from 30 to 100 µm. The details of taste bud morphology vary according to species and an animal's habitat. It is not known how the different taste bud structures relate to the physiology of gustation. The following section will provide only an overview of the diversity of taste bud variation. For a complete review, see Reuter and Witt (1993) and Osculati and Sbarbati (1995).

3.1.1. Lampreys

The taste buds of lampreys, also called "terminal buds" (Baatrup, 1983), lie in the oropharynx being especially prominent on the gill arches. In addition, taste buds line the branchial tube, which serves a purely respiratory function. Despite this location, these taste buds respond to feeding-related substances, for example, amino acids (Baatrup, 1985). The sensory cells of taste buds in lamprey are peculiar in that they are ciliated rather than microvillous.

3.1.2. Bony and Cartilagenous Fishes

Taste buds in these fishes contain three distinct postmitotic cell types: elongate cells bearing a thick microvillus, elongate cells bearing multiple small microvilli, and serotonergic basal cells akin to Merkel cells (Reuter and Witt, 1993). Each of these cell types exhibits synaptic contacts either with another taste cell or with sensory nerve fibers. The taste receptor cells with a single thick apical microvillus are similar to SCCs both in terms of morphology and in expression of arginine-binding proteins (Finger et al., 1996). It is unclear whether a purely supporting cell population exists in taste buds of these fishes.

3.1.3. Amphibia

Although the structure of taste buds of the urodele amphibia (e.g., salamanders) strongly resembles that of fishes, taste buds in anurans are quite derived. Taste buds in frogs take the form of large (500 µm diameter), cylindrical assemblages called "taste disks." Although taste disks contain both elongate sensory cells and "Merkel-like" basal cells, they also contain a distinct type of supporting cell that does not synapse onto sensory nerve fibers (see Osculati and Sbarbati, 1995).

3.1.4. Birds

The few studies detailing the structure of taste buds in birds are predominantly limited to seed-eating species such as chickens and quail. Some avian taste buds are unusual because they are situated deep in the epithelium and have a very long taste pore, called a taste canal (Ganchrow and Ganchrow, 1987). It is

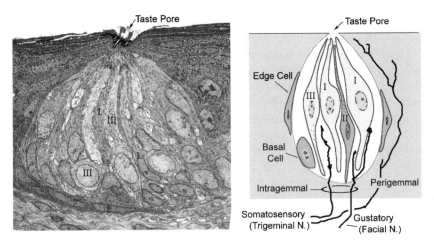

FIG. 12.2. Photomicrograph (A) and semischematic diagram (B) of a longitudinal section through a lingual taste bud of a mammal. **Left:** Electron micrograph of a section through a taste bud of the foliate papillae of a rabbit. The cytoplasmic and nuclear densities are different in the different cell types. The large round nuclei and electron lucent cytoplasm is typical of Type III cells, which in rabbit form obvious basally situated synapses. Numerous electron-dense Type I cells also are evident. Photo courtesy of S. Royer and J. C. Kinnamon (Denver University). **Right:** The principal features of the various cell types of a taste bud of a fungiform papilla in a rat are shown along with the different sorts of innervation. In mice, all three elongate cell types (Type I, Type II, and Type III) are reported to form synapses with the gustatory innervation. Cell types in rats are less clearly distinct than in rabbits (see Table 12.1).

unclear as to whether the unusual structure of taste buds in these species can be generalized to all birds.

3.1.5. Mammals

Although mammalian taste buds share many common features, distinct morphological differences do exist between species, and even within a given species, between taste buds in different areas of the epithelium. Nonetheless, despite this morphological complexity, there is a common plan of taste bud organization. Taste buds consist of proliferative *basal cells*, centrally situated *elongate cells*, and flattened *edge cells* that form the lateral boundary of the taste bud and the transition to extragemmal epithelium (see Fig. 12.2).

Mammalian taste buds contain several morphologically and biochemically distinguishable types of cells. Anatomists of the late nineteenth century described two elongate cell types according to their affinity for basophilic dyes: dark cells (heavier staining) and light cells (lighter staining). With the advent of electron microscopy, these differences in staining were correlated with ultrastructural differences. However, careful analysis of foliate taste buds in rabbits led to the description of a third elongate cell type with prominent synapses, the "Type III" cell. In current usage then, post-mitotic elongate taste cells are divided into three types: Type I (dark cells), Type II (light cells), and Type III (often grouped with light cells on light microscopic criteria).

These morphological classes of cell appear to be clearest in rabbits (Murray, 1986). In brief, Type II cells are described as simple, vesicle-filled elongate cells with a large, round nucleus. Type I cells are more convoluted, and are often observed to embrace or wrap around the light cells. Type I cells also contain characteristic apical, electron-dense, secretory granules (Reutter and Witt, 1993). Some controversy exists as to whether these two cell types form distinct morphological classes (Pumplin et al., 1997), or whether they grade gradually one into the other (Delay et al., 1986). Type III cells superficially resemble Type II cells but are characterized by prominent perinuclear dense-cored vesicles and synapses onto nerve fibers (Table 12.1). Unfortunately, it is not straightforward to apply these ultrastructural criteria for Type III cells to species other than rabbits.

The terminology applied to mammalian taste buds is confusing in part because of interspecies differences. Following the description by Murray for rabbits (Murray, 1969; Murray et al., 1969), many authors writing about rodents use the term "Type I cell" for dark cells and "Type II cell" for light cells. Similarly, the term "Type III cell" has been assigned to the serotonin-accumulating taste cells with dense-cored vesicles situated near the synaptic regions at the base of the cell. Use of this numbered terminology may be more useful than the "light" and "dark" descriptors since the apparent electron density of the cytoplasm may change according to fixative employed (Murray, 1969; Cottler-Fox et al., 1987). However, application of this numbered terminology across taxa may be confusing in that taste cells of one species may not have all of the defining morphological traits of a particular cell type as defined for rabbits. For example, the serotonin-

TABLE 12.1. **Morphological charactristics of the various elongate cells of rodent taste buds**

Cell Type	Characteristics
Type I (Dark)	Dark cytoplasm (electron dense and basophilic)
	Irregular, elongate nucleus; electron dense; often invaginated
	Flattened processes, often wrapping other cells
	Large ($\cong 0.1\ \mu$m), apical dark granules
	Long, slender, filamentous apical microvilli.
Type II (Light)*	Light cytoplasm
	Round or oval nucleus; less electron dense
	Elongate, spindle-shape
	Short, thick apical microvilli
Type III (Intermediate?)*	Light cytoplasm
	Somewhat elongate nucleus, some indentation
	Elongate, spindle-shape
	Perinuclear small dense-core vesicles
	Thick apical process
	Prominent synaptic contacts onto nerve fibers

* Both Type II and Type III cells are included in the category of "Light" cells in Pumplin et al., 1997.

accumulating cells of rodents do not share all of the defining ultrastructural features with Type III, serotonin-accumulating, cells of rabbits. Further, different authors rely on different key characteristics in defining taste cell types, especially in regard to "Type III" cells. As mentioned above, some authors consider "Type III" cells to be those that accumulate biogenic amines (Kim and Roper, 1995). Other authors define "Type III" cells to be those maintaining synaptic contact with nerve fibers (e.g., Suzuki and Takeda, 1983; Kanazawa and Yoshie, 1996; Yoshiei et al., 1997) following the description that Type III cells in rabbits synapse with nerve fibers (Murray, 1986)—although Murray cautions against the assumption that Type III cells would be unique in this regard. Unfortunately, the net result of this confusion in terminology is that the literature is replete with dubious attribution of cell type. Accordingly, one must take caution in interpretation of findings based simply on cell-type descriptions.

The situation in rats and perhaps other rodents is that the elongate taste cells of the bud are divisible on morphological criteria into two distinct types, called "dark" and "light" (Pumplin et al., 1997). The dark cells have flattened process that separate or envelop other elongate cells in the bud, whereas light cells are more regular, spindle-shaped cells. However, the "light" cells defined by Pumplin probably include the class of "intermediate" cells of Delay et al. (1986). Various immunocytochemical studies suggest that the "intermediate" cells of Delay et al. may correspond to Type III cells, but this remains to be established rigorously. Thus the Pumplin et al., (1997) division of taste cells into "light" and "dark" categories seems to combine the Type II and Type III cells of other authors.

Taste cells differ from one another histochemically as well as morphologically. Numerous neurochemicals have been localized to subsets of taste cells. These substances include: calbindin (Miyawaki et al., 1996), NCAM (Takeda et al., 1992; Nelson and Finger, 1993; Smith et al., 1993), PGP 9.5 (ubiquitin C-terminal hydrolase) (Kanazawa and Yoshie, 1996), neuron specific enolase (Yoshie et al., 1989), carbonic anhydrase (Brown et al., 1984; Daikoku et al., 1999), enkephalin (Yoshie et al., 1993), substance P receptors (Chang et al., 1996) and CCK (Herness, 1990; Welton et al., 1992). The staining patterns for each of these substances are complex, with some being co-localized in single taste cells and others being mutually exclusive (Böttger et al., 1997). As of this writing, no simple system of classification has been agreed upon for most of these substances. In rat circumvallate papillae dark (Type I) cells react uniquely with an antibody to a blood group carbohydrate (blood group H-antigen), but whether this is generally applicable to other species is unknown (Pumplin et al., 1997). Nonetheless, this does indicate that the different morphological types of taste cells do display different histochemical features, both in terms of cell surface glycoproteins as well as cytoplasmic contents.

3.2. Cell Types and Specificity

One very important question is whether the morphologically different types of cells respond differentially to chemical stimuli. Functionally, taste cells usually respond to some extent to multiple classes of taste stimuli but respond preferentially to one stimulus class. For example, in rats, taste receptor cells that respond to sucrose tend not respond to the bitter tastant denatonium (Bernhardt et al., 1996). Whether these different functional types correspond to structurally distinct types of taste receptor cells remains unknown. Recent immunocyto-chemical evidence does, however, suggest a correlation between structure and function in taste cells. For example, gustducin, a G-protein implicated in both sweet and bitter transduction (see Chapter 13, Glendinning et al.), occurs only in some Type II cells (Tabata et al., 1995; Boughter et al., 1997) indicating that Type II cells are not homogeneous (or that the anatomical classification is not clearly related to all functional properties). Likewise, the amiloride-sensitive sodium channel (ASNaC), implicated in detection of sodium salts by rats, is present only in Type II cells (Lin et al., 1999). Whereas all gustducin-containing cells also contain ASNaC subunits, not all ASNaC-containing cells exhibit gustducin-immunoreactivity. Similarly, immunocytochemical and lectin-binding studies in catfish (Finger et al., 1996) show that different taste receptor cells have binding sites for different amino acid tastants. Recent molecular evidence (Hoon et al., 1999) also demonstrates some specificity of taste receptor cells. The degree of overlap in expression of two putative taste receptor genes — $\cong 10\%$ of the cells — corresponds well with the reported overlap in arginine and alanine binding sites in the taste buds of catfish (Finger et al., 1996). Taken together, these studies indicate that taste receptor cells show incomplete specificity for tastants (see also Chapter 13, Glendinning et al.). Conversely, single taste cells may express nume-rous members of the family of genes encoding bitter receptors (Adler et al., 2000).

Another line of evidence supporting correlations between function and structure of taste cells is that a given nerve fiber innervates only one type (then called "dark" or "light") of taste cell (Kinnamon et al., 1988). Since gustatory nerve fibers exhibit some specificity of gustatory response (see Chapter 14, Smith and Davis), then the specificity of connectivity lends credence to the idea that receptor cell structure correlates with functional attributes. Thus, both physio-logical and anatomical evidence indicates that taste cells may be preferentially responsive, but not necessarily highly selective, for particular classes of tastants.

In taste buds, small clusters of taste receptor cells (three to four) are functionally coupled via gap junctions (Roper, 1992). These clusters are proposed to represent small functional units within the taste bud (Roper, 1992). It is not presently known whether the clusters of coupled taste cells represent the same morphological cell type and/or the same functional cell type. Coupling of cells with like sensitivity would tend to increase the signal-to-noise ratio in the system, whereas coupling of functionally dissimilar cells would tend to increase the dynamic range of the response of nerve fibers onto which the taste receptor cells synapse (see also Chapter 13, Glendinning et al.).

3.3. Cell Turnover

As noted, taste receptor cells are epithelial in the sense that they have a limited life span and therefore must be replaced to maintain the structure of the epithelium. In rodents, the cells in a taste bud are totally replaced every 10 to 14 days (Beidler and Smallman, 1965; Farbman, 1980; Delay et al., 1986). In poikilothermic vertebrates, for example, fishes, the rate of turnover is dependent on temperature but appears similar to the rate in rodents if the fish were held at an equivalent body temperature (Randerman-Little, 1979). Since a taste bud in a rodent contains approximately 50 to 75 cells, the turnover rate implies that about three to five cells are replaced each day. The three to five cells that are replaced daily corresponds to the number of apoptotic cells observed under normal conditions in lingual taste buds (Takeda et al., 1996). Once a taste cell "dies," it is apparently phagocytosed by neighboring cells in the taste bud (Suzuki et al., 1996).

3.3.1. Role of Innervation—Apoptosis and Differentiation Status

After denervation, the number of apoptotic cells increases dramatically within 24 hours (Suzuki et al., 1996). Whether the nerve fibers provide trophic support for the taste cells, and/or whether the degenerating nerve fiber release "apoptosis-inducing factors" into the taste bud has not been established. In either event, in most vertebrates, and especially in mammals, an intact nerve supply is required for a taste bud to maintain a fully differentiated state (Whitehead et al., 1987; Simon et al., 1993a; Smith et al., 1994). The literature on taste bud degeneration can be confusing. Many older works report that taste buds completely disappear following denervation. These reports were, however, based on the disappearance of taste pores. Careful morphological studies on hamsters indicate that at least in fungiform papillae of that species, despite the absence of a taste pore (which has been taken as evidence for the absence of taste buds), partially differentiated remnant taste buds remain in the epithelium (Whitehead et al., 1987; Barry and Savoy, 1993). If reinnervation occurs, then these taste bud remnant populations appear capable of regenerating differentiated taste buds complete with a taste pore. Even in vallate and foliate papillae, where remnant taste buds are not evident, regeneration of the taste nerves results in reestablishment of differentiated taste buds.

Although the specific gustatory fibers in cranial nerves VII (facial), IX (glossopharyngeal), and X (vagus) appear to provide the bulk of the trophic support for mammalian taste buds, the general cutaneous fibers (carried in the trigeminal nerve to the fungiform taste buds) are capable of supporting some partially differentiated taste cells (Kinnman and Aldskogius, 1991). The transplantation studies carried out by Zalewski in the 1970s also demonstrated the capability of nongustatory nerves to support (partially) differentiated taste buds (see Chapter 15, Barlow). In BDNF (brain-derived neurotrophic factor) and trkB (tyrosine kinase B: the a receptor for BDNF) knockout mice, in which the specific gustatory innervation is essentially lost early in development, remnant taste buds are present (e.g., Fritzsch et al., 1997; Nosrat et al., 1997; see also Chapter 15, Barlow). It is likely that these taste cells receive trophic support from the general somatosensory nerves, which are largely unaffected in the transgenic animals.

3.3.2. Role of Growth and Trophic Factors

The dependence of the taste bud on an intact nerve supply is suggestive of a trophic support of the taste bud by nerve fibers (Olmstead, 1920; Guth, 1971). Even interruption of axonal transport in the taste nerves is sufficient to produce degeneration of the taste buds (Sloan et al., 1983). However, no such neural factor has been identified. In some vertebrates, for example, axolotls, taste buds are not dependent on innervation to maintain a differentiated phenotype (e.g., see Chapter 15, Barlow). In contrast, in rodents, full differentiation of taste buds is dependent on innervation.

4. CELL BIOLOGY OF TASTE EPITHELIUM

4.1. Overall Morphology of Epithelium

In most studies of gustation, the epithelia surrounding the taste buds, if noted at all, is merely considered to be the supporting or protective structure for the taste buds. The various epithelial cells, neurons, and blood vessels present in gustatory mucosa, can however, influence taste receptor cells and the subsequent responses of gustatory neurons. Lingual epithelium, like other epithelia, serves many functions. It protects against physical and chemical trauma, the penetration of chemicals, water loss, entry of microorganisms, and infection. It also provides somatosensory information (mechanical, thermal, itch, pain). In addition, epithelium is self-regenerating and self-healing. Structurally, lingual epithelium must be pliable and able to deform without cracking. In mammals, taste buds lie in stratified, squamous epithelia on the tongue, soft palate, pharynx, and epiglottis where they occupy much less than 1% of the epithelial surface (Holland et al., 1989; Miller, 1995). For this reason alone, it is important to understand how the lingual epithelium interacts with the taste system. To make this broad topic somewhat tractable we will emphasize the characteristics of the epithelium surrounding taste buds rather than those of the heavily cornified filiform and conical papillae that are on the anterior and posterior tongue, respectively (Bradley, 1971; Baratz and Farbman, 1975; Miller, 1995).

Figures 12.3 and 12.4 show schematic diagrams of a taste bud in a gustatory epithelium. The entire epithelial structure consists of an epidermis of variable thickness (depending on the number of layers in the different strata; see Chen and Squire, 1984), a basement membrane that separates the papillary layer from the epidermis, and a dermal layer. The dermis contains connective tissue, blood vessels, neurons, fibroblasts and possibly glands (e.g., von Ebners gland, Fig. 12.3). In the vicinity of taste buds, the epidermis thickness ranges from about 50 to 90 μm depending on the location in the oral cavity, and on the particular species (Farbman, 1965; Miller, 1995). The major features of the epithelium as they relate to cell biology and influence on taste transduction are discussed.

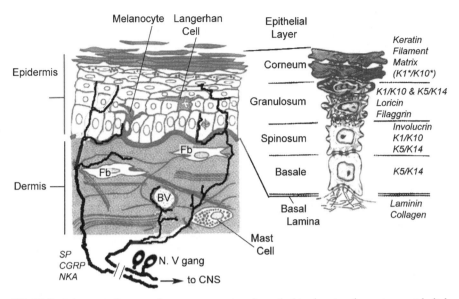

FIG. 12.3. Schematic diagram of transverse section through skin showing the various epithelial strata (labeled), the basement membrane, and the dermis and two neurons whose peripheral processes terminate in the epithelium. The dermal layers contain a blood vessel and a peptiergic nerve fiber (SP, CGRP, NKA) that can induce vasodilation or vasoconstriction. Also shown in the dermal layer are mast cells and fibroblasts. The basal layer in the epidermis contains a dividing stem cell and a melanocyte in addition to basal cells. The epithelium also contains a Langerhans cell without its associated nerve fiber (see text for additional details). The schematic diagram on the right hand side is a expanded view of the epidermis with several of the proteins that are expressed in the different epithelial strata. The diagram on the right hand-side shows the changes in keratiocyte morphology during differentiation. The K's refer to subtypes of keratin (see text for additional details). (Figure modified from Bowden, 1993, and Priestly, 1993).

4.2. Dermis

The dermis is a well-defined layer of connective tissue containing blood vessels that terminate as fine capillary networks below the epidermis (Fig. 12.3). These capillary networks supply nutrients to the epidermis and taste cells and also collect their metabolites. The blood vessels in the dermal layer can constrict or dilate, which not only changes the oxygen supply to the sensory organ, but also can change the tongue's temperature, which, in turn, can alter gustatory responses (Nakamura and Kurihara, 1988; Karita and Izumi, 1993; Wang et al., 1995). The dermis also has mast cells, macrophages, lymphocytes, plasma cells, and polymorphonuclear leukocytes (Fig. 12.3). Many of these cells become activated during inflammation or activation of local peptidergic nerve fibers (see Chapter 4, Bryant and Silver) resulting in release a variety of compounds (e.g., cytokines, histamine, neuropeptides) that can modulate the activity of epithelial

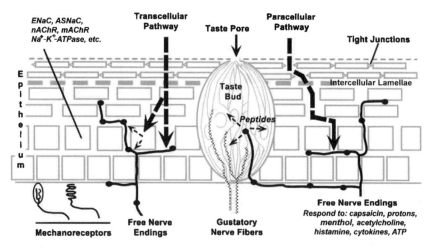

FIG. 12.4. Schematic diagram of a taste bud and the surrounding epithelium illustrating the routes by which potential chemical stimuli can enter lingual epithelium. A gap in the stratified epithelium above the taste bud (taste pore) permits chemical stimuli direct access to the taste cells. Gustatory nerve fibers that synapse with the taste cells are indicated. The keratinocytes in the uppermost layers of the epithelium are joined by cation-selective tight junctions that form the major permeability barrier for passage of small electrolytes (Na^+, Cl^-) into the epithelium. Also shown (filled rectangles) are lipid-dense intercellular lamallae. These extracellular lipids not only prevent water loss from the epithelium but form part of the paracellular pathway for nonpolar molecules that enter the epithelium. Such molecules then can enter the plasma membranes of the epithelial cells or interact with receptors on free nerve endings (direct pathway). The transcellular pathway (left side) is a major pathway for the transport of electrolyte across the epithelium. When chemical stimuli interact with the metabolically active epithelial cells, they may react by altering their transport properties (upper left corner) thereby changing the extracellular environment (e.g., increasing KCl), or by releasing bioactive compounds (e.g., cytokines) that can activate intraepithelial free nerve endings (indirect pathway). The epithelial cells contain a variety of channels, ATPases, and transportors, some of which are indicated. Activation of perigemmal fibers can potentially influence gustatory responses via axon-reflex release of peptides that might interact with taste cells and modulate their activity. The free nerve endings can be activated by a variety of stimuli, some of which are noted. Menthol is included to point out that some of these free nerve endings are cold fibers. Finally, low threshold mechanoreceptors usually are present in the papillary layer.

cells (Barnes, 1991; Lundberg, 1995), and presumably taste receptor cells. For example, mast cells release histamine and other bioactive substances in response to nociceptive stimuli, for example, capsaicin applied to the tongue surface. The chemicals released from mast cells can, in principle, affect the function of taste receptor cells either directly (when taste receptor cells have receptors for these compounds), or indirectly by their vasodilatory effects (Hellekant, 1977; Wang et al., 1995).

4.3. Epidermis

The dermis is separated from the epidermis by a basement membrane. This is a freely permeable membrane, about 0.25 μm thick, containing laminins, collagen (mainly Type IV), integrins, and heparin sulfate. Cells in the epidermis are called keratinocytes because they contain the protein keratin. Different types of keratin are expressed during the process of terminal differentiation (keratinization). Marked changes in the lipid composition also occur during differentiation (see below). Further, different types of epithelia, for example, stratified or columnar, express different cytokeratins. For example, cells of taste buds, which essentially are a columnar epithelium, have a different keratin than the surrounding stratified epithelium (Zhang et al., 1995).

As kerationcytes of the lingual epithelium mature, they change morhpology as they move upward in the epithelium passing from one layer to the next (Fig. 12.3). From the basement membrane up, the epithelial layers are called stratum basale, stratum spinosum, stratum granulosum, and stratum corneum (Fig. 12.3 Holland et al., 1989). After about two weeks (Hill, 1984) the keratinocytes will be sloughed at the surface.

The *stratum basale* consists of a single layer of cuboidal cells (Fig. 12.3). This layer contains mitoticially active stem cells (daughter cells), which, after division, leave this stratum to give rise to mature keratinocytes (see Fig. 12.3). In cases of inflammation and accompanying trauma, the other normally quiescent stem cells can be recruited to activation by cytokines. [1] The cells in the *stratum spinosum* contain several cytoplasmic processes that exhibit a spiny appearance (especially when dehydrated upon fixation). These processes contain desmosomes that bridge adjacent cells. As spinosal cells age and move toward the surface they start to flatten and become orientated parallel to the surface of the epithelium (Fig. 12.4; Holland et al., 1989). The *stratum granulosum* consists of two to four cell layers of long and flattened cells that adhere via mature desmosomes and, in the uppermost layer, tight junctions (Holland et al., 1989). Granulocytes lose their Na^+-K^+-ATPase (Simon et al., 1991) and therefore cannot volume regulate, which may account for their flattened shape and reduced water content. Cells in the stratum granulosum also contain lamellar bodies, which are sheets of lipid bilayers surrounded by a protein envelope. These organelles are secreted into the extracellular space and are bound to the corneocytes via their protein envelope. As granulocytes are displaced upward in the epithelium, they transform into corneocytes, becoming more flattened, with thicker membranes, and losing much of their water (Fig. 12.3). They may retain their nucleus, in which case they are called orthokeratinized, or lose their nucleus, in which case they are keratinized. Obviously, orthokeratinized cells, which are found, for example, near

---

[1]Cytokines are low molecular weight (≥ 5 kd) proteins that bind to receptors on epithelial cells thereby regulating biological responses via second messenger pathways that produce biological responses. Cytokines include interlukins, interferons, cytotoxins, and growth factors. Many (e.g., IL-1, TNFα) are synthesized in the epidermis and are released to act locally in an autocrine or paracrine fashion. Cytokines can affect transporters, protein synthesis, and cell proliferation.

rat taste buds in circumvallate papillae, will be more permeable to compounds than keratinized cells, which are essentially "chemically resistant." The corneocytes are connected to one another through desmosomes and tight junctions. They are subsequently exfoliated as small clusters of cells (Simon and Wang, 1993).

The epithelium also contains two other cell types: *melanocytes*, which produce melanin to protect the basal stem cells from ultraviolet radiation, and *Langerhans cells*, which participate in the skin's immune reaction (Fig. 12.3). Epidermal melanocytes lie on the basement membrane in contact with keratinocytes. Melanocyte production is regulated by a variety of stimuli including cytokines, fibroblast growth factor, UV radiation, and melanocyte stimulating hormone. Langerhans cells originate in the bone marrow, comprise about 3% of the epidermal cells, and are present in most epithelia strata. In skin they behave as nonphagocytic macrophages in that they present antigens to T-lymphocytes. Langerhans cells are modulated by cytokines (e.g., IL-10) and neutropeptides (e.g., substance P and CGRP). They are intimately associated with peptidergic neurons, and it is thought that the Langerhans cell–neuron interaction is both afferent and efferent (Tori et al., 1997). It is not known whether, or how, either of these two cell types can influence taste receptor cells, although the release of cytokines and/or neuopeptides by these two cell types are distinct possibilities. Interactions between the gustatory system and the immune system do occur, perhaps through the agency of cytokines (Phillips and Hill, 1996).

4.4. Epithelial Junctions or Bridges

A primary role of the oral epithelia is that of protecting the tissue from dehydration and from potentially damaging chemical stimuli (e.g., alcohol, hydrocarbons, acid, capsaicin), antigens, temperature extremes, and tissue-damaging mechanical stimuli. Recall that the taste cells and their nerve fibers are protected from most noxious agents because the exposed sensory surface area is small, thereby limiting influx of potential xenobiotics. In addition, the sensory elements are protected by tight junctions that limit the entry of all large molecules and many charged molecules into the epithelium (Holland et al., 1989; Elliott and Simon, 1990). Most of the area of the anterior of the tongue is covered by filiform papillae, with a thick stratum corneum that severely limits the permeability of molecules (Baratz and Farbman, 1975; Simon and Wang, 1993). Between these papillae the epithelium is much thinner and less cornified so that molecules potentially can diffuse more easily into the epithelium. In fungiform papillae, the taste buds are embedded in a stratified squamous epithelium with few cornified layers near the apex of the papilla. The upper surface of fungiform papillae is richly innervated by somatosensory-type neurons. Thus the fungiform papilla serves as both a tactile/thermal organ and a platform for chemical interactions with the underlying chemosensory nerve fibers.

4.5. Permeability of Gustatory Epithelium

When substances penetrate the epithelium, they gain access to free intraepithelial nerve endings and even to basal regions of the taste bud. For a substance to penetrate into the epithelium it must cross many epithelial strata, each having different permeability properties. Compounds can either diffuse *through* the keratinocytes (transcellular pathway), or *around* them (paracellular pathway) (Fig. 12.4). Free diffusion of substances around epithelial cells is precluded by the presence sheets of lipid bilayers in the uppermost epithelial strata and by tight junctions of the zonula type in the stratum corneum and upper layers of the stratum granulosum (Holland et al., 1989).

4.5.1. Paracellular Pathways

For most lingual epithelia, the transepithelial resistance (TER) is approximately equal to the resistance of the paracellular pathway, which, in turn, is mainly that of the resistance of ions diffusing across the tight junctions (Desimone et al., 1984). Despite their name, tight junctions are not totally impermeable but rather act as molecular sieves (like ion channels that select on the basis of charge and size) allowing for the diffusion of small ions into the extracellular space of the epithelium (Salas and Moreno, 1982). Thus, the trivalent cation, lanthanum, cannot permeate tight junctions (Wang et al., 1993), whereas monovalent cations such as Na^+, K^+ and NH_4^+ can diffuse (more rapidly than Cl^- and other anions) across tight junctions in lingual epithelium where they can alter the transepithelial potential and modulate the responses of taste cells to tastants (Ye et al., 1993). Because of this characteristic of tight junctions, lingual trigeminal nerve responses to salts, such as NaCl, can be blocked by the addition of 5–10 mM $LaCl_3$ (Wang et al., 1993, Bryant and Moore, 1995; see also Chapter 4, Bryant and Silver). In contrast, trigeminal responses to hydrophobic molecules, for example, menthol, which penetrate the epithelium by diffusing through the lipids of the cell membranes, are not inhibited by the addition of $LaCl_3$ (Wang et al. 1995).

4.5.2. Transcellular Pathways

The stratum corneum, which comprises thin sheets of corneocytes and a small extracellular space, deserves a special place in this discussion. Corneocytes are composed of a thick, insoluble (15 nm) protein coat surrounding highly aggregated network of folded sheets of keratins and other proteins (e.g., filaggrin). Compared with other types of cells, corneocytes contain little water. In regard to epithelial permeability, corneocytes are like "bricks in a wall" (i.e., impermeable) because of their chemical resistance and mechanical strength. The "mortar" between the bricks is waxy, stratified sheets of lipid bilayers containing ceramides, cholesterol, cholesterol sulfate, and free fatty acid. These extracellular lipids prevent water loss through the epithelium (about 98%), but also provide a pathway for hydrophobic molecules to enter the epithelium. Once molecules diffuse across the lipid layer (which is often the rate-limiting step) they may partition into the plasma membranes of the epithelial cells. In fact, the flux of molecules into epithelium is directly proportional to the permeability, which, in turn, is directly proportional to the membrane–water partition coefficient (Mistretta, 1971; Siegel et al., 1981; see also Chapter 4, Bryant and Silver). Therefore, nonpolar molecules like long chain alcohols, menthol, capsaicin, and nicotine readily enter the epithelium where they can alter epithelial physiology (Simon and Sostman, 1991). In all epithelia, the permeability of uncharged molecules is much greater than for charged molecules (Siegel et al., 1981). Therefore, whether or not weak acids (e.g., aspirin) or weak bases (e.g., nicotine) will be permeable, will depend on their pK_a or pK_b, and the pH in the oral cavity.

4.5.3. Gap Junctions

Gap junctions are involved in intercellular communication in that they provide low resistance pathways for ions and other molecules of molecular weight less than 1 kD to pass between cells by diffusion. Such molecules include ATP, IP_3, and Ca^{2+}. The epithelial cells in all strata, except the stratum corneum, are coupled via gap junctions (Holland et al., 1989). In this sense the living lingual epithelial cells behave functionally as a syncitum or as a single, extended cell (Margolis et al., 1995). This means that, in principle, the effects of cells in one layer can affect cells in other strata. Although we believe it to be highly unlikely, we do not presently know whether epithelial cells are coupled to taste receptor cells. If so, the physiological implications would be quite profound!

4.6. Responses to Chemical Stimuli by Epithelium

Lingual epithelial cells are distinguished from taste cells in terms of electrical behavior (resting potential, input resistance) and sensitivities to various chemical stimuli (Kinnamon and Roper, 1987). Nonetheless, lingual epithelial cells, like those throughout the body, can respond to a variety of substances including acetylcholine, protons, various peptides, and irritants (Fig. 12.4; Sostman and Simon, 1991; Maggi, 1993; Simon and Wang, 1993). Typically, the epithelial cells respond by releasing cytokines and neuropeptides that in turn can influence epithelial cells, intraepithelial free nerve endings (Maggi, 1993; Lundberg, 1995), and potentially even the taste receptor cells themselves.

Like other epithelial cells, the lingual epithelial cells have a variety of transporters to control their volume and intracellular concentrations of electrolytes and nonelectrolytes. This includes, of course, the ubiquitous calcium and sodium ATPases, basolateral potassium channels, and amiloride- sensitive ion channels (Simon and Wang, 1993; Simon et al., 1993b). Since the lingual epithelium can respond to chemical stimuli by changing transport of various ions, its response can indirectly activate intraepithelial nerve fibers simply by increasing the potassium concentration in the extracellular space via the inhibition of the Na^+, K^+- ATPase (thereby depolarizing intraepithelial nerve fibers). For example, inhibition of the Na^+-K^+ ATPase by ethanol will increase the extracellular potassium concentration and may activate a variety of intraepithelial nerve fibers. (See Chapter 4, Bryant and Silver). Similarly, lingual epithelial cells possess both nicotinic and muscarinic acetylcholine receptors (Simon, unpubl. obs.), thus indicating that they are likely to respond to acetylcholine, atropine, and other bitter tasting compounds frequently believed to be the sole domain of taste cells. Keratinocytes normally produce low levels of acetylcholine, which may function as a local signal in the epithelium to regulate proliferation, differentiation, and adhesion (Grando, 1997). Local release of acetylcholine or other local tissue factors is likely to influence cells in taste buds as well.

4.7. Innervation of the Gustatory Epithelium: Afferent and Efferent

The taste epithelium receives an especially dense innervation. Classically, the nerve fibers associated with taste buds are divided into three anatomically defined categories (Fig. 12.2b): intragemmal fibers (those within a taste bud), perigemmal fibers (the dense network surrounding a taste bud), and extragemmal fibers (those ending in the epithelium some distance from taste buds see Fig. 12.2). The intragemmal fibers include mostly, but not entirely, the specific gustatory fibers of the facial, glossopharyngeal, or vagus nerves onto which the taste cells synapse.The intragemmal contingent also may include some general somatosensory nerve fibers, for example, polymodal nociceptors (Ogawa et al., 1968; Finger, 1986; Silverman and Kruger, 1990). The perigemmal and extragemmal innervation serves mechanoreceptors, thermal receptors, and nociceptors in the vicinity of taste buds (Munger, 1993). In addition, there is some evidence for the presence of efferent innervation of taste buds (e.g., Kinnamon et al., 1985; Yoshie et al., 1996), although this is not generally accepted (Royer and Kinnamon, 1988).

Gustatory fibers, that is, those upon which taste cells synapse, arise only from ganglion cells of the facial, glossopharyngeal, or vagus nerves, depending on the location of the taste bud in the oropharynx (see also Chapter 14, Smith and Davis). The facial nerve innervates taste buds at the anterior end of the oral cavity: chorda tympani for the anterior tongue (fungiform papillae and anterior most foliate papillae) and the greater superficial petrosal nerve for the anterior palate. The glossopharyngeal nerve innervates the posterior tongue (vallate and most foliate papillae) and soft palate, while the vagus nerve innervates taste buds in the pharynx, epiglottis, and upper esophagus (when such are present). These latter two nerves also convey general somatosensory fibers to the same area of the epithelium. The general somatosensory innervation for the facial nerve territories is carried by the trigeminal rather than the facial nerve; the anterior two-thirds of the tongue being innervated by the lingual branch of mandibular N and the soft palate by the lesser palatine branch of maxillary N (see also Chapter 4, Bryant and Silver). Thus the anterior taste fields are a natural situation in which one can study the separate functions of the general somatosensory as opposed to gustatory innervation.

The nongustatory oropharyngeal nerve fibers can affect perception of oral stimuli by providing information about mechanical and thermal qualities of substances in the oral cavity. The somatosensory-type innervation of lingual papillae forms a major portion of the innervation. For example, in fungiform papillae, 75% of the nerve fibers entering the papillae originate from the lingual branch of the trigeminal nerve (Farbman and Hellekant, 1978). Sensory fibers may terminate in specialized endorgans such as Merkel disks, Meissners corpuscles, Krause endorgans, or paciniform endings, all of which are involved in mechanoreception and contribute to the perception of the oral quality of food, for example, texture or temperature (Holland, 1984; Munger, 1993). Other nerve fibers terminate in the epithelium as coiled (loose or tight) endings of unknown function (Munger, 1993). A large number of fibers terminate as free nerve endings (Figs. 12.2b and 12.3)—most of which terminate in the dermis, although many extend upward through the basement membrane well into the epithelium (Finger, 1986; Silverman and Kruger, 1990). These free endings are thought to be thermoreceptors and/or nociceptors.

Of the nociceptors, the major subset are polymodal nociceptors (PMNs), so named because they are activated by any of several noxious stimuli including chemical, thermal, and mechanical ones. Activation of the PMNs will result in release of various bioactive substances (e.g., CGRP, SP, NKA, VIP, CCK) from their widely branched terminals (Kruger and Mantyh, 1989; Szallasi and Blumberg, 1999), thereby producing an "Axon reflex" (see Chapter 4, Bryant and Silver and below).

4.8. Interactions Between "Irritants" and Gustation

Psychophysically, the presence of irritants affects taste perception (Karrer and Bartoshuk, 1995; Yangagasawa et al., 1998; see Chapter 16, Breslin). While some of this effect may be due purely to attentional phenomena, stimulation of perigemmal and intragemmal peptidergic PMNs can influence taste transduction and gustation in several ways and at several levels of the sensory neuraxis (Silver et al., 1985; Serova, 1990; Davis and Smith, 1997).

4.8.1. Peripheral Mechanisms

The PMNs have numerous functions including: (1) serving as transducers for nociceptive or strong chemical stimuli, for example, nicotine, menthol, ethanol, or many alkaloids (Silver et al., 1991; Wang et al., 1993, 1995; Bryant and Moore, 1995; Carstens et al., 1998; see also Chapter 4, Bryant and Silver); (2) acting as a source of trophic factors (see above); and (3) serving as modulators of gustatory responses by releasing peptides from their terminals (Fig. 12.4, Finger, 1986; Kinnman and Aldskogius, 1991; Welton et al., 1992; Nagai et al., 1996; Olsen et al., 1998). These functions need not be mutually exclusive. When

PMNs are stimulated, for example, by capsaicin, they release neuropeptides into the surrounding tissue (Holzer, 1991). Some data suggest that neuropeptides can modulate gustatory responses. When peptides like CCK or substance P are injected into rats, they alter chorda tympani responses to NaCl, sucrose, or quinine (Gosnell and Hsiao, 1984; Serova and Esakov, 1985; Serova, 1990). Since taste receptor cells contain receptors for SP (NK1 receptors) (Chang et al., 1996), and presumably for other neuropeptides (e.g., CCK), neuropeptides released from peri- and intragemmal fibers may modulate the response of taste cells to gustatory stimuli. The peripheral release of the bioactive peptides such as CGRP and substance P also causes significant changes in the local tissues surrounding taste buds (Holzer, 1991). Such effects include changes in blood flow, plasma extravazation, ion channel and gap junction permeability, cell kinetics, and activation of mast cells (Holzer, 1991, 1998). For example, activation of PMNs by either electrical and/or chemical stimulation by nociceptive stimuli causes peptide-induced vasodilatation (Karita and Izumi, 1993; Wang et al., 1995) that is obvious from reddening of the tongue, and an increase of several degrees in surface temperature (Wang et al., 1995). This is relevant to gustation since changing the temperature of the epithelium not only alters the perception of chemical quality of foods but also changes the responses of CT neurons to various chemical stimuli (Hellekant, 1977).

4.8.2. Central Mechanisms

In addition to these multiple peripheral effects, activation of peri- and intragemal peptidergic fibers also is likely to affect taste perception by changing the activity of central pathways. The central branches of the intra- and perigemmal fibers project to the spinal trigeminal nucleus and to the nucleus of the solitary tract (Whitehead and Frank, 1983; see also Chapter 14, Smith and Davis). The collateral projections to the nucleus of the solitary tract terminate within the primary gustatory nuclei where they can either directly or indirectly influence transmission and processing of gustatory information (King et al., 1993; Davis and Smith, 1997). There is, however, virtually nothing known about interactions between the taste and nociceptor systems. Higher-order interactions could also take place in the thalamus, cortex, or basal ganglia, but little is known about this potentiality.

5. SUMMARY

All metazoan animals have chemoreceptors associated with mouthparts and involved in regulation of food intake. In invertebrates, these taste receptors are primary sensory neurons with axons extending into the central nervous system. Taste receptor cells of arthropods are associated with uniporus sensilla often located on walking legs as well as on mouthparts. In vertebrates, taste receptor cells are axonless, modified epithelial cells that synapse onto the peripheral processes of the primary gustatory neurons of the facial, glossopharyngeal, and vagual nerve ganglia. The taste receptor cells are aggregated into taste buds lying within the oropharynx. In some aquatic vertebrates, taste buds also occur on the external body surface.

Taste buds comprise several morphological cell types however the detailed structure of taste buds varies in different species. In mammals, taste buds contain three types of cells, one of which appears capable of concentrating biogenic amines, for example, serotonin. The different morphological types of taste cells appear to have some correlation with function, although all types are reported to synapse with gustatory nerve fibers and therefore must be viewed as receptor cells.

The epithelium surrounding the taste buds is likely to play an important role as a conduit by which potential taste stimuli may gain access to the basolateral region of the taste receptor cells and to the nerve fibers supplying both the bud and the surrounding epithelium. Activation of cheminociceptive fibers in the epithelium may play an important role in modulation of taste receptor cell activity as well as in regulation of epithelial dynamics and function.

REFERENCES

Adler, E., M. A. Hoon, K. L. Mueller, J. Chandrashekar, N. J. Ryba, and C. S. Zucker (2000). A novel family of mammalian taste receptors. *Cell* **100**:693–702.

Baatrup, E. (1983). Terminal buds in the branchial tube of the brook lamprey (*Lampetra planeri* (Bloch))—Putative respiratory monitors. *Acta Zool (Stockh)* **64**:139–147.

Baatrup, E. (1985). Physiological studies on the pharyngeal terminal buds in the larval brook lamprey, *Lampetra planeri* (Bloch). *Chem Senses* **10**:549–558.

Bargmann, C. I., E. Hartwieg, and H. R. Horvitz (1993). Odorant-selective genes and neurons mediate olfaction in *C. elegans*. *Cell* **74**:515–527.

Baratz, R. S. and A. I. Farbman (1975). Morphogenesis of rat lingual filiform papillae. *Am J Anat* **143**:283–302.

Barnes, P. J. (1991). Neurogenic inflammation in airways. *Int Arch Allergy Appl Immunol* **94**:303–309.

Barry, M. A. and L. D. Savoy (1993). Persistence and calcium-dependent ATPase staining of denervated fungiform taste buds in the hamster. *Arch Oral Biol* **38**:5–15.

Beidler, L. M. and R. L. Smallman (1965). Renewal of cells within taste buds. *J Cell Biol* **27**:263–272.

Bernhardt, S. J., M. Naim, U. Zehavi, and B. Lindemann (1996). Changes in IP_3 and cytosolic Ca^{2+} in response to sugars and non-sugar sweeteners in transduction of sweet taste in the rat. *J Physiol (Lond)* **490**:325–336.

Böttger, B., K. Reutter and T. E. Finger (1997). Double-label immunocytochemistry of taste cells in the rat: gustducin, NCAM, carbonic anhydrase, and calbindin. *Chem Senses* **22**:647–648.

Boughter, J. D., D. W. Pumplin, C. Yu, R. C. Christy, and D. V. Smith (1997). Differential expression of alpha-gustducin in taste bud populations of the rat and hamster. *J Neurosci* **17**:2852–2858.

Bowden, G. C. (1993). Keratins and other epidermal proteins. In: G. C. Priestly (ed). *Molecular Aspects of Dermatology*. New York: John Wiley and Sons, pp 19–54.

Bradley, R. M. (1971). Tongue topography. In: L. M. Beidler (ed). *Chemical Senses. Part 2: Taste. Handbook of Sensory Physiology, Vol. IV.* pp 1–30.

Braun, C. B. (1998). Schreiner organs: a new craniate chemosensory modality in hagfishes. *J Comp Neurol* **392**:135–163.

Brown, D., L. M. Garcia-Segura, and L. Orci (1984). Carbonic anhydrase is associated with taste buds in rat tongue. *Brain Res* **324**:346–348.

Bryant, B. P. and P. A. Moore (1995). Factors affecting the sensitivity of the lingual trigeminal nerve to acids. *Am J Physiol* **268**:R58–65.

Caprio, J., J. G. Brand, J. H. Teeter, T. Valentincic, D. L. Kalinoski, J. Kohbara, T. Kumazawa, and S. Wegert (1993). The taste system of the channel catfish: from biophysics to behavior. *Trends Neurosci* **16**:192–197.

Carstens, E., N. Kuenzler, and H. O. Handwerker (1998). Activation of neurons in rat trigeminal subnucleus caudalis by different irritant chemicals applied to the oral or ocular mucosa. *J Neurophysiol* **80**:465–492.

Chang, G. Q., S. R. Vigna, and S. A. Simon (1996). Localization of substance P NK-1 receptors in rat tongue. *Regul Pept* **63**:85–89.

Chen, S. Y. and C. A. Squire (1984). The ultrastructure of the oral epithelium. In: J. Meyer, C. A. Squire and S. J. Gerson (eds). *The Structure and Function of Oral Mucosa*. New York: Pergamon Press, pp 7–29.

Cottler-Fox, M., K. Arvidson, E. Hammarlund, and U. Friberg (1987). Fixation and occurrence of dark and light cells in taste buds of fungiform papillae. *Scand J Dent Res* **95**:417–427.

Daikoku H., I. Morisaki, Y. Ogawa, T. Maeda, K. Kurisu, and S. Wakisaka (1999). Immunohistochemical localization of carbonic anhydrase isozyme II in the gustatory epithelium of the adult rat. *Chem Senses* **24**:255–261.

Davis, B. J. and D. V. Smith (1997). Substance P modulates taste responses in the nucleus of the solitary tract of the hamster. *Neuroreport* **8**:1723–1727.

Delay, R. J., J. C. Kinnamon, and S. D. Roper (1986). Ultrastructure of mouse vallate taste buds: II. Cell types and cell lineage. *J Comp Neurol* **253**:242–252.

Desimone, J. A., G. L. Heck, S. Mierson, and S. K. Desimone (1984). The active ion transport properties of canine lingual epithelia in vitro. Implications for gustatory transduction. *J Gen Physiol* **83**:633–656.

Dethier, V. G. (1993). The role of taste in food intake: a comparative view. In: S. A. Simon and S. D. Roper (eds). *Mechanisms of Taste Transduction*. Boca Raton, FL: CRC Press, pp 3–28.

Elliott, E. J. and S. A. Simon (1990). The anion in salt taste: a possible role for paracellular pathways. *Brain Res* **535**:9–17.

Farbman, A. I. (1965). Fine structure of the taste bud. *J Ultrastruct Res* **12**:328–350.

Farbman, A. I. (1980). Renewal of taste bud cells in rat circumvallate papillae. *Cell Tissue Kinet* **13**:349–357.

Farbman, A. I. and G. Hellekant (1978). Quantitative analyses of the fiber population in rat chorda tympani nerves and fungiform papillae. *Am J Anat* **153**:509–521.

Finger, T. E. (1986). Peptide immunocytochemistry demonstrates multiple classes of perigemmal nerve fibers in the circumvallate papilla of the rat. *Chem Senses* **11**:135–144.

Finger, T. E. (1997). Evolution of taste and solitary chemoreceptor cell systems. *Brain Behav Evol* **50**:234–243.

Finger, T. E., B. P. Bryant, D. L. Kalinoski, J. H. Teeter, B. Böttger, W. Grosvenor, R. H. Cagan, and J. G. Brand (1996). Differential localization of putative amino acid receptors in taste buds of the channel catfish, *Ictalurus punctatus*. *J Comp Neurol* **373**:129–138.

Fritzsch, B., P. A. Sarai, M. Barbacid, and I. Silos-Santiago (1997). Mice with a targeted disruption of the neurotrophin receptor trkB lose their gustatory ganglion cells early but do develop taste buds. *Int J Dev Neurosci* **15**:563–576.

Ganchrow, J. R. and D. Ganchrow (1987). Taste bud development in chickens (*Gallus gallus domesticus*). *Anat Rec* **218**:88–93.

Gosnell, B. A. and S. Haiso (1984). The effects of cholectystokinin on taste preference and sensitivity in rats. *Behav Neurosci* **3**:452–460.

Grando, S. A. (1997). Biological functions of keratinocyte cholinergic receptors. *J Invest Dermatol Symp Proc* **2**:41–48.

Grando, S., D. A. Kist, M. Oi, and M. V. Dahl (1993). Human keratinocytes synthesize, secrete, and degrade acetylcholine. *J Invest Dermatol* **101**:32–36.

Guth, L. (1971). Degeneration and regeneration of taste buds. In: L. M. Beidler (ed). *Chemical Senses. Part 2: Taste. Handbook of Sensory Physiology*. Berlin: Springer-Verlag, pp 63–74.

Hellekant, G. (1977). Vasodilator fibres to the tongue in the chorda tympani proper nerve. *Acta Physiol Scand* **99**:292–299.

Herness, M. S. (1990). The digestive hormone cholecystokinin is present in rat taste cells. In: K. J. Dving (ed). *ISOT X; Proceedings of the Tenth International Symposium on Olfaction and Taste*. Oslo: GVS A/S, p 303.

Herrick, C. J. (1905). Central gustatory paths in the brains of bony fishes. *J Comp Neurol* **15**:375–456.

Hill, M. W. (1984). The ultrastructure of the oral epithelium. In: J. Meyer et al. (eds). *The Structure and Function of Oral Mucosa*. New York: Pergamon Press.

Holland, G. R. (1984). Innervation of the oral mucosa. In: J. Meyer et al. (eds). *The Structure and Function of Oral Mucosa*. New York: Pergamon Press, pp 195–218.

Holland, V. F., G. A. Zampighi, and S. A. Simon (1989). Morphology of fungiform papillae in canine lingual epithelium: location of intercellular junctions in the epithelium. *J Comp Neurol* **279**:13–27.

Holzer, P. (1991). Capsaicin: cellular targets, mechanisms of action, and selectivity for thin sensory neurons. *Pharmacol Rev* **43**:143–201.

Holzer, P. (1998) Neural injury, repair and adaption in the GI tract. II: the elusive action of capsaicin on the vagus nerve. *Am J Physiol* **275**:G8–G13.

Hoon M. A., E. Adler, J. Lindemeier, J. F. Battey, N. J. Ryba, and C. S. Zuker (1999). Putative mammalian taste receptors: a class of taste-specific GPCRs with distinct topographic selectivity. *Cell* **96**:541–551.

Kanazawa, H. and S. Yoshie (1996). The taste bud and its innervation in the rat as studied by immunohistochemistry for PGP 9.5. *Arch Histol Cytol* **59**:357–367.

Karita, K. and H. Izumi (1993). Dual afferent pathways of vasodilator reflex induced by lingual stimulation in the cat. *J Auton Nerv Syst* **45**:235–240.

Karrer, T. and L. Bartoshuk (1995). Effects of capsaicin desensitization on taste in humans. *Physiol Behav* **57**:421–429.

Kim, D. J. and S. D. Roper (1995). Localization of serotonin in taste buds: a comparative study in four vertebrates. *J Comp Neurol* **353**:364–370.

King, M. S., L. Wang, and R. M. Bradley (1993). Substance P excites neurons in the gustatory zone of the rat nucleus tractus solitarius. *Brain Res* **619**:120–130.

Kinnamon, J. C., T. A. Sherman, and S. R. Roper (1988). Ultrastructure of mouse vallate taste buds: III. Patterns of synaptic connectivity. *J Comp Neurol* **270**:1–10.

Kinnamon, J. C., B. J. Taylor, R. J. Delay, and S. D. Roper (1985). Ultrastructure of mouse vallate taste buds. I. Taste cells and their associated synapses. *J Comp Neurol* **235**:48–60.

Kinnamon, S. C. and S. D. Roper (1987). Passive and active membrane properties of mudpuppy taste receptor cells. *J Physiol (Lond)* **383**:601–614.

Kinnman, E. and H. Aldskogius (1991). The role of substance P and calcitonin gene-related peptide containing nerve fibers in maintaining fungiform taste buds in the rat after a chronic chorda tympani nerve injury. *Exp Neurol* **113**:85–91.

Knapp, L., A. Lawton, B. Oakley, L. Wong, and C. Zhang (1995). Keratins as markers of differentiated taste cells of the rat. *Differentiation* **58**:341–349.

Kruger, L. and P. Mantyh (1989). Gustatory and related chemosensory systems. In: A. Bjorkland et al. (eds). *Integrated Systems of the CNS, Part II. Handbook of Chemical Neuroanatomy Vol. 7*. Amsterdam: Elsevier, pp 323–410.

Lin W., T. E. Finger, B. C. Rossier, S. C. Kinnamon (1999). Epithelial Na$^+$ channel subunits in rat taste cells: localization and regulation by aldosterone. *J Comp Neurol* **405**:406–420.

Lundberg, J. M. (1995). Tachykinins, sensory nerves, and asthma. *Can J Physiol Pharm* **73**:908–914.

Maggi, C. A. (1993). Tachykinin receptors and airway pathophysiology. *Eur Respir J* **6**: 735–742.

Margolis, L., B. Baibakov, C. Collin, and S. A. Simon (1995). Dye-coupling in three-dimensional histoculture of rat lingual frenulum. *In Vitro Cell Dev Biol Anim* **31**:456–461.

Miller I. J. Jr. (1995). Anatomy of the peripheral taste system. In: R. L. Doty (ed). *Handbook of Taste and Olfaction*. New York: Marcel Dekker, pp 521–547.

Mistretta, C. M. (1971) Permeability of tongue epithelium and its relation to taste. *Am J Physiol* **220**:1162–1167.

Miyawaki, Y., I. Morisaki, M. J. Tabata, K. Kurisu, and S. Wakisaka (1996). Calbindin D28k-like immunoreactivity in the gustatory epithelium in the rat. *Neurosci Lett* **214**:29–32.

Munger, B. L. (1993). The general somatic afferent terminals in oral mucosae. In: S. A. Simon and S. D. Roper (eds). *Mechanisms of Taste Transduction*. Boca Raton, FL: CRC Press, pp 83–104.

Murray, R. G. (1969). Cell types in rabbit taste buds. In: C. Pfaffman (ed). *Olfaction and Taste III*. New York: Rockefeller University Press, pp 331–334.

Murray, R. G. (1986). The mammalian taste bud type III cell: a critical analysis. *J Ultrastruct Mol Struct Res* **95**:175–188.

Murray, R. G., A. Murray, and S. Fujimoto (1969). Fine structure of gustatory cells in rabbit taste buds. *J Ultrastruct Res* **27**:444–461.

Nagai, T., D. J. Kim, R. J. Delay and S. D. Roper (1996). Neuromodulation of transduction and signal processing in the end organs of taste. *Chem Senses* **21**:353–365.

Nakamura, M. and K. Kurihara (1988). Temperature dependence of amiloride-sensitive and -insensitive components of rat taste nerve response to NaCl. *Brain Res* **444**:159–164.

Nelson, G. M. and T. E. Finger (1993). Immunolocalization of different forms of neural cell adhesion molecule (NCAM) in rat taste buds. *J Comp Neurol* **336**:507–516.

Nagai, T., D. Kim, R. J. Delay, and S. D. Roper (1996). Neuromodulation of transduction and signal processing in the end organs of taste. *Chem Sens* **21**:353–365.

Nosrat, C. A., J. Blomlöf, W. M. ElShamy, P. Ernfors, and L. Olson (1997). Lingual deficits in BDNF and NT3 mutant mice leading to gustatory and somatosensory disturbances, respectively. *Development* **124**:1333–1342.

Ogawa, H., M. Sato, and S. Yamashita (1968). Multiple sensitivity of chorda tympani fibres of the rat and hamster to gustatory and thermal stimuli. *J Physiol* **199**:223–240.

Olmstead, J. M. D. (1920). The nerve as a formative influence in the development of taste buds. *J Comp Neurol* **31**:465–468.

Olsen, T. H., M. S. Reidl, L. Vulchanova, X. R. Ortiz-Gonzalez, and R. Elde (1998). An acid sensing ion channel (ASIC) localizes to small primary afferent neurons in rats. *NeuroReport* **9**:1109–1113.

Osculati, F. and A. Sbarbati (1995). The frog taste disc: a prototype of the vertebrate gustatory organ. *Prog Neurobiol* **46**:351–399.

Peters, R. C., K. Kotrschal, and W. D. Krautgartner (1991). Solitary chemoreceptor cells of *Ciliata mustela*, Gadidae, Teleostei, are turned to mucoid stimuli. *Chem Senses* **16**:31–42.

Phillips, L. M. and D. L. Hill (1996). Novel regulation of peripheral gustatory function by the immune system. *Am J Physiol* **271**:R857–R862.

Priestley, G. C. (1993). An introduction to the skin and its disease. In: G. C. Priestley (ed). *Molecular Aspects of Dermatology*. New York: John Wiley and Sons, pp 19–54.

Pumplin, D. W., C. Yu, and D. V. Smith (1997). Light and dark cells of rat vallate taste buds are morphologically distinct cell types. *J Comp Neurol* **378**:389–410.

Randerman-Little, R. (1979). The effect of temperature on the turnover of taste bud cells in catfish. *Cell Tiss Kinet* **12**:269–280.

Reutter, K. and M. Witt (1993). Morphology of vertebrate taste organs and their nerve supply. In: S. A. Simon and S. D. Roper (ed). *Mechanisms of Taste Transduction*. Boca Raton, FL: CRC Press, pp 30–82.

Roper, S. (1983). Regenerative impulses in taste cells. *Science* **220**:1311–1312.

Roper, S. D. (1992). The microphysiology of peripheral taste organs. *J Neurosci* **12**:1127–1134.

Royer S. M. and J. C. Kinnamon (1988). Ultrastructure of mouse foliate taste buds: synaptic and nonsynaptic interactions between taste cells and nerve fibers. *J Comp Neurol* **270**: 11–24, 58–59.

Salas, P. J. and J. H. Moreno (1982). Single-file diffusion multi-ion mechanism of permeation in paracellular epithelial channels. *J Membr Biol* **64**:103–112.

Serova, O. N. (1990). Changes in the activity of the taste receptor apparatus in the perference for a sodium chloride solution in newborn rats receiving capsaicin (in Russian). *Biull Eksper Biol Med* **109**:215–217.

Serova, O. N. and A. I. EsaKov (1985). Activating effect of cholecystokinin-pancreozymin on the taste receptor system on the rat. *Fiziol Zh SSSr* **71**:1271–1277.

Shields, V. D. C. (1994). Ultrastructure of the uniporus sensilla on the galea of larval *Mamestra configurata* (Walker) (Lepidoptera: Noctidae). *Can J Zool* **72**:2016–2031.

Siegel, I. A., K. T. Izutsu, and E. Watson (1981). Mechanisms of non-electrolyte penetration across dog and rabbit oral mucosa in vitro. *Arch Oral Biol* **26**:357–361.

Silver, W. L. and T. E. Finger (1984). Electrophysiological examination of a non-olfactory, non-gustatory chemosense in the searobin, *Prionotus carolinus*. *J Comp Physiol* **154**: 167–174.

Silver, W. L., J. R. Mason, D. A. Marshall, and J. A. Maruniak (1985). Rat trigeminal, olfactory and taste responses after capsaicin desensitization. *Brain Res* **333**:45–54.

Silver, W. L., L. G. Farley, and T. E. Finger (1991). The effects of neonatal capsaicin administration on trigeminal nerve chemoreceptors in the rat nasal cavity. *Brain Res* **561**:212–216.

Silverman, J. D. and L. Kruger (1990). Analysis of taste bud innervation based on glyco-conjugate and peptide neuronal markers. *J Comp Neurol* **292**:575–584.

Simon, S. A. and A. L. Sostman (1991). Electrophysiological responses to non-electrolytes in lingual nerve of rat and lingual epithelia of dog. *Arch Oral Biol* **36**:805–813.

Simon, S. A. and Y. Wang (1993). Chemical responses of lingual nerves and lingual epithelia. In: S. A. Simon and S. D. Roper (eds). *Mechanisms of Taste Transduction*. Boca Raton, FL: CRC Press, pp 225–252.

Simon, S. A., V. F. Holland, and G. Zampighi (1991). Localization of Na,K-ATPase in lingual epithelia. *Chem Senses* **16**:283–293.

Simon, S. A., E. J. Elliott, R. P. Erickson, and V. F. Holland (1993a). Ion transport across lingual epithelium is modulated by chorda tympani nerve fibers. *Brain Res* **615**:218–228.

Simon, S. A., V. F. Holland, D. J. Benos, and G. A. Zampighi (1993b). Transcellular and paracellular pathways in lingual epithelia and their influence in taste transduction. *Microsc Res Tech* **26**:196–208.

Singh, R. N. (1997). Neurobiology of the gustatory systems of *Drosophila* and some terrestrial insects. *Microsc Res Tech* **39**:547–63.

Sloan, H. E., S. E. Hughes, and B. Oakley (1983). Chronic impairment of axonal transport eliminates taste responses and taste buds. *J Neurosci* **3**:117–123.

Smith, D. V., R. A. Akeson, and M. T. Shipley (1993). NCAM expression by subsets of taste cells is dependent upon innervation. *J Comp Neurol* **336**:493–506.

Smith, D. V., R. Klevitsky, R. A. Akeson, and M. T. Shipley (1994). Expression of the neural cell adhesion molecule (NCAM) and polysialic acid during taste bud degeneration and regeneration. *J Comp Neurol* **347**:187–196.

Sostman, A. L. and S. A. Simon (1991). Trigeminal nerve responses in the rat elicited by chemical stimulation of the tongue. *Arch Oral Biol* **36**:95–102.

Suzuki, Y. and M. Takeda (1983). Ultrastructure and monoamine precursor uptake of taste buds in the pharynx, nasopalatine ducts, epiglottis and larynx of the mouse. *Kaibogaku Zasshi* **58**:593–605.

Suzuki, Y., M. Takeda, N. Obara, and Y. Nagai (1996). Phagocytic cells in the taste buds of rat circumvallate papillae after denervation. *Chem Senses* **21**:467–476.

Szallasi, A. and P. M. Blumberg (1999). Vanilloid (capsaicin) receptors and mechanisms. *Pharm Revs* **51**:159–211.

Tabata, S., H. H. Crowley, B. Böttger, T. E. Finger, R. F. Margolskee, and J. C. Kinnamon (1995). Immunoelectron microscopical analysis of gustducin in taste cells of the rat. *Chem Senses* **21**:778.

Tabish, M., Z. K. Siddiqui, K. Nishikawa, and S. S. Siddiqui (1995). Exclusive expression of *C. elegans* osm-3 kinesin gene in chemosensory neurons open to the external environment. *J Mol Biol* **247**:377–389.

Takeda, M., Y. Suzuki, N. Obara and Y. Nagai (1992). Neural cell adhesion molecule of taste buds. *J Electron Microsc* **41**:375–80.

Takeda, M., Y. Suzuki, N. Obara, and Y. Nagai (1996). Apoptosis in mouse taste buds after denervation. *Cell Tissue Res* **286**:55–62.

Torii, H., Z. Yan, J. Hosoi, and R. D. Granstein (1997). Expression of neurotrophic factors and neuropeptide receptors by Langerhans cells and by Langerhans- like cell line XS52: further support for a functional relationship between Langerhans cells and epidermal nerves. *J Invest Dermatol* **109**:588–591.

Wang, Y., R. P. Erickson, and S. A. Simon (1993). Selectivity of lingual nerve fibers to chemical stimuli. *J Gen Physiol* **101**:843–66.

Wang, Y., R. P. Erickson, and S. A. Simon (1995). Modulation of rat chorda tympani nerve activity by lingual nerve stimulation. *J Neurophysiol* **73**:1468–1483.

Welton, J., R. Taylor, A. J. Porter, and S. D. Roper (1992). Immunocytochemical survey of putative neurotransmitters in taste buds from *Necturus maculosus. J Comp Neurol* **324**:509–521.

Whitear, M. (1992). Solitary chemosensory cells. In: T. J. Hara (ed). *Chemoreception in Fishes, 2nd Ed.* London: Elsevier Press, pp 103–125.

Whitehead, M. C. and M. E. Frank (1983). Anatomy of the gustatory system in the hamster: central projections of the chorda tympani and the lingual nerve. *J Comp Neurol* **220**:378–395.

Whitehead, M. C., M. E. Frank, T. P. Hettinger, L. T. Hou, and H. D. Nah (1987). Persistence of taste buds in denervated fungiform papillae. *Brain Res* **405**:192–195.

Yangagisawa, K., L. M. Bartoshuk, F. A. Catalanotto, T. A. Karrer, and J. F. Kveton (1998). Anesthesia of the chorda tympani causes taste phantoms. *Physiol Behav* **63**:329–335.

Ye, Q., R. E. Stewart, G. L. Heck, D. L. Hill, and J. A. DeSimone (1993). Dietary Na($+$)-restriction prevents development of functional Na$^+$ channels in taste cell apical membranes: proof by in vivo membrane voltage perturbation. *J Neurophysiol* **70**:1713–1716.

Yoshie, S., Y. Teraki, T. Iwanaga, and T. Fujita (1989). Immunocytochemistry of neuron-specific proteins and neuropeptides in taste buds and associated nerves. *Arch Histol Cytol* **52**:389–396.

Yoshie S., C. Wakasugi, H. Kanazawa, and T. Fujita (1993) Met-enkephalin-Arg6-Gly7-Leu8-like immunoreactivity in mammalian taste buds. *Arch Histol Cytol* **56**:495–500.

Yoshie, S., H. Kanazawa, and T. Fujita (1996). A possibility of efferent innervation of the gustatory cell in the rat circumvallate taste bud. *Arch Histol Cytol* **59**:479–484.

Yoshie, S., H. Kanazawa, Y. Nishida, and T. Fujita (1997). Occurrence of subtypes of gustatory cells in cat circumvallate taste buds. *Arch Histol Cytol* **60**:421–426.

Zacharuk, R. Y. and V. D. Shields (1991). Sensilla of immature insects. *Annu Rev Entomol* **36**:331–354.

Zhang, C., M. Cotter, A. Lawton, B. Oakley, L. Wong, and Q. Zeng (1995). Keratin 18 is associated with a subset of older taste cells in the rat. *Differentiation* **59**:155–162.

13

Taste Transduction and Molecular Biology

JOHN I. GLENDINNING
Department of Biological Science, Barnard College, Columbia University,
New York, NY

NIRUPA CHAUDHARI
Department of Physiology/Biophysics, University of Miami
School of Medicine, Miami, FL

SUE C. KINNAMON
Department of Anatomy and Neurobiology, Colorado State University,
Fort Collins, CO

1. INTRODUCTION

Among the sensory systems, the taste system is unusual in its capacity to respond to a large number of stimuli that vary greatly in molecular size, lipophilicity, and pH (e.g., salts, amino acids, sugars, acids, vitamins, fatty acids, and many toxic compounds). To accommodate this structural diversity, taste cells utilize a diverse array of transduction mechanisms. However, only a fraction of these mechanisms occurs within any given taste cell; different subsets of taste cells appear to express different transduction mechanisms, and hence different molecular receptive ranges (e.g., Bigiani and Roper, 1993; Cummings et al., 1993; Koganezawa and Shimida, 1997). This feature of the peripheral taste system enables it to accommodate a plethora of stimuli and, at the same time, discriminate between

The Neurobiology of Taste and Smell, Second Edition, Edited by Thomas E. Finger, Wayne L. Silver, and Diego Restrepo.
ISBN 0-471-25721-4   Copyright © 2000 Wiley-Liss, Inc.

different classes of chemical stimuli. Such discrimination helps animals select foods adaptively (Dethier, 1993) and determine the appropriate amounts of saliva, insulin, and specific digestive enzymes to secrete for each meal (e.g., Behrman and Kare, 1968; Teff and Engelman, 1996).

In this chapter, we review the different types of taste transduction mechanisms present in vertebrate and invertebrate taste cells. We begin with an introduction to the general steps in taste transduction. Then we discuss the specific transduction mechanisms that mediate reception of the most common categories of taste stimuli: salts, sweeteners, amino acids, acids, and "bitter" stimuli. Finally, we review the factors known to modulate the response properties of taste cells. Throughout this discussion, we develop the hypothesis that animals with specific feeding habits (e.g., herbivores, omnivores, and carnivores) each have evolved a different collection of taste transduction pathways that best enable them to meet their particular nutritional requirements.

2. STEPS IN TASTE TRANSDUCTION

2.1. Invertebrates

Because most studies of invertebrate taste transduction have involved insects, we limit our discussion to this taxon. The general steps in transduction begin with a sapid compound encountering the tip of a chemosensory hair, and dissolving into the fluid in the terminal pore (for review, Morita, 1992; see also Chapter 12, Finger and Simon). This fluid (or receptor lymph) constitutes the outer ionic milieu for the dendritic process of each taste cell, which are bipolar sensory neurons. Once the sapid compound encounters the distal dendritic membranes, it is thought to interact with specific receptor sites and/or ionic channels, leading to a change in membrane conductance (Fig. 13.1). This change induces a current to flow through the transduction ion channels and down to the base of the dendritic process, where current outflow causes depolarization. While the receptor lymph provides some of the driving electromotive force for this taste cell current, most of the force appears to come from a $^+40$ to $^+60$ mV potential difference between the receptor lymph space and the blood (or hemolymph), called the transepithelial voltage (Thurm and Küppers, 1980; Kijima et al., 1995). The transepithelial voltage is thought to be established by electrogenic ion pumps located in the folded membrane of the supporting cells (tormogen and trichogen cells) that surround the taste cells (see Fig. 12.1, Chapter 12, Finger and Simon).

2.1.1. Methods and Approaches

Two methods are commonly used for recording from insect taste sensilla: tip recording and side-wall recording (review in Frazier and Hanson, 1986). Both

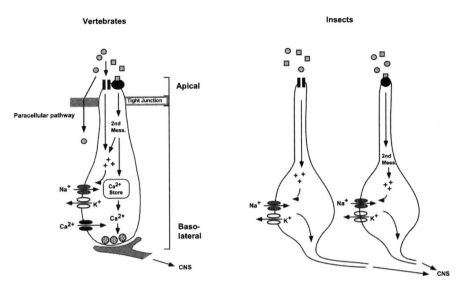

FIG. 13.1. Diagrammatic representation of the general steps in taste transduction for vertebrate taste receptor cells and insect taste receptor neurons. Vertebrate taste cells are polarized cells of epithelial origin, with tight junctions separating the apical, taste receptive membrane from the basolateral membrane, which forms chemical synapses with primary afferent nerve fibers. Taste stimuli usually interact with either ion channels or receptors on the apical membrane, resulting in taste cell depolarization, generation of taste cell action potentials, activation of voltage-gated Ca^{2+} channels, and neurotransmitter release onto sensory afferents. Some monovalent salts and acids can penetrate tight junctions to enter the paracellular pathway, where they interact with basolateral ion channels to depolarize taste cells. In contrast, insect taste cells are bonafide neurons, with taste receptors and taste-modulated ion channels located on the apical dendritic tip. Activation of insect taste cells results in generation of action potentials, which are transmitted to the CNS. Insect taste neurons are highly tuned, with different taste cells responding to different classes of taste stimuli. In contrast, vertebrate taste cells tend to be less narrowly tuned, with an individual taste cell responsive to more than one type of taste stimulus.

methods involve recording potential changes between the body of an insect and the distal dendritic processes of its taste cells. In tip recording, a glass micro-capillary tube, filled with a sapid stimulus in an electrolyte solution, is placed over the tip of a taste sensillum; in this configuration, the glass microcapillary tube serves as the stimulating and recording electrode. Note that this configuration is similar to the loose-patch configuration used for vertebrate taste buds (see below). In side-wall recording, a tungsten recording electrode is inserted into the lumen of a taste sensillum; stimulation is accomplished by contacting the sensillum tip with a microelectrode containing a sapid stimulus. The indifferent electrode (in both configurations) is either inserted directly into the body of the insect or coupled to the insect's intracellular fluid indirectly through an electrolyte bath (Gothilf and Hanson, 1994). Both techniques generate multi-unit traces that consist of trains of action potentials; one can often assign action potentials to different taste cells based on their amplitude and unique temporal pattern of firing using specialized software (e.g., Smith et al., 1990).

Research on transduction mechanisms in insect taste cells has lagged behind that in vertebrate taste cells largely because the taste cells are surrounded by a tough cuticle layer, making them relatively inaccessible. However, three recent reports offer promising approaches for addressing this problem. One approach involves removing the tip of a taste sensillum during the molting process, and patch clamping the distal dendritic process of taste cells as they extrude from the end of the sensillum (Murakami and Kijima, 2000). The second approach involves extracting poly$(A+)$ RNA from tissue samples containing thousands of taste sensilla, and functionally expressing the RNA in *Xenopus* oocytes (Koganezawa et al., 1996). The third approach will include determining the functional significance of a large family of seven transmembrane proteins discovered (through analysis of genomic sequence databases) in a gustatory organ (the labellum) of adult *Drosophila* (Clyne et al., 2000). All evidence to date indicates that these proteins are G protein-coupled taste receptors.

2.2. Vertebrates

Most taste stimuli interact initially with the apical membrane of vertebrate taste cells. This interaction usually leads to a change in membrane conductance, depolarization, Ca^{2+} influx, and transmitter release onto gustatory afferent neurons (Fig. 13.1). Despite their epithelial origin (see Chapter 15, Barlow), vertebrate taste cells express several neuronal properties, including voltage-gated ion channels and the capacity to generate action potentials (Roper, 1983). Even though many taste stimuli act by depolarizing taste cell membranes (e.g., NaCl, sugars, and acids; Behe et al., 1990; Gilbertson et al., 1992; Cummings et al., 1993), others act by hyperpolarizing them (e.g., amino acids and some bitter stimuli; Sato and Beidler, 1983; Hayashi et al., 1996; Zviman et al., 1996; Ogura et al., 1997). Whether changes in transmitter release occur in response to hyperpolarizing receptor potentials is not known.

In addition to participating in apical receptor events, some monovalent salts and acids penetrate the tight junctions at the apex of the taste bud and interact with sites on the basolateral membrane of taste cells (Ye et al., 1991, 1993; see also Chapter 12, Finger and Simon). This pathway is usually referred to as the paracellular pathway, and recent data suggest that it may mediate transduction of some salt and acids in mammals (Ye et al., 1994; DeSimone et al., 1995a; Kloub et al., 1997).

2.2.1. Receptor Mechanisms

Several different types of mechanisms underlie the early events in taste transduction. Simple stimuli (e.g., ions) do not require the presence of specific membrane receptors, but utilize apically (or in some cases basolaterally) located ion channels for transduction. Some stimuli (e.g., Na^+) permeate the channels, while others (e.g., H^+) block channels to depolarize taste cells. More complex stimuli are thought to bind to apically located membrane receptors that may be coupled directly to ligand-gated channels in the apical membrane or G protein-mediated second messenger systems.

Eight different G protein alpha subunits have been cloned from taste tissue, but only a few (gustducin, rod transducin, G_i, and G_{14}) have been localized definitively to taste cells (McLaughlin et al., 1992; Yang et al. 1999). A novel taste-specific G protein gamma subunit is also found in many gustducin-positive taste cells (Huang et al., 1999). A variety of second messenger-modulated ion channels have been identified in taste cells, including a K^+ channel that is closed by protein kinase A (Avenet et al., 1988), a cyclic nucleotide-blocked cation channel (Kolesnikov and Margolskee, 1995), and a cyclic nucleotide-gated cation channel (Misaka et al., 1997). Although their specific roles in taste transduction have yet to be determined, these channels represent likely targets of taste-stimulated second messengers. Finally, some lipophilic bitter stimuli (e.g., quinine) may penetrate the membrane and activate intracellular second messenger cascades directly.

Recent molecular cloning studies have identified a number of G protein-coupled receptors in taste cells. The G protein-coupled receptors underlying transduction of "bitter" and glutamate ("umami") tastes have been cloned and are discussed in sections 3.3.2 and 3.5.2 below, and in Kinnamon (2000). Another pair of receptors (T1R1 and T1R2) are expressed in distinct subsets of taste cells (Hoon et al., 1999), but their functional significance for taste transduction is unclear.

2.2.2. Methods and Approaches

There are several physiological methods for studying taste transduction in vertebrate taste cells. Taste cells can be isolated from the lingual epithelium and studied with either patch-clamp recording techniques (for review, Cummings and Kinnumon, 1995; Gilbertson, 1995) or Ca^{2+} imaging (for review, Restrepo et al., 1995). For Ca^{2+} imaging, taste cells are loaded with a Ca^{2+}-sensitive dye such as fura-2 and changes in intracellular Ca^{2+} are monitored in response to taste stimulation. A loose-patch technique for recording from taste buds in situ has been utilized to study transduction in mammalian fungiform taste buds (for review, Lindemann, 1995). This involves placing a loose-patch pipette over a single fungiform papillae in an intact tongue, and recording action currents (which reflect taste cell action potentials) to taste stimuli or membrane-permeant second messengers. Finally, chorda tympani nerve recordings can be obtained while the lingual field is under voltage-clamp (for review, see DeSimone et al., 1995b). This technique has been useful for evaluating the role of the paracellular pathway in taste transduction. It assumes that if the chorda tympani response to an ionically charged taste stimulus is not affected by applying a voltage field across the lingual epithelium, then the transduction for that chemical must occur on the basolateral membrane via the paracellular pathway.

3. TRANSDUCTION OF THE MOST COMMON CATEGORIES OF TASTE STIMULI

3.1. Salts

All terrestrial animals possess taste cells that respond to salts. In most animals, low levels of excitatory input from these salt-responsive taste cells elicits appetitive responses, and high levels of excitatory input elicits aversive responses (e.g., Dethier, 1968). This concentration-dependent change in affective response may have evolved because salts satisfy the nutritional needs of most animals at low millimolar concentrations, but threaten osmotic equilibrium at higher concentrations.

Because plant tissues (as a general rule) contain low concentrations of sodium, many terrestrial herbivores (and to a lesser extent, omnivores) are chronically sodium deficient. To address this dietary constraint, herbivores and omnivores appear to have evolved several physiological adaptations (for reviews, see Dethier, 1977; Denton, 1982). First, as compared with carnivores, many herbivores constitutively maintain relatively low sodium concentrations in their blood. Second, vertebrate herbivores and omnivores have highly responsive renin-aldosterone systems for scavenging dietary sodium. Third, most herbivores (and some omnivores) have evolved a Na^+-specific appetite, which is enhanced by sodium deprivation. In vertebrates, the basis of this Na^+-specific appetite appears to be the amiloride-sensitive Na^+ channel, as described in detail below.

3.1.1. Invertebrates

Na^+ deprivation causes some insects to exhibit a Na^+-specific appetite (e.g., see Arms et al., 1974). Unfortunately, the mechanistic basis for this phenomenon is unknown. While it is clear that flies can discriminate between Na^+ and non-Na^+ salts (Maes and Bijpost, 1979), flies lack amiloride-sensitive Na^+ channels on their salt-responsive taste cells (Liscia et al., 1997).

Monovalent inorganic cations are the most effective stimulants of salt-responsive taste cells in insects (Dethier, 1976). However, these cations differ in their stimulatory effectiveness (e.g., in the fleshfly, Boettcherisca peregrina, $Li^+ < Na^+ < Cs^+ < Rb^+ < K^+$; den Otter, 1972). den Otter suggested that this order does not vary as a function of ionic size or mobility because each cation interacts differentially with negative groups (probably phosphates) on the surface of the taste cells. However, this hypothesis needs to be evaluated with modern neurobiological techniques.

Salt transduction appears to be accomplished through cation-selective channels in the dendritic membrane (for review, see Dethier, 1976) (Fig. 13.2). At least one species of insect (Drosophila melanogaster) appears to possess two types of cation-selective channels in the same salt-responsive taste cell: one channel accepts only Na^+ while the other accepts Na^+, K^+, and Li^+ (Siddiqi

Na$^+$ Salt

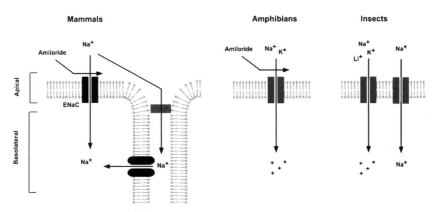

FIG. 13.2. Mechanisms of Na$^+$ salt transduction. Rodents and other herbivores have apically located epithelial Na$^+$ channels (ENaCs) that are blocked by the diuretic drug amiloride; Na$^+$ simply diffuses passively down its concentration gradient into taste cells, resulting in membrane depolarization. Some Na$^+$ also diffuses through tight junctions between taste cells at the apex of the taste bud. This paracellular Na$^+$ enters taste cells through basolateral Na$^+$ channels, but the molecular identity of these channels is unknown. In amphibians, insects, and other mammals (primarily carnivores), Na$^+$ is transduced through Na$^+$ channels that also are permeable to K$^+$. In some but not all vertebrate species, these rather nonselective channels occur primarily on the basolateral membrane, where they are also blockable by amiloride.

et al., 1989). Direct permeation of cations through apical channels is consistent with the observation that pretreatment of salt-responsive taste cells with G protein modulators does not alter their responsiveness to NaCl (Koganezawa and Shimida, 1997).

3.1.2. Vertebrates

The first evidence that amiloride-sensitive Na$^+$ channels are involved in Na$^+$ taste transduction came from the studies of DeSimone and colleagues, who showed that chorda tympani responses to NaCl in rats could be reduced substantially by the diuretic amiloride, which blocks Na$^+$ channels in epithelial cells (Heck et al., 1984). They proposed that NaCl stimulation caused Na$^+$ to flow through apically located amiloride-sensitive Na$^+$ channels, resulting in taste cell depolarization (Fig. 13.2). This seminal finding was confirmed when amiloride-sensitive Na$^+$ currents were identified in whole-cell recordings of isolated taste cells (Avenet and Lindemann, 1988; Gilbertson et al., 1993; Doolin and Gilbertson, 1996; Gilbertson and Zhang, 1998a, 1998b). The amiloride-sensitive Na$^+$ channel in rodent taste cells has properties similar to the well-studied epithelial Na$^+$ channels (ENaC) of colon, kidney, and lung, including inhibition by submicromolar amiloride, apical membrane distribution, high selectivity for Na^+ over K^+, and regulation by hormones such as vasopressin and aldosterone. ENaCs are expressed in rodent taste cells as well as nontaste lingual epithelial cells (Li et al., 1994; Kretz et al., 1999; Lin et al., 1999) and are heterotetrameric channels composed of three distinct homologous subunits (alpha, beta, and gamma; Canessa et al., 1993, 1994). However, the fact that ENaC subunits can be detected in taste cells that are not functionally amiloride-sensitive suggests a complex relationship between ENaC expression and functional amiloride-sensitive Na^+ channels in taste cells.

Additional mechanisms must contribute to Na^+ transduction in mammals because only 80% of the chorda tympani response to NaCl is amiloride-sensitive, and amiloride is ineffective in blocking glossopharyngeal nerve responses to NaCl (Frank, 1991). One of these mechanisms involves the paracellular pathway (Elliott and Simon, 1990; Ye et al., 1991, 1993) (Fig. 13.2). Sodium is thought to permeate the tight junctions at the apex of the taste bud, and enter taste cells through as yet unidentified basolateral channels. This latter mechanism would be amiloride-insensitive because amiloride does not penetrate the tight junctions.

Is ENaC also responsible for Na^+ transduction in carnivorous species, where Na^+ is readily available in the diet? The only mammalian obligatory carnivore that has been studied is the cat, and its salt-best units in the chorda tympani nerve respond equally well to NaCl and KCl (Boudreau et al., 1985). Taste cells in amphibian carnivores (frogs) also appear to express amiloride-blockable Na^+ currents, but these channels have a higher K^+ permeability and a smaller conductance than is typical of ENaC channels (Avenet and Lindemann, 1988, 1990); these channels also appear to be expressed primarily on the basolateral membrane (Kitada and Mitoh, 1998). The relatively high K^+ permeability of Na^+ taste channels in frogs is similar to that of Na^+ taste channels in dogs (Mierson et al., 1988; Nakamura and Kurihara, 1990). The molecular basis for these amiloride-sensitive channels is not known, but it is not likely to be ENaC.

There is clear behavioral evidence that Na^+-specific appetite in rats is mediated by the amiloride-sensitive Na^+ channels (Spector and Grill, 1992; Breslin et al., 1995). However, the story is less clear for humans. Psychophysical studies in humans indicate that even though amiloride reduces the perceived intensity of NaCl applied to the tongue, it does not reduce its perceived "saltiness" (Ossebaard and Smith, 1995, 1996; Ossebaard et al., 1997; see Chapter 15 by Breslin). This indicates that the perception of saltiness is not mediated by amiloride-sensitive Na^+ channels. Instead, Na^+ ions may pass through apical Na^+ channels with low amiloride-sensitivity or Na^+ channels on the basolateral membrane (via the paracellular pathway). It is not known whether human taste cells express ENaC.

3.2. Sweeteners

Many terrestrial plant tissues (e.g., fruits, floral structures, and some leaves) are laden with calorically rich taste stimuli that elicit feeding in virtually all terrestrial animals (Dethier, 1976; Martinez del Rio et al., 1988; Ramirez, 1990). These taste stimuli include sugars, sugar alcohols, D-amino acids, and some proteins. The one exception to this pattern is that domestic and wild cats are indifferent to aqueous solutions of sugar alone (Beauchamp et al., 1977), but will readily drink sucrose solutions in preference to water when they contain low concentrations of electrolytes (Bartoshuk et al., 1971). The strong appetitive response elicited by "sweet" stimuli is thought to represent an adaptation for helping animals identify calorically rich foods. The fact that aquatic vertebrates such as fish and salamanders have a weak (to nonexistent) response to sugars probably reflects the fact that these stimuli are less abundant in aquatic plant tissues.

There are no generally agreed upon molecular features that are essential for a compound to elicit sweet taste. Based on saccharide structures, Shallenberger and Acree (1967) originally proposed that all sweet stimuli contain a hydrogen donor (AH) separated by 0.3 nm from a hydrogen acceptor (B), and that receptor activation was induced by complementary hydrogen bonding between the AH–B sites and the putative "sweetener receptor." This model is now considered to be overly simplistic, and has been expanded to include several hydrophobic bonding sites with the receptor (Walters et al., 1993). The situation is further complicated by the observation that D- and L-glucose are equally sweet to humans, raising the possibility that some sweet stimuli may be detected by means other than a specific receptor.

3.2.1. Invertebrates

Like vertebrates, insects possess taste cells that respond to many naturally occurring saccharides (e.g., glucose, sucrose, and fructose) and sugar alcohols (e.g., sorbitol, mannitol, and myo-inositol) (see review in Dethier, 1976). Whereas flies express transduction mechanisms for both types of sugar in one taste cell (i.e., the sugar-responsive taste cell), most caterpillars express them in two different taste cells (i.e., the sugar-responsive taste cells and sugar alcohol-responsive taste cells) (Glendinning et al., 2000). Owing to this separation of function, caterpillars should be capable of discriminating between sugars and sugar alcohols. Such discrimination may be useful because protein levels correlate with sugar alcohol levels in young leaves (Nelson and Bernays, 1998).

Most studies of "sweet" taste transduction have involved the sugar-responsive taste cell in the labellar taste sensilla of flies (see Fig. 12.1, Chapter 12, Finger and Simon). This taste cell is thought to expresses at least two receptor sites for sugars: one that responds to pyranose sugars (e.g., glucose and sucrose) and another to furanose sugars (e.g., fructose) (for review, see Morita, 1992) (Fig. 13.3). The existence of these separate receptor sites is based on several lines of evidence. Treating the sugar-responsive taste cell with p-chloromercuribenzonate (a specific sulphydryl reagent) almost completely eliminates the response to pyranose sugars, but has no impact on the response to furanose sugars (Shimada et al., 1974). Conversely, treating the sugar-responsive taste cell with amiloride abolishes its response to furanose sugars, but not pyranose sugars (Liscia et al., 1997). Finally, Ozaki and her colleagues used affinity electrophoresis to isolate

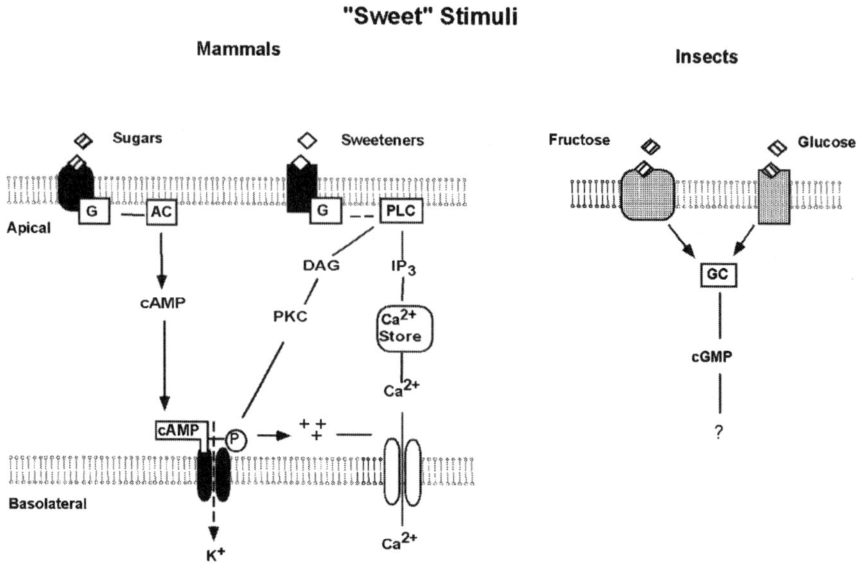

"Sweet" Stimuli

Mammals

Insects

FIG. 13.3. Mechanisms of "sweet" transduction. Sweet tasting stimuli are believe to bind to specific membrane receptors, which in mammals are linked to G proteins. Sugars activate adenylate cyclase (AC), which elevates intracellular cAMP. The cAMP is believed to block a basolateral K^+ conductance directly, leading to membrane depolarization. Synthetic sweeteners and some "sweet" amino acids activate phospholipase C (PLC), leading to the production of inositol trisphosphate (IP_3) and diacylglycerol (DAG). The IP_3 causes release of Ca^{2+} from intracellular stores, while the DAG stimulates protein kinase C (PKC). The PKC phosphorylates and closes the same K^+ channels that are blocked by cAMP. "Sweet" transduction has been studied also in insects, where taste receptor neurons express different receptors for fructose and glucose. Activation of these receptors leads to increases in cGMP, but the downstream target(s) of cGMP have not been identified.

two putative taste receptor molecules; these proteins exhibited sugar binding affinities and specificities that were consistent with separate pyranose and furanose binding site (Ozaki et al., 1993, 1997).

There is growing evidence that both the pyranose and furanose receptors use cGMP, rather than cAMP, as an excitatory intracellular messenger (Fig. 13.3). First, treatment of the sugar-responsive taste cell with inhibitors of protein kinase A and G depress the initial excitatory response to sucrose (Amakawa and Ozaki, 1989). Second, direct stimulation (in situ) with a membrane-permeant cGMP analogue (dibutyryl cGMP) elicits a vigorous excitatory response in the sugar-responsive taste cell, which is comparable to that elicited by sucrose, and much greater than that elicited by dibutyryl cAMP, cGMP, or cAMP (Daley and Vande Berg, 1976; Amakawa et al., 1990). Third, excitatory responses of the sugar-responsive taste cell to mixtures of dibutyryl cGMP and a phosphodiesterase inhibitor (IBMX or theophylline) continue even after stimulation has ended; the duration of this poststimulus response increases with increasing concentrations of the phosphodiesterase inhibitors (Amakawa et al., 1990). Given that phospho-diesterase inhibitors are membrane-permeant, and that elevated intracellular levels of these compounds should slow the rate of dibutyryl cGMP hydrolysis in a concentration-dependent manner, this latter result further bolsters the notion that elevated levels of intracellular cGMP mediate sucrose-induced depolarization of the sugar-responsive taste cell. Fourth, when tissue samples containing thousands of gustatory sensilla from the tarsi of flies are mixed with sucrose (using the quench-flow system), cAMP levels appear to remain at baseline levels during the initial 75 msec of the reaction, which is the time period during which sugar-responsive taste cells normally exhibit their strongest excitatory response (Foster et al., 1998).

3.2.2. Vertebrates

The original model for the transduction of sweet stimuli postulated that the ligands bind to an apical membrane receptor, resulting in a G protein-dependent activation of adenylate cyclase (Fig. 13.3). The increase in cAMP was thought to close basolateral K^+ channels, presumably by activating protein kinase A (PKA). Support for this model came from biochemical studies of membrane preparations containing taste buds from rat, pig, or cow; these studies all reported that sucrose elicited a GTP-dependent increase in cAMP (Striem et al., 1989; Naim et al., 1991; Bernhardt et al., 1996). In addition, electrophysiological studies showed that cAMP caused closure of a resting K^+ current in frog, mouse, rat, and hamster taste cells (Avenet et al., 1988; Behe et al., 1990; Tonosaki and Funakoshi, 1988; Cummings et al., 1996). This model now appears to be essentially correct for the transduction of sucrose in rodents, except recent data suggest that sugar-sensitive K^+ channels may be blocked directly by cAMP, rather than through activation of PKA (Varkevisser and Kinnamon, 1998).

Recent biochemical studies of synthetic sweeteners and D-amino acids have revealed that the cAMP model of "sweet" transduction does not apply for these substances. These two types of sweeteners activate a different pathway, which employs 1,4,5 inositol trisphosphate (IP$_3$) and diacylglycerol (DAG) as second messengers (Bernhardt et al., 1996; Uchida and Sato, 1997a, 1997b) (Fig. 13.3). The complexity of "sweet" transduction is further evidenced by the observation that the IP$_3$ and cAMP pathways co-exist in the same taste cells (Bernhardt et al., 1996; Cummings et al., 1996). Synthetic sweeteners and membrane-permeant cAMP analogues show cross adaptation in patch-clamp studies (Cummings et al., 1996). Recent studies suggest that synthetic sweeteners ultimately activate protein kinase C (PKC) to phosphorylate and close the same K^+ channels that may be directly closed by cAMP (Varkevisser et al., 1997; Varkevisser and Kinnamon, 1998). Since several synthetic sweeteners are amino acid derivatives, it is possible that the natural ligands for synthetic sweetener receptors are sweet amino acids.

Which G proteins are activated by sweet stimuli in rodents? Cyclic AMP is usually produced by activation of adenylate cyclase in response to G_s stimulation.

Diacylglycerol and IP_3, on the other hand, are produced by activation of phospholipase C; several G protein alpha and beta/gamma subunits have been implicated in this activation in other cell types. However, recent experiments also suggest a role for the taste-specific G protein, gustducin, in sweet transduction. Transgenic mice, lacking the gene for gustducin, show reduced preferences for, and afferent nerve responses to, sucrose and synthetic sweeteners (Wong et al., 1996), suggesting that gustducin is involved in the transduction of both types of sweeteners.

Transduction mechanisms for sweet stimuli other than sugars, D-amino acids, and synthetic sweeteners have not been examined in any vertebrates. For example, nothing is known about the transduction of sugar alcohols, even though they: (1) are used extensively as sweeteners by the food and beverage industry, (2) are abundant in many plant tissues, and (3) activate sucrose-best fibers in the chorda tympani nerve of the gerbil, monkey, hamster, and dog (Moskowitz, 1971; Jackinovich and Sugarman, 1988; Harborne and Baxter, 1993).

3.3. Amino Acids

Despite the importance of proteins nutritionally, only a few elicit a distinct taste (e.g., monellin and thaumatin taste sweet to humans). Most animals (including humans), however, can taste the building blocks of proteins: amino acids. In fact, many animals have taste cells that respond strongly and selectively to specific amino acids. While some amino acids taste bitter to humans and inhibit feeding in various animals, many others stimulate feeding, particularly when mixed with other nutrients (Haefeli and Glase, 1990; Mullin et al., 1994). Since free amino acids are often associated with protein-rich foods, animals may have evolved a positive affective response to amino acids because it facilitated the location and identification of protein sources.

One distinctive feature of amino acid-responsive taste cells is that their abundance and sensitivity vary greatly across species. Many aquatic carnivores and detritivores (vertebrate and invertebrate) have large numbers of amino acid-responsive taste cells, and many of these cells are exquisitely sensitive to specific amino acids (Marui and Caprio, 1992). Terrestrial carnivores also have a specialized system for detecting amino acids. For instance, 48% of the geniculate ganglion (facial nerve) taste-responsive neurons in the cat (an obligate carnivore) and 73% in the dog (a facultative carnivore) respond best to amino acids (Boudreau et al., 1985). In contrast, only 9% of the geniculate ganglion taste neurons in the rat (an omnivore) and none in the goat (an herbivore) respond best to amino acids.

3.3.1. Invertebrates

The effects of amino acids on insect taste cells are complex. For example, several investigators have examined the sensory responses to 19 L-amino acids applied to taste cells within the labellar sensilla of flies (Shiraishi and Kuwabara, 1970; Goldrich, 1973). They found that each amino acid elicited one of four classes of response: (1) inhibited all taste cells (aspartic acid, glutamic acid, histidine, arginine, lysine); (2) stimulated the salt-responsive taste cell (proline, hydroxyproline); (3) stimulated the sugar-responsive taste cell (valine, leucine, isoleucine, methionine, phenylalanine, tryptophan); or (4) failed to produce any response (glycine, alanine, serine, threonine, cysteine, tyrosine). These results demonstrate that flies lack taste cells that respond specifically to amino acids. They also show that amino acids constitute a functionally heterogeneous group of taste stimuli for flies.

Caterpillars, on the other hand, have taste cells that respond specifically to amino acids. These amino acid-responsive taste cells exhibit excitatory responses to a large number of amino acids. For example, a single, identified taste cell in two closely related caterpillar species (*Pieris* congeners) can be stimulated by 12 structurally distinct L-amino acids. Moreover, the amino acids that elicited the most vigorous responses are the most nutritionally essential (Dethier, 1973; von Loon and van Eeuwijk, 1989). However, a recent report failed to observe any relationship between stimulatory effectiveness and nutritional importance of amino acids in the spruce budworm (*Choristoneura fumiferana*; Panzuto and Albert, 1998).

Amino acid transduction in the sugar-responsive taste cell of the fly appears to be mediated by at least two molecular receptor sites: one specific for aliphatic, and another for aromatic, amino acids (Shimada and Isono, 1978; Shimada, 1987) (Fig. 13.4). Treating the surface of sugar-responsive taste cells with specific enzymes (e.g., pronase or p-chloromercuribenzonate) differentially abolishes responses to specific chemical classes of amino acids (presumably through differentially altering specific molecular receptor sites). That these putative receptor sites are coupled to a G protein is suggested by the observation that the response of the sugar-responsive cell to an aliphatic and aromatic amino acid is modulated by GTP-γ-S and GDP-β-S (Koganezawa and Shimida, 1997).

The sugar-sensitive taste cell in many species of blood feeding flies also responds to ATP and other noncyclic nucleotides (Furuyama et al., 1999, and references therein). For example, the sugar-sensitive taste cell in the fleshfly (*B. peregrina*) appears to contain at least two classes of receptor sites: one responds to selectively to ADP, and the other to GDP. The location of nucleotide receptor sites on the sugar-sensitive taste cell probably explains why nucleotides stimulate feeding in many species of fly.

3.3.2. Vertebrates

Both ligand-gated channels and G protein-coupled receptors are involved in the transduction of amino acids in vertebrates (Fig. 13.4). These mechanisms have been studied most extensively in the catfish, whose barbels are maximally sensitive to arginine, alanine and proline, but can detect many other free amino acids as well (Marui and Caprio, 1992).

Amino Acids

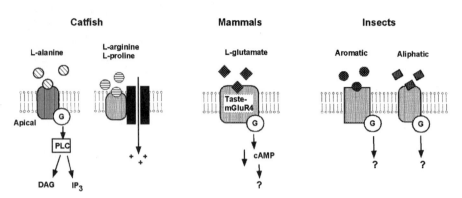

FIG. 13.4. Transduction of amino acids. Taste cells use both G protein-coupled receptors and ligand-gated channels for amino acid transduction. In the catfish, both L-arginine and L-proline utilize ligand-gated cation channels located on the apical membrane. L-alanine binds to G protein-linked receptors that stimulate the PLC pathway. These receptors are located on different subsets of taste cells. In mammals, a taste-specific form of the metabotropic glutamate receptor mGluR4 is believed to transduce glutamate; the receptor causes decreases in cAMP. In insects, there are separate receptor sites for aromatic and aliphatic amino acids; both are believed to be coupled to G proteins, but the second messenger pathways have not been identified.

Transduction of arginine and proline in catfish taste buds involves ligand-gated ion channels. Membrane preparations, originating from barbel epithelia and reconstituted into lipid bilayers, induce the formation of nonselective cation channels, which are selectively gated by arginine (Teeter et al., 1990). Distinct nonselective cation channels, which are gated at higher concentrations by proline, also have been reported (Teeter, 1992). Activation of amino acid-gated channels is thought to generate an inward current, leading to membrane depolarization, calcium entry, and neurotransmitter release. Direct visualization of calcium levels in catfish taste cells has confirmed that calcium concentrations rise in response to arginine (Zviman et al., 1996).

In contrast to arginine and proline, alanine and other short chain, neutral amino acids are transduced via G protein-coupled receptors. These receptors are thought to be coupled to the cAMP and IP$_3$ pathways (Kalinoski et al., 1989b). Histochemical visualization with lectins and antibodies showed that separate taste cells express receptors for alanine or arginine. Taste cells exhibiting the different receptors occur within the same taste bud, demonstrating that although each cell may be tuned to a particular stimulus, taste buds are likely to respond broadly (Finger et al., 1996).

The molecular identity of amino acid receptors in catfish remains elusive in spite of their electrophysiological characterization. High-affinity receptors for L-arginine and L-alanine have been partially purified and appear to be glycoproteins with distinct carbohydrate moieties that exhibit a high specificity for each amino acid (Kalinoski et al., 1987, 1989a, 1992; Grosenenor et al., 1998). One strategy for cloning the cDNAs for such receptors is to express mRNAs from the corresponding tissue in Xenopus oocytes. Indeed, such expression of arginine-activated channels was achieved in an early study (Getchell et al., 1990), but has not yet led to the isolation of cloned cDNAs that could yield more definite answers on transduction mechanisms.

In mammals, the amino acid glutamate (especially as its monosodium salt, MSG) elicits a unique taste termed "umami" (Yamaguchi, 1967). Umami substances are palatable to rodents, carnivores, and humans, especially as components of complex mixtures. They are frequently used to enhance the taste of prepared foods. In behavioral tests and taste nerve recordings, the monophosphates of nucleosides such as guanosine and inosine (GMP and IMP) also generate this same taste quality and potentiate the intensity of glutamate solutions (Kumazawa et al., 1991; Yamaguchi, 1991), a characteristic that is a defining feature of umami taste.

Transduction of glutamate in mammals is thought to involve membrane receptors on the apical surface of taste receptor cells (Fig. 13.4). Molecular modeling, together with electrophysiological measurements on afferent taste fibers, suggested that these receptors resemble the glutamate receptors at central synapses (Faurion, 1991). However, the affinity of glutamate at taste receptors is much lower than that at central synapses: millimolar concentrations are required at taste receptors, and miromolar concentrations at central synapses. Despite this low affinity, biochemical studies demonstrated that radio-labeled glutamate bound to membrane fractions of taste epithelium, and that this binding was enhanced in the presence of GMP and IMP (Torii and Cagan, 1980). The nucleotides caused this enhancement by increasing the number of glutamate-binding sites.

Evidence for glutamate-gated cation channels comes from lipid bilayer studies. Membrane preparations from catfish and mouse taste epithelia induce the formation of glutamate-gated cation channels when incorporated into lipid bilayers (Brand et al., 1991; Teeter et al., 1992). The magnitude of current through these channels increases with glutamate concentration (in the millimolar range), and is further enhanced when GMP is added to glutamate, implying that these channels represent taste receptors. Polymerase chain reaction (PCR) analyses confirmed that several types of glutamate receptors are present in taste epithelia (Chaudhari et al., 1996). These include ionotropic glutamate receptors responsive to NMDA and kainate, and a taste-specific form of one of the metabotropic receptors (mGluR4) that couples to G proteins. Of these, only mGluR4 has been localized unequivocally to taste receptor cells (Chaudhari et al., 1996; Yang et al., 1999). The taste form of mGluR4 is identical to the brain form, except that the n-terminus of the taste form is truncated, resulting in a substantially lower binding affinity for glutamate (Chaudhari et al., 2000). That taste-mGluR4 functions as a taste receptor for glutamate is supported by results of a conditioned-generalization assay (Chaudhari et al., 1996). The rationale for this assay was that when a rodent is conditioned to avoid glutamate, its conditioned aversion should generalize only to taste stimuli that bind to the same receptor as glutamate. The conditioned aversion to glutamate generalized to L-AP4 (a mGluR4 agonist), but not to agonists at inonotropic glutamate receptors. Importantly, the lowest concentrations of glutamate and L-AP4 effective as taste stimuli are also the theshold concentrations necessary to stimulate the truncated taste-mGluR4. Finally, the role of taste-mGluR4 in glutamate transduction is supported by the observation that glutamate decreases cAMP levels in taste cells, an effect expected for activation of taste-mGluR4 (Zhou and Chaudhari, 1997).

Patch-clamp studies of rat taste cells have revealed several types of glutamate-activated currents: an outward current that is mimicked by L-AP4, an inward current mimicked by L-AP4, and an inward current with properties similar to NMDA receptors (Bigiani et al., 1997; Lin and Kinnamon, 1999). Like umami taste, the L-AP4 activated currents can be potentiated by the simultaneous application of GMP (Lin and Kinnamon, 1998). It remains unclear which glutamate receptors are located on the apical membrane, and whether depolarization or hyperpolarization underlies the transduction process. In mouse, direct visualization of membrane potential and cytoplasmic calcium demonstrated that many isolated taste cells depolarize upon glutamate application, leading to calcium increase (Hayashi et al., 1996). Under the same conditions, a few taste cells also hyperpolarize. Whether hyperpolarization in certain cells is an essential component of the code for glutamate taste remains to be determined.

3.4. Acids

In humans, acids elicit the perception of sourness. Although organic acids are generally more sour than inorganic acids at the same pH, the intensity of sourness generally increases with proton concentration. For most vertebrates, acidic solutions are aversive at all detectable concentrations. However, notable exceptions exist. Some species (i.e., deer, sheep, and goats) prefer weak acid solutions over water, but reject more acidic solutions (Bell, 1959; Goatcher and Church 1970; Crawford and Church, 1971; Rice and Church, 1974). One species of frugivorous monkey (*Aotus trivirgatus*) appears to prefer even highly acidic solutions of citric acid (Glaser and Hobi, 1985). In caterpillars, binary mixtures of ascorbic acid (at low concentration) and sucrose strongly stimulate feeding (Ma, 1972).

Organic acids occur in all plant tissues and their concentrations can reach levels over 100 mM (e.g., citric acid from the Krebs cycle; Harborne, 1973). One of the most common and nutritionally significant organic acids, ascorbic acid, occurs in plant tissues at concentrations of 1–10 mM (Schultz and Lechowicz 1986). Since ascorbic acid is an essential vitamin for most herbivorous insects, this observation may explain why caterpillars show appetitive responses to ascorbic acid when it occurs with sugars. The fact that many animals have evolved aversive responses to highly acidic foods makes sense given that highly acidic foods can be toxic (e.g., unripe fruits and spoiled foods) and/or cause excessive depression of gastric pH. In many herbivorous animals, a depression of gastric pH below 4 could destroy the symbiotic organisms that help catabolize cellulose in ingested plant tissues.

Most taste cells respond (in some manner) to acids. This is not surprising because protons modulate ion channel activity (Hille, 1992) and penetrate the paracellular pathway. In general, amphibians utilize apically located channels for acid transduction, while mammals primarily utilize paracellular pathways. There does not appear to be any correlation between food preference and type of transduction mechanism for acid taste, but acid transduction has not been investigated as thoroughly as that of other taste stimuli. The known mechanisms for acid transduction are described below, and are illustrated in Figure. 13.5.

3.4.1. Invertebrates

There is no evidence that insects possess acid-responsive taste cells per se. Instead, acids appear to influence insect taste cells more indirectly: by modulating the responsiveness of taste cells to their respective best stimuli. For example, at low concentrations, acids inhibit the responsiveness of salt-responsive taste

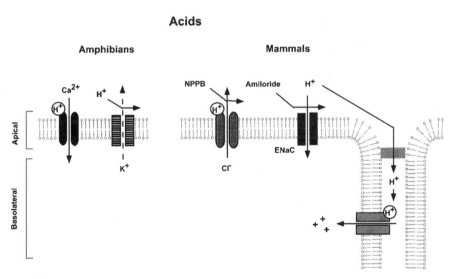

Acids

Amphibians **Mammals**

FIG. 13.5. Mechanisms of acid transduction. Several mechanisms appear to contribute to acid (sour) taste, which is mediated by protons. In amphibians, protons block apically located K^+ channels and also gate apically located Ca^{2+} channels to depolarize taste cells. In mammals, protons gate an apically located Cl^- channel that is sensitive to the drug NPPB, and also permeate apically located amiloride-sensitive Na^+ channels (ENaCs) in some species. The major transduction pathway in mammals is basolateral, since protons readily permeate tight junctions.

cells, whereas at higher concentrations, acids inhibit the responsiveness of sugar- and inositol-responsive taste cells, but stimulate bitter-responsive taste cells (Dethier and Kuch, 1971; Bernays et al., 1998). These modulatory effects of acids are due to pH and not to specific effects of any given acid (Dethier and Kuch, 1971; Bernays et al., 1998).

3.4.2. Vertebrates

Several transduction mechanisms for acids have been described in amphibian taste cells, most involving direct modulation of apical conductances by protons (Fig. 13.5). In the mudpuppy, *Necturus maculosus*, K^+ channels are localized to the apical membrane, where they are exposed directly to protons in the oral cavity (Kinnamon et al., 1988). The apical K^+ conductance consists of a variety of voltage- and Ca^{2+}-activated channels that all have a small open probability at resting potentials (Cummings and Kinnamon, 1992). Potassium efflux through the channel is blocked by protons, thus depolarizing the taste cells. A similar Ca^{2+} activated K^+ conductance is localized to the apical membrane of frog taste cells (Fujiyama et al., 1994), but it is not known if these channels play a role in acid transduction.

Another mechanism for acid transduction in the frog is a proton-gated Ca^{2+} channel, located on the apical membrane (Miyamoto et al., 1988; Okada et al., 1994) (Fig. 13.5). Protons are believed to bind to the channel, allowing Ca^{2+} influx and depolarization. A proton transporter may enhance the response to protons (Okada et al., 1993).

In hamster taste cells, the amiloride-sensitive Na^+ channel is thought to contribute to both acid and Na^+ transduction (Fig. 13.5). In the absence of mucosal Na^+, protons permeate the channel, resulting in taste cell depolarization (Gilbertson et al., 1992; 1993). As would be expected, the response to acids in Na^+-best nerve fibers (Hettinger and Frank, 1990) and brain stem neurons (Boughter and Smith, 1997) is partially inhibited by amiloride, and behavioral experiments suggest that amiloride reduces the aversion to acids in two bottle preference tests (Gilbertson and Gilbertson, 1994). The amiloride-sensitive Na^+ channel does not appear to contribute to acid transduction in rats (DeSimone et al., 1995a), although acids do inhibit responses to NaCl when both stimuli are applied together (Harris et al., 1994).

Another apically located channel that has been shown to participate in the transduction of acids in mouse taste cells is a Cl^- channel. Responses to citric acid in these taste cells reversed at the Cl^- equilibrium potential and were reversibly suppressed by the Cl^- channel blocker NPPB (Miyamoto et al., 1998).

The most important mechanism for transduction of acids in mammalian taste cells appears to be the paracellular pathway, since the chorda tympani response to acids is not influenced by voltage perturbation of the lingual epithelium (DeSimone et al., 1995a; see also Chapter 12 by Finger and Simon). It is thought that protons permeate the tight junctions and interact with sites on the basolateral membrane (Fig. 13.5). One of these sites is likely to be the proton-gated cation channel MDEG1 (also known as BNC1 and BNaC1). Recently, Ugawa et al. (1999) identified this transcript in a taste-specific library and found that antibodies to the channel protein specifically labeled rat taste cells. MDEG1 is also expressed abundantly in the brain and in sensory ganglia, where it mediates transduction of painful acidic stimuli. Rat taste cells express a proton-gated cation conductance (Lin and Kinnamon, 1999), but whether it is mediated by MDEG1 or another proton-gated cation channel remains to be determined. Proton channels and/or pumps are also a potential target of paracellular protons, since pH imaging studies suggest that intracellular pH tracks extracellular pH (Lyall et al., 1997). Another potential target is a newly discovered K^+ conductance that is activated by extracellular K^+ and blocked by H^+; the proton block causes taste cell depolarization (Kolesnikov and Margolskee, 1998). This conductance, identified in frog taste cells, has not yet been found in mammalian taste cells.

3.5. "Bitter" Stimuli

Many different classes of compounds elicit bitter taste in humans and inhibit feeding in animals. These compounds include alkaloids, tannins, methyl-xanthines, peptides, isoprenoids, and many nonsodium salts. Because virtually all naturally occurring poisons taste bitter to humans, this taste quality is thought to have evolved as a mechanism for avoiding toxic substances. This bitterness/toxicity relationship is not always straightforward, however, because many bitter foods are harmless (Glendinning, 1994). Plant tissues in particular contain a diverse range of harmless bitter stimuli (Rouseff, 1990). As a result, animals that include large amounts of plant tissues in their diet (i.e., herbivores and, to a lesser extent, omnivores) appear to have evolved several adaptations for coping with harmless bitter stimuli. One is a reduced overall sensitivity to bitter stimuli; for example, quinine sensitivity (as indicated by two-bottle preference tests) varies systematically with trophic group: carnivores > omnivores > herbivores (Glendinning, 1994). Another adaptation is the ability of animals to habituate selectively to harmless bitter stimuli, while maintaining an aversion to novel (and potentially toxic) ones (Glendinning and Gonzalez, 1995). Finally, there is evidence that several herbivorous and omnivorous animals actually prefer harmless bitter stimuli at low concentrations (the so-called "Schweppes effect"; Glendinning, 1993).

3.5.1. Invertebrates

Many taste stimuli that humans characterize as bitter also rapidly inhibit feeding in insects (e.g., Glendinning et al., 1998, 1999). The mechanistic basis for this rapid feeding inhibition, however, varies across different taxa of insects. For instance, nonherbivorous insects (e.g., flies) lack taste cells that exhibit excitatory responses to bitter stimuli. Instead, bitter stimuli diminish the responsiveness of the sugar-, salt- and water-responsive taste cells to their respective stimuli (Morita and Yamashita, 1959; Dethier and Bowdan, 1992), and in so doing, eliminate most of the stimulatory effect from otherwise preferred foods or fluids. Even though the nature of these inhibitory interactions is unclear, it is known that they are reversible and do not involve competitive or noncompetitive interactions with molecular receptor sites (Dethier and Bowdan, 1989, 1992). From a functional standpoint, this bitter detection system appears to be relatively nondiscriminating because bitter stimuli that differ greatly in toxicity and structure all appear to produce similar effects.

In contrast, herbivorous insects (e.g., caterpillars, grasshoppers, and beetles) have evolved specialized taste cells that exhibit excitatory responses to bitter stimuli (Dethier, 1980; Schoonhoven et al., 1992; Messchendorp et al., 1998). Detailed studies of caterpillars have revealed that their bitter-responsive taste cells usually respond individually to a structurally diverse range of bitter stimuli (Schoonhoven, 1982); this broad molecular receptive range stems in part from the co-occurrence of more than one bitter stimulus transduction pathway within the same taste cell (Glendinning and Hills, 1997). Another feature of caterpillar taste systems is that many of the taste cells responsive to bitter stimuli within the same individual have different molecular receptive ranges (Dethier, 1973; Glendinning et al., 1998, 1999; Waladde et al., 1989; Peterson et al., 1993). This latter feature should permit caterpillars to discriminate between different types of bitter stimuli. Such discrimination would be particularly useful for coping with bitter stimuli that differ in toxicity.

Little is known about how bitter stimuli are transduced by insect taste cells. One promising line of work suggests that a ligand-gated, GABA/glycine chloride channel mediates transduction of several bitter stimuli in a herbivorous beetle (*Diabrotica virgifera*; Mullin et al., 1994) (Fig. 13.6).

3.5.2. Vertebrates

Most bitter stimuli are thought to bind to specific membrane receptors that are coupled to G proteins and second messenger pathways (Fig. 13.6). Phospholipase C (PLC) was the first second messenger to be associated with bitter taste. Calcium imaging experiments showed that denatonium, a synthetic compound bitter at miromolar concentrations, elicited release of Ca^{2+} from intracellular stores in rat taste buds (Akabas et al., 1988). Subsequently, several components of the PLC pathway were identified in taste buds, including the IP_3 receptors (Hwang et al., 1990). Recent biochemical measurements have revealed that several bitter stimuli (denatonium, caffeine, strychnine, sucrose octaacetate, and quinine) elicit IP_3 production in mouse taste tissue, the elevated IP_3 causing release of Ca^{2+} from intracellular Ca^{2+} stores (Spielman et al., 1996; Miwa et al., 1997). Diacylglycerol (DAG) is also produced by activation of PLC, but its function in bitter stimulus transduction has not been examined.

Another enzyme that has been linked to bitter stimulus transduction is phosphodiesterase (PDE). This was first proposed by Price (1973), who showed that several bitter stimuli could activate PDE in biochemical assays. However, this mechanism received little attention until the cloning of gustducin, a

"Bitter" Stimuli

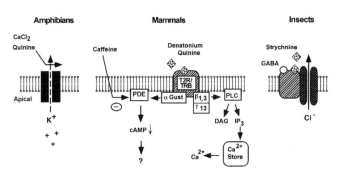

FIG. 13.6. Mechanisms of "bitter" transduction. Bitter stimuli are structurally diverse, and taste cells use several different transduction mechanisms to accommodate them. In amphibians, some "bitter" salts and quinine directly block apically located K^+ channels, the same channels that are also inhibited by protons. In mammals, many bitter stimuli are believed to bind to G protein-coupled receptors that activate both PLC and phosphodiesterase (PDE); the PDE pathway is likely activated by the taste cell-specific G protein gustducin (Gg). Some lipophilic bitter compounds, such as caffeine, may enter taste cells and interact directly with intracellular targets. In insects, some bitter compounds are believed to bind to a GABAergic Cl^- channel.

chemosensory-specific G protein related to the transducins, and expressed in taste cells (McLaughlin et al., 1992). By analogy to phototransduction, it was suspected that gustducin would activate PDE in taste cells, resulting in a decrease in cAMP.

The first direct evidence for a role of gustducin in bitter taste came from biochemical studies, showing that denatonium could activate gustducin (and transducin) in an in vitro assay requiring the presence of taste cell membranes (Ruiz-Avila et al., 1995). More recently, targeted gene replacement was used to generate a null mutation of the gustducin gene, resulting in the gustducin knockout mouse. These mice have reduced responses to several bitter stimuli, suggesting that gustducin is involved in the transduction of bitter stimuli. A potential target for the reduced cAMP is a recently discovered cation channel, which is blocked directly by cyclic nucleotides (Fig. 13.6). Reduction in cyclic nucleotides would relieve block of this channel, resulting in taste cell depolarization (Kolesnikov and Margolskee, 1995; Tsunenari et al., 1996). Since many of the same compounds that activate gustducin also elicit increases in IP_3, the specific role of gustducin in bitter stimulus transduction is unclear. This apparent conundrum was recently resolved by the demonstration (Huang et al., 1999) that many gustducin-positive cells also contain a taste-specific G-gamma subunit, termed $G\gamma13$. This subunit, in combination with $G\beta1$ or $G\beta3$, stimulates phospholipase C activity, independently of the $G\alpha$ subunit. Antibodies to $G\gamma13$ blocked the denatonium-stimulated increase in IP_3, suggesting that $G\gamma13$ mediates the IP_3 response.

Through computerized searches of human and mouse genomic sequence databases, a novel and heterogeneous family of G protein coupled receptors (termed T2Rs or TRBs) was recently identified (Adler et al. 2000; Matsunami, 2000). Several of these receptor genes are clustered at genetic loci that underlie sensitivity to bitter stimuli in humans and mice. Importantly, three of these cloned receptors, when expressed in cultured cells, respond to compounds such as cycloheximide, denatonium and PROP, which generate aversive responses in mice and are bitter tasting to humans. The receptors are surprisingly narrowly tuned, ie. they appear to respond to only one type of stimulus each (Chandrashekhar et al. 2000). Furthermore, they are expressed exclusively in taste cells that also contain the G protein "α-gustducin, suggesting that T2Rs are the gustducin-linked bitter receptors."

Some bitter stimuli do not require G protein-coupled receptors for transduction. These include ionic stimuli, as well as some membrane-permeant and amphiphilic compounds. In amphibians, divalent salts and quinine block apically located K^+ channels, resulting in taste cell depolarization, while K^+ salts permeate the channels, depolarizing taste cells (Kinnamon et al., 1988; Teeter et al., 1989; Cummings and Kinnamon, 1992). In mammals, which appear to lack potassium channels in the apical membrane of taste cells, at least in fungiform taste buds (Ye et al., 1994; Furue and Yoshii, 1997), basolateral K^+ channels could contribute to the transduction of taste stimuli that are either membrane permeant (e.g., quinine) or able to penetrate tight junctions at the apex of the taste bud (e.g., KCl) (Ye et al., 1994). In *Necturus*, quinine enters taste cells and blocks K^+ channels from within (Cummings and Kinnamon, 1992).

Some amphiphilic and membrane permeant bitter stimuli may act independently of membrane receptors, either by changing the phase-boundary potential at the outer surface of the membrane (Kumazawa et al., 1985), by direct activation of G proteins (Naim et al., 1994), or by direct interaction with intracellular proteins, such as enzymes or Ca^{2+} stores. Recent biochemical experiments (Rosenzweig et al., 1999) show that both caffeine and theophylline elicit increases in cGMP. It is presumed that these membrane-permeant bitter stimuli directly inhibit a PDE, causing cGMP levels to increase (Fig. 13.6). The cGMP in turn may open cyclic nucleotide-gated cation channels recently discovered in taste cell membranes (Misaka et al., 1997). In mudpuppy taste cells, the membrane-permeant bitter stimulus dextromethorphan releases Ca^{2+} from intracellular stores by an IP_3-independent mechanism, presumably by entering taste cells and interacting directly with the Ca^{2+} stores (Ogura and Kinnamon, 1999).

4. MODULATION OF TASTE RESPONSES

4.1. Invertebrates

The responses of insect taste cells to binary mixtures of sapid stimuli cannot always be predicted based on the responses to each component alone. Several types of mixture interactions have been discovered (Dethier, 1971; Chapman et al., 1991; Schoonhoven et al., 1992; Shields and Mitchell, 1995). One type involves inhibitory interactions (asymmetrical and symmetrical) between different taste cells within the same sensillum. For the asymmetrical interactions, activation of one taste cell suppresses the responsiveness of another; for the symmetrical interactions, two taste cells mutually suppress each other's responsiveness. The second type of mixture interaction between taste cells is synergistic. For instance, when myo-inositol and sucrose are presented to caterpillars at low concentrations, each of the inositol- and sugar-responsive taste cells respond weakly; however, when the same stimuli are presented at high concentrations, one (as-yet unidentified) taste cell responds vigorously. A final type of mixture interaction occurs within a single taste cell; for example, bitter stimuli can interact directly with the sugar-, salt-, and water-responsive taste cells of flies and reduce their responsiveness.

The mechanistic basis for these peripheral mixture interactions is unclear. Some may involve direct electrical communication between taste cells via gap junctions. In support of this suggestion, two studies have found what appear to be gap junctions between taste cells in flies (Isidoro et al., 1993) and thermo/hygro cells in a moth (Steinbrecht, 1989). However, another study failed to observe any evidence for gap junctions between taste cells in a caterpillar (Shields, 1994).

In addition to these mixture effects, nutrients (e.g., amino acids) in the blood can directly modulate the responsiveness of taste cells in grasshoppers and caterpillars (Simmonds et al., 1992; Simpson and Simpson, 1992). For example, when blood levels of amino acids are low and sugars high, the responsiveness of the taste cells to amino acids is high and that to sugars low. In contrast, when blood levels of amino acids are high and sugars low, then the responsiveness of the taste cells to amino acids is low and that to sugars is high. These modulatory effects of blood nutrients on taste cell responsiveness are mediated peripherally, as they occur in the absence of neural or humoral links between the CNS and taste cells.

4.2. Vertebrates

The responsiveness of vertebrate taste cells to taste stimuli is not fixed. In fact, their responsiveness to taste stimuli can be modulated dramatically by interactions with other chemical substances in foods and by chemical or electrical interactions with other taste cells within the same taste bud (see also Chapters 12, Finger and Simon; Chapter 16, Breslin). Below, we discuss each class of modulation separately.

4.2.1. Direct Effects of Chemical Stimuli on Taste Cell Responsiveness

One of the most robust modulatory effects is produced by cis-polyunsaturated fatty acids, which directly inhibit a class of delayed rectifier K^+ channels in taste cells (Gilbertson et al., 1997). The effect is specific to fatty acids with two or more double bonds in the cis configuration; there is no effect of other fatty acids on taste cells. Ingested fats are hypothesized to be broken down by salivary lipases in the oral cavity, causing release of free fatty acids into the oral cavity. The highly mobile, membrane permeant fatty acids permeate tight junctions and taste cell membranes to interact directly with basolateral K^+ channels. The net effect is taste cell depolarization accompanied by a protracted response to all depolarizing taste stimuli. Recent molecular studies suggest that the target of fatty acid modulation is a Shaker Kv1.5-like K^+ channel, expressed in most taste cells (Kim et al., 1998; Liu et al., 1998). These data indicate that, in addition to the pleasurable "mouth feel" of dietary fat, fatty acids may be detected directly by taste cells. Some taste cells in rat appear to express a fatty acid-transport (FAT) protein in the membrane, which is involved in binding or transporting fatty acids into the cytoplasm (Fukuwatari et al., 1997). Potentially, such a transporter could enhance the fatty acid responsiveness of taste cells.

Another substance that modulates taste cell responsiveness is the citrate anion. Behavioral experiments have shown that citrate enhances preference for amino acids and sweet stimuli, but not for bitter stimuli, in rats. These behavioral changes are correlated with a slight taste cell depolarization together with a prolonged response to sweet stimuli and amino acids (Gilbertson et al., 1997). A recent study found that citrate also enhances responses to amino acids in the glossopharyngeal taste system in the largemouth bass, *Micropterus salmoides* (Ogawa and Caprio, 1999).

Some taste stimuli, when present in a mixture, inhibit the response to other taste stimuli in the mixture. Such is the case for quinine, which inhibits responses of catfish facial nerve fibers to amino acids (Ogawa et al., 1997), and hamster chorda tympani nerve fibers to sucrose (Formaker and Frank, 1996). Despite earlier reports that mixture suppression by quinine occurs as a result of inhibitory interactions among neurons in the brainstem (see Chapter 14, Smith and Davis), the above cited studies establish that mixture suppression also occurs in the periphery. Mixture suppression has not been investigated at the level of single taste cells. Nevertheless, given that quinine can both permeate taste cell membranes and block voltage-gated Na^+, Ca^{2+}, and K^+ currents directly (Kinnamon and Roper, 1988; Cummings and Kinnamon, 1992; Chen and Herness, 1997), it follows that prolonged stimulation with quinine could suppress the voltage-gated currents and overall excitability of taste cells. Quinine may have other modulatory effects as well, such as antagonizing interactions between cell-surface receptors for sweet stimuli or intracellular receptors for sweet-stimulated second messengers. Further studies on isolated taste cells will be required to understand the modulatory effects of quinine on peripheral taste responses.

Protons and K^+ ions can also modulate the amiloride-sensitive Na^+ responses of taste cells (Gilbertson et al., 1992, 1993; Stewart et al., 1996). It is believed that protons and K^+ bind in the channel pore and inhibit Na^+ flux through the channel.

Finally, recent studies suggest that NaCl inhibits the gustatory response to several bitter stimuli. This phenomenon has been observed psychophysically in humans (Breslin and Beauchamp, 1995) and in hamster chorda tympani nerve recordings (Formaker and Frank, 1996). The basis for the suppression of bitterness by NaCl is unknown.

4.2.2. Signal Processing Within Taste Buds

Several lines of evidence indicate that chemical and electrical interactions occur among cells within a taste bud, and that these interactions may alter the subsequent responsiveness of individual taste cells. For example, taste cells in amphibians and fish are modulated by the neurotransmitter serotonin, which is contained in a class of basal cells that in some ways resemble Merkel cells of the skin (see Chapter 12 by Finger and Simon). Taste stimuli elicit release of serotonin, which in turn modulates the voltage-activated Ca^{2+} current in the taste cells (Delay et al., 1997). Serotonin also has been identified in mammalian taste buds (Kim and Roper, 1995). For example, serotonin modulates a Ca^{2+}-dependent K^+ conductance in rat taste cells (Herness and Chen, 1997), but it is not known whether it is released in response to taste stimulation.

In addition to being modulated by neurotransmitters, taste cell responses may be influenced by gap junctions between small groups of taste cells (Yang and Roper, 1987). Although the functional significance of such intercellular coupling is not known, the gap junctional conductance is decreased by intracellular acidification (Bigiani and Roper, 1994). Thus, intercellular interactions may be modulated during acid taste transduction.

5. CONCLUSION

We have attempted to provide a comprehensive overview of taste transduction in the animal world. The resonant message that emerges from this review is that taste cells can respond to a large number of biologically relevant chemicals, and that they do so with a diversity of transduction mechanisms. While our understanding of taste transduction mechanisms is still incomplete, numerous advances have been made over the last decade. We expect that many of the recent findings (e.g. the cloning of G protein-coupled receptors and components of transduction cascades) and technical breakthroughs (e.g., the ability to knock out single genes) will only accelerate this process of discovery.

At this point, it is difficult to determine whether insects and vertebrates share common molecular, biochemical, and physiological mechanisms for responding to taste stimuli. This is because most studies of each taxon have addressed different aspects of taste cell function: vertebrate studies have focused primarily on the nature of the transduction mechanisms themselves (i.e., the cell-surface receptors and second messenger systems), whereas insect studies have focused primarily on the response properties of individual taste cells. This difference in focus stems in large part from methodological limitations that are specific to each taxon, which constrain the types of questions one can address. Nevertheless, among the few transduction systems that have been studied intensively in both taxa (i.e., transduction of sweet and salty stimuli), some common underlying mechanisms exist.

Finally, we have attempted to relate the response properties of the peripheral taste system to the feeding habits of different animal species. Even though the picture is still fragmentary, there are some compelling patterns among vertebrates. For example, many carnivores have evolved taste systems that exhibit a high sensitivity to amino acids, a low sensitivity to sugars, with a limited ability to discriminate between Na^+ and K^+ salts. This pattern of response is well suited for identifying animal tissues, which are high in amino acids and sodium but low in carbohydrates. Many omnivores, in contrast, have a relatively low sensitivity to amino acids, a high sensitivity to sugars, and the ability to discriminate between Na^+ and K^+ salts. Taken together, these results provide support for the hypothesis that species with different feeding habits have each evolved a distinct collection of taste transduction pathways for meeting their particular nutritional requirements.

REFERENCES

Adler, E., M. A. Hoon, K. L. Mueller, J. Chandrashekar, N. J. Ryba, and C. S. Zuker (2000). A novel family of mammalian taste receptors. Cell 100: 693–702.

Akabas, M. H., J. Dodd, and Q. Al-Awqati (1988). A bitter substance induces a rise in intracellular calcium in a subpopulation of rat taste cells. Science 242:1047–1050.

Amakawa, T. and M. Ozaki (1989). Protein kinase C-promoted adaptation of the sugar receptor cell in teh blowfly Phormia regina. J Insect Physiol 35:233–237.

Amakawa, T., M. Ozaki, and K. Kawata (1990). Effects of cyclic GMP on the sugar taste receptor of the fly Phormia regina. J Insect Physiol 36:281–286.

Arms, K., P. Feeny, and R. C. Lederhouse (1974). Sodium: stimulus for puddling behavior by tiger swallowtail bitterflies, Papilio glaucus. Science 185:372–374.

Avenet, P. and B. Lindemann (1988). Amiloride-blockable sodium currents in isolated taste receptor cells. J Membr Biol 105:245–255.

Avenet, P. and B. Lindemann (1990). Fluctuation analysis of amiloride-blockable currents in membrane patches excised from salt-taste receptor cells. J Basic Clin Physiol Pharmacol 1:383–391.

Avenet, P., F. Hofmann, and B. Lindemann (1988). Transduction in taste receptor cells requires cAMP-dependent protein kinase. Nature 331:351–354.

Bartoshuk, L. M., M. A. Harned, and L. H. Parks (1971). Taste of water in the cat: effects on sucrose preference. Science 171:699–701.

Beauchamp, G. K., O. Maller, and J. G. Rogers (1977). Flavor preferences in cats (Felis catus and Panthera sp.). J Comp Physiol Psychol 91:1118–1127.

Behe, P., J. A. DeSimone, P. Avenet, and B. Lindemann (1990). Membrane currents in taste cells of the rat fungiform papilla. Evidence for two types of Ca currents and inhibition of K currents by saccharin. J Gen Physiol 96:1061–1084.

Behrman, H. R. and M. R. Kare (1968). Canine pancreatic secretion in response to acceptable and aversive taste stimuli. *Proc Soc Exp Biol (NY)* **129**:343–346.

Bell, F. R. (1959). Preference thresholds for taste discrimination in goats. *J Agricul Sci* **52**: 125–128.

Bernays, E. A., J. I. Glendinning, and R. F. Chapman (1998). Plant acids modulate chemosensory responses in *Manduca sexta* larvae. *Physiol Entomol* **23**:193–201.

Bernhardt, S. J., M. Naim, U. Zehavi, and B. Lindemann (1996). Changes in IP_3 and cytosolic Ca^{2+} in response to sugars and non-sugar sweeteners in transduction of sweet taste in the rat. *J Physiol (Lond)* **90**:325–336.

Bigiani, A. and S. D. Roper (1993). Identification of electrophysiologically distinct cell subpopulations in *Necturus* taste buds. *J Gen Physiol* **102**:143–170.

Bigiani, A. and S. D. Roper (1994). Reduction of electrical coupling between *Necturus* taste receptor cells, a possible role in acid taste. *Neurosci Lett* **176**:212–216.

Bigiani, A., R. J. Delay, N. Chaudhari, S. C. Kinnamon, and S. D. Roper (1997). Responses to glutamate in rat taste cells. *J Neurophysiol* **77**:3048–3059.

Boudreau, J. C., L. Sivakumar, L. T. Do, T. D. White, J. Oravec, and N. K. Hoang (1985). Neurophysiology of geniculate ganglion (facial nerve) taste systems: species comparisons. *Chem Senses* **10**:89–127.

Boughter, J. D. J. and D. V. Smith (1997). Amiloride suppresses the responses to acids in NaCl-best neurons of the hamster solitary nucleus. *Chem Senses* **22**:648.

Brand, J. G., J. H. Teeter, T. Kumazawa, T. Huque, and D. L. Bayley (1991). Transduction mechanisms for the taste of amino acids. *Physiol Behav* **49**:899–904.

Breslin, P. A. and G. K. Beauchamp (1995). Suppression of bitterness by sodium: variation among bitter taste stimuli. *Chem Senses* **20**:609–623.

Breslin, P. A. S., A. C. Spector, and H. J. Grill (1995). Sodium specificity of salt appetite in Fisher-344 and Wistar rats in impaired by chorda tympani nerve transection. *Am J Physiol* **269**:R350–R356.

Canessa, C. M., J. D. Horisberger, and B. C. Rossier (1993). Epithelial sodium channel related to proteins involved in neurodegeneration [see comments]. *Nature* **361**: 467–470.

Canessa, C. M., A. M. Merillat, and B. C. Rossier (1994). Membrane topology of the epithelial sodium channel in intact cells. *Am J Physiol* **267**:C1682–1690.

Chandrashekar, J., K. L. Mueller, M. A. Hoon, E. Adler, L. Feng, W. Guo, C. S. Zuker, and N. J. Ryba (2000). $T2R_S$ function as bitter taste receptors. *Cell* **100**:703–711.

Chapman, R. F., A. Ascoli-Christensen, and P. R. White (1991). Sensory coding for feeding deterrence in the grasshopper *Schistocerca americana*. *J Exper Biol* **158**:241–259.

Chaudhari, N., H. Yang, C. Lamp, E. Delay, C. Cartford, T. Than, and S. Roper (1996). The taste of monosodium glutamate: membrane receptors in taste buds. *J Neurosci* **16**: 3817–3826.

Chaudhari, N., A. M. Landin, and S. D. Roper (2000). A metabotropic glutamate receptor variant functions as a taste receptor. *Nat Neurosci* **3**:113–119.

Chen, Y. and M. S. Herness (1997). Electrophysiological actions of quinine on voltage-dependent currents in dissociated rat taste cells. *Pflugers Arch* **434**:215–26.

Clyne, P. J., C. G. Warr, and J. R. Carlson (2000). Candidate taste receptors in *Drosophila*. *Science* **287**:1830–1834

Crawford, J. C. and D. C. Church (1971). Responses of black-tailed deer to various chemical taste stimuli. *J Wildl Manag* **35**:210–215.

Cummings, T. A. and S. C. Kinnamon (1992). Apical K+ channels in *Necturus* taste cells. Modulation by intracellular factors and taste stimuli. *J Gen Physiol* **99**:591–613.

Cummings, T. A. and S. C. Kinnamon (1995). Patch-clamping of taste cells in mudpuppy. In: A. I. Spielman and J. G. Brand, (ed). *Experimental Cell Biology of Taste and Olfaction*. Boca Raton, FL: CRC Press, pp 299–308.

Cummings, T. A., J. Powell, and S. C. Kinnamon (1993). Sweet taste transduction in hamster taste cells: evidence for the role of cyclic nucleotides. *J Neurophysiol* **70**:2326–2336.

Cummings, T. A., C. Daniels, and S. C. Kinnamon (1996). Sweet taste transduction in hamster: sweeteners and cyclic nucleotides depolarize taste cells by reducing a K+ current. *J Neurophysiol* **75**:1256–1263.

Daley, D. L. and J. S. Vande Berg (1976). Apparent opposing effects of cyclic AMP and dibutyryl-cyclic GMP on the neuronal firing of the blowfly chemoreceptors. *Biochin Biophys Acta* **437**:211–220.

Delay, R. J., R. Taylor, and S. D. Roper (1993). Merkel-like basal cells in *Necturus* taste buds contain serotonin. *J Comp Neurol* **335**:606–613.

Delay, R. J., A. Mackay-Sim, and S. D. Roper (1994). Membrane properties of two types of basal cells in *Necturus* taste buds. *J Neurosci* **14**:6132–6143.

Delay, R. J., S. C. Kinnamon, and S. D. Roper (1997). Serotonin modulates voltage-dependent calcium current in *Necturus* taste cells. *J Neurophysiol* **77**:2515–2524.

den Otter, C. J. (1972). Differential sensitivity of insect chemoreceptors to alkalai cations. *J Insect Physiol* **8**:109–131.

Denton, D. A. (1982). The Hunger for Salt. New York: Springer-Verlag.

DeSimone, J. A., E. M. Callaham, and G. L. Heck (1995a). Chorda tympani taste response of rat to hydrochloric acid subject to voltage-clamped lingual receptive field. *Am J Physiol* **268**:C1295–1300.

DeSimone, J. A., G. L. Heck, and Q. Ye (1995b). Taste nerve recording from a voltage-clamped receptive field. In: A. I. Spielman and J. G. Brand (ed). *Experimental Cell Biology of Taste and Olfaction*. Boca Raton, FL: CRC Press, pp 227–234.

Dethier, V. G. (1968). Chemosensory input and taste discrimination in the blowfly. *Science* **161**:389–391.

Dethier, V. G. (1971). A surfeit of stimuli: a paucity of receptors. *Am Sci* **59**:706–715.

Dethier, V. G. (1973). Electrophysiological studies of gustation in Lepidopterous larvae. II. Taste spectra in relation to food-plant discrimination. *J Comp Physiol* **82**:103–134.

Dethier, V. G. (1976). *The Hungry Fly: A Physiological Study of the Behavior Associated with Feeding*. Cambridge, MA: Harvard University Press.

Dethier, V. G. (1977). The taste of salt. *Am Sci* **65**:744–751.

Dethier, V. G. (1980). Evolution of receptor sensitivity to secondary plant substances with special reference to deterrents. *Am Nat* **115**:45–66.

Dethier, V. G. (1993). The role of taste in food intake: a comparative view. In: S. A. Simon and S. D. Roper (eds). *Mechanisms of Taste Transduction*. Boca Raton, FL: CRC Press, pp 3–25.

Dethier, V. G. and E. Bowdan (1989). The effect of alkaloids on the sugar receptors of the blowfly. *Physiol Ent* **14**:127–136.

Dethier, V. G. and E. Bowdan (1992). Effects of alkaloids on feeding by *Phormia regina* confirm the critical role of sensory inhibition. *Physiol Entomol* **17**:325–330.

Dethier, V. G. and J. H. Kuch (1971). Electrophsyiological studies of gustation in lepidopterous larvae. I. Comparative sensitivity to sugars, amino acids, and glycosides. *Z Vergl Physiol* **72**:343–363.

Doolin, R. E. and T. A. Gilbertson (1996). Distribution and characterization of functional amiloride-sensitive sodium channels in rat tongue. *J Gen Physiol* **107**:545–554.

Elliott, E. J. and S. A. Simon (1990). The anion in salt taste: a possible role for paracellular pathways. *Brain Res* **535**:9–17.

Faurion, A. (1991). Are umami taste receptor sites structurally related to glutamate CNS receptor sites? *Physiol Behav* **49**:905–912.

Finger, T. E., B. P. Bryant, D. L. Kalinoski, J. H. Teeter, B. Böttger, W. Grosvenor, R. H. Cagan, J. G. Brand, and B. Bottger (1996). Differential localization of putative amino acid receptors in taste buds of the channel catfish, *Ictalurus punctatus. J Comp Neurol* **373**:129–138.

Formaker, B. K., and M. E. Frank (1996). Responses of the hamster chorda tympani nerve to binary component taste stimuli: evidence for peripheral gustatory mixture interactions. *Brain Res* **727**:79–90.

Foster, K. D., A. I. Spielman, and L. M. Kennedy (1998). Rapid kinetics of receptor cell firing and second messenger formation in response to sucrose. *Chem Senses* **23**:549 (abstract).

Frank, M. E. (1991). Taste-responsive neurons of the glossopharyngeal nerve of the rat. *J Neurophysiol* **65**:1452–1463.

Frazier, J. L. and F. E. Hanson (1986). Electrophysiological recording and analysis of insect chemosensory responses. In: J. R. Miller and T. A. Miller (eds). *Insect–Plant Interactions.* New York: Springer-Verlag, pp 285–330.

Fujiyama, R., T. Miyamoto, and T. Sato (1994). Differential distribution of two Ca(2+)-dependent and -independent K+ channels throughout receptive and basolateral membranes of bullfrog taste cells. *Pflugers Arch* **429**:285–290.

Fukuwatari, T., T. Kawada, M. Tsuruta, T. Hiraoka, T. Iwanaga, E. Sugimoto, and T. Fushiki (1997). Expression of the putative membrane fatty acid transporter (FAT) in taste buds of the circumvallate papillae in rats. *FEBS Lett* **414**:461–464.

Furue, H. and K. Yoshii (1997). In situ tight-seal recordings of taste substance-elicited action currents and voltage-gated Ba currents from single taste bud cells in the peeled epithelium of mouse tongue. *Brain Res* **776**:133–139.

Furuyama, A., M. Koganezawa, and I. Shimida (1999). Multiple receptor sites for nucleotide reception in the labellar taste receptor cells of the fleshfly *Boettcherisca peregrina. J Insect Physiol* **45**:249–255.

Getchell, T. V., M. Grill., S. S. Tate, R. Urade, J. H. Teeter, and F. L. Margolis (1990). Expression of catfish amino acid taste receptors in *Xenopus* oocytes. *Neurochem Res* **15**:449–456.

Gilbertson, D. M. and T. A. Gilbertson (1994). Amiloride reduces the aversiveness of acids in preference tests. *Physiol Behav* **56**:649–654.

Gilbertson, D. M., W. T. Monroe, J. R. Milliet, J. Caprio, and T. A. Gilbertson (1997). Citrate ions enhance behavioral and cellular responses to taste stimuli. *Physiol Behav* **62**: 491–500.

Gilbertson, T. A. (1995). Patch-clamping of taste cells in hamster and rat. In: A. I. Spielman and J. G. Brand (ed). *Experimental Cell Biology of Taste and Olfaction.* Boca Raton, FL: CRC Press, pp 317–328.

Gilbertson, T. A. and H. Zhang (1998a). Characterization of sodium transport in gustatory epithelia from the hamster and rat. *Chem Senses* **23**:283–293.

Gilbertson, T. A. and H. Zhang (1998b). Self-inhibition in amiloride-sensitive sodium channels in taste receptor cells. *J Gen Physiol* **111**:667–677.

Gilbertson, T. A., P. Avenet, S. C. Kinnamon, and S. D. Roper (1992). Proton currents through amiloride-sensitive channels in hamster taste cells: role in acid transduction. *J Gen Physiol* **100**:803–824.

Gilbertson, T. A., S. D. Roper, and S. C. Kinnamon (1993). Proton currents through amiloride-sensitive Na+ channels in isolated hamster taste cells: enhancement by vasopressin and cAMP. *Neuron* **10**:931–942.

Gillary, H. L. (1966). Stimulation of the salt receptor of the blowfly. I. NaCl. *J Gen Physiol* **50**:337–350.

Glaser, D. and G. Hobi (1985). Taste responses in primates to citric and acetic acid. *Int J Primatol* **6**:395–398.

Glendinning, J. I. (1993). Preference and aversion for deterrent chemicals in two species of *Peromyscus* mouse. *Physiol Behav* **54**:141–150.

Glendinning, J. I. (1994). Is the bitter rejection response always adaptive? *Physiol Behav* **56**:1217–1227.

Glendinning, J. I. and N. A. Gonzalez (1995). Gustatory habituation to deterrent compounds in a grasshopper: concentration and compound specificity. *Anim Behav* **50**:915–927.

Glendinning, J. I. and T. T. Hills (1997). Electrophysiological evidence for two transduction pathways within a bitter-sensitive taste receptor. *J Neurophysiol* **78**:734–745.

Glendinning, J. I., S. Valcic, and B. N. Timmermann (1998). Maxillary palps can mediate taste rejection of plant allelochemicals by caterpillars. *J Comp Physiol A* **183**:35–44.

Glendinning, J. I., M. Tarre, and K. Asaoka (1999). Contribution of different bitter-sensitive taste cells to feeding inhibition in a caterpillar. *Behav Neurosci* **113**:840–854.

Glendinning, J. I., N. Nelson, E. A. Bernays (2000). How does inositol and glucose modulate feeding in *Manduca sexta* caterpillars. *J Exper Biol* **199**:1522–1534.

Goatcher, W. D. and D. C. Church (1970). Taste responses in ruminants. II. Reactions of sheep to acids, quinine, urea and sodium hydroxide. *J Anim Sci* **30**:784–790.

Goldrich, N. R. (1973). Behavioral responses of *Phormia regina* (Meigen) to labellar stimulation with amino acids. *J Gen Physiol* **61**:74–88.

Gothilf, S. and F. E. Hanson (1994). A technique for electrophysiologically recording from chemosensory organs of intact caterpillars. *Entomol Exper Appl* **72**:304–310.

Grosvenor, W., A. M. Feigin, A. I. Spielman, T. E. Finger, M. R. Wood, A. Hansen, K. L. Kalinoski, J. H. Teeter, and J. G. Brand (1998). The arginine taste receptor: physiology, biochemistry and immunocytochemistry. *Ann N Y Acad Sci* **855**: 134–142.

Haefeli, R. J. and D. Glase (1990). Taste responses and threshold obtained with the primary amino acids in humans. *Food Sci Technol* **23**:523–527.

Harborne, J. B. (1973). *Phytochemical Methods*. London: Chapman and Hall.

Harborne, J. B. and H. Baxter (1993). *Phytochemical Dictionary: A Handbook of Bioactive Compounds from Plants*. Washington, DC: Taylor and Francis, 791 pp.

Harris, D. E., D. M. Gilbertson, W. T. Monroe, S. C. Kinnamon, and T. A. Gilbertson (1994). Contribution of amiloride-sensitive pathways to acid transduction in rats. *Chem Senses* **19**:481–482.

Hayashi, Y., M. M. Zviman, J. G. Brand, J. H. Teeter, and D. Restrepo (1996). Measurement of membrane potential and $[Ca^{2+}]_i$ in cell ensembles: application to the study of glutamate taste in mice. *Biophys J* **71**:1057–1070.

Heck, G. L., S. Mierson, and J. A. DeSimone (1984). Salt taste transduction occurs through an amiloride-sensitive sodium transport pathway. *Science* **223**:403–405.

Herness, S. and Y. Chen (1997). Serotonin inhibits calcium-activated K+ current in rat taste receptor cells. *NeuroReport* **8**:3257–3261.

Hettinger, T. P., and M. E. Frank (1990). Specificity of amiloride inhibition of hamster taste responses. *Brain Res* **513**:24–34.

Hille, B. (1992). *Ionic Channels of Excitable Membranes.* Sunderland, MA: Sinauer Associates.

Hoon, M. A., E. Adler, J. Lindemeier, J. F. Battey, N. J. Ryba, and C. S. Zuker (1999). Putative mammalian taste receptors: a class of taste-specific GPCRs with distinct topographic selectivity. *Cell* **96**:541–551.

Huang, L., Y. G. Shanker, J. Dubauskaite, J. Z. Zheng, W. Yan, S. Rosenzweig, A. I. Spielman, M. Max, and R. F. Margolskee (1999). Gγ13 colocalizes with gustducin in taste receptor cells and mediates IP3 responses to bitter denatonium. *Nat Neurosci* **2**: 1055–1062.

Hwang, P. M., A. Verma, D. S. Bredt, and S. H. Snyder (1990). Localization of phosphatidylinositol signaling components in rat taste cells: role in bitter taste transduction. *Proc Natl Acad Sci U S A* **87**:7395–7399.

Isidoro, N., M. Solinar, R. Baur, P. Roessingh, and E. Städler (1993). Functional morphology of a tarsal sensillum of *Delia radicum* L (Diptera, Anthomyiidae) sensitive to important host-plant compounds. *Int J Insect Morph Embryol* **39**:275–281.

Jackinovich, W. J. and D. Sugarman (1988). Sugar taste reception in mammals. *Chem Senses* **13**:13–31.

Kalinoski, D. L., R. C. Bruch, and J. G. Brand (1987). Differential interaction of lectins with chemosensory receptors. *Brain Res* **418**:24–40.

Kalinoski, D. L., B. P. Bryant, G. Shaulsky, J. G. Brand, and S. Harpaz (1989a). Specific L-arginine taste receptor sites in the catfish, *Ictalurus punctatus*: biochemical and neurophysiological characterization. *Brain Res* **488**:163–173.

Kalinoski, D. L., T. Huque, V. J. LaMorte, and J. G. Brand (1989b). Second messenger events in taste. In: J. G. Brand, J. H. Teeter, R. H. Cagan, and M. R. Kare (eds). *Chemical Senses: Receptor Events and Transduction in Taste and Olfaction.* New York: Marcel Dekker, pp 85–101.

Kalinoski, D. L., L. C. Johnson, B. P. Bryant, and J. G. Brand (1992). Selective interactions of lectins with amino acid taste receptor sites of the channel catfish. *Chem Senses* **7**: 381–390.

Kijima, H., Y. Okada, S. Oiki, S. Goshima, K. Nagata, and T. Kazawa (1995). Free ion concentration in receptor lymph and a role of transepithelia voltage in the fly labellar taste receptor. *J Comp Physiol A* **177**:123–133.

Kim, D. J. and S. D. Roper (1995). Localization of serotonin in taste buds: a comparative study in four vertebrates. *J Comp Neurol* **353**:364–370.

Kim, I., L. Liu, and T. A. Gilbertson (1998). Inhibition of K+ channels by fatty acids may represent a common mechanism for the chemoreception of fat in both pre- and post-ingestive targets. *Chem Senses* **23**:612 (abstract).

Kinnamon, S. C. and S. D. Roper (1988). Membrane properties of isolated mudpuppy taste cells. *J Gen Physiol* **91**:351–371.

Kinnamon, S. C., V. E. Dionne, and K. G. Beam (1988). Apical localization of K+ channels in taste cells provides the basis for sour taste transduction. *Proc Natl Acad Sci U S A* **85**:7023–7027.

Kinnamon, S. C. (2000). A plethora of taste receptors. *Neuron* **25**:507–510.

Kitada, Y. and Y. Mitoh (1998). Amiloride does not affect the taste responses of the frog glossopharyngeal nerve to NaCl. In: *International Symposium on Olfaction and Taste XII.* San Diego: CME, 73 pp.

Kloub, M. A., G. L. Heck, and J. A. DeSimone (1997). Chorda tympani responses under lingual voltage clamp: implications for NH4 salt taste transduction. *J Neurophysiol* **77**:1393–1406.

Koganezawa, M. and I. Shimida (1997). The effects of G protein modulators on the labellar taste receptor cells of the fleshfly (*Boettcherisca pereprina*). *J Insect Physiol* **43**:225–233.

Koganezawa, M., R. Shingai, K. Isono, and I. Shada (1996). Expression of taste reception of fleshfly in *Xenopus* oocytes. *NeuroReport* **7**:2063–2067.

Kolesnikov, S. S. and R. F. Margolskee (1995). A cyclic-nucleotide-suppressible conductance activated by transducin in taste cells. *Nature* **376**:85–88.

Kolesnikov, S. S. and R. F. Margolskee (1998). Extracellular K+ activates a K(+)- and H(+)-permeable conductance in frog taste receptor cells. *J Physiol (Lond)* **507**:415–432.

Kretz, O., P. Barbry, R. Bock, and B. Lindemann (1999). Differential expression of RNA and protein of the three pore-forming subunits of the amiloride-sensitive epithelial sodium channel in taste buds of the rat. *J Histochem Cytochem* **47**:51–64.

Kumazawa, T., M. Kashiwayanagi, and K. Kurihara (1985). Neuroblastoma cell as a model for a taste cell: mechanism of depolarization in response to various bitter substances. *Brain Res* **333**:27–33.

Kumazawa, T., M. Nakamura, and K. Kurihara (1991). Canine taste nerve responses to umami substances. *Physiol Behav* **49**:875–881.

Li, X. J., S. Blackshaw, and S. H. Snyder (1994). Expression and localization of amiloride-sensitive sodium channel indicate a role for non-taste cells in taste perception. *Proc Natl Acad Sci U S A* **91**:1814–1818.

Lin, W. and S. C. Kinnamon (1998). Responses to monosodium glutamate and guanosine 5'-monophosphate in rat fungiform taste cells. *Ann N Y Acad Sci* **855**:407–411.

Lin, W. and S. C. Kinnamon (1999). Proton-activated currents in taste cells of rat vallate papilla. *Chem Senses* **24**: 570.

Lin, W., T. E. Finger, B. C. Rossier, and S. C. Kinnamon (1999). Immunolocalization of epithelial sodium channel subunits in rat tongue: possible transduction pathways for Na^+ salt. *J Comp Neurol* **405**:406–420.

Lindemann, B. (1995). Loose-patch recording from taste buds in situ. In: A. I. Spielman and J. G. Brand (eds). *Experimental Cell Biology of Taste and Olfaction.* Boca Raton, FL: CRC Press, pp 333–340.

Liscia, A., P. Solari, R. Majone, I. Tomassini Barbarossa, and R. Crnjar (1997). Taste reception mechanisms in the blowfly: evidence of amiloride sensitive and insensitive receptor sites. *Physiol Behav* **62**:875–879.

Liu, L., I. Kim, S. Hu, H. Wang, H. Zhang, and T. A. Gilbertson (1998). Identification of a Shaker Kv1.5-like K+ channel in taste cells: the primary target for fatty acid inhibition. *Chem Senses* **23**:612(abstract).

Lyall, V., G. M. Feldman, G. L. Heck, and J. A. DeSimone (1997). Effects of extracellular pH, PCO_2, and HCO_3^- on intracellular pH in isolated rat taste buds. *Am J Physiol* **273**:C1008–1019.

Ma, W.-C. (1972). Dynamics of feeding responses in *Pieris brassicae* Linn as a function of chemosensory input: a behavioral and electrophysiological study. *Meded Laudbouwhogesch Wageningen* **72-11**:1–162.

Maes, F. W. and S. C. A. Bijpost (1979). Classical conditioning reveals discrimination of salt taste quality in the blowfly *Calliphora vicina*. *J Comp Physiol* **133**:53–62.

Martinez del Rio, C., B. R. Stevens, D. E. Daneke, and P. T. Andreadis (1988). Physiolgoical correlates of preference and aversion for sugars in three species of birds. *Physiol Zool* **61**:222–229.

Marui, T. and J. Caprio (1992). Teleost gustation. In: T. J. Hara (ed). *Fish Chemoreception*. London: Chapman and Hall, pp 171–198.

Matsunami, H., J.-P. Montmayeur, and L. B. Buck (2000). A family of candidate taste receptors in human and mouse. Nature **404**:601–604

McLaughlin, S. K., P. J. McKinnon, and R. F. Margolskee (1992). Gustducin is a taste-cell-specific G protein closely related to the transducins. *Nature* **357**:563–569.

Messchendorp, L., H. M. Smid, and J. J. A. von Loon (1998). The role of an epipharyngeal sensillum in the perception of feeding deterrents by *Leptinotarsa decemlineata* larvae. *J Comp Physiol A* **183**:255–264.

Mierson, S., S. K. DeSimone, G. L. Heck, and J. A. DeSimone (1988). Sugar-activated ion transport in canine lingual epithelium. Implications for sugar taste transduction. *J Gen Physiol* **92**:87–111.

Misaka, T., Y. Kusakabe, Y. Emori, T. Gonoi, S. Arai, and K. Abe (1997). Taste buds have a cyclic nucleotide-activated channel, CNGgust. *J Biol Chem* **272**:22623–22629.

Miwa, K., F. Kanemura, and K. Tonosaki (1997). Tastes activate different second messengers in taste cells. *J Vet Med Sci* **59**:81–83.

Miyamoto, T., Y. Okada, and T. Sato (1988). Ionic basis of receptor potential of frog taste cells induced by acid stimuli. *J Physiol (Lond)* **405**:699–711.

Miyamoto, T., R. Fujiyama, Y. Okada, and T. Sato (1998). Sour transduction involves activation of NPPB-sensitive conductance in mouse taste cells. *J Neurophys* **80**:1852–1859.

Morita, H. (1992). Transduction precess and impulse initiation in insect contact chemoreceptor. *Zool Sci* **9**:1–16.

Morita, H., and S. Yamashita (1959). Generator potential of insect chemoreceptor. *Science* **130**:922.

Moskowitz, H. R. (1971). The sweetness and pleasantness of sugars. *Am J Psychol* **84**: 387–405.

Mullin, C. A., S. Chyb, H. Eichenseer, B. Hollister, and J. L. Frazier (1994). Neuroreceptor mechanisms in insect gustation: a pharmacological approach. *J Insect Physiol.* **40**: 913–931.

Murakami, M. and H. Kijima (2000). Transduction ion channels directly gated by sugars on the insect taste cell. *J. Gen. Physiol.* **115**:455–466.

Naim, M., T. Ronen, B. J. Striem, M. Levinson, and U. Zehavi (1991). Adenylate cyclase responses to sucrose stimulation in membranes of pig circumvallate taste papillae. *Comp Biochem Physiol B* **100**:455–458.

Naim, M., R. Seifert, B. Nurnberg, L. Grunbaum, and G. Schultz (1994). Some taste substances are direct activators of G-proteins. *Biochem J* **297**:451–454.

Nakamura, M., and K. Kurihara (1990). Non-specific inhibition by amiloride of canine chorda tympani nerve responses to various salts: do Na(+)-specific channels exist in canine taste receptor membranes? *Brain Res* **524**:42–48.

Nelson, N., and E. A. Bernays (1998). Inositol in two host plants of *Manduca sexta*. *Entomol Exper Appl* **88**:189–193.

Ogawa, K. and J. Caprio (1999). Citrate ions enhance taste responses to amino acids in the largemouth bass. *J Neurophys* **81**:1603–1607.

Ogawa, K., T. Marui, and J. Caprio (1997). Quinine suppression of single facial taste fiber responses in the channel catfish. *Brain Res* **769**:263–272.

Ogura, T. and S. C. Kinnamon (1999). IP$_3$- independent release of Ca^{2+} from intra-cellular stores: A novel mechanism for bitter taste transduction. *J Neurophys* **82**: 2657–2666.

Ogura, T., A. Mackay-Sim, and S. C. Kinnamon (1997). Bitter taste transduction of denatonium in the mudpuppy *Necturus maculosus*. *J Neurosci* **17**:3580–3587.

Okada, Y., T. Miyamoto, and T. Sato (1993). Contribution of proton transporter to acid-induced receptor potential in frog taste cells. Comp. Biochem. Physiol. *Comp Physiol* **105**:725–728.

Okada, Y., T. Miyamoto, and T. Sato (1994). Activation of a cation conductance by acetic acid in taste cells isolated from the bullfrog. *J Exp Biol* **187**:19–32.

Ossebaard, C. A. and D. V. Smith (1995). Effect of amiloride on the taste of NaCl, Na-gluconate and KCl in humans: implications for Na+ receptor mechanisms. *Chem Senses* **20**:37–46.

Ossebaard, C. A. and D. V. Smith (1996). Amiloride suppresses the sourness of NaCl and LiCl. *Physiol Behav* **60**:1317–1322.

Ossebaard, C. A., I. A. Polet, and D. V. Smith (1997). Amiloride effects on taste quality: comparison of single and multiple response category procedures. *Chem Senses* **22**: 267–275.

Ozaki, M., T. Amakawa, K. Ozaki, and F. Tokunaga (1993). Two types of sugar-binding protein in the labellum of the fly: putative taste receptor molecules for sweetness. *J Gen Physiol* **102**:201–216.

Ozaki, M., W. Idei, and F. Tokunga (1997). Antogonistic bindings of polysaccharides to the sugar receptor proteins and their inhibitory effects on teh feeding behavior of the blowfly, *Phormia regina*: two types of putative sugar receptor proteins in the leg of the fly. *Chem Senses* **22**:766.

Panzuto, M. and P. J. Albert (1998). Chemoreception of amino acids by female fourth- and sixth-instar larvae of the spruce budworm. *Entomol Exp Appl* **86**:89–96.

Peterson, S. C., F. E. Hanson, and J. D. Warthen Jr. (1993). Deterrence coding by a larval *Manduca* chemosensory neurone mediating rejection of a non-host plant, *Canna generalis* L. *Physiol Entmol* **18**:285–295.

Price, S. (1973). Phosphodiesterase in tongue epithelium: activation by bitter taste stimuli. *Nature* **241**:54–55.

Ramirez, I. (1990). Why do sugars taste good? *Neurosci Biobehav Rev* **14**:125–134.

Restrepo, D., M. M. Zviman, and N. E. Rawson (1995). Imaging of intracellular calcium in chemosensory receptor cells. In: A. I. Spielman and J. G. Brand (eds). *Experimental Cell Biology of Taste and Olfaction.* Boca Raton, FL: CRC Press, pp 387–398.

Rice, P. R. and D. C. Church (1974). Taste responses of deer to browse extracts, organic acids, and odors. *J Wildl Manage* **38**:830–836.

Roper, S. (1983). Regenerative impulses in taste cells. *Science* **220**:1311–1312.

Rosenzweig S., W. Yan, M. Dasso, and A. I. Spielman (1999). Possible novel mechanism for bitter taste mediated through cGMP. *J Neurophysiol* **81**:1661–1665.

Rouseff, R. L. (1990). *Bitterness in Foods and Beverages.* New York: Elsevier Science Publishers.

Ruiz-Avila, L., S. K. McLaughlin, D. Wildman, P. J. McKinnon, A. Robichon, N. Spickofsky, and R. F. Margolskee (1995). Coupling of bitter receptor to phosphodiesterase through transducin in taste receptor cells. *Nature* **376**:80–85.

Sato, T. and L. M. Beidler (1983). Dependence of gustatory neural response on depolarizing and hyperpolarizing receptor potentials of taste cells in the rat. *Comp Biochem Physiol A* **75**:131–137.

Schoonhoven, L. M. (1982). Biological aspects of antifeedants. *Ent Exp Appl* **31**:57–69.

Schoonhoven, L. M., W. M. Blaney, and M. S. J. Simmonds (1992). Sensory coding of feeding deterrents in phytophagous insects. In: E. A. Bernays (ed). *Insect-Plant Interactions, Vol. IV.* Boca Raton, FL: CRC Press, pp 59–79.

Schultz, J. C. and M. J. Lechowicz (1986). Host-plant, larval age, and feeding behavior influence midgut pH in the gypsy moth (*Lymantria dispar*). *Oecologia* **71**:133–137.

Shallenberger, R. S. and T. E. Acree (1967). Molecular theory of sweet taste. *Nature* **216**: 480–482.

Shields, V. D. C. (1994). Ultrastructure of the uniporous sensilla on the galea of larval *Mamestra configurata* (Walker) (Lepidoptera: Noctuidae). *Can J Zool* **72**: 2016–2031.

Shields, V. D. C. and B. K. Mitchell (1995). The effect of phagostimulant mixtures on deterrent receptor(s) in two crucifer-feeding lepidopterous species. *Phil Trans R Soc Lond B* **347**:459–464.

Shimada, I. (1987). Stereospecificity of the multiple receptor sites in the sugar taste receptor cell of the fleshfly. *Chem Senses* **12**:235–244.

Shimada, I. and K. Isono (1978). The specific receptor site for aliphatic carboxylate anion in the labellar sugar receptor in the fleshfly. *J Insect Phyiol* **24**:807–811.

Shimada, I., A. Shiraishi, H. Kijima, and H. Morita (1974). Separation of two receptor sites in a single labellar sugar receptor of the flesh-fly by treatment with p-chloromercuribenzoate. *J Insect Physiol* **20**:605–621.

Shiraishi, A., and M. Kuwabara (1970). The effect of amino acids on the labellar hair chemosensory cells of the fly. *J Gen Physiol* **56**:768–782.

Siddiqi, O., S. Joshi, K. Arora, and V. Rodrigues (1989). Genetic investigation of salt reception in *Drosophila melanogaster. Genome* **31**:646–651.

Simmonds, M. S. J., S. J. Simpson, and W. M. Blaney (1992). Dietary selection behaviour in *Spodoptera littoralis*: the effects of conditioning diet and conditioning period on neural responsiveness and selection behavior. *J Exper Biol* **162**:73–90.

Simpson, S. J. and C. L. Simpson (1992). Mechanisms controlling modulation by haemolymph amino acids of gustatory responsiveness in the locust. *J Exp Biol* **168**:269–287.

Smith, J. J. B., B. K. Mitchell, B. M. Rolseth, A. T. Whitehead, and P. J. Albert (1990). SAPID Tools: microcomputer programs for an analysis of multi-unit nerve recordings. *Chem Senses* **15**:253–270.

Spector, A. C. and H. J. Grill (1992). Salt taste discrimination after bilateral section of the chorda tympani nerve or the glossopharyngeal nerve. *Am J Physiol* **263**:R169–R176.

Spielman, A. I., H. Nagai, G. Sunavala, M. Dasso, H. Breer, I. Boekhoff, T. Huque, G. Whitney, and J. Brand (1996). Rapid kinetics of second messenger production in bitter taste. *Am J Physiol* **39**:C926–C931.

Steinbrecht, R. A. (1989). The fine structure of thermohygrosensitive sensilla in the silkmoth *Bombyx mori*: receptor membrane structure and sensory cell contacts. *Cell Tissue Res* **255**:49–57.

Stewart, R. E., G. L. Heck, and J. A. Desimone (1996). Taste-mixture suppression: functional dissection of cellular and paracellular origins. *J Neurophysiol* **75**:2124–2128.

Striem, B. J., U. Pace, U. Zehavi, M. Naim, and D. Lancet (1989). Sweet tastants stimulate adenylate cyclase coupled to GTP-binding protein in rat tongue membranes. *Biochem J* **260**:121–126.

Teeter, J. H., K. Sugimoto, and J. G. Brand (1989). Ionic currents in taste cells and reconstituted taste epithelial membranes. In: J. G. Brand, J. H. Teeter, R. H. Cagan, and M. R. Kare (eds). *Chemical Senses: Receptor Events and Transduction in Taste and Olfaction*. New York: Marcel Dekker, pp 151–170.

Teeter, J. H., J. G. Brand, and T. Kumazawa (1990). A stimulus-activated conductance in isolated taste epithelial membranes. *Biophys J* **58**:253–259.

Teeter, J. H., T. Kumazawa, J. G. Brand, D. L. Kalinoski, E. Honda, and G. Smutzer (1992). Amino acid receptor channels in taste cells. In: D. P. Corey and S. D. Roper, (eds). *Sensory Transduction* New York: Rockefeller University Press, pp 292–306.

Teff, K. L. and K. Engelman (1996). Oral sensory stimulation improves glucose tolerance in humans: effects on insulin, C-peptide, and glucagon. *Am J Physiol* **270**:R1371–1379.

Thurm, U. and J. Küppers (1980). Epithelial potential of insect sensilla. In: M. Locke and D. Smith (eds). *Insect Biology of the Future*. New York: Academic Press, pp 735–763.

Tonosaki, K., and M. Funakoshi (1988). Cyclic nucleotides may mediate taste transduction. *Nature* **331**:354–356.

Torii, K. and R. H. Cagan (1980). Biochemical studies of taste sensation. IX. Enhancement of L-[3H]glutamate binding to bovine taste papillae by 5′-ribonucleotides. *Biochim Biophys Acta* **627**:313–323.

Tsunenari, T., Y. Hayashi, M. Orita, T. Kurahashi, A. Kaneko, and T. Mori (1996). A quinine-activated cationic conductance in vertebrate taste receptor cells. *J Gen Physiol* **108**:515–523.

Uchida, Y. and T. Sato (1997a). Changes in outward $K+$ currents in response to two types of sweeteners in sweet taste transduction of gerbil taste cells. *Chem Senses* **22**:163–169.

Uchida, Y. and T. Sato (1997b). Intracellular calcium increase in gerbil taste cell by amino acid sweeteners. *Chem Senses* **22**:83–91.

Ugawa S, Y. Minami, W. Guo, Y. Saishin, K. Takatsuji, T. Yamamoto, M. Tohyama, and S. Shimada (1999). Receptor that leaves a sour taste in the mouth. *Nature* **395**:555–556.

Varkevisser, B. and S. C. Kinnamon (1998). Is PKA involved in sweet taste transduction? *Chem Senses* **23**:613(abstract).

Varkevisser, B., T. Ogura, and S. C. Kinnamon (1997). Evidence for a role of protein kinase C in sweet taste transduction. *Chem Senses* **22**:815.

von Loon, J. J. A. and F. A. van Eeuwijk (1989). Chemoreception of amino acids in larvae of two species of *Pieris*. *Physiol Entomol* **14**:459–469.

Waladde, S. M., A. Hassanali, and S. A. Ochieng (1989). Taste sensilla responses to limonoids, natural insect antifeedants. *Insect Sci Applic* **10**:301–308.

Walters, D. E., G. E. DuBois, and M. S. Kellog (1993). Design of sweet and bitter tastants. In: S. A. Simon and S. D. Roper (eds). *Mechanisms of Taste Transduction*. Boca Raton, FL: CRC Press, pp 463–478.

Wong, G. T., K. S. Gannon, and R. F. Margolskee (1996). Transduction of bitter and sweet taste by gustducin [see comments] [published erratum appears in *Nature* 1996 Oct 10; 383(6600):557]. *Nature* **381**:796–800.

Yamaguchi, S. (1967). The synergistic taste effect of monosodium glutamate and disodium 5' inosinate. *J Food Sci* **32**:473–478.

Yamaguchi, S. (1991). Basic properties of umami and effects on humans. *Physiol Behav* **49**: 833–841.

Yang, J. and S. D. Roper (1987). Dye-coupling in taste buds in the mudpuppy, *Necturus maculosus*. *J Neurosci* **11**:3561–3565.

Yang, H., I. B. Wanner, S. D. Roper and N. Chaudhari (1999). An optimized model for in situ hybridization with signal amplification that allows detection of rare mRNAs. *J Histochem Cytochem* **47**:431–446.

Ye, Q., G. L. Heck, and J. A. DeSimone (1991). The anion paradox in sodium taste reception: resolution by voltage- clamp studies [see comments]. *Science* **254**:724–726.

Ye, Q., R. E. Stewart, G. L. Heck, D. L. Hill, and J. A. DeSimone (1993). Dietary Na(+)- restriction prevents development of functional Na+ channels in taste cell apical membranes: proof by in vivo membrane voltage perturbation. *J Neurophysiol* **70**:1713–1716.

Ye, Q., G. L. Heck, and J. A. DeSimone (1994). Effects of voltage perturbation of the lingual receptive field on chorda tympani responses to Na+ and K+ salts in the rat: implications for gustatory transduction. *J Gen Physiol* **104**:885–907.

Zhou, X. and N. Chaudhari (1997). Modulation of cAMP levels in rat taste epithelia following exposure to monosodium glutamate. *Chem Senses* **22**:834.

Zviman, M. M., D. Restrepo, and J. H. Teeter (1996). Single taste stimuli elicit either increases or decreases in intracellular calcium in isolated catfish taste cells. *J Membr Biol* **149**: 81–88.

14

Neural Representation of Taste

DAVID V. SMITH and BARRY J. DAVIS
Department of Anatomy and Neurobiology, and Program in Neuroscience,
University of Maryland School of Medicine, Baltimore, MD

1. INTRODUCTION

The sense of taste provides a gateway to the body's internal environment by monitoring the types of substances an organism ingests. The qualities of taste and their positive and negative hedonic aspects provide signals important for decisions about ingestion and rejection. Gustatory neurobiologists generally agree that humans experience four basic taste qualities (sweet, salty, sour, and bitter), although they disagree about the possibility of others (see Chapter 13, Glendinning et al.; Chapter 16, Breslin). These qualities and the behaviors associated with them provide the means by which an animal makes ingestive decisions. Through these qualities, taste helps to ensure the animal's energy supply (sweet), maintain the proper electrolyte balance (salt), and avoid the ingestion of toxic substances (sour, bitter). Of considerable debate is the way in which taste quality is represented in the activity of gustatory neurons. Many investigators accept the idea that peripheral gustatory nerve fibers and central neurons can be classified meaningfully into groups on the basis of similarities and differences in their response profiles. Increasingly, the description of the physiology of taste neurons has been characterized by analyses that employ a neuron classification scheme in an attempt to impose some order on these data.

The Neurobiology of Taste and Smell, Second Edition, Edited by Thomas E. Finger, Wayne L. Silver, and Diego Restrepo.
ISBN 0-471-25721-4   Copyright © 2000 Wiley-Liss, Inc.

There is evidence in mammals that these neuron types play a critical role in defining differences in the across-neuron response patterns elicited by highly discriminable stimuli. In this chapter, we describe the neuroanatomy of the gustatory system, discuss the basic response properties of gustatory neurons, present the arguments about the nature of taste quality coding, and discuss the role of neurotransmitters in the processing and modulation of gustatory afferent information.

2. ANATOMY OF THE GUSTATORY SYSTEM

2.1. Peripheral Gustatory Nerves

The facial (VIIth), glossopharyngeal (IXth), and vagus (Xth) nerves innervate taste receptor cells, located throughout the oropharyngeal cavity. In mammals, the facial nerve contains two sensory components related to taste, the chorda tympani (CT) and the greater superficial petrosal (GSP) nerves. The CT innervates taste buds in the fungiform papillae of the anterior tongue and the more rostral foliate papillae along the lateral aspect of the posterior tongue. Palatine branches of the GSP innervate taste buds on the soft palate, nasoincisor ducts, and, in the rat, along the "Geschmacksstreifen" (taste stripe) at the junction of the soft and hard palate. The cell bodies of both the CT and GSP lie in the geniculate ganglion. Together, the CT and GSP innervate 30–35% of all taste buds in the rat and hamster oral cavity (Miller and Smith, 1984). The lingual-tonsillar branch of IX, with the cell bodies lying in the petrosal ganglion, innervates taste receptors in the vallate and remaining foliate papillae. In rodents, the IXth nerve innervates about 55% of all taste buds. The superior laryngeal branch of X innervates taste buds of the larynx and upper esophagus; the cell bodies lie in the nodose ganglion.

The gustatory fibers of CT, GSP, IX, and X project topographically into the brainstem to terminate in the rostral portion of the nucleus of the solitary tract (NST), the location of the first central synapse in the taste pathway. These projections overlap rostrocaudally so that the target zones of the CT/GSP and IX, and IX and X overlap at their margins. The most caudal projections of IX and X are outside of the gustatory zone and consist of fibers innervating arterial baro- and chemoreceptors, atrial volume receptors, and osmoreceptors. The oral mucosa and lingual epithelium are richly innervated by general somatosensory fibers mediating touch, pain, and thermal sensitivity, carried within the trigeminal nerve (V), IX, and X (See also Chapter 12, Finger and Simon). The lingual branch of V provides somatosensory innervation to the anterior tongue. Centrally, in addition to projections to the trigeminal complex, the lingual branch of V also terminates in the gustatory portion of the NST, overlapping rostrocaudally with the gustatory inputs of the CT and GSP nerves (see Chapter 4, Bryant and Silver).

2.2. Organization of the Central Gustatory Pathway

Although this chapter concentrates on vertebrates, some features of the gustatory connections in invertebrates are noteworthy even given the striking dissimilarities in brain organization between these groups. The perioral taste receptors in arthropods (uniporus sensilla, see Chapter 12, Finger and Simon), the best-studied invertebrate group, send axons into the subesophogeal ganglion (see Singh, 1997, for an excellent review). This area is relatively far-removed from the locus of termination of olfactory (antennal) inputs to the superesophogeal ganglion. Thus in invertebrates as well vertebrates, areas processing primary taste information are essentially independent of those dealing with primary olfactory input.

For vertebrates, the central taste pathway begins in the medulla and projects rostrally in parallel with the general visceral afferent system. In the gustatory system, first-order afferent input is associated with three cranial nerves (VII, IX, and X), which project into the rostral portions of the visceral sensory column of the medulla, which, in tetrapod vertebrates, is the nucleus of the solitary tract (NST) (see Fig. 14.1). The VIIth nerve is primarily motor (to the facial musculature), with a small gustatory component, and IX and X are mixed motor and sensory nerves supplying gustatory and somatosensory innervation to the posterior tongue, pharynx, and larynx. Second-order projections arising from the NST ascend mostly ipsilaterally. In mammals, this ascending tract has not been

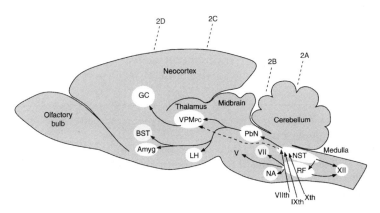

FIG. 14.1. Schematic diagram of the ascending gustatory pathway; descending projections are not shown. Solid lines show connections of the rodent gustatory system within the CNS; a dashed line indicates the projection from NST to VPMPC in primates. *Abbreviations:* Amyg, central nucleus of the amygdala; BST, bed nucleus of the stria terminalis; GC, gustatory cortex; LH, lateral hypothalamus; NA, nucleus ambiguus; NST, nucleus of the solitary tract; PbN, parabrachial nuclei; VPMPC, ventral posterior medial nucleus of the thalamus, parvicellular division; V, VII, and IX, trigeminal, facial, and hypoglossal motor nuclei; VIIth, IXth, and Xth, axons of peripheral gustatory fibers in the facial, glossopharyngeal, and vagal cranial nerves. The approximate level of each section shown in Fig. 14.2 is indicated by dashed lines (2A–2D).

named; it travels diffusely as a component of the central tegmental tract through the dorsal tegmentum of the pons and terminates in the parabrachial nuclei (PbN) of the dorsal pons. In primates, there is a direct projection from the medulla to the thalamus (Pritchard, 1991; dashed line in Fig. 14.1), bypassing the dorsal pons, as in catfish (Finger, 1987). The PbN projects directly to the parvicellular division of the ventral posterior medial nucleus (VPMpc) of the thalamus, which, in turn, projects to agranular insular (primary gustatory) cortex (GC). In addition, the PbN projects to limbic structures that are typically associated with visceral functions, including the lateral hypothalamus (LH), the central nucleus of the amygdala (Amyg), and the bed nucleus of the stria terminalis (BST). In addition to its ascending connections, the NST makes local medullary connections with a number of brainstem motor nuclei (V, VII, XII, nucleus ambiguus, inferior and superior salivatory nuclei), either directly or through interneurons in the reticular formation (RF; Travers and Norgren, 1983). In summary, the gustatory pathway involves three cranial nerves, a diffuse, primarily ipsilateral projection through the brainstem, a pontine relay nucleus, a thalamocortical projection to neocortex, and substantial connections to limbic regions and oromotor nuclei (Fig. 14.1).

2.2.1. Nucleus of the Solitary Tract (NST)

The NST is a heterogeneous collection of subnuclei that extends rostrally from the level of the spinal cord/medulla junction near the IVth ventricle to a more lateral (and ventral) position in the pons at the level of the dorsal cochlear and facial nuclei. Historically, the NST has been divided into functional and morphological divisions that are associated with visceral afferent activity, a combination of somatic motor and visceromotor activity, and gustation (Norgren, 1995). The rostral half of the NST (Fig. 14.2A) receives the gustatory fibers of

$$\longrightarrow$$

FIG. 14.2. A series of Nissl-stained coronal sections of the major components of the ascending gustatory pathway in the hamster.

(a) The rostral pole (gustatory portion) of the NST.

(b) A section through the caudal one-third of the PbN, showing the pontine taste area.

(c) The taste relay in the thalamus, VPMpc.

(d) Agranular insular (gustatory) cortex. *Abbreviations:* ac, anterior commissure; ACB, nucleus accumbens; AGC, agranular insular cortex; AQ, cerebral aqueduct; br, brachium conjunctivum; CP, caudate and putamen; cpd, cerebral peduncle; DCN, dorsal cochlear nucleus; fr, fornix; icp, inferior cerebellar peduncle; IV, fourth ventricle; LC, locus coeruleus; LV, lateral ventricle; lot, lateral olfactory tract; mcp, middle cerebellar peduncle; ml, medial lemniscus; MN, medial nucleus of the PbN; MV, motor nucleus of V; mVN, medial vestibular nucleus; NST, nucleus of the solitary tract; PbN, parabrachial nuclei; PC, piriform cortex; PFN, parafascicular nucleus of the thalamus; pyr, pyramid; RF, reticular formation; SN, substantia nigra; spV, spinal tract of V; spVN, spinal vestibular nucleus; sV, sensory nucleus of V; VII, facial nucleus; VPM-VPL, ventral posterior medial and lateral nuclei of the thalamus; VPMpc, ventral posterior medial nucleus of the thalamus, parvicellular division; ZI, zona incerta. Calibration bar: 1 mm.

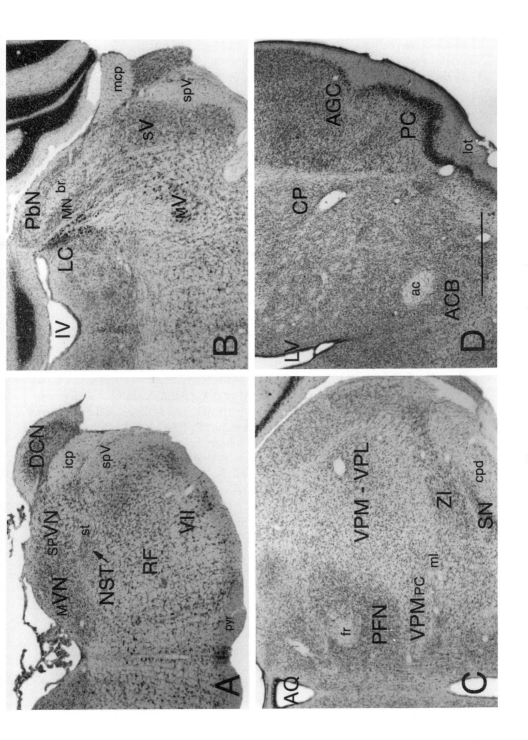

357

cranial nerves VII, IX, and X, and some somatosensory fibers of V and possibly IX and X. This portion of the NST has been given several functional monikers such as the "gustatory recipient zone," "gustatory zone," or "gustatory NST." However, an unexpectedly high number of neurons in the rostral NST is activated only by tactile stimulation (Travers and Norgren, 1995). The caudal NST receives afferent input from branches of IX and X that carry baro-, osmo-, and chemo-receptor information regulating cardiovascular and respiratory reflexes, from subdiaphragmatic branches of X regulating gastric motility and feeding, and from regions of the dorsal horn of the spinal cord mediating pain. In rodents, the gustatory NST consists of lateral, central, medial, dorsal, and ventral subnuclei, defined according to their cytoarchitecture and their positions relative to the solitary tract (Whitehead, 1988). The prefix *rostral* is sometimes used to avoid confusion with similarly named nuclei of the caudal NST. The rostral central, rostral lateral, and rostral ventral subnuclei are the largest and most discernible.

The gustatory NST contains several morphologically distinct classes of neurons. The nomenclature for these cells varies, but the neurons are typically fusiform (elongated or bipolar), multipolar (stellate), or ovoid in shape, with relatively simple dendritic branching patterns. The dendrites extend distances that often exceed the boundaries of the gustatory zone. Unfortunately, electro-physiological studies have uncovered no obvious relationship between a variety of morphological measures (e.g., somal size, number of spines, complexity of dendritic arborizatons) and intrinsic firing patterns or taste-response profiles (Renehan et al., 1996). Nonetheless, it is likely that many of the small ovoid neurons are γ-aminobutyric acid- (GABA) or enkephalin-containing local circuit neurons (Lasiter and Kachele, 1988; Davis, 1993). The neurons of the gustatory NST show a preferred orientation in the horizontal plane, with dendrites oriented in parallel with or perpendicular to the solitary tract, the source of peripheral afferent input (Davis, 1988). These dendrites share the same orientation as the incoming gustatory afferent fibers, which is functionally significant because the terminations of the gustatory fibers are exclusively axodendritic (Whitehead, 1986; Davis, 1998).

The ascending gustatory pathway originates primarily from neurons in the rostral central and rostral lateral subnuclei of the NST (Whitehead, 1990). The ascending pathway is presumably responsible for the conscious awareness of taste quality (sweet, salty, sour, or bitter), taste hedonics (pleasantness or unpleasant-ness), and the integration of sensory activity (gustatory and autonomic activity) associated with eating and drinking. As described below, the ascending projection from the NST terminates in the so-called "pontine taste area" of the parabrachial nuclei, which actually corresponds to two to three subnuclei of the PbN. About one-third of the neurons in the gustatory NST contribute to this ascending pathway (Whitehead, 1990). Some neurons, mostly in the rostral ventral subnucleus, project ventrally and/or caudally to the reticular formation, to more caudal nuclei of the NST, or to the hypoglossal nucleus (Travers and Norgren, 1983). These medullary projections to adjacent somatic and visceral premotor/motor areas are responsible for the modulation of several reflexes (e.g., swallowing, salivation, preabsorptive insulin release) associated with ingestion.

2.2.2. Parabrachial Nuclei (PbN)

The parabrachial complex is located in the dorsolateral aspect of the pons (Fig. 14.2B). The original definition of the "pontine taste area" was based on the pattern of degeneration in the parabrachial nuclei (PbN) following electrolytic lesions of the gustatory NST and corresponded to the medial nucleus of the PbN (Norgren and Leonard, 1973). Electrophysiological studies also showed that neurons of the medial nucleus responded to taste stimulation of the tongue (Norgren and Pfaffmann, 1975).

The PbN is physically divided into medial and lateral subdivisions by the obliquely directed fibers of the brachium conjunctivum (superior cerebellar peduncle). Each major subdivision is further divisible into distinct nuclei on the bases of electrophysiological data, cytoarchitectonic criteria, or efferent projection patterns. A more expanded definition of the taste-responsive area of the PbN corresponds to the medial and ventrolateral nuclei in the caudal third of the PbN (Travers, 1993). Ascending general visceral afferent projections arising from caudal, nongustatory regions of the NST and the ventrolateral medulla terminate primarily in nuclei of the lateral subdivision of the PbN.

The PbN projects bilaterally (heavier ipsilaterally) to the parvicellular division of the ventral posterior medial nucleus (VPMPC) of the thalamus. Other projections from the PbN go to the ventral forebrain, most notably to the lateral hypothalamus, central nucleus of the amygdala, bed nucleus of the stria terminalis, and substantia innominata (Norgren, 1976). It appears that different populations of PbN neurons contribute to the thalamic and ventral forebrain projections (Travers, 1993). In monkey, the bulk of the projections from PbN are directed toward the ventral forebrain. In contrast, the monkey PbN projects lightly to VPMPC, which receives gustatory input directly from the NST (Pritchard, 1991). In both rodents and monkeys, the PbN projections to limbic areas are extensive and reflect the substantial interactions between the gustatory and general visceral afferent systems.

2.2.3. Gustatory Thalamus: VPMPC

The gustatory thalamic relay is located in the parvicellular division of the ventral posterior medial nucleus (VPMPC) of the dorsal thalamus, which is a medial extension of VPM, the thalamic relay for somatosensory information arising from the head region via the trigeminal system (Fig. 14.2C). The VPMPC contains a topographic representation of the oral cavity that is oriented mediolaterally with overlapping gustatory and somatosensory (lingual) receptor fields. Neurons in VPMPC are responsive to combinations of gustatory, oral tactile, and thermal stimulation (Ogawa and Nomura, 1988) and, therefore, the activity of neurons in the VPMPC encompasses all forms of sensory stimulation of the oral cavity. The neurons of the VPMPC are generally small and densely packed, hence the commonly used term *parvicellular*. Neuronal size and packing density can vary across species and the term *paucicellular* (sparse cells) is often substituted in cases where cell packing appears less dense than in the adjacent VPL. In the rat and hamster, the ascending gustatory fibers arising from the PbN also cross over at the level of VPM to terminate in the contralateral VPMPC (Norgren, 1976). These bilateral projections to the VPMPC account for the bilateral responses of thalamic neurons to sapid stimulation (Blomquist et al., 1962).

2.2.4. Gustatory Neocortex

The nomenclature describing gustatory neocortex is species dependent and somewhat confusing. In rodents, VPMPC projects to agranular insular cortex, which is exposed on the lateral convexity of the brain just dorsal to the rhinal fissure. Agranular insular cortex is cytoarchitectonically distinct from the granular cortex (Fig. 14.2D). Granular cortex has a prominent layer IV or granule cell layer and is considered primary somatosensory cortex. Oral somatosensory (trigeminal) input to primary somatosensory cortex is targeted to the most ventral aspect of the granular layer, immediately dorsal to the agranular cortex that receives gustatory input. There is an ongoing debate about the cytoarchitectonic definition and boundaries of the rodent agranular, transitional dysgranular (i.e., contains some granule cells), or granular cortices. This issue should be resolvable in the future with more precise mapping studies of taste-responsive neurons, better electrode tract reconstruction, and stimulation of anterior and posterior oral receptive fields with combinations of sapid and somatosensory stimuli. In primates, the primary gustatory neocortex corresponds to both the frontal operculum and buried insular cortex toward the rostral end of the lateral fissure. Both cortical areas receive major inputs from the VPMPC. A secondary gustatory cortex is located within orbitofrontal cortex of primates but this area does not receive direct projections from VPMPC (Pritchard, 1991).

In some rodents, there are direct projections from the PbN to cortex that bypass the thalamus (Shipley and Sanders, 1982). In addition, there are projections from the lateral hypothalamus and amygdala to the cortex that overlap the inputs from VPMPC (Allen et al., 1991). Consequently, gustatory neocortex receives converging limbic and gustatory input, consistent with the intimate relationship between the gustatory and general visceral afferent systems that is observed throughout the ascending gustatory pathway.

2.2.5. Descending Corticofugal and Limbic Projections

The gustatory system, like other sensory systems, is organized hierarchically and numerous descending projections from higher levels impinge upon relay points in the ascending pathway. Insular cortex has widespread projections to subcortical areas that include the VPMPC, central nucleus of the amygdala, PbN, and NST (Saper, 1982). The VPMPC projects to the central and lateral nuclei of the amygdala and the central nucleus of the amygdala projects heavily to the ipsilateral medial and lateral divisions of the PbN and has widespread projections throughout the NST. The descending projections of the PbN are destined for the spinal cord, reticular formation, and oromotor nuclei (Travers and Norgren, 1983). Neither the VPMpc nor the gustatory regions of the PbN appear to project caudally to the gustatory NST. Although it is tempting to assume that these descending projections modulate the processing of ascending gustatory information, many of these connections only have been defined anatomically (but see below).

2.3. Taste and Visceromotor Integration

The processing of gustatory information is often discussed solely in terms of sensory coding and taste discrimination. These aspects of taste processing usually receive the greatest attention in introductory textbooks that cover sensory systems. However, it is becoming more difficult to discuss the gustatory system outside the context of visceromotor integration. Norgren (1995) has provided compelling arguments that a more contemporary view of gustation and the autonomic nervous system should include considerations of the numerous anatomical interconnections between visceral afferent and efferent systems and the brainstem gustatory relay nuclei. Gustation is represented in the rostral pole of a functional column that also receives anatomically separate inputs related to respiration, blood pressure, blood pH, gastrointestinal activity, and pain. Anatomically, this functional column corresponds to the NST, with the gustatory zone located in its rostral half. This visceral afferent column is closely associated with both somatic and branchial motor nuclei (i.e., the retrofacial area, motor nucleus of V, nucleus ambiguus, and hypoglossal nucleus) and visceromotor (salivatory and vagal preganglionic parasympathetic neurons) columns containing neurons that initiate chewing, tongue movement, salivation, swallowing, gastrointestinal motililty, acid secretion, and other reflexes associated with eating, drinking, and painful stimulation (Fig. 14.1). The gustatory and general visceral afferent systems provide the requisite sensory input to effectively control these assorted somatic and visceromotor reflexes.

Activation of the gustatory system is necessary for an organism to assess the palatability of food. Taste input generates activity in the ascending gustatory pathway and presumably evokes a perceptual component that involves an awareness of taste quality (sweet, salty, sour, or bitter). Simultaneously, taste stimulation activates the afferent limb of a variety of visceromotor reflexes (appetitive or aversive) accompanying eating and drinking. These stereotypic behaviors are characterized by well-defined somatic motor and visceromotor responses (Grill and Norgren, 1978b). The circuitry needed to initiate such reflexes exists in the brainstem, since chronic decerebrate rats display both acceptance and rejection behaviors (Grill and Norgren, 1978c). A multisynaptic pathway(s) supports these ingestive and aversive reflexes and probably involves intrinsic projections of the gustatory NST to other parts of the NST and to premotor brainstem areas involved in swallowing, salivation, and chewing (Travers and Norgren, 1983).

The gustatory and general visceral afferent systems are anatomically distinct in fish. In some species (e.g., catfish), external taste buds are located on the head, body, lips, and specialized structures, such as barbels. These taste buds are innervated by the facial nerve and are important for locating food (Atema, 1971). In contrast, internal taste buds are located in the mouth, pharynx, and gill chambers and, in some species, a palatal organ in the posterior oral cavity. The oropharyngeal taste buds, innervated by the IXth and Xth nerves, are critical for the initiation of swallowing and the determination of palatability. The facial, glossopharyngeal, and vagus nerves project to the corresponding facial, glosso-pharyngeal (generally quite small), and vagal lobes in the medulla, which are most prominent in highly gustatory fishes such as catfish and carp. The central projections of the facial and vagal lobes remain discrete. The neurons in the facial lobe project caudally to trigeminal sensory nuclei where gustatory and somatosensory inputs probably converge; in contrast, the caudal projections of the vagal lobe terminate directly on the dendrites of motor neurons in nucleus ambiguus that innervate pharyngeal musculature and initiate swallowing (Finger and Morita, 1985). Obviously, the ingestion of food requires the integration of both sensory and motor components. In fish, the neural substrates that support such integration remain distinguishable in the medulla; in rodents it is more difficult to define precisely the anatomical substrate for such integration.

The rostral projections of the facial and vagal lobes also remain relatively distinct (Finger, 1987). The neurons of the facial lobe project to the ipsilateral thalamus, as well as to the superior secondary gustatory nucleus in the dorsal pons. In contrast, the neurons of the vagal lobe project primarily to the gustatory nucleus in the dorsal pons. The superior secondary gustatory nucleus projects to both the thalamus and limbic regions of the inferior lobe in a fashion analogous to the efferent projections of the PbN in mammals. These rostral projection patterns indicate that gustatory and visceral afferent information ascend in parallel but separate pathways to eventually reach thalamic and limbic structures, as in mammals.

3. BASIC RESPONSE PROPERTIES OF GUSTATORY NEURONS

A universal characteristic of mammalian gustatory neurons is their responsive-ness to stimuli representing more than one of the classic four taste qualities. The responses of these neurons are modulated also by changes in stimulus concen-tration and often by tactile and thermal stimuli as well. Therefore, understanding the role of these neurons in the coding of stimulus information must consider this lack of stimulus specificity. Even when gustatory neurons can be shown to be relatively narrowly tuned across taste qualities, as occurs in cells of the primate orbitofrontal cortex, these neurons still can be modulated by stimulus concentration and are likely to be multimodal in their responsiveness. This multiple sensitivity raises a serious issue with regard to taste quality coding by labeled lines, as discussed below.

3.1. Multiple Sensitivity of Peripheral Taste Fibers

The chemical responsiveness of peripheral taste fibers varies with the organism, the location of the taste receptor organ within a single organism, and may depend on natural diet as well as taxon. The peripheral taste fibers of vertebrates are neurons that integrate information from multiple taste cells and relay it to the nucleus of the solitary tract. In contrast, the peripheral taste "fibers" of invertebrates are the taste cells themselves; these bipolar sensory neurons project directly (in most cases) to the subesophageal ganglion of the CNS (see Fig. 12.1, Chapter 12, Finger and Simon). Taste fibers—whether in invertebrates or vertebrates—can be restricted in terms of breadth of chemical responsiveness or more broadly "tuned." For example, the silkworm, *Bombyx mori*, has several taste sensilla, and each contains four identified taste "fibers" with well-characterized molecular receptive ranges. The most sharply tuned fiber is depolarized exclusively by specific isomers of the sugar alcohol, inositol; the other three fibers are less sharply tuned and are depolarized respectively by sucrose and related sugars; D-glucose and related sugars; or various salts, (Jakinovich and Agranoff, 1971). Other species of insect have even more broadly tuned taste fibers. For instance, many flies have an identified sugar-sensitive taste fiber that is excited by many sugars and amino acids, and is inhibited by a variety of compounds that humans describe as bitter (Dethier and Bowdan, 1989; Pollack and Balakrishnan, 1997). Despite the specificity of response of many peripheral taste fibers in these invertebrates, discrimination between stimuli, for example, discriminating a weak salt solution from pollen extract, may require comparison of activity across the different sensory cells (Dethier, 1976; Hansen et al., 1998).

Even the earliest electrophysiological recordings of the activity of single fibers in the CT nerve showed that mammalian taste neurons respond to stimuli of different qualities. In fact, this multiple sensitivity of CT fibers in the rat, cat, and rabbit first led Carl Pfaffmann (1955, 1959) to propose that taste quality is represented by the pattern of activity across these afferent nerve fibers. Since mammalian gustatory nerve fibers are not specifically tuned to a single taste quality, many investigators have attempted to classify them into functional groups in an effort to impose order on the organization of gustatory afferent information.

The feeding habits of rodents are quite different from other species whose taste systems have been characterized electrophysiologically. For example, taste fibers of channel catfish are extremely sensitive to L-amino acids, especially L-arginine and L-alanine, which stimulate different fiber types in the facial nerve (Davenport and Caprio, 1982). Amino acids are potent feeding stimulants for catfish, so it is not surprising that their gustatory systems are especially tuned to these stimuli. Analysis of single facial taste fibers in the sea catfish demonstrated several broadly tuned groups of fibers, which responded differentially to various amino acids (Michel and Caprio, 1991), although the role of these fiber types in coding information about amino acids is not clear.

3.1.1. Responses to Basic Taste Stimuli: Response Profiles and Fiber Types

Although there have been several schemes for classifying taste fibers into meaningful groups, the most common are those based on similarities and differences in the fibers' response profiles. Of these, the most widely used has been the "best-stimulus" classification, first used by Frank (1973) for grouping hamster CT nerve fibers. When the anterior tongue is stimulated by the mid-range concentrations of 0.1 M sucrose, 0.03 M NaCl, 0.003 M HCl, and 0.001 M quinine hydrochloride (QHCl), fibers of the hamster's CT nerve can be categorized into one of three groups: sucrose (S)-best, NaCl (N)-best, or HCl (H)-best. In Frank's study, only one CT fiber out of 79 responded best to QHCl (Q-best). Interestingly, when the stimuli are arranged in order of taste preference (i.e., S, N, H, and Q), the response profile of each best-stimulus group peaks at a single point (the best stimulus), with the responses to the "side-band" stimuli diminishing on either side (see Fig. 14.3). The response profiles within each best-stimulus group are relatively similar, suggesting a good deal of homogeneity of responsiveness within a group. Similar groups are seen at central gustatory nuclei of the hamster (Fig. 14.3) when the same stimuli are applied to the anterior portion of the tongue (Travers and Smith, 1979; Van Buskirk and Smith, 1981).

The homogeneity of the response profiles within a best-stimulus group makes it relatively easy to predict the responsiveness of a taste neuron to other stimuli within the same quality class. That is, if a neuron responds vigorously to sucrose it usually gives strong responses to other stimuli that are sweet to humans and that are classified as similar to sucrose by hamsters in behavioral experiments. Similar observations follow for the responses of N-best neurons to other sodium and lithium salts and of H-best neurons to other acids and nonsodium salts. Hierarchical cluster analyses of the responses of taste fibers in the CT nerve (Frank et al., 1988) and neurons in the NST and PbN (Smith et al., 1983b) provide strong support for the classification of these neurons into types, based on similarities and differences in their response profiles. The dendrogram of Figure 14.4 depicts the cluster solution for hamster PbN neurons in response to 18 stimuli. The response profiles of S-, N-, and H-best neurons are strikingly different from each other, even when the responses to as many as 18 stimuli are included in the profiles and when the cells are broadly tuned (see below). There is enough similarity within each neuron type to warrant considering each to be a distinct class of neurons. Hierarchical cluster analysis has been applied in many other species and at several levels of the gustatory system (Van Buskirk and Smith, 1981; Frank et al., 1988; Scott and Giza, 1990; Rolls, 1995) and most investigators agree on the usefulness of classifying neurons into functional types. At the very least, describing taste neurons by their best stimulus provides a

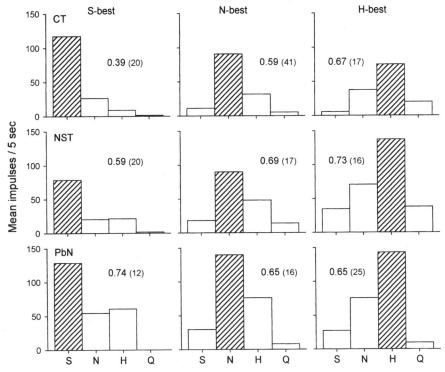

FIG. 14.3. Mean neural response functions for S-, N-, and H-best neurons in the CT nerve, the NST, and the PbN of the hamster. The response of each class of neurons to its best stimulus is shaded. The mean breadth of tuning (H) is given above each profile; the number in parenthesis is the number of neurons comprising each mean profile. (Data are from Frank (1973) for the CT, Travers and Smith (1979) for the NST, and Van Buskirk and Smith (1981) for the PbN.)

convenient method for imposing some order on the system. Beyond that, there are a number of studies suggesting that these neuron types are biologically relevant, for example, studies showing that only N-best neurons receive input from the amiloride-sensitive Na^+ transduction mechanism (Hettinger and Frank, 1990; Scott and Giza, 1990). The role that these neuron types play in the coding of taste quality is addressed below.

3.1.2. Breadth of Responsiveness

Because taste neurons typically respond to stimuli representing more than one taste quality, many investigators have attempted to describe their breadth of responsiveness across these stimuli. Early attempts generally employed some criterion for what constituted a response, and because these criteria varied across laboratories, it was difficult to compare cells in different species or at different levels of the gustatory system. Smith and Travers (1979) introduced a measure of

Hamster PbN neurons

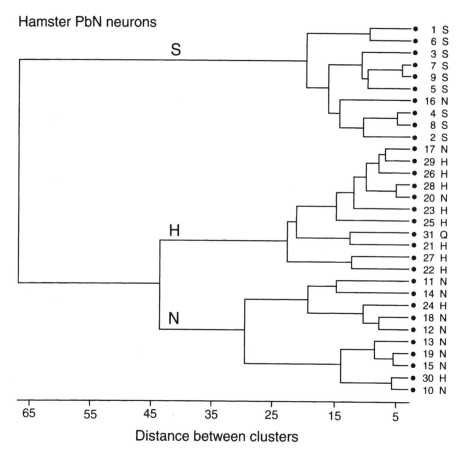

FIG. 14.4. Cluster dendrogram depicting the results of a hierarchical cluster analysis of 31 hamster PbN neurons based on similarities in their neural response profiles across 18 gustatory stimuli. The neuron numbers and the best-stimulus designations (S, H, or N) are shown on the right of the figure. The distance between neurons or clusters of neurons joined at each step is shown along the abscissa. S, H and N indicate the three major clusters. (From Smith et al. (1983b).)

the breadth of tuning based on the entropy equation used in information theory to measure information transmission. In this procedure, the response of a neuron to each stimulus is expressed as a proportion (p_i) of the response to all four. The breadth of tuning (H) is given by the equation:

$$H = -K \Sigma p_i \log p_i,$$

where H=breadth of responsiveness (entropy) and K is a scaling constant (1.661 for four stimuli), which sets the value of H between 0.0, indicating a neuron that responds exclusively to one of the four stimuli, and 1.0, indicating a cell that responds equally to all four. Therefore, larger values of H reflect a greater breadth of tuning across the four basic stimuli. This measure has been used to compare the breadth of tuning of gustatory neurons across different levels of the nervous system (Van Buskirk and Smith, 1981; Rolls, 1995) and across species (Travers, 1993).

Because gustatory neurons respond to different stimulus classes and are modulated by stimulus concentration, the response profile of a cell can vary with stimulus intensity. Generally, taste neurons increase their firing rates as stimulus concentration is increased, but the slopes of these concentration-response functions often vary across stimuli, even in the same neuron. Consequently, the shapes of the response profiles may be considerably different across the range of effective stimulus concentrations. As seen in auditory tuning curves and in other sensory modalities, increasing stimulus intensity generally produces greater breadth of responsiveness (Smith and Travers, 1979). When stimulus concentrations are reduced, cells become more narrowly tuned until they only respond to a single (best) stimulus. Therefore, measures of the breadth of tuning of gustatory neurons, regardless of the method used to quantify this concept, can be greatly confounded by the relative concentrations of the test stimuli. Making comparisons across neural levels, across experiments from different laboratories, and especially across species, is a difficult task. There are instances, however, where stimulus concentrations and other factors have been held constant across experiments, allowing a direct comparison of the breadth of tuning of taste neurons under different conditions (Van Buskirk and Smith, 1981; Rolls, 1995).

3.1.3. Neuron Types in Different Nerves and Different Species

The response profiles of gustatory nerve fibers have been characterized in several species, but those in the hamster are the easiest to compare because the data were collected under similar conditions in each experiment. Within the hamster CT nerve there are three distinct fiber types, including the S-, N-, and H-best (Frank, 1973). Mean response profiles for 20 S-, 41 N-, and 17 H-best CT fibers are shown in Figure 14.3 (top). In the CT nerve, the S-best fibers are clearly the most specifically tuned (see Fig. 14.3, top), with a mean entropy of 0.39, compared to 0.59 and 0.67 for the N- and H-best fibers, respectively (Van Buskirk and Smith, 1981). The responses of the S-best fibers are quite narrowly tuned to sucrose and other sweet-tasting stimuli, with some sensitivity to NaCl and very little response to HCl or QHCl (Fig. 14.3, top). In comparison to the rat, the CT nerve of the hamster is considerably more responsive to sweet stimuli. The greater breadth of tuning of the N- and H-best fibers in the hamster CT occurs largely because these fiber types respond to both NaCl and HCl, albeit to different degrees.

There is considerably more sensitivity to sweet stimuli in the GSP nerve of both the rat (Nejad, 1986) and hamster (Harada and Smith, 1992) than in the CT nerve of either species. In addition, single-neuron recordings have been made from cells in the rat NST that respond to stimulation of the nasoincisor ducts (Travers and Norgren, 1991), showing a considerable number of S-best neurons, many of which receive converging input from both GSP and CT nerves. Behavioral experiments have shown that taste receptors innervated by the GSP contribute substantially to the rat's ability to respond to sweet-tasting stimuli (Spector et al., 1993, 1997).

Bitter stimuli, such as QHCl, are relatively ineffective in driving fibers of either the CT or the GSP, at least at mid-range concentrations. Fibers of the glossopharyngeal (IXth) nerve are relatively more responsive to QHCl in both the rat and hamster (Hanamori et al., 1988; Frank, 1991). In comparison to the hamster's CT nerve, where there are few Q-best and many S- and N-best fibers, the distribution of these fiber types is reversed in the IXth nerve. The magnitudes of the responses to these chemicals are also reversed between the CT, where sucrose and NaCl produce much larger responses, and the IXth nerve, where the response to QHCl is considerably greater than that to either sucrose or NaCl. In the rat, there are many Q-best and a number of S-best fibers in the IXth nerve and no N-best fibers at all (Frank, 1991). Consequently, the information projected to the NST in both of these species by the VIIth nerve is dominated by preferred stimuli (sucrose and NaCl), whereas that from the IXth nerve is more concerned with aversive stimuli (QHCl), although the rat's IXth nerve (but not the hamster's) is also relatively sensitive to sucrose.

Taste buds within the laryngeal mucosa are innervated by fibers of the internal branch of the superior laryngeal nerve (SLN), a branch of the vagus (Xth) nerve. Unlike gustatory fibers of the VIIth and IXth nerves of the hamster, those of the SLN do not discriminate among taste qualities (Smith and Hanamori, 1991). These fibers respond well to water, NaCl, and HCl, but poorly to sucrose and QHCl. A hierarchical cluster analysis shows no distinct fiber types in the SLN. Instead, the sensitivities of SLN fibers are graded along a continuum, with no distinct differences among fibers responding best to different stimuli. Laryngeal nerve fibers appear to serve a role in protection of the airway from foreign substances but not in taste discrimination.

The tendency for different nerves to encode slightly different information obtains in nonmammalian as well as mammalian vertebrates. Even though amino acids are the predominantly effective stimuli for catfish, the fibers of the IXth and Xth nerves are more sensitive to QHCl than VIIth nerve taste fibers (Kanwal and Caprio, 1983). The IXth nerve fibers of amphibians are also highly responsive to bitter substances (Gordon and Caprio, 1985). Thus, the preferential sensitivity of the IXth nerve to aversive stimuli is characteristic of fish, amphibians, and mammals.

Taste fibers recorded from primates show sensitivities similar to those of rodents, except that they appear to be somewhat more narrowly tuned across the four taste qualities. Responses of CT fibers of the squirrel monkey are divisible into distinct S-, N-, or H-best categories, very similar to those observed in the hamster CT nerve (Pfaffmann et al., 1976). Although the breadth of tuning has not been quantified for these fibers, the squirrel monkey CT fibers are somewhat more narrowly tuned to their best stimulus than are those of the hamster. Fibers of the macaque CT nerve are divisible into four relatively distinct groups, responding best to sucrose, NaCl, HCl, or QHCl (Sato et al., 1975). Like the CT fibers in squirrel monkey, those in the macaque also appear to be more specifically tuned to their best stimulus than are those in either rat or hamster. Recent data from the chimpanzee also suggest a high degree of specificity of CT afferent fibers (Hellekant and Ninomiya, 1991). Thus, in comparison to rodents, the peripheral gustatory afferent fibers of primates are more narrowly tuned to the basic taste stimuli.

3.2. Convergence and Breadth of Tuning of Brainstem Neurons

Gustatory neurons in the CNS are typically more broadly tuned than peripheral nerve fibers in the same species. This increased breadth of tuning is likely due to convergence of peripheral fibers onto brainstem neurons. Neurons in the NST, for example, often can be driven by stimulation of taste receptors in anatomically separate regions of the oral cavity. In addition to greater breadth of tuning to gustatory stimuli, central taste neurons often respond to thermal and tactile stimuli, and at cortical levels to visual and olfactory stimuli.

3.2.1. Increase in Breadth of Tuning in NST and PbN

Experiments in which CT fibers and neurons of the NST and PbN of the hamster were stimulated by applying the same stimuli to the anterior tongue have revealed a systematic increase in the breadth of responsiveness of these cells from the periphery to the pons. A comparison of the mean response profiles of S-, N-, and H-best neurons of the hamster CT, NST, and PbN is shown in Fig. 14.3. The mean breadth of tuning (H) significantly increases from the CT to the PbN. In particular, the breadth of tuning of the S-best cells, which are the most specifically tuned fibers in the CT nerve, increases from 0.39 in the CT to 0.59 in the NST to 0.74 in the PbN (see Fig. 14.3). The S-best neurons are more broadly tuned in the pons than the very broadly tuned H-best fibers of the CT nerve (Van Buskirk and Smith, 1981). This reflects a change from an S-best CT fiber that responds predominantly to sucrose with only slight sideband sensitivity to one other stimulus (usually NaCl) to an S-best pontine neuron that responds well to at least three of the four stimuli. The broad tuning of hamster PbN neurons is quite evident in Figure 14.5, which shows the mean responses of S-, N-, and H-best PbN neurons to an array of 18 stimuli.

An increase in breadth of responsiveness from periphery to brainstem has been noted in other mammals as well (Travers, 1993), although the comparisons have been made across data collected in different laboratories and not necessarily employing stimuli at the same concentrations. Across five CT fiber experiments on rat, hamster, and monkey, the mean breadth of responsiveness was 0.56 (range = 0.52–0.61), whereas the mean breadth across 10 studies of NST cells in rat, hamster, and monkey was 0.74 (range = 0.62–0.86). Travers (1993) also compared these data to those from cells in insular cortex of rats and insular-opercular cortex of monkeys, where the mean breadth of tuning across six experiments was an intermediate level of 0.63 (range = 0.54–0.76). Given the

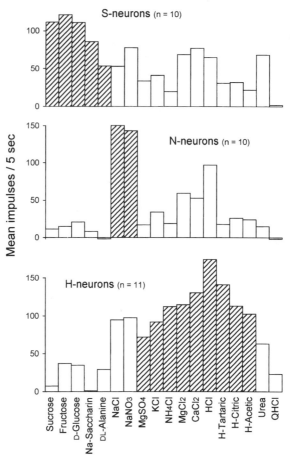

FIG. 14.5. Mean responses (impulses in 5 sec) of each neuron class in the hamster PbN to each of 18 stimuli. The sweet-tasting stimuli are shaded in the profile for the S-neurons, the sodium salts are shaded in the profile for the N-neurons, and the nonsodium salts and acids are shaded in the response profile for the H-neurons. (Modified from Smith et al. (1983a).)

caveat that these studies are not completely comparable across laboratories or species, these data suggest that the trend of increasing breadth of responsiveness between periphery and brainstem appears to hold across several species. They further suggest no additional increase in the breadth of tuning of taste cells between brainstem and cortex and perhaps even a slight return toward the narrower tuning characteristic of peripheral fibers. Data from several experiments on the macaque (Rolls, 1995) show that central gustatory neurons in this primate decrease their breadth of tuning between the NST and the cortex; NST cells have a breadth of tuning value of 0.87, frontal operculum cells a value of 0.67, neurons of the insula a value of 0.56, and the cells in the secondary cortical taste area of the orbitofrontal cortex a breadth of tuning of 0.39. In comparison, CT fibers in the macaque have a mean breadth of responsiveness of 0.57 (Sato et al., 1975), showing that the effect of synaptic processing from the CT to the insula is to increase the breadth of tuning and then return it to peripheral levels. Although cells in the orbitofrontal cortex are quite narrowly tuned to tastants, they are often multimodal, responding also to olfactory and visual stimuli (Rolls and Baylis, 1995).

3.2.2. Convergence onto Single NST Cells from Different Fields

There is some convergence of input onto single CT fibers, since each one branches to innervate several fungiform papillae. The sensitivities of each branch of a CT axon appear to be the same, suggesting little consequence of this peripheral convergence on taste information processing other than might accrue through spatial summation (Oakley, 1975). On the other hand, there is considerable convergence of peripheral fiber inputs onto second-order gustatory cells of the NST. The receptive fields of taste-responsive cells in the NST of the sheep are considerably larger than those of the CT nerve, suggesting that CT fibers converge onto secondary neurons in the NST (Vogt and Mistretta, 1990). About half of the taste-responsive cells of the rat NST receive converging input from separate taste bud regions. In a careful series of studies, Travers and colleagues demonstrated that the predominant patterns of convergence of peripheral fibers onto NST neurons involve taste buds of either the anterior oral cavity (i.e., anterior tongue, nasoincisor ducts, or sublingual organ), or those of the posterior oral cavity (i.e., foliate papillae, soft palate, or retromolar mucosa) (Travers et al., 1986). However, there is little evidence that anterior and posterior receptive fields combine in the responsiveness of an NST neuron. These patterns of convergence reflect the orderly anterior-to-posterior progression of food through the oral cavity during ingestion, making the simultaneous stimulation of the converging inputs likely. Although convergence has been examined mostly in the NST, it has also been noted at other levels of the central gustatory pathway. The available evidence suggests no greater convergence of separate taste bud subpopulations at pontine or thalamic levels, although the increased breadth of responding of S-best neurons in the hamster PbN (Fig. 14.3) suggests that within a receptor population (anterior tongue) there may be further convergence at higher levels.

3.2.3. Multimodal Sensitivity: Touch and Temperature

In addition to their extensive breadth of responsiveness across stimuli of different taste quality, gustatory neurons at all levels often respond to other sensory modalities, most often to intraoral somatosensory stimuli. Fibers of the CT nerve of rats and hamsters respond to both gustatory and thermal stimulation (Ogawa et al., 1968) and this sensitivity to warming and cooling is maintained in central gustatory neurons. There is a strong correlation at both peripheral and central levels between the responses to warming the tongue and to sucrose and between

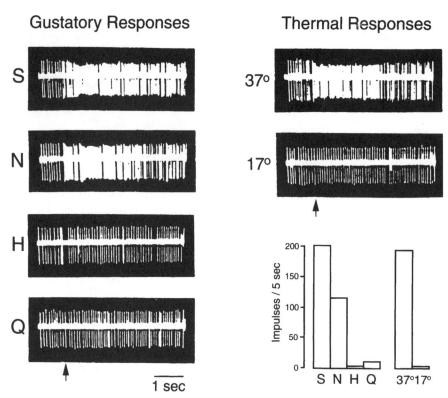

FIG. 14.6. Oscillographic records for the responses of a hamster PbN neuron to the four basic gustatory stimuli: sucrose (S), NaCl (N), HCl (H), and QHCl (Q) and to warming (37°C) and cooling (17°C) of the tongue. The arrows indicate the approximate onsets of the responses. The histograms in the lower right show the mean responses of this neuron to the six stimuli for three tests of the taste and two tests of the thermal stimuli. (From Travers and Smith (1984).)

the responses to cooling and to NaCl or HCl. The responses of a hamster PbN neuron to gustatory and thermal stimulation are shown in Figure 14.6. This S-best neuron responded also to NaCl and to 37°C distilled water (Travers and Smith, 1984). In the rat NST, one half of the taste-responsive cells with receptive fields in the anterior oral cavity and all of those with receptive fields in the posterior oral cavity are responsive also to mechanical stimulation of their receptive fields (Travers and Norgren, 1995). In fact, a number of studies have shown that about 64% of taste-responsive NST neurons receive converging input from gustatory and tactile receptors. The few studies that have examined this question at the level of the PbN and VPMpc show an average amount of gustatory/mechanical convergence of 77% and 88%, respectively (Travers, 1993). The reports of convergence between gustatory and tactile stimuli on cells of the rat insular cortex vary from 67% (Ogawa et al., 1990) to none (Yamamoto,

1984), which may reflect the intensity of the tactile stimuli used in the various experiments. In awake, freely ingesting rats, there is some convergence of both thermal and olfactory stimuli onto cortical gustatory neurons (Yamamoto et al., 1989). Multimodal convergence has been examined in primates most extensively in the secondary cortical taste area in the orbitofrontal cortex, where there are cells responding to combinations of taste and olfactory, visual, or somatosensory stimuli (Rolls and Baylis, 1995). Thus it is clear that taste-responsive cells in the central nervous system are driven often by stimuli differing in taste quality, intensity, and sensory modality. The contribution of the responses of these multimodal cells to the coding of any particular attribute, such as taste quality, must be assessed with this multiple sensitivity in mind.

3.2.4. Ambiguity in the Response of One Cell

Individual gustatory neurons, both peripheral and central, typically respond to stimuli representing several different taste qualities (Pfaffmann, 1955; Smith et al., 1983a, 1983b; Frank et al., 1988; Scott and Giza, 1990). Hamster taste neurons are more broadly tuned in the NST and PbN than in the CT nerve (see Fig. 14.3), and even in the periphery, taste fibers become more broadly responsive as stimulus concentration increases. Because both quality and intensity can modulate the responses of taste neurons, the response of any one neuron is entirely ambiguous with respect to either parameter (Pfaffmann, 1955, 1959). In addition, gustatory neurons are often responsive to thermal and tactile stimuli (Ogawa et al., 1968; Travers and Smith, 1984; Hanamori et al., 1987; Travers, 1993). Thus impulse traffic in a single neuron may be related to several stimulus modalities, making the unambiguous interpretation of that signal impossible without comparing it to activity in other cells (Crick, 1979; Erickson, 1982). Therefore, in thinking about how sensory information is coded in the gustatory system, it is important to remember that cells at all levels of the pathway are broadly responsive to stimuli that vary in perceptual quality, are more broadly responsive at high than at low intensities, and often are sensitive to other modalities, such as touch and temperature.

3.3. Temporal and Spatial Aspects of Taste Processing

The responses of taste neurons typically are analyzed by examining the impulse frequency during some defined period, such as the 1-sec or 5-sec interval after the arrival of the stimulus at the tongue. The responses to tastants, however, vary over time and to some extent the temporal parameters of the responses to different stimuli are relatively distinct. Consequently, some investigators have suggested that temporal characteristics may be important in coding information about taste quality. In addition, there is some evidence that cells responding to different stimuli are differentially distributed in central gustatory nuclei, suggesting the possibility of a chemotopic code for taste quality. Overall, however, the data supporting either a strictly temporal or topographic spatial code for gustatory quality are not compelling.

3.3.1. Time Course of Taste Responses: Differences Among Stimuli

The temporal characteristics of gustatory neural responses are often ignored in studies of taste physiology. Where they have been examined, investigators have noted that different classes of stimuli produce responses with different time courses. For example, a brief phasic response followed by a slowly declining tonic component characterizes responses to NaCl. The responses to sucrose, on the other hand, have a much less pronounced, if not absent, phasic component and are sometimes characterized by periodic bursts of impulses. There are differences among NaCl, HCl, QHCl, and sucrose in the time it takes the rat CT nerve to reach its maximum firing rate, the effects of stimulus concentration on that rise time, and the ratio of the phasic to the tonic component of the response (Harada et al., 1983). Responses of rat CT nerve fibers to different tastants are correlated with different temporal patterns, but this correlation is not precise enough to serve as an unambiguous code for taste quality. The time course of the response is largely a function of the stimulus, but also depends on which neurons are activated. Differences in time course, however, can add to the distinctions that occur in the activity across cells to enhance the neural distinction among stimuli, resulting in a kind of spatiotemporal pattern that provides a unique signature for each stimulus (Scott, 1987). Although differences in the temporal patterns of activity elicited by different stimuli may enhance the resolution among stimuli, the neural representation of taste quality is provided primarily by the pattern of activity across neurons, as discussed below.

3.3.2. Response to Constant Stimulation: Rate of Onset Versus Adaptation

The temporal characteristics of the response to a taste stimulus reflect a number of underlying factors, including the rate of stimulus application and the process of adaptation. The response to NaCl in the rat CT nerve has two distinct components: an initial phasic discharge followed by a tonic response component. This phasic/tonic response pattern is common to many stimuli and occurs in the gustatory responses of all species that have been investigated. Analysis of the time course of the NaCl response in the rat CT nerve shows it to decrease from the initial phasic peak in two exponential stages. The initial phasic response declines with a time constant of about 0.5 sec and this is followed by a tonic response component that decays much more slowly, with a time constant of about 30 sec (Smith et al., 1975). The ratio of the phasic and tonic components of the CT response to NaCl varies directly with the rate of stimulus onset. The magnitude of the phasic response is a power function of the rate of stimulus rise and if the stimulus is applied slowly enough the phasic response is completely eliminated (Smith and Bealer, 1975). These data show that the gustatory system is exquisitely sensitive to the rate of stimulus change. Taste neurons respond appropriately to increasing stimulus concentration, but the rate at which that concentration changes is a particularly salient feature of the stimulus. Therefore, the time course of the neural response to a tastant reflects not only the particular stimulus but also the rate at which it flows over the tongue. Beyond this initial rapid decline of the phasic response, there is a secondary decline during the tonic component of the response, which probably reflects adaptation at the level of the taste receptor cell, where a similar slow decline is observed.

3.3.3. Chemotopic Correlates of Taste in the CNS: Orotopic Representation

The idea that taste quality might be represented in the nervous system by a topographic pattern of activity has been around for many years, although it does not have strong experimental support. Topographic distributions of responsiveness have been noted in the NST of the rat (Halpern and Nelson, 1965), hamster (Dickman and Smith, 1989), and monkey (Scott et al., 1986). Generally, NaCl and sucrose responses are more prevalent in the most rostral portion of the NST and responses to HCl and QHCl are prominent at more caudal levels. Largely, these differences reflect the differential sensitivity of the CT and IXth nerves, since these nerves terminate within the NST in a rostral to caudal order, projecting an orotopic map onto the NST. However, even with only anterior tongue stimulation, there is some indication of a differential distribution of best-stimulus categories of cells in the hamster PbN. Seventy-five percent of N- and S-best cells are located in the caudal half and 65% of H-best neurons in the rostral half of the pontine taste area (Van Buskirk and Smith, 1981). A similar crude topographic organization of chemosensitivity occurs in the macaque NST, but not in the insular-opercular cortex of that species (Smith-Swintosky et al., 1991). In the rat cortex, however, distinct spatial patterns of responsiveness are generated by taste stimuli, due to differential input from the various cranial nerves. Sucrose responses are most dominant in the anterodorsal region, quinine responses in the posterior region, NaCl responses in the central and ventral regions, and HCl responses evenly distributed throughout the cortical taste area (Yamamoto et al., 1985). These "across-region patterns" have been proposed as a code for the representation of taste quality (Yamamoto, 1989). Even if such patterns can be reliably related to taste quality, their generation is dependent to a large extent on the topographic distribution of afferent inputs from different parts of the oral cavity, which serves to create the regional differences in responsiveness seen at various levels of the CNS. Since taste qualities can be reliably identified by humans following stimulation of a single fungiform papilla (Bealer and Smith, 1975), it is hard to imagine how such crude topographic representations of the entire oral cavity could serve as a neural code for gustatory quality.

4. TASTE INFORMATION PROCESSING: BEHAVIOR AND NEURAL CODING

Although we have already addressed the possibility of temporal and chemotopic coding, the principal debate over taste quality coding has focused on two competing theories: the *across-fiber pattern* theory and the *labeled-line* hypothesis (see also Chapter 1, Finger et al.). There has been considerable disagreement over the past two decades about the merits of these two ways of representing taste quality. The labeled-line hypothesis proposes that gustatory neuron types are specific coding channels for taste quality. The across-fiber pattern theory, on the other hand, proposes that quality is coded by the pattern of activity across neurons. In the following discussion, we examine the role of gustatory neuron types in the definition of the across-neuron pattern. From the result of such an examination, we see that the separate neuron types establish and define the distinctions among the patterns that are necessary for taste discrimination. Therefore, the same neurons are critical for coding taste quality, regardless of whether they are considered a "labeled line" or a necessary part of the "across-fiber pattern." Taste discrimination is impossible without the contribution of each neuron type. In this section we first consider the role of taste in behaivior and then examine the mechanisms used for the representation of taste information in the nervous system.

4.1. What Role Does Taste Play in Behavior?

Before examining how information about taste stimuli is coded in the nervous system, it is instructive to consider what functions taste serves in the life of an organism. Taste information is important to help ensure the animal's energy supply (sweet), maintain the proper electrolyte balance (salt), and avoid the ingestion of toxic substances (sour, bitter). As such, the gustatory system serves as the gatekeeper of the internal milieu, acting to guide ingestive and avoidance behaviors (see also Chapter 13, Glendinning et al.). Because the anatomical organization of the central taste pathway parallels that of other visceral sensory input, and because taste stimulation triggers visceral responses, it is useful to consider taste as a component of the visceral system. Taste serves both a discriminative and an affective function, allowing animals to make distinctions among potential foods and providing information that guides decisions about the ingestion or rejection of palatable and unpalatable substances.

4.1.1. Perception Versus Visceral Function

As humans, it is natural for us to think about the gustatory system in terms of our own experience. Consequently, much of the research on the taste system has been directed toward understanding the neural basis of the human sensations of sweet, salty, sour, and bitter. These qualities are associated with stimuli that are important for organisms to recognize as nutrients or potential toxins. Preference and subsequent ingestion, for example, characterize the response to sweet-tasting carbohydrates, whereas bitter-tasting alkaloids and other toxins are avoided (Pfaffmann, 1964). Certainly these perceptual categories are important to our investigation of the taste system, but input from gustatory receptors also directly controls an array of reflexive motor responses involved in food ingestion and a range of autonomic responses important for digestive and other homeostatic processes, such as electrolyte balance. Gustatory stimulation produces effects on the digestion, absorption, and utilization of nutrients by eliciting a range of neurally mediated reflexes (Friedman and Mattes, 1991). Salivary, gastric, and pancreatic secretions and cardiovascular and thermoregulatory responses resulting from gustatory stimulation are referred to as cephalic phase reflexes, which prepare the body for the arrival and utilization of nutrients. Stimulation of the laryngeal epithelium with water produces diuresis, mediated by changes in the release of antidiuretic hormone (Shingai et al., 1988). These and other data demonstrate that gustatory stimulation produces a variety of responses beyond taste perception. Therefore, understanding the neural organization of the gustatory system requires consideration of the many different roles that taste plays in the physiology of the organism, from perception to autonomic function.

4.1.2. Ingestion and Rejection: Taste Reactivity

Grill and Norgren (1978b) first described the reflexive behaviors in the rat resulting from gustatory stimulation. They termed these ingestive and aversive response sequences, which largely reflect stimulus palatability, "taste reactivity." Sucrose and other palatable stimuli evoke a pattern of ingestive responses that begins with rhythmic mouth movements followed by midline tongue protrusions, lateral tongue flicks, and swallowing. Unpalatable stimuli, such as QHCl, produce a response sequence that includes oral gapes and a number of somatic motor responses associated with rejection, such as chin rubbing, paw flailing, head shaking, and face washing. These response sequences occur in decerebrate rats (Grill and Norgren, 1978c), suggesting that the neural substrate for these reflexive behaviors is intact within the brainstem. In normal animals, but not in decerebrate ones, these responses are modifiable by learning, so that a normally ingestive sequence becomes an aversive one after a conditioned taste aversion (Grill and Norgren, 1978a). In general, the results from taste reactivity tests correspond with other short-term palatability measures such as lick-rate tests. These experiments demonstrate that much of the neural distinction among stimuli of different qualities, as evidenced by specific patterns of motor output, can occur within the hindbrain and does not require the contribution of more rostral levels of the gustatory system.

4.1.3. Generalization and Discrimination

In order to make decisions about ingestion or rejection, the ability to discriminate among different taste stimuli is essential. Behavioral studies have employed operant conditioning techniques or conditioned taste aversion generalization to determine the behavioral capacity of the gustatory system of experimental animals. A taste aversion can be conditioned to a particular tastant by pairing that stimulus with gastrointestinal malaise produced by drugs or radiation. Following such learning, the degree to which the aversion generalizes to other stimuli can serve as an indication of the gustatory similarity among a number of tastants. The results of these kinds of studies show, for example, that hamsters and rats generalize a learned sucrose aversion to several other stimuli described as sweet by humans, including fructose, glucose, and Na-saccharin (Smith et al., 1979; Nowlis et al., 1980). These techniques also demonstrate that a number of sodium and lithium salts are considered similar by rodents and distinct from nonsodium salts such as KCl or NH_4Cl, which are behaviorally similar to acids. Finally, bitter-tasting substances, such as QHCl and $MgSO_4$, show cross generalization in these types of experiments. To a considerable extent, such generalization studies produce groupings of gustatory stimuli that are similar to those produced by analyzing the responses of peripheral or central gustatory neurons of the same species. That is, stimuli that produce similar patterns of neural activity or that drive the same sets of neurons, are grouped together in behavioral generalization tests.

Discriminative operant conditioning has been used to examine both the generalization and the discriminability among gustatory stimuli. Using shock avoidance, Erickson (1963) demonstrated that rats conditioned to avoid KCl would also avoid NH_4Cl but not NaCl. A similar pattern of generalization occurred with an appetitive operant task, in which rats were trained to respond on one of two levers, depending upon the taste of a previously conditioned stimulus (Morrison, 1967). Transection of the CT nerve but not the IXth nerve severely impairs the ability of rats to discriminate between NaCl and KCl (Spector and Grill, 1992). These and other data suggest strongly that information arriving via the CT nerve is important for discriminating NaCl from nonsodium salts. Treatment of the tongue with amiloride blocks the response in N-best fibers of the CT nerve to NaCl and eliminates the ability of rats to make a discrimination between NaCl and KCl (Spector et al., 1996). In addition, when rats are conditioned to avoid NaCl in the presence of amiloride, they generalize that aversion to KCl (Hill et al., 1990). Since rats do not have an amiloride-sensitive response to NaCl in the IXth nerve (Formaker and Hill, 1991) nor even any N-best IXth nerve fibers (Frank, 1991), it is undoubtedly the comparison between N-best and other nerve fibers in the CT and the GSP that allows this discrimination. Through such careful behavioral dissection of the response to tastants, a role for the differential inputs of the various cranial nerves subserving taste is beginning to emerge.

4.2. Theories of Quality Coding

Prior to the development of neurophysiological recording methods, a long tradition of human psychophysical research had provided considerable support for the notion that taste experience could be reduced to a few basic qualities, although not necessarily the traditional four (see also Chapter 16, Breslin). This idea, combined with Meuller's doctrine of specific nerve energies, led to the expectation that the perception of taste quality would arise from the activation of a few neuron types, each coding a single taste quality. This strict "labeled-line" theory was discounted by early neurophysiological recordings showing that peripheral taste fibers in several species are responsive to stimuli representing more than one taste quality (Pfaffmann, 1955). In response, an "across-fiber pattern" theory was proposed, which held that taste quality is coded by the relative activity across the population of responsive neurons (Pfaffmann, 1959; Erickson, 1968, 1982). This theory accommodates the multiple sensitivity of taste fibers and requires neither specific fiber types nor taste primaries. However, the prevailing view of the importance of primary tastes led to a modification of the labeled-line theory, which proposed that taste quality is coded by the activity in a few "best-stimulus" channels, that is, by neurons that respond best, but not specifically, to one of the basic taste qualities (Pfaffmann, 1974; Pfaffmann et al., 1976).

Although each coding theory has its strengths, both strain to encompass the full range of data. For example, accumulating evidence suggests that there are indeed functional classes of neurons that correspond in some way to primary taste qualities (Frank, 1973, 1991; Smith et al., 1983b; Frank et al., 1988; Hanamori et al., 1988; Ninomiya and Funakoshi, 1988; Hettinger and Frank, 1990; Scott and Giza, 1990; Smith and Frank, 1993). On the other hand, analysis of the relative activity of these neuron groups shows that no single class of cells in isolation can reliably discriminate between different taste qualities (Smith et al., 1983b; Smith and Frank, 1993). The neural coding problem essentially rests on whether the activity in a given taste neuron is an unambiguous representation of the quality of the stimulus applied to its receptors (i.e., a labeled line) or whether this activity is meaningful only in the context of activity in other afferent neurons (i.e., in the pattern of responses). Here we review the neurophysiological data that bear on this issue and conclude that taste quality is coded in the relative activity across several neuron types.

4.2.1. Response Profiles: Labeled Lines

Although mammalian taste neurons are broadly tuned, many investigators have attempted to group them into functionally meaningful categories (Pfaffmann, 1955; Frank, 1973; Travers and Smith, 1979; Van Buskirk and Smith, 1981; Smith et al., 1983b; Frank et al., 1988; Hanamori et al., 1988). Hierarchical cluster analyses of the similarities and differences in the shapes of their response profiles result in distinct groups of CT (Frank et al., 1988) and IXth nerve (Hanamori et al., 1988) fibers that correspond to these "best-stimulus" groups. The implication of distinct fiber types in the coding of taste quality began with the categorization of hamster CT fibers into best-stimulus groups (Frank, 1973). This type of categorization became the focus of an ensuing controversy over the neural representation of taste quality when Pfaffmann (1974, 1976), in a reversal of his earlier ideas about pattern coding, proposed that these fiber types code taste quality in a labeled-line fashion. This hypothesis suggests that "sweetness" is coded by activity in S-best neurons, "saltiness" by activity in N-best neurons, etc. Thus, in contrast to a "population" approach to taste coding (across-neuron patterns), this labeled-line position advocates a "feature extraction" approach, in which particular neurons (or groups of neurons) play specific roles in the representation of taste quality. A labeled-line code, by definition, implies that activity in that "line" carries a specific message, without any reliance on the activity in other "lines" (Perkel and Bullock, 1968).

4.2.2. Breadth of Tuning: Across-Fiber Patterns

The multiple sensitivity of fibers in the CT nerve originally led Pfaffmann (1995, 1959) to propose that taste quality is coded by the pattern of activity across taste fibers. With this coding hypothesis, taste quality remains invariant with increased intensity, although any single neuron may increase its breadth of responsiveness. The pattern of activity generated across the entire array of taste neurons at a higher concentration is similar in shape, but varies in amplitude (Erickson, 1968, 1982). Stimuli with similar tastes, such as sweeteners or sodium salts, generate highly similar patterns of activity across afferent taste neurons. The similarities among these patterns are typically measured by calculating the across-neuron correlation between pairs of stimuli (Erickson, 1963), although other indices have been proposed. Behavioral investigations show that stimuli evoking highly correlated neural patterns are judged by experimental animals to have similar tastes (Erickson, 1963; Morrison, 1967; Smith et al., 1979; Nowlis et al., 1980). This across-fiber pattern view of quality coding incorporates the multiple sensitivity of gustatory neurons as an essential part of the neural code for taste quality. This theoretical view stresses that the code for quality is given in the response of the entire population of cells, placing little or no emphasis on the role of an individual neuron. Erickson (1968) has argued that pattern coding is used by many sensory systems, particularly nontopographic modalities (such as color vision and taste) employing neurons that are broadly tuned across their stimulus array. Even in insects, where the peripheral taste receptor cells are rather narrowly tuned, across-fiber patterns are thought to play a role in the discrimination among potential food sources (Dethier, 1976).

When the across-neuron correlations among the responses to an array of gustatory stimuli are calculated across either peripheral or central neurons, stimuli with similar tastes correlate highly and those with different tastes correlate less (Pfaffmann et al., 1976; Travers and Smith, 1979; Van Buskirk and Smith, 1981; Erickson, 1982; Smith et al., 1983a; Frank et al., 1988; Scott and Giza, 1990). Often these across-neuron correlations serve as input to a multivariate statistical procedure, used to a generate a "taste space" representing the neural similarities and differences among the stimuli. A taste space for 18 stimuli is shown in Figure 14.7; this three-dimensional space was derived from correlations among the responses of 31 hamster PbN neurons and was generated using multidimensional scaling (Smith et al., 1983a). Within this space, there is a clear separation between the sweet-tasting stimuli, the sodium salts, the nonsodium salts and acids, and the two bitter-tasting stimuli. This arrangement of stimuli based on similarities among their across-neuron patterns suggests that there is

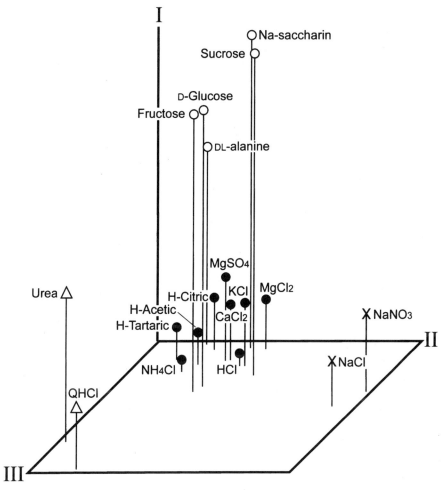

FIG. 14.7. Three-dimensional space showing the locations of 18 stimuli, obtained from multidimensional scaling (KYST) of the correlations in the responses evoked across 31 hamster PbN neurons. Four groups of stimuli are indicated by different symbols: sweet-tasting (open circles), sodium salts ($\times$), nonsodium salts and acids (solid circles), and bitter-tasting (open triangles). (Modified from Smith et al. (1983a).)

sufficient information within these patterns to discriminate among these four groups of stimuli, even though any one cell in the hamster PbN is likely to respond to stimuli of more than one group (see above, Fig. 14.5).

4.3. The Role of Neuron Types in Taste Quality Coding

The labeled-line hypothesis requires the existence of neuron types for the coding of taste quality, whereas the across-neuron pattern theory does not. The number of labeled lines would equal the number of discrete taste qualities, which would each be signaled by activity in separate afferent channels. Consequently, the existence of gustatory neuron types has been sharply contested, based on the assumption that their existence somehow implicates them as labeled lines. However, the mere existence of neuron or fiber types (defined by their best stimulus, similarities in their response profiles, or by other criteria such as their amiloride sensitivity) does not necessarily imply that these classes of cells comprise labeled lines. A classic example where receptor types are evident but where there is general agreement about the existence of a pattern code is in vertebrate color vision (Erickson, 1968, 1982; Crick, 1979; Smith and Frank, 1993).

4.3.1. Neural Discrimination of Taste Inputs

Because peripheral taste fibers and central neurons are broadly tuned, it is difficult for the activity in a single neuron or even in a single neuron type to discriminate between stimuli that differ in taste quality. Several years ago, Smith and colleagues (Smith et al., 1983a, 1983b) examined the contribution of separate neuron types in the hamster brainstem to the across-neuron patterns and the ability of these neuron types alone or in combination to discriminate between stimuli. These investigators showed that each neuron type is necessary for determining the similarities among stimuli for which they are "best," but that the neural discrimination between stimuli of different classes depends upon the activity in more than one neuron type. For example, within a multivariate taste space based on the across-neuron patterns generated from the responses of neurons in the hamster PbN, sodium salts and nonsodium salts are readily distinguishable (see Fig. 14.7). However, if either the N-best or the H-best class of cells, which are preferentially sensitive to sodium salts and nonsodium salts, respectively, are missing from the data matrix, the across-neuron patterns for the remaining cells cannot distinguish these two classes of salts. Thus, the neural discrimination between stimuli with different tastes depends upon the relative activity in different neuron types; one neuron type alone is insufficient to discriminate between stimuli with different taste qualities.

This "across-neuron type" code is simply a restatement of the across-fiber pattern theory first proposed by Pfaffmann (1959), except that it puts an emphasis on the activity in recognizable neuron types. This coding mechanism is similar to the coding of stimulus wavelength by vertebrate visual systems, where three types of broadly sensitive photoreceptor pigments are involved. The color of the wavelength of light falling on the retina can be accurately encoded by considering the relative activity in these three photoreceptors, that is, by a pattern (Erickson, 1968; Boynton, 1971). Deficiencies in one or more of the photoreceptor pigments result in various forms of visual chromatic deficiency, or "color blindness." Similarly, the neural discrimination among different classes of gustatory stimuli depends upon the relative activity in different neuron types (Smith and Frank, 1993).

4.3.2. Comparison of Activity Across Neuron Types

There are no known analogies to color blindness in taste, but the experiments of Scott and his colleagues (Scott and Giza, 1990) on the rat NST provide an experimental demonstration of an analogous phenomenon. Blocking the NaCl response of the N-best neurons with lingual application of amiloride results in the inability of cells in the NST to discriminate between sodium and nonsodium salts; that is, their across-neuron patterns are not distinct without the activity of N-best cells. As shown by Smith and colleagues in the hamster (Smith et al., 1983a), without the differential response of two neuron types (N-best and H-best) to these two classes of stimuli, they are coexistent within the taste space of the amiloride-treated rat. These results are compatible with the interpretation that taste quality discrimination depends on a comparison of activity across broadly tuned neuron types, comparable to the coding of color vision by broadly tuned photoreceptors (Smith et al., 1983a; Smith and Frank, 1993). In both taste and color vision, elimination of any one cell type results in a lack of separation between stimuli of different quality within multidimensional space (Scott and Giza, 1990; Smith and Frank, 1993) and a loss of behavioral discrimination among the same stimuli (Graham et al., 1961; Boynton, 1971; Hill et al., 1990; Spector et al., 1996).

The data cited above show that no one gustatory neuron type alone is capable of providing information that can reliably distinguish the across-neuron patterns evoked by dissimilar-tasting compounds. More than one neuron type must contribute to the pattern in order for the patterns evoked by unlike stimuli to be distinct. Thus, in that sense, the various neuron types (S-, N-, H-, or Q-best) are critically important for the discrimination of taste quality. The N-best cells or H-best cells can define the similarities among sodium salts, but activity in both neuron types is required to distinguish sodium salts from nonsodium salts and acids. Behavioral and neural data in rats support this requirement, but there is no evidence to date that N-best cells alone are labeled lines signaling "saltiness." On the contrary, N-best cells provide the critical part of a pattern across neuron types that codes saltiness; these cells by themselves cannot distinguish sodium salts from other stimuli. It is the activity in these particular neuron types that largely defines the unique patterns that represent taste quality. In that sense, whether these neuron types are viewed as a "labeled line" or the critical part of an "across-fiber pattern" is an irrelevant question. Taste discrimination depends on the differential activity in these separate neuron classes.

5. MODULATION OF AFFERENT GUSTATORY SIGNALS

Responses to gustatory stimulation recorded from brainstem cells in rodents are subject to several modulatory influences, including the effects of gastric distension, blood glucose, and insulin levels, and conditioned taste aversion learning.

That is, the responses of rodent brainstem cells are modulated not only by stimulus quality, intensity, and modality, but also by the animal's physiologic state and prior experience. In primates, taste neurons in the orbitofrontal cortex but not in the brainstem are modulated by the animal's state of satiety. Although the circuitry underlying such modulatory effects on taste processing are not well understood, in rodents there are descending connections to brainstem taste nuclei from all of the forebrain areas that receive gustatory input, including the cortex, amygdala, and lateral hypothalamus. Descending influences on brainstem taste responses can be either excitatory or inhibitory. Neurophysiological studies have implicated a number of neurotransmitters and peptides in the modulation of NST neuronal activity.

5.1. Effects of Physiologic State on Taste Processing

Taste information is important for guiding food ingestion and it contributes to autonomic regulation by evoking cephalic phase responses. Therefore, it is not surprising that visceral afferent information might feed back onto the gustatory system to alter its response to food-related substances. Although the number of studies is relatively small, there is evidence from rodents that a number of physiologic factors can modulate the responses of brainstem taste neurons.

5.1.1. Physiologic Factors

Several physiologic factors can alter the responsiveness of gustatory neurons. In rodents, these effects are seen in neurons of the NST. In acute, anesthetized preparations, intravenous administration of blood glucose (Giza and Scott, 1983), insulin (Giza and Scott, 1987), or pancreatic glucagon (Giza et al., 1992) decreases the response of NST neurons to glucose, with little or no effect on the responses to other taste stimuli. The selective effect of these satiety factors on the response to sweet-tasting stimuli also occurs following gastric distension (Glenn and Erickson, 1976). On the other hand, systemic administration of cholecy-stokinin (CCK), which has been implicated in satiety in a number of studies, has no appreciable effect on taste responses in the rat NST (Giza et al., 1990). All of these factors appear to preferentially suppress responses to stimuli that normally drive ingestion, suggesting that decreasing the gustatory effect of these stimuli helps to promote the termination of feeding. In contrast, recordings from small groups of NST neurons in behaving monkeys show that the responses to glucose, NaCl, HCl, QHCl, or blackcurrant juice are completely unchanged when the animals feed to satiety (Yaxley et al., 1985). On the other hand, neurons of the monkey's orbitofrontal cortex demonstrate the effects of sensory-specific satiety, decreasing their response during feeding, but only to the consumed substance (Rolls et al., 1989). Such effects do not occur in primary taste cortex of the insula (Yaxley et al., 1998) and frontal operculum (Rolls et al., 1988). In both primates (Rolls et al., 1986) and rats (Norgren, 1970), the state of satiety can modulate the responsiveness of lateral hypothalamic neurons to taste stimuli. These differences between rodents and primates in the influence of physiologic state on taste responsiveness may reflect differences in the anatomy of the ascending and descending gustatory pathways, which bypass the PbN in primates, or differences between awake, behaving animals and acute, anesthetized preparations. Nevertheless, these data show that the level of satiety can have a regulatory effect on gustatory afferent processing.

5.1.2. Descending Modulation

The brainstem gustatory nuclei receive descending projections from their forebrain targets, including insular cortex, central nucleus of the amygdala, bed nucleus of the stria terminalis, and lateral hypothalamus, all areas known to be involved in visceral regulation and in the control of ingestive and motivated behavior. Although it is well known that the discriminative capacity of the gustatory system is intact within the hindbrain of decerebrate animals (Grill and Norgren, 1978c), there is also evidence that taste responses in brainstem nuclei are altered followng decerebration (DiLorenzo, 1988; Mark et al., 1988). Generally, descending influences from the forebrain on taste neurons are excitatory, although there is some evidence for inhibitory effects as well. Reversible suppression of the gustatory cortex with procaine results in altered responses to taste stimulation in rat PbN (DiLorenzo, 1990) and NST (DiLorenzo and Monroe, 1995) neurons, showing that these cells can be either facilitated or inhibited by descending corticofugal inputs. Stimulation of the hamster gustatory cortex produces either excitatory or inhibitory corticofugal modulation of at least one-third of taste-responsive cells sampled from the NST; the inhibition is mediated by $GABA_A$ receptors (Smith et al., 1998a). Lateral hypothalamic stimulation can either excite (Matsuo et al., 1984) or inhibit (Murzi et al., 1986) the activity of gustatory cells in the NST. Such descending circuits may be involved in the modification of taste responsiveness in the NST that has been shown to follow a conditioned taste aversion (Chang and Scott, 1984), since such learned behavior depends upon an intact forebrain (Grill and Norgren, 1978a). Given the complexity of the central gustatory pathway, involving both thalamocortical and ventral forebrain afferent projections and descending connections, much remains to be learned about how taste information processing is influenced by these circuits.

5.2. Synaptic Modulation of Gustatory Activity

Until recently, little was known about the neurotransmitters and neuromodulators involved in gustatory processing. Several neuroactive substances have been identified in the gustatory NST, including substance P (SP), tyrosine hydroxylase (TH, probably reflecting the dopamine phenotype), methionine enkephalin, GABA, excitatory amino acids, and calbindin (Davis and Jang, 1988; Lasiter and Kachele, 1988; Davis, 1993; Davis and Kream, 1993).

In vitro experiments on brainstem slice preparations show that cells in the gustatory NST of the rat are excited by glutamate agonists (Wang and Bradley, 1995). Electrical stimulation of the solitary tract produces excitatory postsynaptic potentials in rostral NST neurons, which are blocked or reduced by both NMDA and non-NMDA glutamate receptor antagonists, suggesting that glutamate is an excitatory neurotransmitter between peripheral fibers and NST neurons. A similar conclusion follows from experiments on in vitro slices through the goldfish gustatory lobe (Smeraski et al., 1999). The responses of NST neurons in vivo to either gustatory stimuli or anodal current applied to the anterior tongue and the cells' spontaneous activity are completely blocked by glutamate receptor antagonists (Li and Smith, 1977), suggesting that the majority of excitatory input to these cells, including any potential descending ones, is glutamatergic.

GABA, acting at both $GABA_A$ and $GABA_B$ receptors, mediates inhibition within the rostral NST of both the rat (Wang and Bradley, 1993) and hamster (Liu et al., 1993). GABA inhibits approximately 68% of neurons in the rostral NST of both species in vitro. Similarly, the application of GABA in vivo suppresses the activity of about 60% of taste-responsive NST neurons (Smith and Li, 1998). Gustatory NST neurons are maintained under a tonic GABAergic inhibition, as shown by the excitatory effect of microinjections of the $GABA_A$ receptor antagonist, bicuculline methiodide (Smith and Li, 1998). Tonic inhibition is also evident in 400-µm thick brainstem slices (Liu et al., 1993; Wang and Bradley, 1993), suggesting that much of this inhibition arises within the brainstem. One of the roles of GABA in this system, as in other sensory systems, is to modulate the breadth of tuning of the cells. Following local administration of bicuculline, there is a significant increase in the breadth of responsiveness of cells in the hamster NST (Smith and Li, 1998).

Microinjection of substance P into the hamster NST produces enhancement of the responses of more than one-half of taste-responsive cells, whether stimulated with anodal current or chemical stimulation of the tongue (Davis and Smith, 1997). Substance P suppresses the responses of a small number of cells, as has also been shown in vitro (King et al., 1993). When gustatory cells in the NST are classified into best-stimulus categories, the enhanced response to NaCl following Substance P occurs predominantly in N-best neurons (Smith et al., 1998a), although this analysis has not yet been attempted for other tastants.

Systemic administration of morphine produces a suppression of gustatory activity in the PbN of the rat, which is reversible by naloxone (Hermann and Novin, 1980). Microinjection of met-enkephalin into the hamster NST produces a dose-dependent inhibition of the activity of some taste-responsive neurons (Smith et al., 1998b). Conversely, stimulation of taste receptors by sucrose produces an opioid-mediated analgesia in newborn rats (Blass et al., 1987). These data and others on the effects of opioids on food intake, suggest a two-way interaction between the gustatory and nociceptive systems. Although additional experiments are needed to further characterize the modulation of taste activity by opioid peptides, these data and those on glutamate, GABA and Substance P show that taste-responsive neurons in the NST are subject to modulation by a variety of mechanisms. Understanding how these various systems function to shape the responsiveness of NST cells will require careful analysis of the anatomical substrates that control these systems, their neuropharmacology, both in vivo and in vitro, and the way in which they modulate the responses of the different gustatory neuron types to different stimuli.

6. SUMMARY

The sense of taste is represented by neural activity in cells that are relatively broadly responsive to perceptually distinct stimuli. Although peripheral taste neurons in invertebrates and some species of fish are sometimes responsive to restricted classes of stimuli, peripheral taste fibers are relatively broadly tuned in mammals. This breadth of tuning increases through convergence onto central gustatory neurons, which are also often responsive to tactile and thermal stimuli. The response profiles of gustatory neurons at all levels of the nervous system have been used to classify neurons into types, based on their best stimulus. However, the broad responsiveness of mammalian gustatory neurons strongly suggests that gustatory quality cannot be represented by the activity in particular neural groups, such as NaCl- or sucrose-best, but that it is the pattern of activity across these broadly tuned cells that represents taste quality. In addition to providing information important for the discrimination of taste quality, gustatory input also drives the initial components of food ingesion (and rejection). Both the anatomical substrate and the functional role of taste emphasize its relationship to other visceral sensory inputs. There is evidence that visceral events, such as the state of sodium balance, blood glucose levels, or gastric distension, can feed back onto gustatory cells to modulate the flow of information. Synaptic modulation of central gustatory neurons is effected by a number of neurotransmitters and neuropeptides, including glutamate, GABA, substance P, and enkephalin. Thus, the processing of gustatory infomation through the taste pathway is carried out by broadly tuned cells that are subject to synaptic modulation in response to metabolic need.

REFERENCES

Allen, G. V., C. B. Saper, K. M. Hurley, and D. F. Cechetto (1991). Organization of visceral and limbic connections in the insular cortex of the rat. *J Comp Neurol* **311**:1–23.

Atema, J. (1971). Structures and functions of the sense of taste in the catfish (*Ictalurus natalis*). *Brain Behav Evol* **4**:273–294.

Bealer, S. L. and D. V. Smith (1975). Multiple sensitivity to chemical stimuli in single human taste papillae. *Physiol Behav* **14**:795–799.

Blass, E., E. Fitzgerald, and P. Kehoe (1987). Interactions between sucrose, pain and isolation distress. *Pharmacol Biochem Behav* **26**:483–489.

Blomquist, A. J., R. M. Benjamin, and R. Emmers (1962). Distribution of thalamic units responsive to thermal, mechanical and gustatory stimulation of the tongue of the albino rat. *Fed Proc* **21**:343.

Boynton, R. M. (1971). Color vision. In: J. W. Kling and L. A. Riggs (eds). *Woodworth and Schlosberg's Experimental Psychology.* New York: Holt, Rinehart and Winston, pp 315–368.

Chang, F.-C. T., and T. R. Scott (1984). Conditioned taste aversions modify neural responses in the rat nucleus tractus solitarius. *J Neurosci* **4**:1850–1862.

Crick, F. H. C. (1979). Thinking about the brain. *Sci Am* **241**:219–232.

Davenport, C. J. and J. Caprio (1982). Taste and tactile recordings from the ramus recurrens facialis innervating flank taste buds in the catfish. *J Comp Physiol* **147**:217–229.

Davis, B. J. (1988). Computer-generated rotation analyses reveal a key three-dimensional organizational feature of the nucleus of the solitary tract. *Brain Res Bull* **20**:545–548.

Davis, B. J. (1993). GABA-like immunoreactivity in the gustatory zone of the nucleus of the solitary tract in the hamster: light and electron microscopic studies. *Brain Res Bull* **30**:69–77.

Davis, B. J. (1998). Synaptic relationships between the chorda tympani and tyrosine hydroxylase-immunoreactive dendritic processes in the gustatory zone of the nucleus of the solitary tract in the hamster. *J Comp Neurol* **392**:78–91.

Davis, B. J. and T. Jang (1988). Tyrosine hydroxylase-like and dopamine-β-hydroxylase-like immunoreactivity in the gustatory zone of the nucleus of the solitary tract in the hamster: light- and electron-microscopic studies. *Neuroscience* **27**:949–964.

Davis, B. J., and R. M. Kream (1993). Distribution of tachykinin- and opioid-expressing neurons in the hamster solitary nucleus: an immuno- and in situ hybridization histochemical study. *Brain Res* **616**:6–16.

Davis, B. J. and D. V. Smith (1997). Substance P modulates taste responses in the nucleus of the solitary tract of the hamster. *Neuroreport* **8**:1723–1727.

Dethier, V. G. (1976). *The Hungry Fly: A Physiological Study of the Behavior Associated with Feeding.* Cambridge, MA: Harvard University Press, 489 pp.

Dethier, V. G. and Bowdan, E. (1989). The effect of alkaloids on the sugar receptors of the blowfly. *Physiol Ent* **14**:127–136.

Dickman, J. D., and D. V. Smith (1989). Topographic distribution of taste responsiveness in the hamster medulla. *Chem Senses* **14**:231–247.

DiLorenzo, P. M. (1988). Taste responses in the parabrachial pons of decerebrate rats. *J Neurophysiol* **59**:1871–1887.

DiLorenzo, P. M. (1990). Corticofugal influence on taste responses in the parabrachial pons of the rat. *Brain Res* **530**:73–84.

DiLorenzo, P. M., and S. Monroe (1995). Corticofugal influence on taste responses in the nucleus of the solitary tract in the rat. *J Neurophysiol* **74**:258–272.

Erickson, R. P. (1963). Sensory neural patterns and gustation. In: Y. Zotterman (ed). *Olfaction and Taste.* Oxford: Pergamon, pp 205–213.

Erickson, R. P. (1968). Stimulus coding in topographic and non-topographic afferent modalities: on the significance of the activity of individual sensory neurons. *Psychol Rev* **75**:447–465.

Erickson, R. P. (1982). The "across-fiber pattern" theory: an organizing principle for molar neural function. In: W. D. Neff (ed). *Contributions to Sensory Physiology, Vol. 6.* New York: Academic Press, pp 79–110.

Finger, T. E. (1987). Gustatory nuclei and pathways in the central nervous system. In: T. E. Finger and W. L. Silver (eds). *Neurobiology of Taste and Smell.* New York: John Wiley, pp 331–354.

Finger, T. E., and Y. Morita (1985). Two gustatory systems: facial and vagal gustatory nuclei have different brainstem connections. *Science* **227**:776–778.

Formaker, B. K. and D. L. Hill (1991). Lack of amiloride sensitivity in SHR and WKY glossopharyngeal taste responses to NaCl. *Physiol Behav* **50**:765–769.

Frank, M. E. (1973). An analysis of hamster afferent taste nerve response functions. *J Gen Physiol* **61**:588–618.

Frank, M. E. (1991). Taste-responsive neurons of the glossopharyngeal nerve of the rat. *J Neurophysiol* **65**:1452–1463.

Frank, M. E., S. L. Bieber, and D. V. Smith (1988). The organization of taste sensibilities in hamster chorda tympani nerve fibers. *J Gen Physiol* **91**:861–896.

Friedman, M. I. and R. D. Mattes (1991). Chemical senses and nutrition. In: T. V. Getchell, R. L. Doty, L. M. Bartoshuk, and J. B. Snow, Jr. (eds). *Smell and Taste in Health and Disease.* New York: Raven Press, pp 391–404.

Giza, B. K. and T. R. Scott (1983). Blood glucose selectively affects taste-evoked activity in rat nucleus tractus solitarius. *Physiol Behav* **31**:643–650.

Giza, B. K. and T. R. Scott (1987). Intravenous insulin infusions in rats decrease gustatory-evoked responses to sugars. *Am J Physiol* **252**(*Regul Integrat Comp Physiol 21*):R994–R1002.

Giza, B. K., T. R. Scott, and R. F. Antonucci (1990). Effect of cholecystokinin on taste responsiveness in rats. *Am J Physiol* **258**(*Regul Integrat Comp Physiol 27*): R1371–R1379.

Giza, B. K., T. R. Scott, and D. A. Vanderweele (1992). Administration of satiety factors and gustatory responsiveness in the nucleus tractus solitarius of the rat. *Brain Res Bull* **28**: 637–639.

Glenn, J. F. and R. P. Erickson (1976). Gastric modulation of gustatory afferent activity. *Physiol Behav* **16**:561–568.

Gordon, K. D. and J. Caprio (1985). Taste responses to amino acids in the southern leopard frog, *Rana sphenocephala. Comp Biochem Physiol A* **81**:525–530.

Graham, C. H., H. G. Sperling, Y. Hsia, and A. H. Coulson (1961). The determination of some visual functions of a unilaterally color-blind subject: methods and results. *J Psychol* **51**: 3–32.

Grill, H. J. and R. Norgren (1978a). Chronically decerebrate rats demonstrate satiation but not bait shyness. *Science* **201**:267–269.

Grill, H. J. and R. Norgren (1978b). The taste reactivity test. I. Mimetic responses to gustatory stimuli in neurologically normal rats. *Brain Res* **143**:263–280.

Grill, H. J. and R. Norgren (1978c). The taste reactivity test. II.: Mimetic responses to gustatory stimuli in chronic thalamic and decerebrate rats. *Brain Res* **143**:281–297.

Halpern, B. P. and L. M. Nelson (1965). Bulbar gustatory responses to anterior and to posterior tongue stimulation in the rat. *Am J Physiol* **209**:105–110.

Hanamori, T., N. Ishiko, and D. V. Smith (1987). Multimodal responses of taste neurons in the frog nucleus tractus solitarius. *Brain Res Bull* **18**:87–97.

Hanamori, T., I. J. Miller Jr., and D. V. Smith (1988). Gustatory responsiveness of fibers in the hamster glossopharyngeal nerve. *J Neurophysiol* **60**:478–498.

Hansen K., S. Wacht, H. Seebauer, and M. Schnuch (1998). New aspects of chemoreception in flies. *Ann N Y Acad Sci* **855**:143–147.

Harada, S. and D. V. Smith (1992). Gustatory sensitivities of the hamster's soft palate. *Chem Senses* **17**:37–51.

Harada, S., T. Marui, and Y. Kasahara (1983). Analysis of the initial taste responses from rat chorda tympani nerve. *Jpn J Oral Biol* **25**:566–570.

Hellekant, G. and Y. Ninomiya (1991). On the taste of umami in chimpanzee. *Physiol Behav* **49**:927–934.

Hermann, G. and D. Novin (1980). Morphine inhibition of parabrachial taste units reversed by naloxone. *Brain Res Bull* **5**(Suppl. 4):169–173.

Hettinger, T. P. and M. E. Frank (1990). Specificity of amiloride inhibition of hamster taste responses. *Brain Res* **513**:24–34.

Hill, D. L., B. K. Formaker, and K. S. White (1990). Perceptual characteristics of the amiloride-suppressed sodium chloride taste response in the rat. *Behav Neurosci* **104**:734–741.

Jakinovich, W. J. and B. W. Agranoff (1971). The stereospecificity of the inositol receptor of the silkworm *Bombyx mori. Brain Res* **33**:173–180.

Kanwal, J. S. and J. Caprio (1983). An electrophysiological investigation of the oro-pharyngeal (IX-X) taste system in the channel catfish, *Ictalurus punctatus. J Comp Physiol* **150**:345–357.

King, M. S., L. Wang, and R. M. Bradley (1993). Substance P excites neurons in the gustatory zone of the rat nucleus tractus solitarius. *Brain Res* **619**:120–130.

Lasiter, P. S. and D. L. Kachele (1988). Organization of GABA and GABA-transaminase containing neurons in the gustatory zone of the nucleus of the solitary tract. *Brain Res Bull* **21**:623–636.

Li, C.-S., and D. V. Smith (1997). Glutamate receptor antagonists block gustatory afferent input to the nucleus of the solitary tract. *J Neurophysiol* **77**:1514–1525.

Liu, H., M. M. Behbehani, and D. V. Smith (1993). The influence of GABA on cells in the gustatory region of the hamster solitary nucleus. *Chem Senses* **18**:285–305.

Mark, G. P., T. R. Scott, F.-C. T. Chang, and H. J. Grill (1988). Taste responses in the nucleus tractus solitarius of the chronic decerebrate rat. *Brain Res* **443**:137–148.

Matsuo, R., N. Shimizu, and K. Kusano (1984). Lateral hypothalamic modulation of oral sensory afferent activity in nucleus tractus solitarius neurons of rats. *J Neurosci* **4**:1201–1207.

Michel, W. and J. Caprio (1991). Responses of single facial taste fibers in the sea catfish, *Arius felis*, to amino acids. *J Neurophysiol* **66**:247–260.

Miller, I. J. Jr. and D. V. Smith (1984). Quantitative taste bud distribution in the hamster. *Physiol Behav* **32**:275–285.

Morrison, G. R. (1967). Behavioural response patterns to salt stimuli in the rat. *Can J Psychol* **21**:141–152.

Murzi, E., L. Hernandez, and T. Baptista (1986). Lateral hypothalamic sites eliciting eating affect medullary taste neurons in rats. *Physiol Behav* **36**:829–834.

Nejad, M. S. (1986). The neural activities of the greater superficial petrosal nerve of the rat in response to chemical stimulation of the palate. *Chem Senses* **11**:283–293.

Ninomiya, Y. and M. Funakoshi (1988). Amiloride inhibition of responses to rat single chorda tympani fibers to chemical and electrical tongue stimulations. *Brain Res* **451**:319–325.

Norgren, R. (1970). Gustatory responses in the hypothalamus. *Brain Res* **21**:63–71.

Norgren, R. (1976). Taste pathways to hypothalamus and amygdala. *J Comp Neurol* **166**:17–30.

Norgren, R. (1995). The gustatory system. In: G. Paxinos (ed). *The Human Nervous System.* New York: Academic Press, pp 845–861.

Norgren, R. and C. M. Leonard (1973). Ascending central gustatory pathways. *J Comp Neurol* **150**:217–238.

Norgren, R. and C. Pfaffmann (1975). The pontine taste area in the rat. *Brain Res* **91**:99–117.

Nowlis, G. H., M. E. Frank, and C. Pfaffmann (1980). Specificity of acquired aversion to taste qualities in hamsters and rats. *J Comp Physiol Psychol* **94**:932–942.

Oakley, B. (1975). Receptive fields of cat taste fibers. *Chem Sens Flav* **1**:431–442.

Ogawa, H. and T. Nomura (1988). Receptive field properties of thalamocortical taste relay neurons responsive to natural stimulation of the oral cavity in rats. *Exp Brain Res* **73**:364–370.

Ogawa, H., M. Sato, and S. Yamashita (1968.) Multiple sensitivity of chorda tympani fibres of the rat and hamster to gustatory and thermal stimuli. *J Physiol (Lond)* **199**: 223–240.

Ogawa, H., S. Ito, N. Murayama, and K. Hasegawa (1990). Taste area in granular and dysgranular insular cortices in the rat identified by stimulation of the entire oral cavity. *Neurosci Res* **9**:196–201.

Perkel, D. H. and T. H. Bullock (1968). Neural coding. *Neurosci Res Program Bull* **6**:221–348.

Pfaffmann, C. (1955). Gustatory nerve impulses in rat, cat and rabbit. *J Neurophysiol* **18**:429–440.

Pfaffmann, C. (1959). The afferent code for sensory quality. *Am Psychol* **14**:226–232.

Pfaffmann, C. (1964). Taste, its sensory and motivating properties. *Am Sci* **52**:187–206.

Pfaffmann, C. (1974). Specificity of the sweet receptors of the squirrel monkey. *Chem Senses Flav* **1**:61–67.

Pfaffmann, C., M. Frank, L. M. Bartoshuk, and T. C. Snell (1976). Coding gustatory information in the squirrel monkey chorda tympani. In: J. M. Sprague and A. ￼. Epstein (eds). *Progress in Psychobiology and Physiological Psychology, Vol. 6.* New York: Academic Press, pp 1–27.

Pollack, G. S. and R. Balakrishnan (1997). Taste sensilla of flies: function, central neuronal projections, and development. *Microsc Res Techn* **39**:532–546.

Pritchard, T. (1991). The primate gustatory system. In: T. V. Getchell, R. L. Doty, L. M. Bartoshuk, and J. B. J. Snow (eds). *Smell and Taste in Health and Disease.* New York: Raven, pp 109–125.

Renehan, W. E., Z. G. Jin, X. G. Zhang, and L. Schweitzer (1996). Structure and function of gustatory neurons in the nucleus of the solitary tract. 2. Relationships between neuronal morphology and physiology. *J Comp Neurol* **367**:205–221.

Rolls, E. T. (1995). Central taste anatomy and neurophysiology. In: R. L. Doty (ed). *Handbook of Olfaction and Gustation.* New York: Marcel Dekker, pp 549–573.

Rolls, E. T. and L. L. Baylis (1995). Gustatory, olfactory and visual convergence within the primate orbitofrontal cortex. *J Neurosci* **14**:5437–5452.

Rolls, E. T., E. Murzi, S. Yaxley, S. J. Thorpe, and S. J. Simpson (1986). Sensory-specific satiety: food-specific reduction in responsiveness of ventral forebrain neurons after feeding in the monkey. *Brain Res* **368**:79–86.

Rolls, E. T., T. R. Scott, Z. J. Sienkiewicz, and S. Yaxley (1988). The responsiveness of neurones in the frontal opercular gustatory cortex of the macaque monkey is independent of hunger. *J Physiol (Lond)* **397**:79–86.

Rolls, E. T., J. Sienkiewicz, and S. Yaxley (1989). Hunger modulates the responses to gustatory stimuli of single neurons in the orbitofrontal cortex. *Eur J Neurosci* **1**:53–60.

Saper, C. B. (1982). Reciprocal parabrachial-cortical connections in the rat. *Brain Res* **242**: 33–40.

Sato, M., H. Ogawa, and S. Yamashita (1975). Response properties of macaque monkey chorda tympani fibers. *J Gen Physiol* **66**:781–810.

Scott, T. R. (1987). Coding in the gustatory system. In: T. E. Finger and W. L. Silver (eds). *Neurobiology of Taste and Smell*. New York: John Wiley, pp 355–378.

Scott, T. R. and B. K. Giza (1990). Coding channels in the taste system of the rat. *Science* **249**:1585–1587.

Scott, T. R., S. Yaxley, Z. J. Sienkiewicz, and E. T. Rolls (1986). Gustatory responses in the nucleus tractus solitarius of the alert cynomolgus monkey. *J Neurophysiol* **55**:182–200.

Shingai, T., Y. Miyaoka, R. Ikarashi, and K. Shimada (1988). Diuresis mediated by the superior laryngeal nerve in rats. *Physiol Behav* **44**:431–433.

Shipley, M. T. and M. S. Sanders (1982). Special senses are really special: evidence for a reciprocal, bilateral pathway between insular cortex and nucleus parabrachialis. *Brain Res Bull* **8**:493–501.

Singh R. N. (1997). Neurobiology of the gustatory systems of *Drosophila* and some terrestrial insects. *Microsc Res Tech* **39**:547–63.

Smeraski, C. A., T. V. Dunwiddie, L. Diao, and T. E. Finger (1999). NMDA and non-NMDA receptors mediate responses in the primary gustatory nucleus in goldfish. *Chem Senses* **24**:37–46.

Smith, D. V. and S. L. Bealer (1975). Sensitivity of the rat gustatory system to the rate of stimulus onset. *Physiol Behav* **15**:303–314.

Smith, D. V. and M. E. Frank (1993). Sensory coding by peripheral taste fibers. In: S. A. Simon and S. D. Roper (eds). *Mechanisms of Taste Transduction*. Boca Raton, FL: CRC Press, pp 295–338.

Smith, D. V. and T. Hanamori (1991). Organization of gustatory sensitivities in hamster superior laryngeal nerve fibers. *J Neurophysiol* **65**:1098–1114.

Smith, D. V. and C.-S. Li (1998). Tonic GABAergic inhibition of taste-responsive neurons in the nucleus of the solitary tract. *Chem Senses* **23**:159–169.

Smith, D. V. and J. B. Travers (1979). A metric for the breadth of tuning of gustatory neurons. *Chem Sens Flav* **4**:215–229.

Smith, D. V., J. W. Steadman, and C. N. Rhodine (1975). An analysis of the time course of gustatory neural adaptation in the rat. *Am J Physiol* **229**:1134–1140.

Smith, D. V., J. B. Travers, and R. L. Van Buskirk (1979). Brainstem correlates of gustatory similarity in the hamster. *Brain Res Bull* **4**:359–372.

Smith, D. V., R. L. Van Buskirk, J. B. Travers, and S. L. Bieber (1983a). Coding of taste stimuli by hamster brainstem neurons. *J Neurophysiol* **50**:541–558.

Smith, D. V., R. L. Van Buskirk, J. B. Travers, and S. L. Bieber (1983b). Gustatory neuron types in the hamster brainstem. *J Neurophysiol* **50**:522–540.

Smith, D. V., C.-S. Li, and B. J. Davis (1998a). Excitatory and inhibitory modulation of taste responses in the hamster brainstem. In: C. Murphy (ed). *Olfaction and Taste XII*. New York: New York Academy of Sciences, pp 450–456.

Smith, D. V., C.-S. Li, and B. J. Davis (1998b). Opioid modulation of gustatory responses in the solitary nucleus: electrophysiological and immunohistochemical evidence. *Chem Senses* **23**:578.

Smith-Swintosky, V. L., C. R. Plata-Salaman, and T. R. Scott (1991). Gustatory neural coding in the monkey cortex: stimulus quality. *J Neurophysiol* **66**:1156–1165.

Spector, A. C. and H. J. Grill (1992). Salt taste discrimination after bilateral section of the chorda tympani or glossopharyngeal nerves. *Am J Physiol* **263**:R169–R176.

Spector, A. C., S. P. Travers, and R. Norgren (1993). Taste receptors on the anterior tongue and nasoincisor ducts of rats contribute synergistically to behavioral responses to sucrose. *Behav Neurosci* **107**:694–702.

Spector, A. C., N. A. Guagliardo, and S. J. St. John (1996). Amiloride disrupts NaCl versus KCl discrimination performanace: implications for salt taste coding in rats. *J Neurosci* **16**:8115–8122.

Spector, A. C., S. Markison, S. J. St. John, and M. Garcea (1997). Sucrose vs. maltose taste discrimination by rats depends on the input of the seventh cranial nerve. *Am J Physiol* **272**(*Regul Integr Comp Physiol 41*):R1210–R1218.

Travers, S. P. (1993). Orosensory processing in neural systems of the nucleus of the solitary tract. In: S. A. Simon and S. D. Roper (eds). *Mechanisms of Taste Transduction*. Boca Raton, FL: CRC Press, pp 339–394.

Travers, J. B. and R. Norgren (1983). Afferent projections to the oral motor nuclei in the rat. *J Comp Neurol* **220**:280–298.

Travers, S. P. and R. Norgren (1991). Coding the sweet taste in the nucleus of the solitary tract: differential roles for anterior tongue and nasoincisor duct gustatory receptors in the rat. *J Neurophysiol* **65**:1372–1380.

Travers, S. P. and R. Norgren (1995). Organization of orosensory responses in the nucleus of the solitary tract of the rat. *J Neurophysiol* **73**:2144–2162.

Travers, J. B. and D. V. Smith (1979). Gustatory sensitivities in neurons of the hamster nucleus tractus solitarius. *Sens Processes* **3**:1–26.

Travers, S. P. and D. V. Smith (1984). Responsiveness of neurons in the hamster parabrachial nuclei to taste mixtures. *J Gen Physiol* **84**:221–250.

Travers, S. P., C. Pfaffmann, and R. Norgren (1986). Convergence of lingual and palatal gustatory neural activity in the nucleus of the solitary tract. *Brain Res* **365**:305–320.

Van Buskirk, R. L. and D. V. Smith (1981). Taste sensitivity of hamster parabrachial pontine neurons. *J Neurophysiol* **45**:144–171.

Vogt, M. B. and C. M. Mistretta (1990). Convergence in mammalian nucleus of the solitary tract during development and functional differentiation of salt taste circuits. *J Neurosci* **10**:3148–3157.

Wang, L. and R. M. Bradley (1993). Influence of GABA on neurons of the gustatory zone of the rat nucleus of the solitary tract. *Brain Res* **616**:144–153.

Wang, L. and R. M. Bradley (1995). In vitro study of afferent synaptic transmission in the rostral gustatory zone of the rat nucleus of the solitary tract. *Brain Res* **702**: 188–198.

Whitehead, M. C. (1986). Anatomy of the gustatory system in the hamster: synaptology of facial afferent terminals in the solitary nucleus. *J Comp Neurol* **244**:72–85.

Whitehead, M. C. (1988). Neuronal architecture of the nucleus of the solitary tract in the hamster. *J Comp Neurol* **276**:547–572.

Whitehead, M. C. (1990). Subdivisions and neuron types of the nucleus of the solitary tract that project to the parabrachial nucleus in the hamster. *J Comp Neurol* **301**: 554–574.

Yamamoto, T. (1984). Taste responses of cortical neurons. *Progr Neurobiol* **23**:273–315.

Yamamoto, T. (1989). Role of the cortical gustatory area in taste discrimination. In: R. H. Cagan (ed). *Neural Mechanisms in Taste*. Boca Raton, FL: CRC Press, pp 197–219.

Yamamoto, T., N. Yayuma, T. Kato, and Y. Kawamura (1985). Gustatory responses of cortical neurons in rats. III. Behavioral vs. neural correlates of taste quality. *J Neurophysiol* **53**:1370–1386.

Yamamoto, T., R. Matsuo, Y. Kiyomitsu, and R. Kitamura (1989). Taste responses of cortical neurons in freely ingesting rats. *J Neurophysiol* **61**:1244–1258.

Yaxley, S., E. T. Rolls, Z. J. Sienkiwicz, and T. R. Scott (1985). Satiety does not affect gustatory activity in the nucleus of the solitary tract of the alert monkey. *Brain Res* **347**:85–93.

Yaxley, S., E. T. Rolls, and Z. J. Sienkiewicz (1988). The responsiveness of neurones in the insular gustatory cortex of the macaque monkey is independent of hunger. *Physiol Behav* **42**:223–229.

15

Gustatory System Development

LINDA A. BARLOW

Department of Biological Sciences, University of Denver, Denver, CO

1. INTRODUCTION

The taste system of vertebrates consists of peripheral taste buds innervated by branches of three cranial nerves, which in turn transmit gustatory information to the central nervous system. How these individual components develop and connect with one another is the focus of this chapter. This review is not exhaustive, since several complete and excellent reviews of taste system development are available (Mistretta, 1991; Oakley, 1993a; Mistretta and Hill, 1995). Rather, I emphasize new findings that address embryological aspects of the taste system. In that light, I have included a section on chemosensory organ genesis in *Drosophila*. The developmental pattern of these organs is reminiscent of that of taste buds, and may be instructive for investigations of the developing taste periphery of vertebrates. I have also distinguished between initial formation of the taste system versus growth and elaboration of the embryonic pattern, since these two phases of development may occur via different mechanisms.

The Neurobiology of Taste and Smell, Second Edition, Edited by Thomas E. Finger, Wayne L. Silver, and Diego Restrepo.
ISBN 0-471-25721-4   Copyright © 2000 Wiley-Liss, Inc.

2. EMBRYONIC DEVELOPMENT OF THE TASTE SYSTEM OF VERTEBRATES

2.1. Development of the Taste Periphery

The taste system of vertebrates develops during embryogenesis, so that a functional taste system is present by hatching or birth. In general, taste buds arise directly from oral and pharyngeal epithelia (Barlow and Northcutt, 1995; Stone et al., 1995). Morphological differentiation of taste buds has been described in a variety of species (for review see Mistretta and Hill, 1995), and is summarized below.

Taste bud primordia appear as aggregates of irregularly oriented cells within a cuboidal epithelium (Fig. 15.1D). These cells then elongate and orient perpendicularly to the basement membrane, but remain covered by superficial, squamous epithelial cells (Fig. 15.1E). Cell apices subsequently break through the epithelium (Fig. 15.1F), and some or all of the fusiform cells differentiate as taste cells (Kinnamon et al., 1985). In mammals and birds, the apical microvilli of taste cells are protected within a taste pore or pit. The pore develops last, and characterizes a fully mature bud in mammals and birds. Fishes and aquatic amphibians, in contrast, lack taste pores; instead, taste cell apices simply extrude as a bundle between epithelial cells (also see Chapter 12, Finger and Simon).

In many vertebrates, taste buds occur in papillae (for review see Reutter and Witt, 1993). In mammals, taste buds on the anterior tongue reside on fungiform papillae, whereas those on the posterior tongue are located in circumvallate, and in some species, foliate papillae. In early embryos, papillae first appear as thickenings in the tongue epithelium (Fig. 15.1B) (Farbman, 1965; Mistretta, 1972). As development progresses, the thickenings protrude, acquire a mesenchymal core (Fig. 15.1C), and subsequently, taste bud primordia appear within papillae (Fig. 15.1D–F). The development of circumvallate and foliate papillae is comparable to fungiform ones, except that deep trenches form the walls of these papillae (Mistretta, 1972). Taste buds develop within these epithelial walls, with their apices accessing the lumen of the trench.

While taste bud primordia develop after papillae differentiate, papillae are not necessarily precursors of taste buds since some buds, for example, in the palate, larynx, and epiglottis, lack papillae (Mistretta, 1972; Bradley et al., 1980; Belecky and Smith, 1990). In fishes and terrestrial amphibians, taste buds also are either embedded in the epithelium or located on papillae, while differentiated taste buds of aquatic amphibians are generally not associated with papillae (for review see Reutter and Witt, 1993).

2.2. Development of the Innervation of Taste Buds

Taste buds are innervated by branches of the VIIth (facial), IXth (glossopharyngeal), and Xth (vagal) cranial nerves. Typically, innervation of the epithelium

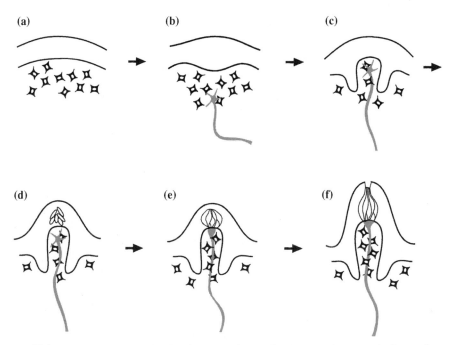

FIG. 15.1. Diagram depicting the development of generalized mammalian taste buds, papillae and their innervation (see text for full details).

(a) Early in embryonic development, cranial mesenchymal cells (black cells with white nuclei) are found adjacent to the lingual epithelium, which initially is uniform in its thickness.

(b) The first indications of papillar formation are small thickenings in the lingual epithelium. At this stage, cranial nerves have invaded the mesenchyme of the tongue (gray axon with star-shaped growth cone).

(c) Papillae develop as epithelial evaginations, which possess a mesenchymal core. At this stage, nerve fibers typically have reached the lingual epithelium.

(d) Taste bud primordia develop within papillae, and appear first as aggregates of loosely organized cells (hollow elliptical cells). The taste papillae continue to grow.

(e) Cells within the developing bud become elongate, with their apices oriented toward the luminal surface (onion-shaped aggregate of elongate cells).

(f) The taste pore develops, allowing taste cells access to the lingual environment. Taste cells now have formed afferent synapses on cranial nerve fibers.

precedes taste bud differentiation (Landacre, 1907; Farbman and Mbiene, 1991; Whitehead and Kachele, 1994; Barlow et al., 1996; Witt and Reutter, 1996) Gustatory fibers follow nerve-specific trajectories from their origin within cranial ganglia to the tongue (Mbiene and Mistretta, 1997) and invade the tongue mesenchyme when papillae are first forming (Fig. 15.1B). However, fibers do not contact papillae until later, when the fibers grow into the mesenchymal core of each papilla and ultimately reach the epithelium (Fig. 15.1C) well before taste buds differentiate (Fig. 15.1D–F) (Farbman and Mbiene, 1991; Whitehead and Kachele, 1994).

2.3. The Role of Innervation in the Development of Taste Buds

Communication between taste buds and nerves is clearly important during development. The adult size of fungiform taste buds in rats is positively correlated with the number of nerve fibers innervating each bud at birth (Krimm and Hill, 1998). In addition, the number of taste buds per fungiform papilla in sheep increases during gestation and decreases postnatally; these changes are paralleled by an increase and subsequent decrease in nerve fibers innervating each papilla (Mistretta et al., 1988). However, the data are correlative and do not reveal if nerves or taste buds or both dictate the developmental patterns observed.

Because innervation precedes the appearance of differentiated taste buds (Fig. 15.1) and because mature taste buds are maintained by nerve contact (Guth, 1958), it has been proposed that nerves induce taste buds. The majority of tests of this neural induction hypothesis have used postnatal rats, where most circumvallate taste buds fail to differentiate fully, that is, develop taste pores, when innervation of papillae was disrupted postnatally (for review see Oakley, 1993b). However, these experiments were carried out long after nerve fibers had initially innervated papillae (Mbiene et al., 1997) and when immature taste buds already were present (Mistretta, 1972; Oakley et al., 1991), and thus these experiments do not represent a test of induction.

The term induction refers to specific signaling between cells, typically during embryogenesis, where recipient cells acquire a new fate (see Slack, 1991). By definition, in the absence of induction, the new developmental pathway is not embarked upon. There are two components of induction: (1) Signals emanate from inducing cells to change the fate of responding cells; and (2) Recipient cells are able, or *competent*, to respond to inductive signals. Applying these definitions to hypothesis of taste bud development, the *inductive* signaling from nerves should trigger the development of taste buds within a *competent* epithelium.

To test explicitly the neural induction of taste buds, one must determine if nerve-naïve epithelia can give rise to taste buds. Stone (1940) initially tested this hypothesis by grafting the oropharyngeal region prior to innervation and taste bud differentiation, from one amphibian embryo to the flank of another embryo, and found that grafted lower jaws developed taste buds. Stone concluded that these ectopic buds had differentiated independently of specific gustatory innervation, but did not determine if the grafted taste buds might have been induced by spinal nerves. Since adult taste buds can be maintained by inappropriate nerves (Kinnman and Aldskogius, 1988), the requirement of nerves for taste bud induction might also be more general. To resolve this ambiguity, the embryonic oropharyngeal region was raised in culture in the total absence of nerves (Barlow et al., 1996). These nerve-free explants developed taste buds morphologically indistinguishable from controls, confirming Stone's original claim that taste buds are not induced by nerves.

To test the neural induction hypothesis in mammals, similar culture experiments have been performed. Explanted tongues from rat embryos differentiate taste papillae in the absence of innervation (Farbman and Mbiene, 1991; Mbiene et al., 1997), consistent with papillar formation in vivo before nerves arrive (Fig. 15.1), but explants failed to exhibit differentiated taste buds. Whether this is because nerves are necessary for taste bud induction, or because culture conditions precluded survival of explanted tongues long enough for taste bud differentiation is not clear. Mammalian tissue in general develops more slowly in vitro, and thus taste bud formation may have been delayed. Alternatively, nerves may be required for differentiation of mammalian taste buds.

Recently, genetically altered mice have been used to test the neural induction hypothesis (Fritzsch et al., 1997; Nosrat et al., 1997; Zhang et al., 1997; Oakley et al., 1998). Specific neurotrophins or their receptors have been "knocked out" and the effect on taste buds examined. Neurotrophins are a family of secreted growth factors expressed by a variety of target tissues and are essential for the survival of neurons that innervate those targets (for review see Davies, 1994). Different neurons express different neurotrophin receptors, and thus are specifically responsive to a subset of neurotrophins (Table 15.1).

In the case of taste buds and their innervation, brain-derived neurotrophic factor (BDNF) and its receptor, tyrosine receptor kinase B (trkB), are crucial. In rodents, *BDNF* mRNA is expressed in all taste papillae when they are first forming, and expression persists in adult taste buds (Nosrat et al., 1996). It has been postulated that gustatory neurons express trkB and require contact with taste buds expressing BDNF to survive (Nosrat et al., 1997). Results from both BDNF and trkB null mice support this hypothesis; cranial ganglion neurons are greatly reduced in number in both mutants and axons do not reach the tongue (Jones et al., 1994; Conover et al., 1995; Fritzsch et al., 1997). BDNF and trkB knockout mice survive postnatally and, when examined several weeks after birth, have disrupted taste buds and papillae (Nosrat et al., 1997; Zhang et al., 1997; Oakley et al., 1998). Yet because these observations were made weeks after taste buds and their innervation develop in normal mice, these data do not address whether nerves induce taste buds.

In one study (Fritzsch et al., 1997), trkB knockout mice were examined when taste buds first appear. In contrast to postnatal animals, taste bud primordia and papillae in these embryos were indistinguishable from controls, at the level of

TABLE 15.1. Genes expressed in developing taste epithelium in rodents

Ligand	Primary Receptor	Transduction Elements
Brain-derived neurotrophin (BDNF)	Tyrosine receptor kinase B (trkB)	
Neurotrophin-3 (NT3)	Tyrosine receptor kinase C (trkC)	
Sonic hedgehog (Shh)	Patched (Ptc)	Gli1 (homolog of *Drosophila* cubitus interruptus)

normal histology. At postnatal stages, however, papillae and taste buds were progressively reduced in number, consistent with the findings of others. This result suggests that formation of taste buds is nerve-independent, but that differentiation of taste buds may require innervation. This view is reinforced by recent findings in mice genetically engineered to express BDNF in inappropriate tissues (Ringstedt et al., 1999). These transgenic animals express BDNF in the tongue musculature through which developing gustatory nerves must travel to reach the lingual epithelium. When these mice are examined at embryonic day 19, gustatory fibers are arrested in the tongue mesenchyme and are not found in the epithelium. Interestingly, taste buds are also severely reduced as judged by the absence of immunocytochemical markers for differentiated taste buds. These data are consistent with the idea that taste bud induction is nerve-independent, but that in rodents, terminal differentiation of taste buds requires neural contact.

The situation is further complicated since lingual taste buds are innervated by gustatory or intragemmal fibers, whereas the papillar epithelium is innervated by general cutaneous or perigemmal fibers. In BDNF or trkB knockout mice, only intragemmal innervation is disrupted, while the perigemmal fibers are undisturbed (Fritzsch et al., 1997; Nosrat et al., 1997; Ringstedt et al., 1999). Thus, the initial development of taste buds and papillae in trkB knockout mice (Fritzsch et al., 1997) could reflect taste bud induction by perigemmal fibers. These latter neurons are dependent upon neurotrophin-3 (NT-3) (ElShamy and Ernfors, 1996), which is expressed in the epithelium of the papillae surrounding the BDNF-expressing area (Nosrat et al., 1996). When NT-3 is knocked out, perigemmal innervation is abolished, although taste buds and papillae are unperturbed (Nosrat et al., 1997). trkB/trkC (the NT-3 receptor) double knockout mice have been generated, and geniculate ganglion neurons, a subset of which innervate taste buds, are reduced to 2% of controls (Silos-Santiago et al., 1997). Unfortunately the effect of this neuronal loss on taste bud development was not reported.

Clearly, neural induction of taste buds has not been explicitly tested in mammals, since nerve-naïve lingual epithelia that survive until taste buds would develop have not yet been generated. Nonetheless, data from mammals and amphibians indicate that early taste bud and papilla development is nerve-independent. In amphibians, both taste bud induction and terminal differentiation occur in the absence of innervation (Barlow et al., 1996). In mammals while papillae (Farbman and Mbiene, 1991; Mbiene et al., 1997) and taste bud primordia (Fritzsch et al., 1997) can form without nerves, terminal differentiation of taste buds, that is, acquisition of a taste pore, may rely upon innervation. These species differences may reflect an early manifestation of the variability seen across taxa with respect to neural maintenance of taste buds (for review see Barlow, 1998). Alternatively, real species differences may exist in the induction of taste buds. Until the role of nerves is elucidated in mammalian embryos, we cannot distinguish between these two possibilities.

2.4. The Role of Mesenchyme in the Development of Taste Buds and Taste Papillae

Cranial mesenchyme is another candidate tissue potentially involved in induction of papillae and taste buds. Embryonic interactions between mesenchyme (a loosely arranged meshwork of irregularly shaped cells) and epithelium result in numerous epithelial specializations, including teeth (Thesleff and Sharpe, 1997), hair follicles (Oliver, 1980), and feather buds (Dhouailly, 1984). Epithelial–mesenchymal interactions may be involved in papilla formation (Farbman and Mbiene, 1991; Mbiene et al., 1997), and possibly in the induction of taste buds.

Long before taste buds differentiate, paraxial mesoderm (the mesoderm found bilaterally along the axial notochord) and neural crest cells (initially located dorsal to the neural tube), which comprise cranial mesenchyme, migrate from their dorsal positions to lateral and ventral locations within the head (Noden and De Lahunta, 1985). Mesenchyme thus contacts the epithelium before taste buds differentiate. To test the inductive capacity of mesenchyme, the presumptive oropharyngeal epithelium was removed from amphibian embryos prior to mesenchyme migration (Barlow and Northcutt, 1997). These cultured explants, devoid of mesenchyme, still gave rise to well-differentiated taste buds, demonstrating that taste bud differentiation in amphibians is independent of mesenchymal signals and, surprisingly, is intrinsic to the oropharyngeal epithelium. Data from other mammalian tissues indicate some aspects of epithelial patterning also are independent of mesenchyme. For example, while teeth require extensive reciprocal interactions between epithelium and mesenchyme to complete differentiation (for review see Lumsden, 1987), the epithelium dictates the pattern of tooth development (Lumsden, 1988). Similarly, while whisker formation requires epithelial–mesenchymal interactions (Oliver, 1980), *Sonic hedgehog* (*Shh*, a secreted factor) is expressed in the epithelium in future whisker follicle sites before mesenchymal contact (Iseki et al., 1996).

Shh is also expressed in developing taste papillae (Bitgood and McMahon, 1995; Hall et al., 1999) and may represent intrinsic patterning of the tongue. *Patched* (*Ptc*) the receptor for *Shh*, and another gene downstream in the *Shh* cascade, *Gli1*, are also expressed in developing papillae (Table 15.1). Initially, expression of all three genes is broad, but becomes restricted to sites of future papillae. Narrowing of expression occurs before innervation of the tongue epithelium (Hall et al., 1999), and thus, like papillar formation, is nerve-independent (Farbman and Mbiene, 1991; Mbiene et al., 1997). Furthermore, focal *Shh* expression before papillae form suggests that *Shh* signaling may pattern the epithelium. However, mesenchymal involvement in this *Shh* mediated patterning is implicated by the early expression of *patched* in the papillar mesenchyme.

While *Shh* expression becomes restricted to regions of taste papillae that will give rise to taste buds (Hall et al., 1999), it is not known if *Shh* signaling controls taste bud development. *Shh* is expressed often in epithelial–mesenchymal outpouchings, such as teeth, follicles, and limb buds (Thesleff and Sharpe, 1997), and thus *Shh* expression in papillae may reflect involvement in evagination of papillae rather than in taste bud patterning. A test of this idea would be to examine *Shh* expression in developing taste buds devoid of papillae, such as those in the larynx, epiglottis, or palate. If *Shh* is expressed in these buds, then *Shh* is likely involved in taste bud formation. However, if *Shh* is not expressed in nonpapillary taste buds, an alternative must be considered: The action of earlier genes may specify taste bud pattern within the epithelium, which could then regulate *Shh* signaling in only a subset of taste bud precursors that reside in papillae. The fact that *Shh* is not expressed in developing foliate papillae (Hall et al., 1999) would tend to support this latter hypothesis.

2.5. A New Model of Taste System Development

In light of recent findings, a new model for development of the taste periphery has been constructed (Barlow and Northcutt, 1998; Northcutt and Barlow, 1998). Rather than focusing on perinatal events, our model presumes that very early embryonic events are pivotal for development of taste buds (Fig. 15.2). In brief: (1) During gastrulation, cells above the blastopore are specified as the presumptive oropharyngeal epithelium (Fig. 15.2a). This means that, at this early stage, these cells are already destined to become the oropharyngeal epithelium, (2) After gastrulation, within this specified epithelial field, local cell–cell interactions give rise to taste bud progenitors (Fig. 15.2b), (3) The oropharyngeal epithelium attracts ingrowing gustatory fibers with long-range chemoattractants (Fig. 15.2c), (4) Once afferent fibers reach the oropharyngeal epithelium, signa-

FIG. 15.2. Diagram of a series of working hypotheses for the development of the taste system of vertebrates.

(a) A sagittal view of an early amphibian gastrula. The presumptive oropharyngeal endoderm is indicated in gray and sits at the dorsal lip of the blastopore (bp) at the onset of gastrulation. These cells will involute first, and follow the trajectory indicated by the dashed gray line, to their position at the anterior of the embryo (head of gray arrow). Here they will give rise to the oropharyngeal epithelium and taste buds.

(b) Most likely after gastrulation, cells within the oropharyngeal endoderm signal to one another (asterisks) such that some cells become taste bud progenitors (dark gray cylinders), while the majority remain epithelial (light gray cylinders).

(c) The oropharyngeal epithelium produces a long-range chemical cue that is found in a gradient (dark gray to light gray shading with +'s) to attract ingrowing cranial nerves from a distance.

(d) Once nerve fibers have reached the vicinity of the epithelium, local cues (small black arrows) produced by taste bud progenitors (dark gray cylinders) serve to establish specific contacts between taste cells and growing nerve fibers (black axons with star-shaped growth cones).

(e) Having received the proper signals for target recognition, cranial nerve fibers collapse their growth cones (club shapes at the end of black axons) and stop growing.

(f) Taste buds differentiate fully (dark gray onion shapes) and form synapses with gustatory afferents.

ling between taste bud primordia and sensory neurites results in target recognition and maintained contact between these cell types (Fig. 15.2d–e) (5) Taste receptor cells mature and form synaptic contacts with afferent nerves (Fig. 15.2f).

2.5.1. The Oropharyngeal Epithelium is Specified as the only Embryonic Region to Later Generate Taste Buds

Oropharyngeal specification occurs at least by the time gastrulation is complete since explants of this region taken from late gastrulae produce taste buds in vitro

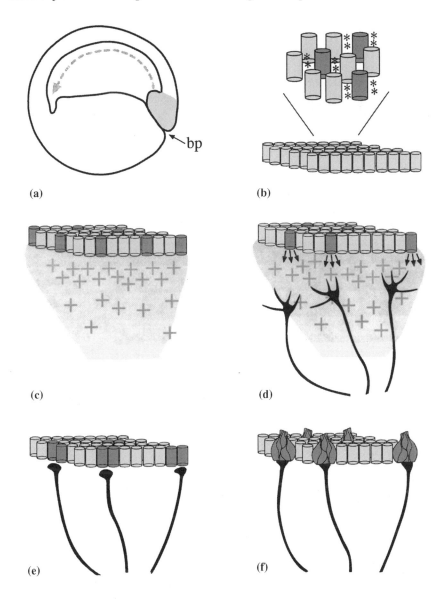

(a)　　　　　　　　　　　(b)

(c)　　　　　　　　　　　(d)

(e)　　　　　　　　　　　(f)

(Barlow and Northcutt, 1997). However, specification may actually occur earlier. Before gastrulation begins, the presumptive oropharyngeal epithelial cells are located at the dorsal lip of the blastopore (Fig. 15.2a) (Vogt, 1929; Cleine and Slack, 1985). These cells involute first during gastrulation and are part of Spemann's Organizer, which "organizes" the body axes (Spemann and Mangold, 1924; for review see Harland and Gerhart, 1997). This embryonic region has received extraordinary attention from developmental biologists for over 70 years, and recently numerous organizer-specific genes have been identified (for review see Harland and Gerhart, 1997). Expression of these genes is restricted to the organizer prior to gastrulation (Vodicka and Gerhart, 1995; Bouwmeester et al., 1996), indicating that these cells are already molecularly distinct from the rest of the embryo. Among these cells are the presumptive oropharyngeal epithelial cells, some of which will give rise to taste buds (Vogt, 1929; Barlow and Northcutt, 1995). These molecular data attest to the uniqueness of these cells early on, and corroborate the hypothesis that the presumptive oropharyngeal epithelium is specified during gastrulation.

2.5.2. Local Cell–Cell Interactions Are Responsible for the Genesis of Taste Bud Primordia

One possible scenario is that taste bud precursors arise via lateral inhibitory signaling between taste bud primordia and epithelial cells (Fig. 15.2b), as occurs in *Drosophila* neural development (for reviews see Ghysen et al., 1993; Artavanis-Tsakonas et al., 1995). In *Drosophila*, all cells within a competent subset of epithelial cells express both an inhibitor of neuronal differentiation and its receptor. Via stochastic fluctuations in inhibitor protein levels, one cell within the competency group expresses more inhibitor than its neighbors, causing neighboring cells to downregulate their own inhibitor expression and thus they no longer inhibit the initial signaler from differentiating as a neural precursor. Concurrently, neighbors differentiate as epithelial cells, which results in distributed neural primordia within an epithelial field. However, it remains to be tested whether a comparable cellular strategy is used by vertebrate embryos to set up an array of taste buds.

2.5.3. The Oropharyngeal Epithelium Attracts Neurites via Long-Range Chemoattractants

The developing cranial nerves extend a large distance before neurites reach the oropharynx. Generally, several mechanisms can operate during neuronal outgrowth, ranging from contact cues within the extracellular matrix to long-range diffusible signals (for review see Goodman and Tessier-Lavigne, 1997). Our conjecture that taste fibers are attracted to the oropharynx via long-range chemical cues (Fig. 15.2c) was based on similarities with the developing trigeminal innervation of whisker pads, where Lumsden and Davies (1983) demonstrated that developing trigeminal axons were attracted from a distance to an appropriate peripheral target. Recent data, however, indicate that the chemorepellant semaphorin also plays a role in guiding ingrowing cranial nerves in the tongue (Rochlin and Farbman, 1998). Trigeminal axons are restricted initially to the lateral margins of the developing tongue, but with time progressively invade the medial tongue. The early repellant activity of the medial tongue can be abolished when growing axons are exposed to a function blocking antibody against neuropilin-1, a receptor for semaphorin. This result, and the gradual restriction of semaphorin expression from a broad domain to a narrow medial strip in the tongue during development (Giger et al., 1996), indicate that semaphorins are important for guidance of trigeminal axons in the tongue. Interestingly, these authors (Rochlin and Farbman, 1998) report preliminarily that geniculate axons are also repelled by semaphorins, suggesting that gustatory fibers may also use chemorepellant cues to find their targets.

2.5.4. Specific Signals from Taste Bud Primordia to Sensory Neurites Result in Maintained Contact Between Taste Bud Primordia and Gustatory Neurites

Once gustatory nerves reach the epithelium, they must find and innervate appropriate targets (Fig. 15.2d). This can result from growth directly to target cells, or pruning of fibers that fail to encounter the proper target (for review see Goodman and Tessier-Lavigne, 1997). In developmental studies of fungiform papillae, facial and trigeminal nerves grew directly to papillae (Farbman and Mbiene, 1991; Whitehead and Kachele, 1994), supporting the hypothesis that local chemoattractants guide ingrowing fibers. While the predominant view has been that neurotrophins do not have a tropic action in vivo (for review see Goodman and Tessier-Lavigne, 1997), some recent findings indicate neurotrophins may indeed be involved in axon guidance in the short range (see McFarlane and Holt, 1997). Thus, in addition to their classic role in neurotrophic support, neurotrophin expression in developing taste papillae and buds may serve to guide gustatory axons to their appropriate targets (Nosrat et al., 1996).

2.5.5. Taste Receptor Cells Mature and Form Synaptic Contacts with Nerve Fibers

Once afferent fibers have reached their targets, synaptic contacts develop, and taste buds and sensory neurons likely become reciprocally dependent (Fig. 15.2e,f): Taste buds and papillae require neural contact for maintenance (Fritzsch et al., 1997; Nosrat et al., 1997; Oakley et al., 1998), while sensory neurons obtain trophic support from the periphery (Davies, 1994). Cell turnover becomes a feature of mature taste buds (Beidler and Smallman, 1965; Delay et al., 1986; Ganchrow et al., 1994), and this constant turnover requires the continuous renewal of synaptic contacts. Persistent BDNF expression in adult taste buds may act to maintain innervation of taste buds with an ever-changing receptor cell population (Nosrat et al., 1996).

3. DEVELOPMENT OF *DROSOPHILA* SENSORY BRISTLES

The peripheral nervous system of flies consists of chemosensory and mechan-osensory bristles (also termed hairs or sensilla) on the wings, legs, and feeding appendages (reviewed by Stocker, 1994; Pollack and Balakrishnan, 1997). Among the chemoreceptors is a class of uniporous chemosensory sensilla, which are often referred to as taste sensilla (see Chapter 12, Finger and Simon). While it is unlikely that uniporous bristles and vertebrate taste buds are homologous given their divergent morphology and evolutionary history, the recent finding that mechanisms intrinsic to the epithelium produce taste buds in amphibians converges on the well-documented intrinsic patterning of the sensory bristles of flies. Portions of the genetic network responsible have been elucidated in flies (for reviews see Artavanis-Tsakonas et al., 1995; Jan and Jan, 1998), and many *Drosophila* genes have vertebrate homologs that function during vertebrate neurogenesis (for review see Anderson and Jan, 1997). Thus an examination of genetic mechanisms that generate sensory bristles in flies, which has shed substantial light on the molecular genetics of vertebrate neurogenesis, may provide insight into mechanisms of taste bud development.

3.1. Morphological Development of Sensory Bristles in *Drosophila*

Uniporous chemosensory sensilla are present in both larval and adult flies, and are distinguished from mechanosensory sensilla based on a curved shaft that tapers to a distal pore (see Figure 12.1, Chapter 12, Finger and Simon), while mechanosensory hairs are straight and lack a pore (Pollack and Balakrishnan, 1997). Each uniporous chemosensory hair is multicellular, comprising single socket, shaft, and sheath cells, and up to five sensory neurons (see Chapter 12, Finger and Simon; Chapter 13, Glendinning et al.). All cells within a sensilla arise from a single cell, the sensory organ precursor (Bate, 1978). The sensory organ precursor differentiates within the epithelium as a result of an elaborate gene cascade, discussed below, and divides asymmetrically to produce two daughter cells. Each daughter then divides asymmetrically to produce the cells of the sensillum.

3.2. Genetic Cascades Involved in Sensory Bristle Formation

Long before bristles form, cells acquire spatial identity within broad domains within the epithelium. As development proceeds, this map is increasingly refined to produce smaller, more focal patches of bristle-competent cells within larger patches destined to a non-neural fate. Ultimately one sensory organ precursor arises within a small cluster of competent cells. The sensory organ precursor is then specified as to bristle type, that is, mechanosensory or chemosensory, and gives rise to the appropriate lineage. This sequence of events involves repression and activation of multiple gene products (for review see Anderson and Jan, 1997; Jan and Jan, 1998) (Table 15.2).

TABLE 15.2. Genes expressed in developing sensory organs in *Drosophila melanogaster*

Gene Category	Genes Drosophila (Mammalian Homologs)*	Protein Class	Expression Pattern
	Prepattern genes		
Iroquois complex (IROC)	Auracan, Caupolican (*Irx1, Irx2, Irx3*)	Homeobox transcription factors	Discrete ectodermal domains[1]
	Proneural genes		
Achaete-scute complex (AS-C)	Achaete, Scute (*Mash1, Mash2, NeuroD*, neurogenin1, neurogenin 2)	basic helix-loop-helix (bHLH) transcription factors	Proneural clusters in ectoderm[2]
	Neurogenic genes		
	Notch (*Notch1, Notch2, Notch3, Notch4*)	Membrane receptor	Throughout proneural cluster, then restricted to SOP neighboring epithelial cells[3]
	Delta (*Delta1, Delta2*)	Membrane-bound ligand	Throughout proneural cluster, then restricted to SOPs[3]
	Neuronal selector genes		
	Paired-box neuro (poxn) (*Pax2, Pax5, Pax8*)	Paired-box transcription factor	Chemosensory SOPs and daughter cells[4]
	Sensory lineage fate decisions		
	Notch (*Notch1, Notch2, Notch3, Notch4*)	Membrane receptor	Non-neuronal daughters of SOP lineage[5]
	Delta (*Delta1, Delta2*)	Membrane-bound ligand	Neuronal daughters of SOP lineage[5]
	Numb (*Numb*)	Cytoplasmic protein associated with the membrane	SOPs and neuronal daughter cells[6]

*Some mammalian homologs of *Drosophila* genes are listed, but expression patterns refer only to *Drosophila*.
[1]Gomez-Skarmeta et al., 1996 Gomez-Skarmeta and Modolell, 1996
[2]Dambly-Chaudiere and Ghysen, 1987; Dambly-Chaudiere and Vervoort, 1998.
[3]Artavanis-Tsakonas et al., 1995; Ghysen et al., 1993.
[4]Dambly-Chaudiere et al., 1992.
[5]Jan and Jan, 1998.
[6]Rhyu et al., 1994; Uemura et al., 1989.

3.2.1. Development and Refinement of a Spatial Map

In fly embryos, as in all animals, the embryonic axes are established early on through the complex interaction of many genes, and subregions within the polarized embryo are then set up via further gene interactions (see Gilbert, 1997). Subsequently, smaller patches of molecularly distinct cells are generated. Genes involved in this process have been named prepattern genes, and thus far consist of members of the Iroquois Complex (IROC): *araucan* and *caupolican*. These are novel homeodomain proteins whose expression is repressed by *engrailed* and *wingless* (genes involved in regionalization of embryos) which restrict *IROC* expression (Gomez-Skarmeta and Modolell, 1996).

3.2.2. Formation of Proneural Cell Clusters

Groups of cells competent to differentiate as sensory organ precursors are termed proneural clusters, and express genes of the *Achaete-Scute Complex*, (AS-C), *achaete* and *scute* (Dambly-Chaudiere and Ghysen, 1987). The locations of proneural clusters roughly prefigure the zones in which sensory organs will form. Prepattern genes, *araucan* and *caupolican*, directly regulate expression of AS-C genes (Gomez-Skarmeta et al., 1996). Araucan protein binds an enhancer sequence upstream of the AS-C promoter region, and loss-of-function alleles of *araucan* or *caupolican* cause a loss of proneural gene expression. However, the proneural, AS-C-expressing cluster is a subset of *IROC* expressing cells, and therefore other genes must be involved in restriction of proneural gene expression within the domain of *IROC* expression.

3.2.3. A Single Neural Precursor Arises Within Each Proneural Cluster

Within each proneural cluster, a single sensory organ precursor develops via lateral inhibition. This interaction is mediated by neurogenic genes, including *Notch* and *Delta*. Initially, both the signaling protein, Delta, and its receptor, Notch, are globally expressed by all cells within a proneural cluster (for review see Ghysen et al., 1993; Artavanis-Tsakonas et al., 1995). Subsequently, via random fluctuations (but see Seugnet et al., 1997) in Delta expression, it is hypothesized that a particular cell will express more Delta signal than its neighbors, which enhances proneural gene expression within that cell and causes it to become an sensory organ precursor. Delta secreted by a sensory organ precursor activates Notch receptors in adjacent cells, which causes the latter to decrease their own expression of Delta. Notch-expressing neighboring cells decrease proneural gene expression and commit to an epithelial fate. When Notch or Delta expression is genetically abolished, lateral inhibitory signaling is lost, and overabundant sensory bristles are produced since all cells within a proneural cluster persist in expressing proneural genes.

3.2.4. Neuronal Selector Genes Determine the Type of Sensory Organ Formed

In broad terms, sensory bristles are either mechanosensory or chemosensory, and as sensory organ precursors are specified, each expresses a selector gene specific for the type of bristle it will produce. *paired-box neuro (poxn)*, a transcription factor, is expressed in chemoreceptive sensory organ precursors (Dambly-Chaudiere et al., 1992). If *poxn* is expressed ectopically in predicted mechano-receptive sensory organ precursors (based on stereotypic position), they are transformed into chemoreceptors; the shaft develops with a pore, and multiple sensory neurons are generated. Central connections also are transformed from a mechanosensory to a chemosensory neuron topography (Nottebohm et al., 1992, 1994). The result is a fully functional chemosensory hair that when stimulated with water elicits drinking behavior in thirsty flies, just as occurs when chemo-sensory bristles are stimulated in wild-type flies (Pollack and Balakrishnan, 1997).

3.2.5. Determination of Cell Fates Within a Sensory Organ Precursor Lineage

Each mechanosensory sensory organ precursor divides asymmetrically to produce two nonequivalent daughter cells (Fig. 15.3). The first daughter divides asymmetrically to give rise to shaft and socket cells, while the second daughter produces a sensory neuron and a sheath cell (Ghysen and Dambly-Chaudiere, 1989). The cellular basis for acquisition of cell fate within a lineage is based both on unequal localization of cytoplasmic determinants during cell division, and on cell signaling between daughters (for review see Anderson and Jan, 1997; Jan and Jan, 1998).

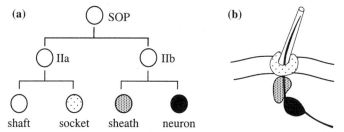

FIG. 15.3. Schematic diagrams of the cell lineage and morphology of mechanosensory bristles of *Drosophila melanogaster*.

(a) Sensory organ precursor cells (SOP) divide to produce two daughter cells, IIa and IIb. The IIa daughter divides to produce the shaft (white) and socket (low-density stippling) cells, both support cells. The IIb daughter divides, and gives rise to a neuron (black) and another support cell, the sheath cell (high-density stippling).

(b) The shaft and socket cells are situated externally, and house the apical dendrite of the sensory neuron. Fills are as in (a).

The cytoplasmic protein Numb is localized basally in a sensory organ precursor during mitosis, and confers a neuron/sheath cell progenitor fate on the daughter that contains the protein (Uemura et al., 1989; Rhyu et al., 1994). In mutants lacking *numb*, no neurons form within sensory organ precursor lineages; instead all cells assume a socket or shaft fate (Uemura et al., 1989). Numb localization requires some type of linkage to the cytoskeleton for proper positioning during mitosis. *inscuteable* acts upstream of Numb, and may control mitotic spindle position, and thus the position of neural determinants (Kraut et al., 1996).

Interactions between daughter cells via Notch-Delta signaling also bias cell fate (for review see Jan and Jan, 1998). This role in sensory organ precursor lineage is distinct from Notch-Delta signaling involved in sensory organ precursor specification. Loss of Notch or Delta during sensory organ precursor or daughter cell division results in overabundant neurons within a bristle at the expense of support cells. This effect on cell lineage is due to direct repression of Notch by Numb; the daughter cell that receives Numb represses its own Notch, which reinforces a neural fate (Guo et al., 1996). Daughters that lack Numb maintain Notch expression, and differentiate as support cells.

This abridged story of fly sensory neurogenesis as a series of discrete genetic steps is an oversimplification. As evidenced by Notch, single genes can act at many stages of development, and influence a broad spectrum of other genes. This complexity in gene action is exacerbated in vertebrates.

3.3. Genes Involved in Vertebrate Neurogenesis: Possible Roles in Taste Bud Development

Vertebrate homologs of many of the genes discussed above act during vertebrate neurogenesis (for reviews see Anderson and Jan, 1997; Jan and Jan, 1998). Typically, several vertebrate homologs exist for each *Drosophila* gene (see Table 15.2), and thus the complexity of neurogenic pathways is enhanced in vertebrates. Furthermore, while some of these genes appear to interact in the same way at a molecular level in flies and vertebrates, for example, Delta is the ligand for Notch, not surprisingly their exact roles in neurogenesis have diverged during evolution (Anderson and Jan, 1997). Interestingly, a few of these genes have been localized to taste buds and may be involved in their development.

Elements of the *Shh* signaling cascade are expressed in developing taste papillae, and expression precedes both innervation of papillae and differentiation of taste receptor cells (Hall et al., 1999). These data suggest that *Shh* signaling may pattern taste papillae and perhaps specify taste bud primordia. In addition, *engrailed* is expressed in the taste buds of zebrafish (Hatta et al., 1991). These findings parallel results in flies where interactions between *engrailed* and *hedgehog*, the insect homolog of *Shh*, create spatial domains in developing epithelia (Gilbert, 1997).

Recently *Mash1*, a mammalian homolog of *achaete-scute complex* (AS-C) basic helix–loop–helix transcription factors (Table 15.2), has been found expressed in the basal cells of taste buds in rats (Seta et al., 1999). These cells are likely stem cells responsible for generating the receptor cells of taste buds (see Chapter 12, Finger and Simon). Numerous basal cells in each adult taste bud express *Mash1*, which is consistent with the finding that the cells within a taste bud arise embryonically from multiple progenitors (Stone et al., 1995). Thus, as in fly sensory organ precursor development, members of the AS-C family of genes may be expressed in taste bud progenitor cells.

Two homologs of *Notch* also have been found in rodent circumvallate papillae (Hoon and Ryba, 1997). In *Drosophila, Notch* is involved in specification of neural precursors and daughter cells within a sensory lineage; developmental events that likely occur in taste buds. For example, *Notch* expression in taste papillae may indicate that lateral inhibitory signaling is occurring to specify taste bud stem cells. Alternatively, *Notch* may serve to assign cell fate within a taste bud lineage.

Certainly, the role of these genes, and possibly other neurogenic genes, deserve further attention in our studies of taste bud development.

4. MATURATION OF THE TASTE SYSTEM

The taste system of vertebrates continues to develop after the basic system forms embryonically (for reviews see Hill, 1987a; Oakley, 1993a; Mistretta and Hill, 1995). In rodents, which are precocial, that is, the timing of their birth is early with respect to the completion of development, a functional, albeit immature, taste system is present at birth. As postnatal life progresses, the majority of taste buds appear, and nervous components of the taste system alter extensively. During this time, early taste preferences and aversions gradually transmute into the mature responses of adults. In altricial mammals, such as humans, other primates, and sheep, birth occurs late with respect to overall development, so that although the sequence of development is comparable to rodents, altricial taste systems have matured substantially before birth.

4.1. Maturation of Taste Bud Morphology

While most rat fungiform papillae have immature taste buds at birth, taste pores only develop postnatally (Farbman, 1965; Mistretta, 1972). Few immature taste buds are evident in rodent circumvallate and foliate papillae at birth; instead these papillae proliferate taste buds predominantly during postnatal development. In addition, taste buds grow postnatally, due to an increase in cell number per bud. In sheep fetuses, the sequence of taste bud development is comparable; immature taste buds are first present in well-formed, innervated papillae, and the number, size, and maturational state of taste buds increase as gestation progresses (Bradley and Mistretta, 1973).

Data from descriptive studies indicate that taste bud proliferation may occur via mechanisms distinct from taste bud induction (Stone, 1940; Bradley et al.,

1980; Miller and Smith, 1988). During embryonic development, immature taste buds are first evident as aggregates of elongate, randomly oriented cells. Subsequently, cell orientation becomes uniform, apical processes appear, and a single taste pore differentiates within each bud. However, during postnatal development (or late gestational development), immature taste buds are encountered less frequently or not at all. Instead, large taste buds with multiple pores are evident. This finding suggests that taste bud proliferation is mediated by a process of growth and then fission of larger taste buds (Mistretta and Hill, 1995). These data suggest that taste bud proliferation may be secondarily nerve-dependent; only trophically maintained taste buds would grow large enough to fission. This could explain why postnatal taste buds do not increase in number when the epithelium is denervated, yet early taste bud induction is nerve independent.

4.2. Maturation of Taste Bud Physiology

Taste cells possess voltage-dependent membrane channels (see Chapter 13, Glendinning et al.), and like neurons (Spitzer, 1979), acquisition of excitability is one component of taste cell development. In mudpuppies, immature taste cells that lack apical processes also lack sodium currents that drive the action potential in mature cells (Mackay-Sim et al., 1996). The full complement of voltage-dependent currents are present only in taste cells with apical processes. In contrast, fungiform taste cells of two-day-old rat pups that have not yet developed a taste pore already have all the voltage-dependent currents found in mature receptor cells (Kossel et al., 1997).

Taste cells also contain the cellular machinery to transduce a wide variety of chemical stimuli (see Chapter 13, Glendinning et al.). While much is known about transduction in mature taste cells, little is known about how transduction cascades develop. Behavioral studies and whole-nerve recordings have shown that sodium sensitivity increases in postnatal rats and gestational sheep (Hill, 1987a). This change was thought to arise via the acquisition of functional amiloride-sensitive sodium channels, since the increased sodium response could be blocked by amiloride (Hill and Przekop, 1988). This turned out not to be the case: Amiloride-sensitive sodium channels are present (Stewart et al., 1995) and functional (Kossel et al., 1997) in immature taste receptor cells. What seems to be more likely is that the onset of salt sensitivity is due to the opening of the taste pore during development.

4.3. Maturation of Gustatory Innervation

The increased behavioral sensitivity of rats to sodium is reflected in an increased aversion to sodium, and maturation of physiological taste responses coincides with the acquisition of this behavior (for reviews see Hill, 1987a; Mistretta and Hill, 1995). In both sheep and rats, recordings from the chorda tympani (CT) branch of the VIIth nerve show responses to sodium increase dramatically during development when compared with responses to ammonium, which increase only slightly. While most CT fibers respond to sodium at birth, the increase in whole nerve salt sensitivity is caused by both an increase in response frequency and a reduction in response latency of individual sodium sensitive fibers (Hill et al., 1982). The decrease in response latency is correlated with progressive myelination of the CT (Ferrell et al., 1985). Increased response frequencies to sodium do not reflect an increase in the number of taste buds innervated by a single CT fiber. Just the contrary; receptive fields of sodium responsive CT fibers actually decrease postnatally in sheep (Mistretta et al., 1988; Nagai et al., 1988). Thus, changes in membrane components of individual taste cells may increase a cell's enhanced sensitivity to sodium and thus would be responsible for increased sodium sensitivity measured in the CT (Nagai et al., 1988; see Chapter 13, Glendinning et al.). Alternatively, opening of the taste pore may substantially impact salt sensitivity.

4.4. Maturation of Brain Stem Gustatory Nuclei

Gustatory sensory neurons project to the rostral nucleus of the solitary tract (NST) in the medulla, and NST neurons in turn ascend to the parabrachial nucleus (PBN) within the pons (see Chapter 14, Smith and Davis). The response profiles and morphology of neurons in both nuclei mature during development.

4.4.1. Nucleus of the Solitary Tract

Initially, neurons of the NST respond to a limited array of taste stimuli (for review see Hill, 1987b). In rat pups, NST cells acquire the ability to respond to all stimuli within the first two postnatal weeks, and sodium sensitivity of individual NST neurons gradually increases until the adult response profile is reached. Developmental changes in NST neurons do not reflect precisely changes in CT responses: (1) the increase in NST sodium response lags behind that of the CT; (2) unlike the CT where the number of sodium responding fibers increases developmentally, the percentage of sodium-best cells does not change in the NST; and (3) receptive fields of individual NST sodium-sensitive neurons increase is size (Vogt and Mistretta, 1990), whereas those of CT neurons decrease. Growth of postnatal NST receptive fields may result from the caudal expansion of CT projections within the NST (Lasiter, 1992), as well as from changes in NST neuron morphology and excitability (Bao et al., 1995).

Three classes of neurons are found in the NST: ovoid, fusiform, and multipolar. The latter two are thought to receive gustatory input, and project to higher order centers (Whitehead, 1986; Lasiter, 1992). While initially all cell types increase their dendritic volume and branching, only multipolar neurons grow through adulthood in rats or birth in sheep (Lasiter et al., 1989; Mistretta and Labyak, 1994). Some studies indicate that reductive remodeling of dendritic spines and branches occurs in projection neurons of the NST to refine gustatory processing (Bao et al., 1995; Renehan et al., 1997). Specifically, Renehan and colleagues have recorded from NST neurons and examined retrospectively the morphology of NST neurons with known response profiles. In young rats, NST salt-sensitive neurons are tuned more broadly with respect to taste stimuli and have larger dendritic arbors than salt-sensitive neurons of older rats.

4.4.2. Parabrachial Nucleus

Response profiles of individual neurons within the PBN also change during postnatal development (Hill, 1987b). Within the first postnatal week, PBN cells respond to all stimuli, and response frequencies to all taste stimuli increase with age. Changes in PBN responses follow temporally changes observed first in CT neurons, and second in neurons of the NST. Concomitant with physiological changes, dendrites of multipolar and fusiform cells within the PBN increase in length (Lasiter and Kachele, 1988).

4.5. Plasticity of Taste System Development

In general, sensory system organization is sensitive to manipulations of sensory inputs during critical periods in development. The taste system is no exception. Sodium-deprived rat dams give birth to pups that fail to develop normal sodium sensitivity (Hill, 1987c). Morphological changes in the taste system are induced by maternal sodium deprivation: organization of CT projections within the NST is disrupted (King and Hill, 1991) and individual NST neurons have longer dendrites than age-matched controls (King and Hill, 1993), perhaps reflecting a lack of normal dendritic pruning. Functional and behavioral recovery of sodium responsiveness can be triggered if deprived rats at any age are allowed a single drink of isotonic NaCl (Przekop et al., 1990).

Unlike other sensory systems, however, the effect of salt deprivation on the taste system is not direct. First, consider the timing of development of the taste system with respect to the critical period during which salt deprivation has its effect. Pregnant rats must be deprived of sodium before their embryos reach embryonic day 8 (E8). However, papillae and taste buds first appear on E12 and E19, respectively, and gustatory ganglion cells and NST neurons do not develop until E12-E15 (Altman and Bayer, 1982). Thus, none of the affected cell types are present during the critical period for a subsequent effect of sodium deprivation on the morphology of these structures. Further, although pregnant females are salt-deprived, their litters are not; the mother's homeostatic mechanisms protect both the developing embryos (Kirksey et al., 1962) and nursing pups (Stewart et al., 1993) from salt deprivation. These findings indicate, rather than a direct effect of salt deprivation on taste buds, that hormonal factors may affect taste system development (Hill, 1987c). The hormone aldosterone is a potential candidate since it is well known that it regulates amiloride-sensitive sodium channels in adult taste receptor cells (Lin et al., 1999). However, aldosterone increases salt sensitivity in adults (Herness, 1992), which is in direct contrast with the decrease in salt sensitivity documented in embryonically salt-

deprived rat pups. Thus either aldosterone functions differently early in development, or other hormonal factors may be involved.

Many of the effects of maternal sodium deprivation can be replicated by
postnatal sensory deprivation. For example, destruction of taste buds or feeding
via intragastric cannula during a critical period between postnatal days 2–8
results in morphological deficits in the CT and NTS (Lasiter and Kachele, 1990;
Lasiter, 1995) comparable to those obtained with maternal salt deprivation (King
and Hill, 1991, 1993). Identical treatment at later postnatal stages has no effect
on CT or NTS morphology. These parallel results from dramatically different
experimental treatments are puzzling. While postnatal manipulation of taste
buds follows the rubric of critical period sensitivity, results from maternal salt
manipulations indicate early hormonal factors are important for normal taste
system development. Whether hormonal mechanisms will prove to be
widespread for the developing organization of other sensory systems, or unique
to the taste system remains to be seen.

5. SUMMARY

The vertebrate gustatory system comprises taste buds with associated papillae,
the sensory neurons that innervate them, and nuclei within the central nervous
system that receive and process gustatory information. Each of these components
arises independently in discrete regions of a vertebrate embryo, but through
development taste buds, cranial ganglion neurons and central neurons become
linked. The intriguing aspect of this embryology is how each of these elements
arises, and how these components contact one another properly to form a
functioning sensory system.

In particular, the development of taste buds historically has received much
attention. Recent findings, however, contradict the conventional view that taste
buds are induced to form by cranial nerve contact, and indicate rather that taste
buds (or their primordia) develop autonomously within local epithelia. How an
epithelium is patterned such that only some cells acquire a receptor cell or
neuronal fate has fascinated developmental biologists for some time, but these
researchers have for the most part focused their attention on invertebrate model
systems. Now the developing taste bud array appears to be an ideal and relatively
simple vertebrate system for examining such patterning and its underlying
mechanisms. Those investigating taste bud development will gain substantial
insight from perusing the *Drosophila* neurogenesis literature, given the continual
new discoveries of vertebrate homologs of developmentally regulated fly genes,
some of which are now known to be expressed in the taste periphery.

Another interesting facet of taste system development is the question of how
gustatory nerves find target taste buds. Again, the taste periphery is a relatively
simple array of receptors that offers an ideal situation to study axon guidance,
pathfinding, and target recognition. A third important aspect of taste system
development is how gustatory sensory neurons find their target neurons within

the central nervous system. While substantial progress has been made in our understanding of the plasticity of this level of the taste system, the developmental mechanisms that create and regulate these neuronal connections have not been determined.

In sum, recent experimental findings have begun to change our ideas of taste system development. As a consequence, the developing taste system offers a host of new and intriguing questions for developmental biologists and neurobiologists.

ACKNOWLEDGMENTS

I thank Lisa Nagy, Heather Eisthen, Tom Finger, and Sue Kinnamon for helpful discussions, and all three editors for concise editorial expertise.

REFERENCES

Altman, J. and S. Bayer (1982). Development of the cranial nerve ganglia and related nuclei in the rat. *Adv Anat Embryol Cell Biol* **74**:1–90.

Anderson, D. J. and Y. N. Jan (1997). The determination of the neuronal phenotype. In: W. M. Cowan, T. M. Jessell, and S. L. Zipursky (eds). *Molecular and Cellular Approaches to Neural Development.* Oxford: Oxford University Press, pp 26–63.

Artavanis-Tsakonas, S., K. Matsuno, and M. E. Fortini (1995). Notch signaling. *Science* **268**:225–232.

Bao, H., R. M. Bradley, and C. M. Mistretta (1995). Development of intrinsic electrophysiological properties in neurons from the gustatory region of rat nucleus of solitary tract. *Dev Brain Res* **86**:143–154.

Barlow, L. A. (1998). The biology of amphibian taste. In: H. Heatwole and E. Dawley (eds). *Amphibian Biology: Sensory Perception.* Chipping Norton: Surrey Beatty & Sons; pp 743–782.

Barlow, L. A., and R. G. Northcutt (1995). Embryonic origin of amphibian taste buds. *Dev Biol* **169**:273–285.

Barlow, L. A. and R. G. Northcutt (1997). Taste buds develop autonomously from endoderm without induction by cephalic neural crest or paraxial mesoderm. *Development* **124**:949–957.

Barlow, L. A. and R. G. Northcutt (1998). The role of innervation in the development of taste buds: insights from studies of amphibian embryos. *Ann N Y Acad Sci* **855**:58–69.

Barlow, L. A., C.-B. Chien, and R. G. Northcutt (1996). Embryonic taste buds develop in the absence of innervation. *Development* **122**:1103–1111.

Bate, C. M. (1978). Development of sensory systems in arthropods. In: M. Jacobson (ed). *Handbook of Sensory Physiology.* Berlin: Springer Verlag, pp 1–53.

Beidler, L. M. and R. L. Smallman (1965). Renewal of cells within taste buds. *J Cell Biol* **27**:263–272.

Belecky, T. L. and D. V. Smith (1990). Postnatal development of palatal and laryngeal taste buds in the hamster. *J Comp Neurol* **293**:646–654.

Bitgood, M. J. and A. P. McMahon (1995). *Hedgehog* and *Bmp* genes are coexpressed at many diverse sites of cell–cell interaction in the mouse embryo. *Dev Biol* **172**:126–138.

Bouwmeester, T., S.-H. Kim, Y. Sasai, B. Lu, and E. M. De Robertis (1996). Cerberus is a head-inducing secreted factor expressed in the anterior endoderm of Spemann's organizer. *Nature* **382**:595–601.

Bradley, R. M. and C. M. Mistretta (1973). The gustatory sense in foetal sheep during the last third of gestation. *J Physiol* **231**:271–282.

Bradley, R. M., M. L. Cheal, and Y. H. Kim (1980). Quantitative analysis of developing epiglottal taste buds in sheep. *J Anat* **130**:25–32.

Cleine, J. H. and J. M. W. Slack (1985). Normal fates and states of specification of different regions in the axolotl gastrula. *J Embryol Exp Morphol* **86**:247–269.

Conover, J. C., J. T. Erickson, D. M. Katz, L. M. Bianchi, W. T. Poueymirou, J. McClain, L. Pan, M. Helgren, N. Y. Ip, P. Boland, B. Friedman, S. Wiegand, R. Vejsada, A. C. Kato, T. M. DeChiara, and G. D. Yancopoulos (1995). Neuronal deficits, not involving motor neurons, in mice lacking BDNF and/or NT4. *Nature* **375**:235–238.

Dambly-Chaudiere, C. and A. Ghysen (1987). Independent subpatterns of sense organs require independent genes of the *achaete-scute* complex in *Drosophila* larvae. *Genes Dev* **1**:297–306.

Dambly-Chaudiere, C. and M. Vervoort (1998). The bHLH genes in neural development. *Int J Dev Biol* **42**:269–273.

Dambly-Chaudiere, C., E. Jamet, M. Burri, D. Bopp, K. Basler, E. Hafen, N. Dumont, P. Spielman, A. Ghysen, and M. Noll (1992). The paired box gene *pox neuro*: a determinant of poly-innervated sense organs in *Drosophila*. *Cell* **69**:159–172.

Davies, A. M. (1994). The role of neurotrophins in the developing nervous system. *J Neurobiol* **25**:1334–1348.

Delay, R. J., J. C. Kinnamon, and S. D. Roper (1986). Ultrastructure of mouse vallate taste buds: II. Cell types and cell lineage. *J Comp Neurol* **253**:242–252.

Dhouailly, D. (1984). Specification of feather and scale patterns. In: G. M. Malacinski and S. V. Bryant (eds). *Pattern Formation*. New York: Macmillan Publishing Co., pp 581–602.

ElShamy, W. M. and P. Ernfors (1996). Requirement of neurotrophin-3 for the survival of proliferating trigeminal ganglion progenitor cells. *Development* **122**:2405–2414.

Farbman, A. I. (1965). Electron microscope study of the developing taste bud in rat fungiform papilla. *Dev Biol* **11**:110–135.

Farbman, A. I. and J.-P. Mbiene (1991). Early development and innervation of taste bud-bearing papillae on the rat tongue. *J Comp Neurol* **304**:172–186.

Ferrell, F., T. Tsuetaki, and R. A. Chole (1985). Myelination in the chorda tympani of the postnatal rat: a quantitative electron microscope study. *Acta Anat* **123**:224–229.

Fritzsch, B., P. A. Sarai, M. Barbacid, and I. Silos-Santiago (1997). Mice lacking the neurotrophin receptor trkB lose their specific afferent innervation but do develop taste buds. *Intl J Dev Neurosci* **15**:563–576.

Ganchrow, D., J. R. Ganchrow, R. Romano, and J. C. Kinnamon (1994). Ontogenesis and taste bud cell turnover in the chicken. I. Gemmal cell renewal in the hatchling. *J Comp Neurol* **345**:105–114.

Ghysen, A. and C. Dambly-Chaudiere (1989). Genesis of the *Drosophila* peripheral nervous system. *Trends Genet* **5**:251–255.

Ghysen, A., C. Dambly-Chaudiere, L. Y. Jan, and Y.-N. Jan (1993). Cell interactions and gene interactions in peripheral neurogenesis. *Genes Dev* **7**:723–733.

Giger, R., D. P. Wolfer, G. M. De Wit, and J. Verhaagen (1996). Anatomy of rat semaphorin III/ collapsin-1 mRNA expression and relationship to developing nerve tracts during neuroembryogenesis. *J Comp Neurol* **375**:378–392.

Gilbert, S. F. (1997). *Developmental Biology*. Sunderland, MA: Sinauer.

Gomez-Skarmeta, J. L. and J. Modolell (1996). *Auracan* and *caupolican* provide a link between compartment subdivision and patterning of sensory organs and veins in the *Drosophila* wing. *Genes Dev* **10**:2935–2945.

Gomez-Skarmeta, J. L., R. D. del Corral, E. de la Calle-Mustienes, D. Ferre-Marco, and J. Modolell (1996). Araucan and caupolican, two members of the novel iroquois complex, encode homeoproteins that control proneural and vein-forming genes. *Cell* **85**:95–105.

Goodman, C. S. and M. Tessier-Lavigne (1997). Molecular mechanisms of axon guidance and target recognition. In: W. M. Cowan, T. M. Jessell, and S. L. Zipursky (eds). *Molecular and Cellular Approaches to Neural Development*. New York: Oxford University Press, pp 108–178.

Guo, M., L. Y. Jan, and Y. N. Jan (1996). Control of daughter cell fates during asymmetric division: interaction of Numb and Notch. *Neuron* **17**:27–41.

Guth, L. (1958). Taste buds on the cat's circumvallate papilla after reinnervation by glossopharyngeal, vagus and hypoglossal nerves. *Anat Rec* **130**:25–37.

Hall, J. M., J. E. Hooper, and T. E. Finger (1999). Expression of *Sonic hedgehog, Patched* and *Gli1* in developing taste papillae of the mouse. *J Comp Neurol* **406**:143–155.

Harland, R. and J. Gerhart (1997). Formation and function of Spemann's organizer. *Annu Rev Cell Dev Biol* **13**:611–667.

Hatta, K., R. Bremiller, M. Westerfield, and C. B. Kimmel (1991). Diversity of expression of Engrailed-like antigens in zebrafish. *Development* **112**:821–832.

Herness, M. S. (1992). Aldosterone increases the amiloride-sensitivity of the rat gustatory neural response to NaCl. *Comp Biochem Physiol* **103A**:269–273.

Hill, D. L. (1987a). Development and plasticity of the gustatory system. In: T. E. Finger and W. L. Silver (eds). *Neurobiology of Taste and Smell*. New York: John Wiley, pp 379–400.

Hill, D. L. (1987b). Development of taste responses in the rat parabrachial nucleus. *J Neurophys* **57**:481–495.

Hill, D. L. (1987c). Susceptibility of the developing rat gustatory system to the physiological effects of dietary sodium deprivation. *J Physiol* **393**:413–424.

Hill, D. L. and P. R. Przekop (1988). Influences of dietary sodium on functional taste receptor development: a sensitive period. *Science* **241**:1826–1827.

Hill, D. L., C. M. Mistretta, and R. M. Bradley (1982). Developmental changes in taste response characteristics of rat single chorda tympani fibers. *J Neurosci* **2**:782–790.

Hoon, M. A, and N. J. P. Ryba (1997). Analysis and comparison of partial sequences of clones from a taste-bud-enriched cDNA library. *J Dent Res* **76**:831–838.

Iseki, S., A. Araga, H. Ohuchi, T. Nohno, H. Yoshioka, F. Hayashi, and S. Noji (1996). *Sonic hedgehog* is expressed in epithelial cells during development of whisker, hair, and tooth. *Biochem Biophys Res Commun* **218**:688–693.

Jan, Y. N. and L. Y. Jan (1998). Asymmetric cell division. *Nature* **392**:775–778.

Jones, K. R., I. Farinas, C. Backus, and L. F. Reichardt (1994). Targeted disruption of the BDNF gene perturbs brain and sensory neuron development but not motor neuron development. *Cell* **76**:989–999.

King, C. T. and D. L. Hill (1991). Dietary sodium chloride deprivation throughout development selectively influences the terminal field organization of gustatory afferent fibers projecting to the rat nucleus of the solitary tract. *J Comp Neurol* **303**:159–169.

King, C. T. and D. L. Hill (1993). Neuroanatomical alterations in the rat nucleus of the solitary tract following early maternal NaCl deprivation and subsequent NaCl repletion. *J Comp Neurol* **333**:531–542.

Kinnamon, J. C., B. J. Taylor, R. J. Delay, and S. D. Roper (1985). Ultrastructure of mouse vallate taste buds. I. Taste cells and their associated synapses. *J Comp Neurol* **235**:48–60.

Kinnman, I. and H. Aldskogius (1988). Collateral reinnervation of taste buds after chronic sensory denervation: a morphological study. *J Comp Neurol* **270**:569–574.

Kirksey, A., R. L. Pike, and J. A. Callahan (1962). Some effects of high and low sodium intakes during pregnancy in the rat. II. Electrolyte concentration of maternal plasma, muscle, bone and brain and of placenta, amniotic fluid, fetal plasma and total fetus in normal pregnancy. *J Nutr* **77**:42–51.

Kossel, A. H., M. McPheeters, W. Lin, and S. C. Kinnamon (1997). Development of membrane properties in taste cells of fungiform papillae: functional evidence for early presence of amiloride-sensitive sodium channels. *J Neurosci* **17**:9634–9641.

Kraut, R., W. Chia, L. Y. Jan, Y. N. Jan, and J. A. Knoblich (1996). Role of *inscuteable* in orienting asymmetric cell divisions in *Drosophila*. *Nature* **383**:50–55.

Krimm, R. F. and D. L. Hill (1998). Quantitative relationships between taste bud development and gustatory ganglion cells. *Ann N Y Acad Sci* **855**:70–75.

Landacre, F. L. (1907). On the place of origin and method of distribution of taste buds in *Ameirus melas*. *J Comp Neurol* **17**:1–66.

Lasiter, P. S. and D. L. Kachele (1990). Effects of early postnatal receptor damage on development of gustatory recipient zones within the nucleus of the solitary tract. *Dev Brain Res* **55**:57–71.

Lasiter, P. S. (1992). Postnatal development of gustatory recipient zones within the nucleus of the solitary tract. *Brain Res Bull* **28**:667–677.

Lasiter, P. S. (1995). Effects of orochemical stimulation on postnatal development of gustatory recipient zones within the nucleus of the solitary tract. *Brain Res Bull* **38**:1–9.

Lasiter, P. S. and D. L. Kachele (1988). Postnatal development of the parabrachial gustatory zone in rat: dendritic morphology and mitochondrial enzyme activity. *Brain Res Bull* **21**:79–94.

Lasiter, P. S., D. M. Wong, and D. L. Kachele (1989). Postnatal development of the rostral solitary nucleus in rat: dendritic morphology and mitochondrial enzyme activity. *Brain Res Bull* **22**:313–321.

Lin, W., T. E. Finger, B. C. Rossier, and S. C. Kinnamon (1999). Epithelial sodium channel subunits in rat taste cells: localization and regulation by aldosterone. *J Comp Neurol* **405**:406–420.

Lumsden, A. G. S. (1987). The neural crest contribution to tooth development in the mammalian embryo. In P. F. A. Maderson (ed). *Developmental and Evolutionary Aspects of the Neural Crest*. New York: John Wiley and Sons, pp 261–300.

Lumsden, A. G. S. (1988). Spatial organization of the epithelium and the role of neural crest cells in the initiation of the mammalian tooth germ. *Development* **103**:155–169.

Lumsden, A. G. and A. M. Davies (1983). Earliest sensory nerve fibres are guided to peripheral targets by attractants other than nerve growth factor. *Nature* **306**:786–788.

Mackay-Sim, A., R. J. Delay, S. D. Roper, and S. C. Kinnamon (1996). Development of voltage-dependent currents in taste receptor cells. *J Comp Neurol* **365**:278–288.

Mbiene, J.-P. and C. M. Mistretta (1997). Initial innervation of embryonic rat tongue and developing taste papillae: nerves follow distinctive and spatially restricted pathways. *Acta Anat* **160**:139–158.

Mbiene, J.-P., D. K. MacCallum, and C. M. Mistretta (1997). Organ cultures of embryonic rat tongue support tongue and gustatory papilla morphogenesis in vitro without intact sensory ganglia. *J Comp Neurol* **377**:324–340.

McFarlane, S. and C. E. Holt (1997). Growth factors: a role in guiding axons? *Trends Cell Biol* **7**:424–430.

Miller, I. J. and D. V. Smith (1988). Proliferation of taste buds in the foliate and vallate papillae of postnatal hamsters. *Growth Dev Aging* **52**:123–131.

Mistretta, C. M. (1972). Topographical and histological study of the developing rat tongue, palate and taste buds. In: J. F. Bosma (ed). *Third Symposium on Oral Sensation and Perception. The Mouth of the Infant.* Springfield, IL: Charles C. Thomas, pp 163–187.

Mistretta, C. M. (1991). Developmental neurobiology of the taste system. In: T. V. Getchell, R. L. Doty, L. M. Bartoshuk, and J. B. Snow (eds). *Smell and Taste in Health and Disease.* New York: Raven Press, pp 35–64.

Mistretta, C. M. and D. L. Hill (1995). Development of the taste system. Basic neurobiology. In: R. L. Doty (ed). *Handbook of Olfaction and Gustation.* New York: Marcel Dekker, pp 635–668.

Mistretta, C. M. and S. E. Labyak (1994). Maturation of neuron types in nucleus of solitary tract associated with functional convergence during development of taste circuits. *J Comp Neurol* **345**:359–376.

Mistretta, C. M., S. Gurkan, and R. M. Bradley (1988). Morphology of chorda tympani receptive fields and proposed neural rearrangements during development. *J Neurosci* **8**:73–78.

Nagai, T., C. M. Mistretta, and R. M. Bradley (1988). Developmental decrease in size of peripheral receptive fields of single chorda tympani nerve fibers and relation to increasing NaCl taste sensitivity. *J Neurosci* **8**:64–72.

Noden, D. M. and A. De Lahunta (1985). *The Embryology of Domestic Animals. Developmental Mechanisms and Malformations.* Baltimore: Williams and Wilkins.

Northcutt, R. G. and L. A. Barlow (1998). Amphibians provide new insights into taste bud development. *Trends Neurosci* **21**:38–42.

Nosrat, C. A., T. Ebendal, and L. Olson (1996). Differential expression of brain-derived neurotrophic factor and neurotrophin 3 mRNA in lingual papillae and taste buds indicates roles in gustatory and somatosensory innervation. *J Comp Neurol* **376**: 587–602.

Nosrat, C. A., J. Blomlöf, W. M. ElShamy, P. Ernfors, and L. Olson (1997). Lingual deficits in BDNF and NT3 mutant mice leading to gustatory and somatosensory disturbances, respectively. *Development* **124**:1333–1342.

Nottebohm, E., C. Dambly-Chaudiere, and A. Ghysen (1992). Connectivity of chemosensory neurons is controlled by the gene *poxn* in *Drosophila. Nature* **359**:829–832.

Nottebohm, E., A. Usui, S. Therianos, K. Kimura, C. Dambly-Chaudiere, and A. Ghysen (1994). The gene *poxn* controls different steps of the formation of chemosensory organs in *Drosophila. Neuron* **12**:25–34.

Oakley, B. (1993a). Control mechanisms in taste bud development. In: S. A. Simon and S. D. Roper (eds). *Mechanisms of Taste Transduction.* Boca Raton: CRC Press, pp 105–125.

Oakley, B. (1993b). The gustatory competence of the lingual epithelium requires neonatal innervation. *Dev Brain Res* **72**:259–264.

Oakley, B., D. E. LaBelle, R. A. Riley, K. Wilson, and L.-H. Wu (1991). The rate and locus of development of rat vallate taste buds. *Dev Brain Res* **58**:215–221.

Oakley, B., A. Brandemihl, D. Cooper, D. Lau, A. Lawton, and C. Zhang (1998). The morphogenesis of mouse vallate gustatory epithelium and taste buds requires BDNF-dependent taste neurons. *Dev Brain Res* **105**:85–96.

Oliver, R. F. (1980). Local interactions in mammalian hair growth. In: R. I. C. Spearman and P. A. Riley (eds). *The Skin of Vertebrates.* London: Academic Press, pp 199–210.

Pollack, G. S. and R. Balakrishnan (1997). Taste sensilla of flies: Function, central neuronal projections, and development. *Microsc Res Tech* **39**:532–546.

Przekop, P., D. G. Mook, and D. L. Hill (1990). Functional recovery of the gustatory system after sodium deprivation during development: how much sodium and where. *Am J Physiol* **259**:R786–791.

Renehan, W. E., J. Massey, Z. Jin, X. Zhang, Y.-Z. Liu, and L. Schweitzer (1997). Developmental changes in the dendritic architecture of salt-sensitive neurons in the nucleus of the solitary tract. *Dev Brain Res* **102**:231–246.

Reutter, K. and M. Witt (1993). Morphology of vertebrate taste organs and their nerve supply. In: S. A. Simon and S. D. Roper (eds). *Mechanisms of Taste Transduction.* Boca Raton, FL: CRC Press, pp 29–82.

Rhyu, M. S., L. Y. Jan, and Y. N. Jan (1994). Asymmetric distribution of Numb protein during division of the sensory organ precursor cell confers distinct fates to daughter cells. *Cell* **76**:477–491.

Ringstedt, T., C. F. Ibanez, and C. A. Nosrat (1999). Role of brain-derived neurotrophic factor in target invasion in the gustatory system. *J Neurosci* **19**:3507–3518.

Rochlin, M. W. and A. I. Farbman (1998). Trigeminal ganglion axons are repelled by their presumptive targets. *J Neurosci* **18**:6840–6852.

Seta, Y., T. Toyono, S. Takeda, and K. Toyoshima (1999). Expression of Mash1 in basal cells of rat circumvallate taste buds is dependent upon gustatory innervation. *FEBS Lett* **444**:43–46.

Seugnet, L., P. Simpson, and M. Haenlin (1997). Transcriptional regulation of *Notch* and *Delta*: requirement for neuroblast segregation in *Drosophila. Development* **124**:2015–2025.

Silos-Santiago, I., A. M. Fagan, M. Garber, B. Fritzsch, and M. Barbacid (1997). Severe sensory deficits but normal CNS development in newborn mice lacking trkB and trkC tyrosine protein kinase receptors. *Eur J Neurosci* **9**:2045–2056.

Slack, J. M. W. (1991). *From Egg to Embryo.* Cambridge: Cambridge University Press.

Spemann, H. and H. Mangold (1924). Über Induktion von Embryonen anlagen durch Implantation artfremder Organisatoren. *Arch Microscop Anat Entwicklungsmechanik* **100**:599–638.

Spitzer, N. C. (1979). Ion channels in development. *Annu Rev Neurosci* **2**:363–397.

Stewart, R. E., H. Tong, R. McCarty, and D. L. Hill (1993). Altered gustatory development in Na+ − restricted rats is not explained by low Na+ levels in mothers' milk. *Physiol Behav* **53**:822–826.

Stewart, R. E., P. S. Lasiter, D. J. Benos, and D. L. Hill (1995). Immunohistochemical correlates of peripheral gustatory sensitivity to sodium and amiloride. *Acta Anat* **153**:310–319.

Stocker, R. F. (1994). The organization of the chemosensory system in *Drosophila melanogaster*: a review. *Cell Tissue Res* **275**:3–26.

Stone, L. M., T. E. Finger, P. P. L. Tam, and S.-S. Tan (1995). Taste receptor cells arise from local epithelium, not neurogenic ectoderm. *Proc Natl Acad Sci U S A* **92**:1916–1920.

Stone, L. S. (1940). The origin and development of taste organs salamanders observed in the living condition. *J Exp Zool* **83**:481–506.

Thesleff, I. and P. Sharpe (1997). Signalling networks regulating dental development. *Mech Dev* **67**:111–123.

Uemura, T., S. Shepherd, L. Ackerman, L. Y. Jan, and Y. N. Jan (1989). *Numb*, a gene required in determination of cell fate during sensory organ formation in *Drosophila* embryos. *Cell* **5**:349–360.

Vodicka, M. A. and J. C. Gerhart (1995). Blastomere derivation and domains of gene expression in the Spemann organizer of *Xenopus laevis*. *Development* **121**:3505–3518.

Vogt, M. B. and C. M. Mistretta (1990). Convergence in mammalian nucleus of solitary tract during development and functional differentiation of salt taste circuits. *J Neurosci* **10**:3148–3157.

Vogt, W. (1929). Gestaltungsanalyse am Amphibienkeim mit örtlicher Vitalfärbung. II. Gastrulation und Mesodermbildung bei Urodelen und Anuren. *Wilhelm Roux' Arch Entwicklungmech Org* **120**:385–706.

Whitehead, M. C. (1986). Anatomy of the gustatory system in the hamster: synaptology of facial afferent terminals in the solitary nucleus. *J Comp Neurol* **244**:72–85.

Whitehead, M. C. and D. L. Kachele (1994). Development of fungiform papillae, taste buds, and their innervation in the hamster. *J Comp Neurol* **340**:515–530.

Witt, M. and K. Reutter (1996). Embryonic and early fetal development of human taste buds: a transmission electron microscopical study. *Anat Rec* **246**:507–523.

Zhang, C. X., A. Brandemihl, D. Lau, A. Lawton, and B. Oakley (1997). BDNF is required for the normal development of taste neurons in vivo. *Neuroreport* **8**:1013–1017.

16

Human Gustation

PAUL A. S. BRESLIN
Monell Chemical Senses Center, Philadelphia, PA

1. INTRODUCTION

The great pioneers of psychophysics, including Newton, Helmholtz, and Maxwell, began their careers as physicists. They wanted to understand how the physical energies they studied, like electromagnetic waves, results in our everyday sensory experiences. Their interests lay not only in the mundane sensations, but also in the aesthetics of experience. The pursuits of these pioneers were founded on the empirical analysis of the most fundamental sensory research questions and were grounded in a solid appreciation of the physical stimuli.

Their goals were lofty and worthy of emulation. If psychophysicists in the chemical senses today are to understand such complex and desirable experiences as gourmet dishes, we should understand the senses from the most basic perceptual phenomena of single stimuli and build from there, just as Newton and Helmholtz did when they attempted to understand sensory experience. Taste researchers have already begun and should continue with the questions of simple physical stimuli, their combinations, and their resultant sensations.

The following chapter is not a review of psychophysical laws, scaling techniques, or power functions, nor is it a survey of taste psychophysical methods (see O'Mahony, 1986, for a review). Rather, I will redirect attention to the fundamental questions of human taste perception that were asked circa the turn of last century. In the process of discussing the status of these questions, I will

The Neurobiology of Taste and Smell, Second Edition, Edited by Thomas E. Finger, Wayne L. Silver, and Diego Restrepo.
ISBN 0-471-25721-4   Copyright © 2000 Wiley-Liss, Inc.

HUMAN GUSTATION

determine if the recent gustatory literature has followed in this vein, and if not, whether these "old" ideas are worth pursuing again.

2. THE PHILOSOPHY OF TERMINOLOGY AND HOW WE FRAME OUR IDEAS

2.1. Definitions

To begin, we shall identify and define what is meant by human taste in three ways: anatomically, psychologically (sensory modality), and teleologically.

2.1.1. Anatomical

Taste, in humans, is the unique modality of sensation usually elicited by chemical stimulation of specialized epithelial "taste receptor cells" in the oro-pharynx and laryngeal areas (see Chapter 12, Finger and Simon). In humans, tastes have been elicited from all the edges of the tongue as well as the back of the tongue, the soft palate, and the throat (Collings, 1974). Unlike other mammals, humans do not have taste sensitivity on the anterior hard palate (Miller and Bartoshuk, 1991; see Chapter 14, Smith and Davis).

2.1.2. Psychological (Modality)

The sense of taste is characterized by qualitative descriptors such as salty, sweet, sour, bitter, and, as will be discussed later, umami (a Japanese term used to describe the taste quality of certain protein-hydrolysates, amino acids-especially gltamate, and 5′-ribonucleotide combinations; the closest English words are "savory," "meaty," or "brothy" (see also Chapter 13, Glendinning et al.). Taste is *not* characterized by descriptors such as hot, putrid, pungent, astringent, oily, greasy, minty, gritty, or prickling. These characteristics represent other oral sensory modalities mediated, in part, by nongustatory nerves (see Chapter 4, Bryant and Silver). Taste, which is most commonly experienced during ingestion, is difficult to tease apart from other food-related sensory modalities, particularly oral somatosensation and olfaction. Even for single stimuli this task can be difficult. For example, one could ask what the taste of menthol is at a moderate concentration. To a typical observer it seems primarily bitter in taste—but also smells minty, irritates the oral cavity, and elicits cold (or warm) sensations (Green, 1992; Cliff and Green, 1994). All four of these sensations may interact to make the complete analytical separation of the individual modalities difficult. It is for this reason that the flavor, the total sensory impact on the upper-airways (mouth, nose, throat), of a food is often confused with the taste sensations of the food (Tichener, 1909). This is also why patients who find that they suddenly cannot smell due to nasal congestion often report they cannot "taste." The multimodal sensations arising from a complex, oral stimulus are usually localized within the oral cavity, despite the fact that reception occurs as far away as the olfactory epithelium. Thus the Gestalt of flavor is a strongly unified percept that can not always be deconstructed into distinct modalities.

2.1.3. Teleological: Why Do We Need to Taste Food; What Function Does Taste Serve?

Teleologically, taste may inform us about the *evolutionarily key features* of potential ingesta placed into the mouth (or of what is being brought into the mouth, as in the case of licking) that cannot otherwise be detected by other modality systems (e.g., olfactory and somatosensory). In addition, taste cues may prepare the GI tract for substances that are about to be ingested so that the ingesta may be processed efficiently by the digestive system (metabolized and absorbed or regurgitated) (e.g., anticipatory metabolic responses; Powley and Berthoud, 1985; Teff and Engelman, 1996; see also Chapter 13, Glendinning et al.).

A "key gustatory feature" is any sapid chemical that is commonly found in plants and animals that may be either: (a) critical to sustain life (such as ions, nutrients, or calories), (b) critical to avoid (such as toxins), or (c) closely associated with (a) and (b) but neither clearly toxic nor nutritious, such as acids/protons. Unlike olfaction, taste in terrestrial animals is a proximal sense; that is, the primary source of the stimuli must be in physical contact with the subject. Many animals including humans have evolved innate acceptance and rejection responses to tastes that characteristically identify beneficial or harmful foods. In general, compounds that taste predominantly and strongly sweet elicit ingestive reflexes while compounds that taste predominantly and strongly bitter elicit rejection reflexes. Both reflex types are present prenatally, perinatally, and postnatally in normal and anencephalic infants (Steiner, 1973, 1974). As in most behaviors, however, these reflexes may change with conditioning (Pelchat and Rozin, 1982).

2.2. The Dimensions of Human Taste

Taste sensations consist of five principal attributes: quality, intensity, temporal dynamics, spatial topography, and hedonics.

2.2.1. Quality

Quality is usually the dominant feature by which any modality is subdivided into different classes of sensation. Qualities are the logical subdivisions of the sensory modalities. Some selected examples from various modalities are: sweet, blue, salty, C-flat, cool, camphorous, tingly, and fruity. The commonly recognized qualities of taste are most often broken into four or five major categories: salty, sweet, bitter, sour, and umami (or savory/protein-like taste). Whether taste has many other *major* qualitative subdivisions than these five categories seems unlikely given the global consensus. Under a major taste quality classification, however, there may be subsubdivisions of distinguishable qualities, such as multiple qualities that can all be categorized as sweet or as bitter, etc.

2.2.2. Intensity

Intensity is the magnitude of the qualitative sensation at any particular time (or integrated over some time period). At absolute detection-threshold levels, a taste sensation can exhibit intensity without any quality. Intensity is used here as a psychological term and should not be confused with changes in the physical dimension. As is known for other senses, for example, the Bezold-Brücke shift in vision (Purdy, 1931), changes in the concentration of a stimulus may both alter the perceived intensity and also be accompanied by subtle changes in quality, for example, NaCl or LiCl (Dzendolet and Meiselman, 1967; Pfaffmann et al., 1971).

Generally, perceived intensity increases exponentially with increases in the physical concentration of the compound. This principle is labeled Stevens' (1969) power law, where the exponent of the intensity function varies depending upon the compound. For taste this law can be represented as

$$I = kC^n$$

where I is perceived intensity, k is a constant, C is the concentration of the stimulus, and n is the exponential variable associated with the concentration-intensity function for the compound in question. In log-log plots these functions are linear and "n" represents the slope of a line. Across most suprathreshold concentration ranges tested, taste compounds are saturating ($n < 1$)—unless other high-threshold modalities, like somatosensation, become activated—and will appear negatively decelerating in linear coordinates. At very low concentration ranges many taste compounds take on the characteristics of an accelerating system ($n > 1$). Thus the behavior of taste power functions vary both with chemical identity as well as with ranges of concentration (Breslin, 1996). Many taste psychometric functions, however, are sigmoidally shaped and take on properties of both an accelerating and a decelerating function depending upon where in the function one looks (Fig. 16.1).

2.2.3. Temporal Dynamics

The temporal dynamics of taste are an important feature of taste but are seldom the focus of either scientific or popular attention (see Section 4.7 below for notable exceptions). Taste experience, as a matter of course, occurs as a function of time. For example, if one focuses attention on how long a taste lasts, one can note that some tastes are short lived and some are very long lived. Since tastes are usually short lived, the long-lived tastes are more noticeable and are commonly referred to as aftertastes (usually with a negative connotation). Certain

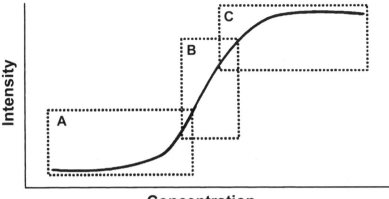

Concentration

FIG. 16.1. A typical psychometric concentration-response function for taste, shaped roughly like a Weibul or logistic function. In rectangle **(A)** the low concentration portion of the function is outlined to show that it is expansive. This means that two solutions of low concentration should combine hyperadditively. In rectangle **(B)** the moderate concentration portion of the function is outlined to show that it is linear. Two moderate solutions should combine additively. In the rectangle **(C)** the high concentration portion of the function is outlined to show that it is compressive. This means that two solutions of high concentration should combine subadditively. Note that all three of these interactions may be synonymous with a linear shift of the entire function along the X-axis.

intensive sweeteners (e.g., thaumatin) and certain bitter tasting compounds (e.g., denatonium) are notorious for the long duration of taste sensation they can elicit. Taste qualities may change over time as well, either as a result of biophysical events occurring at the receptor or in the receptor cells or due to chemical events in the oral cavity altering the compound being tasted (von Skramlik, 1921a, 1921b).

2.2.4. Spatial Topography

The spatial topography of taste refers to both (1) the heterogeneity of the receptor fields in the oral cavity as a function of location as well as (2) the localizability of the taste sensations, the position in the oral cavity from which the taste appears to originate. Some modalities of sensation are sharply localizable, such as tactile sensations (where you were touched) (Weinstein, 1968). Others clearly are not, for example, olfaction (which nostril is stimulated by an odorant—lateralization) (Kobal et al., 1989). And others are intermediate, such as warmth (Hensel, 1982; Stevens, 1991). Localization of taste may help in the identification of a food, the manipulation of a food bolus, or in the removal of select portions of an oral bolus based upon its taste quality. The facts that: (1) the receptive fields of different oral locations are not uniform in sensitivity (Collings, 1974) and (2) that humans can localize tastes (Shikata et al., 2000), both logically result in humans' ability to localize stimulation of an area of high sensitivity to a compound despite experiencing whole mouth stimulation. Thus, if we imagine we have more receptors for sucrose octaacetate on the foliate papillae of the tongue than elsewhere, and we also have the ability to localize where a strong bitter taste originates when placed on the tongue (Shikata et al., 2000), then when we experience sucrose octaacetate (SOA) applied to the whole mouth we will report that it is most strong on the sides of the back of the tongue. For example, when tasting a liquid medicine, it is common to perceive that the medicine tastes bitter in the back of the throat, although the medicine has actually stimulated taste buds over the entire oral cavity to varying degrees (Collings, 1974). Such spatial abilities may even create obstacles for the development of substitute sweeteners and salts that may differ in local stimulating properties from the target tastes of sucrose or NaCl.

2.2.5. Hedonics

The hedonics of taste are the affective evaluations of whether a taste is liked or disliked. The hedonic evaluation of a taste is arguably the most malleable of the sensory attributes of a taste. The quality and intensity (and to a lesser degree the spatial and temporal properties) of a taste dictate our hedonic response to it, whether determined by cognitive, associative, or reflexive means. A single quality of taste from a single compound may vary in an individual from seeming delicious to disgusting depending upon the particular circumstances of tasting, internal state of the taster, and both immediate and long-term experience with the taste. Certain compounds and their associated qualities, especially sweet and bitter, elicit innate hedonic reactions of like and dislike, respectively. These reactions may be reversed as a function of experience with the tastes or their use, for good reason. Many bitter-tasting compounds are associated with nutritive benefit from the foods in which they originate; thus our responses to these bitter tastes needs to be adjustable (Gerber et al., 1999). It may be difficult, however, to alter hedonics so that a relatively high-intensity taste will elicit no affective response; that is, there likely exists limits to the role of experience at overcoming innate affective responses.

3. THE PAST

3.1. What Were the Old Research Questions Asked 80 to 100 Years Ago?

3.1.1. Are There Four Basic Tastes (Salt, Sweet, Sour, Bitter)?

Four central questions in taste asked approximately 100 years ago were, "Do all bitter stimuli elicit the same bitterness; do all sweet stimuli elicit the same sweetness; do all acids elicit the same sourness; and do all salt stimuli elicit the same saltiness?" The question as applied to single compounds with complex tastes became a critical one in the history of taste psychophysics, resulting in questions such as, "Is the salty taste component of KCl, which overall tastes both salty and bitter, the same as the salty taste of NaCl or NH_4Cl?" One must be careful, however, to attend to reasons for apparent differences in taste. For example, do the sweeteners saccharin and aspartame taste differently because they elicit different qualities of sweetness or because saccharin has a bitter side taste and aspartame stimulates taste for slightly longer?

3.1.2. Synthesis Versus Analysis

One of the hottest early discussions in taste was whether a taste sensation could be mentally subdivided into component subsensations (analytically) or whether each taste sensation had its own immutable quality (see Bartoshuk and Gent, 1984, for review). For example, when sugar (sweet) and NaCl (salty) were added together, is the resulting quality of taste novel, as if the two component sensations merged synthetically to create something indivisible and new? Researchers have tried to fit taste between two quintessential representatives of each category: audition for the analytical sense and color vision for the synthetic sense. The great difficulty with this classification scheme is that it is somewhat ambiguous. We all recognize colors that appear blue-green as an analytical mixture of blue and green (Hurvich and Jameson, 1955). Similarly, because audition is so highly nonlinear and many tones have harmonics and residue tones, it becomes very difficult to tease apart certain overlaid tones analytically when presented together, especially if separated by sixths or eighths (Green, 1988).

Since the categorizations of analytic versus synthetic sensations appear to be fuzzy, I hold that the question of whether taste is analytic or synthetic is unclear and therefore difficult to answer. Just as the question was proposed above (3.1.1) whether there are basic tastes (salt, sweet, bitter, and sour), so too may we ask whether there are basic colors of the synthetic color sense (red, green, blue, and yellow). It seems that all senses display aspects of analysis and synthesis depending upon the stimuli involved; taste is surely no exception.

3.1.3. Four Primary Tastes: "Primary" as Percept or Stimulus

There is a long-standing ambiguity in the taste literature regarding the term "basic tastes." The term "basic tastes" is used to refer to the idea that when salt, sweet, sour, and bitter percepts are combined, all taste sensations can be created (independent of whether or not they are synthetic, see Section 3.1.2). This usage confounds the term "basic taste"—(i) a fundamental element of taste *perception*—with the term "primary"—(ii) a *physical stimulus* that can be used to generate all tastes.

(i) Historically, the idea of primary or basic tastes as perceptions holds that salt, sweet, bitter, and sour are introspectively the basis of all gustatory sensations; note that umami is a relatively new qualitative taste label and was not discussed then.

(ii) The idea of a taste primary as a physical stimulus holds that there are sets of N-chemical primaries (historically $N=4$) whereby combinations of these primaries can be used to match the taste of other taste stimuli when mixed at the appropriate proportions. For example, 460 nm, 530 nm, and 650 nm lights can be mixed to produce most colors, as in color television. The reader must be clear that when a "basic taste" is mentioned in the literature it refers to psychological percepts (e.g., salty), and when a "taste primary" is mentioned it refers to physical stimuli that can be used to match other compounds in taste (e.g., NaCl).

3.1.4. Coding Dimensionality of Taste

The "formation of taste," as theorized by von Skramlik (1921a, 1921b), must comprise more than the simple elicitation of salty, sweet, sour, and bitter alone, as there are many stimuli which elicit complex multi-quality sensations. He referred to these four taste qualities, reinstated into the study of taste by von Fick in 1864, as representative qualities from the "pathways" of the sensing organ, by which he presumably meant that each quality has its own independent coding pathway. Von Skramlik was interested in "whether these four qualities could not be further subdivided," and furthermore whether combinations of the four would give rise to all tastes.

This work is based upon the hypothesis that all possible taste experiences can be elicited by a small number, N, of taste stimuli, presented either alone or in combination (von Skramlik, 1921a, 1921b; 1926; cf. Erickson, 1963, 1982; Pfaffmann et al., 1971; Schiffman et al., 1979). This *N-geusia hypothesis* is analogous to the *trichromatic theory* of color vision, which states that all color experiences can be elicited by suitable combinations of three "primaries" ($N=3$) (e.g., 460, 530, and 650 nm light)(Hurvich, 1981; Pugh, 1988). The N-geusia hypothesis states that all taste experiences can be elicited by combinations of N "primary stimuli" (e.g., sodium chloride (NaCl), sucrose, quinine HCl, citric acid, etc.). A primary stimulus is defined as one of a set of stimuli of which each constituent member-stimulus or the combinations of these constituents are sufficient to elicit all taste experiences. This hypothesis remains to be fully tested.

4. THE PRESENT

4.1. What Are the New Questions that Have Been Asked in the Last Two Decades?

This section reviews contemporary works that address measurement of the five attributes of taste (see Section 2.2). Because of space constraints, I am omitting a large body of contemporary taste literature that only tangentially touches upon these issues, for example, comparing discrimination testing methods (O'Mahony, 1986), testing various scaling measures (Marks et al., 1992; Green et al., 1993, 1996), or examining the validity of several mathematical taste–taste mixture interaction models (Frijters and De Graaf, 1987; De Graaf and Frijters, 1988; McBride, 1989; Schifferstein, 1996; Schifferstein and Kleykers, 1996).

4.2. Perceiving Intensity

4.2.1. Intensity Discrimination—Weber Fractions

Humans can assign, either verbally (weak, moderate, strong) or numerically, a direct scale value with great consistency to the perceived intensity of a sensation (Stevens, 1975). Although such direct scaling of perceived taste intensity is today the norm (Stevens, 1969), indirect scaling techniques can be used to measure subjects' sensitivity to increases or decreases in the concentration of a compound. In many sensory systems, the ability to detect changes in intensity is described by a fixed function of the ratio of the magnitude of the change to the starting concentration of the stimulus, in other words a percent increase or decrease (see Weber fractions—Stevens, 1975). For example, a subject may be able to just noticeably detect an increase or a decrease of 10% in concentration regardless of the starting concentration (just noticeable difference—j.n.d.). Weber fractions may be constant over several orders of magnitude in concentration (above threshold levels) (Breslin et al., 1996), but exceptions to this rule occur at very low and very high concentrations (Holway and Hurvich, 1938). Importantly, McBride (1983) has shown that measurement of just noticeable differences yields convergent data with the direct scaling of intensity using methods such as category scaling, thus validating the two techniques.

The rate at which intensity grows with concentration, the exponent of Steven's power function ($I = kC^n$, where I is intensity and C is concentration), helps to determine the input-output intensity function of the compound under study. One might surmise that the exponent for a compound's intensity power function bears some relation to the quality of the sensation; within a single quality, however, the exponents (n) for several compounds' power functions do not form natural groupings. Therefore, the rate at which intensity increases is a function of peripheral events in the epithelium (e.g., individual molecule's binding kinetics) and not psychological events such as the qualitative categorization of sweetness or sourness.

4.2.2. Threshold Effects

In most sensory modalities, much of the psychophysical work concerns threshold measurements. Such studies can show additivity properties of subthreshold stimuli as well as masking effects of suprathreshold backgrounds, for example, the detection of stimulus A is measured in the presence of background levels of suprathreshold B (Pugh and Kirk, 1986; Green, 1988).

In taste, relatively little threshold work has been conducted. J. Stevens has reported both additivity and masking effects (Stevens, 1991, 1995, 1996, 1997; Stevens and Traverzo, 1997). That is, taste is a highly integrative sense across all of the taste qualities. For example, when two chemicals (e.g., sucrose and quinine) are presented at half threshold concentrations they yield a detectable solution (Stevens, 1995). Likewise, if each of 24 solutions is decreased to 1/24th of its detection concentration, and they are all combined, the mixture yields a detectable solution (Stevens, 1997). These results imply the presence of a neural integrator across quality channels. This is surprising given electrophysiological evidence from chimpanzees, *Pan troglodyte*, that the qualitative taste channels appear to code quality independently (Hellekant and Ninomiya, 1991, 1994; Hellekant et al., 1997, 1998; c.f., Chapter 14, Smith and Davis for an alternative interpretation).

Stevens also studied subjects' abilities to detect tastes in the presence of suprathreshold background tastes (maskers)(Stevens and Traverzo, 1997). In everyday experience, we often need to detect a taste in the presence of a background taste or flavor—such as when seasoning any complex food or dish or adding salt to a plate of food. The ability of suprathreshold stimuli to mask weak stimuli also helps to address issues of interquality taste integration.

Interestingly, there has never been a strong relationship established between absolute sensitivity for various compounds via detection thresholds and the suprathreshold intensity at higher concentrations. Two notable exceptions to this are the phenylthiocarbamide/propylthiouracil (PTC/PROP) thresholds and suprathreshold intensity correlations (Bartoshuk et al, 1994), and the threshold and suprathreshold ratings for quinine and urea (Yokomukai et al., 1993; Cowart et al., 1994). It is worth noting that the latter study suggests that quinine and urea stimulate different bitter taste mechanisms, an observation also made by McBurney et al. (1972) (see Section 4.4.4 below).

A table of selected thresholds for common test substances is provided as a reference tool (Table 16.1).

4.2.3. Is Taste Chronically Stable over Time?

Taste studies, like those in other modalities, have not generally been associated with measures of reliability. Taste responses, however, may be more variable than responses in vision or audition studies due to salivary changes, adaptational states, and hormone levels that vary both acutely and chronically. Repeated measures over time are often laborious and, in taste, often yield poor results. Poor reliability can call into question both the techniques and the utility of the data. To alleviate this problem, some researchers have taken the time to retest their subjects repeatedly (Faurion, 1987a; Stevens et al., 1995). Reliability can sometimes decay over time (weeks) (Faurion, 1987a). Thus only by averaging over several repeated measures can a true estimate of sensitivity be obtained for

TABLE 16.1. Detection thresholds for various common taste substances. These tastants are listed according to their major taste quality although several have additional side tastes

Quality	Compound	Threshold (in M)	Ref	Quality	Compound	Threshold (in M)	Ref
BITTER	Caffeine	$5 \times exp -4$	4	SWEET	Aspartame	$(1.76-2.08) \times exp -5$	9
	MgSO$_4$	$3.85 \times exp -4$	7		Fructose	$8.9 \times exp -4$	1, 5
	Quinine HCl	$1.4 \times exp -6$	9		Glucose	$7.33 \times exp -3$	1, 5
	Sucrose Octaacetate	$3.58 \times exp -6$	7		Glycine	$3.09 \times exp -2$	8
	Urea	$(1.07-1.72) \times exp -2$	9		Saccharin	$(8.58-10.1) \times exp -6$	9
	Prop Taster	$2 \times exp -5$	2		Sucrose	$6.5 \times exp -4$	4
	Non-taster	$6 \times exp -4$	2				
SALTY	NH$_4$Cl	$8.39 \times exp -4$	9	SOUR	Acetic Acid	$(1.07-1.12) \times exp -4$	9
	CaCl$_2$	$8 \times exp -6$	10		Citric Acid	$7 \times exp -5$	3
	LiCl	$(0.9-4) \times exp -2$	1, 6		HCl	$(1.6-1.63) \times exp -4$	9
	NaCl	$1.02 \times exp -3$	9		Malic Acid	$(7.38-7.16) \times exp -5$	9
	KCl	$(6.31-6.49) \times exp -3$	9		Tartaric Acid	$4.78 \times exp -5$	9
UMAMI	Monosodium glutamate (MSG)	$5 \times exp -4$	3	UMAMI	L-arginine	$1.23 \times exp -3$	8
					L-glutamine	$9.77 \times exp -3$	8

References: (1) Amerine, M. A., R. M. Pangborn, and E. B. Roessler (1965). The sense of taste. In: *Principles of Sensory Evaluation of Food.* New York: Academic Press, pp 28–144; (2) Bartoshuk L. M., Duffy V. B., Miller I. J. (1994). PTC/PROP tasting: anatomy, psychophysics, and sex effects. *Physiol Behav* **56**:1165–1171; (3) Horio, T. and Y. Kawamura (1998). Influence of physical exercise on human preferences for various taste solutions. *Chem Senses* **23**:417–421; (4) James, C. E., D. G. Laing, and N. Oram (1997). A comparison of the ability of 8–9-year-old children and adults to detect taste stimuli. *Physiol Behav* **62**:193–197; (5) Pangborn, R. M. (1963). Relative taste intensities of selected sugars and organic acids. *J Food Sci* **28**:726–733; (6) Pfaffmann, C. (1959) The sense of taste. In: H. W. Magoun (ed) *Handbook of Physiology, Sect.1. Neurophysiology: Vol. I.* Washington, DC: American Physiological Society, pp 507–534; (7) Schiffman, S. S., L. A. Gatlin, A. E. Frey, S. A. Heiman, W. C. Stagner, and D. C. Cooper (1994). Taste perception of bitter compounds in young and elderly persons: relation to lipophilicity of bitter compounds. *Neurobiol Aging* **15**:743–750; (8) Schiffman, S. S., K. Sennewald, and J. Gagnon (1981). Comparison of taste qualities and thresholds of D- and L-amino acids. *Physiol Behav* **27**:51–59; (9) Stevens, J. C. (1997). Detection of very complex taste mixtures: generous integration across constituent compounds. *Physiol Behav* **62**:1137–1143; (10) Tordoff, M. G. (1996). Some basic psychophysics of calcium salt solutions. *Chem Senses* **21**:417–424.

taste (Stevens et al., 1995). Certain qualities of taste seem more reliable than others. For example, sweeteners yield low reproducibility using threshold measures (Faurion, 1987a; Stevens, 1995). At the extreme of variability, some women have been categorized as taste blind for antithyroid drugs in one part of their menstrual cycle and sensitive to the same compounds in another phase (Glanville and Kaplan, 1965; see also Duffy et al., 1998). In general, the quality of taste psychophysical data is only as good as its replicability. Future work should emphasize repeated testing, despite its logistical difficulties.

4.3. Does the Taste System Appear Linear or Nonlinear?

In the cases of audition and vision, linear systems analyses (LSA) have yielded important data. LSA allows for the measurement of an input/output function (or transfer function) given regular variation in the spatial or temporal domain. McBurney (1976) has been the only researcher to attempt such analyses with taste stimuli. This may in part be due to the difficulty of manipulating taste solutions and adjusting their concentrations rapidly and smoothly during testing. A standard LSA technique is to input into the system a sinusoidal gradient and/ or a step function (square wave) and measure the perceptual output. The transfer function is a complete theoretical description of the temporal or spatial properties of the system under study. McBurney attempted to describe the *temporal* transfer function for taste; to date no one has attempted to study the *spatial* transfer function for taste. The taste system appears efficient at detecting changes in the rate of change of concentration rather than the absolute change in concentration, perhaps as a result of rapid adaptation at slower rates of change (McBurney, 1976; see also Smith and Bealer, 1975). Under optimal conditions, subjects can detect the equivalent of a 3.4% change (Weber fraction) in NaCl or sucrose concentration, suggesting high sensitivity. The taste system also approaches linearity for slow frequency modulation ($\sim$0.5 Hz) and is log-linear at greater frequencies, which was the majority of the perceptual range tested. This observation supports Weber's law that the sensitivity of the system is proportional to the intensity of the standard stimulus.

4.4. Perception of Quality

4.4.1. Umami as a Distinct Quality

The most typical methods of measuring taste quality involve either: (1) the direct scaling of qualities' intensities by subjects using a variety of scaling techniques and a variable number of scales presented per trial, or (2) the rating of total intensity and the subsequent division of the total intensity into various percent portions of salty, sweet, bitter, and sour. Occasionally, the quality umami is included in these techniques.

    Umami is the Japanese term to describe the "savory" taste elicited by certain foods, including, mushrooms (shiitake), seaweeds (sea tangle), fish (bonito), and vegetables (tomato). The prototypical chemical elicitors of umami are monosodium glutamate mixed with 5'-ribonucleotides like inosine monophosphate (IMP) or guanosine monophosphate (GMP) (see Chapter 13, Glendinning et al.). All the foods described as umami are rich in these compounds. This quality of taste is easily perceived by people of many different cultures. Therefore, there must be cultural reasons that umami is not generally included with the standard four taste qualities: salt, sour, bitter, and sweet. When offered MSG with 5'-ribonucleotides (especially in warm water), Americans or Europeans describe it as brothy, soupy, meaty, and savory. While the term savory has not been included in our parlance of qualitative taste descriptors, the term umami has been accepted in Japan (at least within food industry). The term is relatively new, however, first coined in 1908 by Ikeda for the taste of a broth made of sea tangle and bonito fish (cited in Yamaguchi, 1987).

4.4.2. Qualitative Shifts with Changes in Chemical Concentration

Several compounds shift in their perceived quality as a function of changes in concentration. Most notable among these are the salts. NaCl is often observed to taste distinctly sweet at very low concentrations. Concentrations of salt that are weaker than the levels in saliva may give rise to bitter tastes, which may, in fact, be water tastes (Bartoshuk et al., 1964; McBurney and Shick, 1971; Bartoshuk, 1968). Water tastes occur when the mouth is completely or nearly completely adapted to any suprathreshold concentration of a compound and then water is sampled.

Shifts in quality that occur with increasing concentration may be a function of high threshold receptors being activated. For example, maltose is indiscriminable from glucose at low concentrations but at higher concentrations becomes increasingly easier to discriminate (Breslin et al., 1996). Presumably, maltose and glucose are activating similar perceptual channels at low concentrations, but at higher concentrations maltose activates an additional channel. Note that reference to perceptual channels here does *not* mean ion channels, rather it means the combination of peripheral and/or central neural activity that uniquely gives rise to a particular taste quality.

4.4.3. Adaptation

As mentioned above, a fundamental characteristic of taste is the adaptation to background levels of sapid compounds, with the exception of some acids and bitter tasting compounds. For example, saliva is usually tasteless even though it contains many ions and other potentially sapid chemicals. In psychophysics, adaptation is defined as the decrement in intensity or sensitivity to a compound under constant stimulation by this compound.

Several reports suggest that adaptation has both a peripheral epithelial basis as well as a central neural component. For example, a small portion of the tongue can be adapted to a stimulus, then neighboring patches of taste epithelium can be tested. Adaptation impacts the nonstimulated epithelium both across the midline of the tongue and on the same side (Kroeze and Bartsohuk, 1985). These effects cannot be attributed solely to peripheral adaptation of the receptor cells.

4.4.4. Cross-Adaptation

Cross-adaptation occurs when the perceived intensity of a solution is decreased following adaptation to a different compound, relative to adaptation to water. Cross-adaptation is usually not as complete as adaptation and may not be symmetrical. These phenomena are likely due to the complex (multiquality) tastes of most all taste stimuli. While select qualities of a stimulus may be cross-adapted, the stimulus as a whole will not.

Various salts do not appear to cross-adapt when measuring estimates of total stimulus strength, but do cross-adapt if only the salty quality is rated (Smith and McBurney, 1969; Smith and van der Klaauw, 1995). This suggests that the saltiness of the different salts share a common perceptual pathway. There often is little cross-adaptation between compounds that differ in quality such as between sucrose and quinine, although cross-quality adaptation can occur.

More interestingly, there are stimuli that elicit the same quality of taste sensation but do not cross-adapt one another. Most notable among these are different groupings of intensive sweeteners and the various groupings of bitter tasting compounds. Thus, three groupings of bitter compounds have been identified as a function of cross-adapting within group but not between groups (McBurney et al., 1972; c.f., Yokomukai et al., 1993). For example, quinine is believed to stimulate multiple transduction mechanisms, but it does not cross-adapt phenyl-thio-carbamide (PTC), another bitter compound. Thus, PTC is believed to have a transduction mechanism separate from that of quinine. PTC detection is believed to be dependent upon a single recessive gene yielding a bi- or trimodal threshold distribution for all sample populations (Bartoshuk et al., 1994; Reed et al., 1999).

4.4.5. Blockade of a Taste Quality with Topical Oral Agents: (i) Amiloride, (ii) Chlorhexidine, (iii) Gymnema sylvestre, (iv) Lactisole, PA-LG (Bitter Taste Blockers)

Many studies in recent years have examined whether a topical pharmacological agent can be administered to the mouth to block elicitation of a particular quality. These psychophysico-pharmacological techniques have attracted attention because they hold the potential both to elucidate transduction mechanisms as well as to implicate the number of potential mechanisms within a quality of taste (by blocking some but not all compounds within a class). The greatest attention has been devoted to amiloride, an epithelial sodium-channel blocker (see Chapter 13, Glendinning et al.; see also Halpern, 1998, for review). This compound was studied because of its direct implications for the transduction mechanism of NaCl's salty taste and because of its successful use in selected rodent species and strains (Bernstein and Hennessy, 1987; Hill et al., 1990; McCutcheon, 1991; Spector et al., 1996; Harada et al., 1997; Miyamoto et al., 1998; Roitman and Bernstein, 1999). Amiloride's pharmacological effects are relatively specific; if it blocked perception of salty taste, one could infer that salty taste was conveyed via amiloride-sensitive epithelial sodium channels. In humans, many reports suggested that topical amiloride altered the taste of NaCl, partially decreasing its intensity on the tip of the tongue (Schiffman et al., 1983; McCutcheon, 1992; Tennissen, 1992; Smith and Ossebaard, 1995; Tennissen and McCutcheon, 1996; Anand and Zuniga, 1997). More careful analysis showed that while amiloride does reduce the overall perceived intensity of NaCl, amiloride does not reduce the perceived saltiness. Rather, amiloride appears to decrease the subtle sour side-quality of NaCl (Ossebaard and Smith, 1996, 1997; Ossebaard et al., 1997; c.f., Halpern and Darlington, 1998). Despite the success of amiloride at blocking the unique characteristics of sodium taste in selected rodents, a potent selective blocker of salty taste in humans remains to be discovered. There are reports, however, that chronic use of the topical anti-gingival agent chlorhexidine selectively reduces the salty taste of NaCl (Lang et al., 1988; Helms et al., 1995). More recently, Breslin and Thorp (2000) have shown that acute oral ringes with chlorhexidine reduce the saltiness of a variety of salts, as well as the bitterness of several compounds, without affecting sweetness, sourness or savoriness. The inhibition of salty taste from several ions suggests their saltiness is transduced by a common mechanism.

Sweet taste in humans, on the other hand, is blockable by at least two potent agents. The first, *Gymnema sylvestre*, or its component gymnemic acid, works as a pretreatment. If subjects hold a tea made of *Gymnema* leaves in the mouth, then any sweet tasting compound presented afterwards will be greatly diminished in sweetness (Meiselman and Halpern, 1970). Interestingly, *Gymnema* tea has been used as a folk remedy for diabetes in India for thousands of years as it appears to slow the absorption of sugar in the gut (Shimizu et al., 1997). The second compound, lactisole (2-(4-methoxyphenoxy)propanoic acid, sodium salt) is a fast-acting competitive antagonist to almost any compound tasting sweet (Lindley, 1991; Johnson et al., 1994; Sclafani and Perez, 1997). Although, its inhibition of high-potency sweeteners has been debated, lactisole appears to block the sweet taste of high-potency sweeteners initially but not after the sweetener has been in the mouth a few seconds (G. Dubois [personal communication]; Seitz, 1998; c.f., Schiffman et al., 1999). The ability to block the sweet taste of virtually any sweet-eliciting compound (or at least all moderately potent sweeteners) suggests that there may be a common transduction element to all of them, possibly a receptor.

Although bitter taste blockers have been much sought after, only a few have been found that have widespread bitter taste blocking abilities. Foremost among these is the simple sodium ion. Sodium salts block the bitter taste of a large variety of bitter tasting compounds (Schifferstein and Frijters, 1992; Breslin and Beauchamp, 1995). These salts work to varying degrees, however, depending upon the bitter tasting compound in question. Some may be blocked almost completely while others appear to be blocked not at all. The mechanism for sodium ions' ability to block bitter taste remains to be elucidated. Another blocker of bitter tasting compounds is a combination of protein (especially lactoglobulin) bound to phosphatidic acid (PALG) (Katsuragi et al., 1997). PALG is especially good at blocking the taste of quinine, although this may be because it precipitates quinine, rendering it insoluble (Katsuragi and Kurihara, 1997 [p. 275]). Caffeine, on the other hand, is suppressed well by PALG and is not removed from solution (Katsuragi and Kurihara, 1997). The general observation that no bitter taste blocker works on *all* bitter compounds the way sweet taste blockers work on all sweet tasting compounds suggests that there may be several bitter taste transduction mechanisms (see also Chapter 13, Glendinnning et al.). More recently, Ming et al. (1999) have suggested that adenosine monophosphate (AMP) may block bitterness of several bitter tasting agents.

There are no known blocking agents for acid sourness or umami in humans at present.

4.5. Mixture Interactions for Different and Same Quality Taste Compounds

When taste compounds are mixed together in solution, they often interact with one another so that each tastes different than it would were it presented alone at the same concentration. A more thorough discussion of these phenomena can be found in a review (Breslin, 1996). The following examples describe only suprathreshold stimuli; but related phenomena occur for threshold measurements as well (see Section 4.2.2). In almost all mixture studies, tastes have been combined in binary mixtures, and occasionally as trinary mixtures (O'Mahony et al., 1983; Breslin and Beauchamp, 1997). Considering just the binary combinations of taste qualities, only half have been studied to any degree, with sweetness mixtures being the most studied and umami mixtures being the least. If one considers that different chemical representatives of any given quality, say bitter, do not interact with all other compounds in the same manner (Breslin and Beauchamp, 1995), then the amount of research that remains to be done in this area of human taste alone is vast.

4.5.1. Enhancement

When two (or more) compounds are mixed together and a particular quality of taste is increased in intensity, seen as a leftward shift of the concentration–intensity curve, then enhancement has occurred. That is, every point along the concentration axis is perceived as being more intense when in the presence of a fixed concentration of a second compound. Since the concentration–intensity function is generally sigmoidal rather than linear, the magnitude of the effect will be dependent upon which point along the concentration axis the second compound is added. Interactions that occur at low concentrations of compounds tend to show enhancement, where the curve is expansive (concave looking);

where the curve is linear (in the middle) there tend to be small linear interactions, and at high concentrations, where the curve is compressive (convex looking), there tend to be suppressive effects (Fig. 16.1) (Breslin, 1996; see also Bartoshuk, 1975). (These statements are also true for all of the interactions discussed below in Section 4.5) Examples of enhancement are relatively rare for cross-quality compounds but are the general rule for compounds that elicit the same quality. One example of a cross-quality enhancement is seen with the salty taste of NaCl and the bitter taste of the amino acid arginine. Although arginine has no salty taste on its own, it enhances the salty taste of NaCl (Riha et al., 1997).

4.5.2. Synergy

Synergy is similar to enhancement but is a more potent form of positive interaction. When two (or more) compounds are mixed together and a particular quality of taste is increased both as a leftward shift of the concentration–intensity curve and *as a steepening of its slope*, then there is synergy. That is, every point along the concentration axis is perceived as being more intense when in the presence of a fixed concentration of a second compound. Synergy is very rare in taste, however. There are two prototypical examples, both involving same quality mixtures. The first and most clear example comes from the combination of MSG with 5′-ribonucleotides (Rivkin and Bartoshuk, 1980). These two compounds synergize their respective umami tastes. The second is seen with the certain intensive sweeteners such as aspartame and acesulfame-K. Together their sweetnesses synergize (Schiffman et al., 1995; Lawless, 1998). It remains to be shown that cross-quality synergy exists.

4.5.3. Suppression

Suppression is the counterpart to enhancement. When two (or more) compounds are mixed together and a particular quality of taste is decreased as a rightward shift of the concentration–intensity curve, then there is suppression (Breslin, 1996). That is, every point along the concentration axis is perceived as being less intense when in the presence of a fixed concentration of a second compound. Suppression is highly common, especially among compounds of different qualities. For example, when making lemonade, the sourness of the lemons will be suppressed slightly by the sweetness of the sugar and the sweetness of the sugar will be suppressed by the sourness of the lemons. Suppression is usually symmetrical but it can also be asymmetrical (Kamen et al., 1961; Breslin and Beauchamp, 1995).

4.5.4. Masking

Masking is the counterpart to synergy. Masking is similar to suppression but is a more potent form of negative interaction. When two (or more) compounds are mixed together and a particular quality of taste is decreased both as a rightward shift of the concentration–intensity curve and as a *shallowing of its slope*, then there is masking. That is, every point along the concentration axis is perceived as being less intense when in the presence of a fixed concentration of a second compound. Examples of masking come from the taste blocking/ inhibiting literature discussed above (Section 4.3.5). NaCl decreases the bitterness of urea not only as a rightward shift of the curve but also as a shallowing of the bitterness function slope (Breslin and Beauchamp, 1995). In general, both masking and synergy involve peripheral pharmacological effects on the taste cells. The more general phenomena of enhancement and suppression tend to involve more cognitive interactions, although there could also be a peripheral component to these mixture phenomena.

4.5.5. Release from Suppression

Because interaction effects can be asymmetrical, and compounds almost always interact with one another by one of the four mechanisms mentioned above, there are interesting results when adding a third compound to a binary mixture. For example, if a bitter-tasting compound (urea) is mixed with a sweetener (sucrose) there is likely to be mutual suppression whereby the sweet suppresses the bitter and vice versa (see Section 4.5.3). If a sodium salt is added to the binary bitter-sweet mixture, the sodium will have a large masking effect upon the bitter taste but only a very weak suppressive effect upon the sweet. What remains perceptually is predominantly sweet and very slightly bitter, a large relative change. Since the bitter taste suppresses the sweetness of the sucrose, the sweetness will increase in intensity when released from the suppression by the bitter (Breslin and Beauchamp, 1997) (Fig. 16.2). Similar phenomena have been discussed in the context of employing either sequential adaptation stimuli or taste blockers (Lawless, 1979, 1982).

4.6. Individual Differences as Evidence of the Range of Potential Transduction Sequences

Widespread individual differences in taste are observed for most compounds, especially for bitter compounds, intensive sweeteners, and multiquality salts. For some compounds the range in sensitivity across the population can be several orders of magnitude. Of particular interest is the observation that sensitivity to one bitter-tasting compound does not necessarily predict sensitivity to another bitter-tasting compound. For example, one subject may be somewhat sensitive to quinine and relatively insensitive to urea, another subject may be the reverse, while a third person may be highly sensitive to both (Yokomukai et al., 1993).

The most famous example of individual differences in taste comes from the anti-thyroid compounds propylthiouracil (PROP) and phenylthiocarbamide (PTC) which contain an $N-C=S$ bond (Kalmus, 1971). A frequency histogram of sensitivity to these compounds reveals a distribution that appears bimodal (Bartoshuk et al., 1994; see Table 16.1) (or statistically trimodal; Reed et al.,

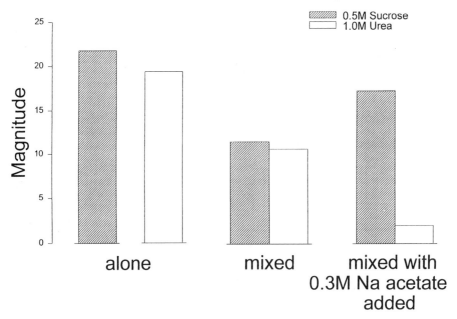

FIG. 16.2. The sweetness (hatched bars) and the bitterness (open bars) are plotted for a high concentration of sucrose (0.5 M) and a high concentration of urea (1.0 M) either alone (left bars), when mixed together (middle bars) or when mixed together with Na acetate (0.3 M) added to the mixture (right bars). The bitterness of urea and the sweetness of sucrose significantly suppress one another mutually (middle bars cf. left bars). As Na acetate is added to the urea-sucrose mixtures the suppression of bitterness increases. The addition of Na acetate to solutions that contained both urea and sucrose simultaneously decreased the bitterness and increased sweetness to levels that approximated the sweetness of sucrose in pure deionized water.

1995). This has captured the attention of the taste community because it suggests a single recessive gene that is responsible for sensitivity to PROP and PTC, thus opening the way for identifying a human gene coding for a bitter taste receptor (Reed et al., 1999).

Another technique that may help elucidate the basis for bitter taste variability is to examine the populations represented by the two tail ends of a broad unimodal distribution of sensitivity to a compound like denatonium benzoate. It is possible that by examining sensitivities to many compounds in a population, the manner in which sensitivities tend to cluster will help identify the number of bitter taste mechanisms (e.g., one for urea and tryptophan (Guinard et al., 1994) and another for quinine and SOA) (see also Chapter 3, Sengupta and Carlson).

4.7. Temporal Dynamics: Time-Intensity

The temporal dynamics of taste are obvious when people report that some food or medicine has a bitter aftertaste (meaning after they swallowed or expectorated).

The temporal dimension of taste, however, has been long overlooked by taste researchers (but not by food industry). Halpern has been a leader in the study of the temporal aspects of taste, including reaction times and the time-intensity profiles of taste compounds (Kelling and Halpern, 1988). The temporal aspects of tastes are critical when distinguishing between stimuli such as bitter-tasting compounds or intensive sweeteners. For example, some sweeteners such as acesulfame-k and neohesperidine dihydrochalcone (NHDC) have a very rapid sweet taste onset (a fraction of a second), but the sweet taste of acesulfame-k decays relatively rapidly while the sweet taste of NHDC lingers for up to three minutes. For bitter-tasting compounds, similar time-intensity profiles can be described but over a much longer time span (i.e., Leach and Noble, 1986). Denatonium benzoate at higher concentrations may be tasted by some people 20 minutes after sampling the solution.

An interesting question arises from the fact that people can track intensity over time (time-intensity functions (T-I)) from the first 100 msecs at onset to the offset, dozens of seconds later. The question is how does a simple, point intensity rating of a compound relate to the shape of the time-intensity function. Clearly, the peak amplitude of a time-intensity function does not translate linearly to single point intensity ratings, nor to the amplitude of the T-I function at the time the intensity rating is given. A complex semi-integrative process goes into each intensity rating based upon the time of onset, peak intensity, area under the curve, duration of the taste, and/or possibly average intensity rating over the interval between onset and response (Kelling and Halpern, 1988). How we integrate the temporal-intensity experience to determine the overall intensity rating is an important area that remains to be investigated. One parameter that has a clear influence is duration. The longer a taste is experienced the stronger it is perceived, doubling in intensity from 50 msec to 2000 msec (Kelling and Halpern, 1988). Such temporal summation (and spatial summation as discussed in Section 4.8.4) demonstrates that taste and somatosensation share many common features.

4.8. The Spatial Dimension of Taste: Sensory Heterogeneity, Localization, Resolution, and Spatial Summation

4.8.1. Heterogeneity of Receptive Fields

The heterogeneity of the receptive fields in the mouth was first discussed in 1901 by Hänig. He reported that the thresholds for four chemicals (representing four qualities of taste) were lowest on certain portions of the tongue (e.g., sugar on the tip and quinine in the back). This statement has been interpreted by some to mean that all bitter tastes are tasted only at the back of the tongue, sweet at the tip, etc. This belief is false. Virtually every area of the mouth that is taste sensitive will respond to all sapid stimuli if the concentration is adjusted appropriately. The fact remains, however, that the different oral taste areas differ in their relative sensitivities to stimuli (Collings, 1974). Furthermore, it is difficult to make generalizations about where qualities of taste are strongest (e.g., sweet at the tip), because each quality is represented by thousands of stimuli and each stimulus may have a different spatial profile, for example, many intense sweeteners (such as aspartame) appear strongest at the back of the tongue when swished briefly and swallowed. Also, Collings (1974) demonstrated that the greatest absolute sensitivity to quinine occurs at the tip of the tongue, while the posterior tongue gives rise to greater sensations than the tip with suprathreshold concentrations of quinine. This heterogeneity may have a functional purpose as the posterior tongue in rodents seems to serve a different functional role (acceptance versus rejection) than the anterior tongue (stimulus discrimination) (Frank, 1991; St. John and Spector, 1998).

4.8.2. Taste Localization

People's ability to localize taste stimuli in the mouth was first studied by painting spots of taste solutions on the tongue simultaneously with fine brushes and asking subjects to indicate where certain qualities were located (Von Skramlik, 1924). People are good at determining where four different taste stimuli are located in the oral cavity. Furthermore, the spatial resolution of taste localization was good; subjects could localize different stimuli presented adjacently to one another on opposite, or same, sides of the tongue. Unfortunately, von Skramlik's painting technique does not control perfectly for temporal discordance in the delivery of the stimuli; thus there is discordance in the onset of the taste and tactile sensations. More recently Shikata et al. (2000) have developed a technique that controls for temporal discriminative cues by delivering two taste stimuli at precisely the same time. They demonstrated that subjects are very good at lateralizing taste cues when either one taste is compared to water (Shikata et al., 2000), or when one taste is presented simultaneously with another taste (personal observation) (Fig. 16.3). These data strongly suggest that there is a central neural map of the location of all taste buds in the oral cavity, and that this map can be used to determine the location of where each quality is originating in the oral cavity.

4.8.3. Tactile Capture of Taste

Subjects will localize a taste to an area in the oral cavity where they are being touched, even if that area contains no taste buds (Todrank and Bartoshuk, 1991). This is a strong phenomenon and may account for common experiences such as why sweet taste is localized to a hard candy as it is whorled around in the mouth, despite being placed in areas where there are no taste buds, such as in the pocket of a cheek. The localization of tastes and the tactile capture of tastes can be competing phenomena depending on where in the oral cavity a stimulus is located. Under these conditions the dominant trait will depend largely upon whether attention is drawn to gustatory or tactile cues (Delwiche et al., 2000).

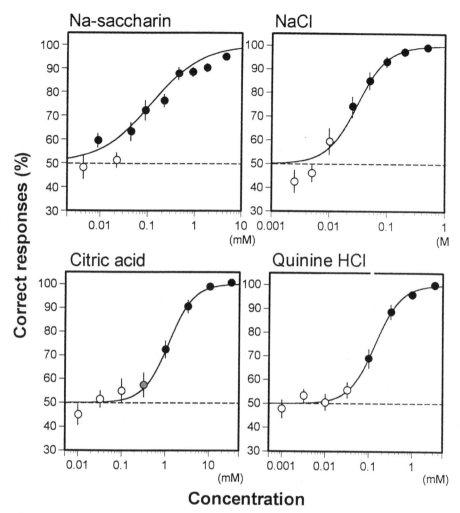

FIG. 16.3. The response curves of lateralization for each compound. Each plot represents mean values and standard errors across >200 trials. Filled circles identify data points significantly different from chance (P<0.01); gray circles identify points weakly different from chance (0.01<p<0.05); open circles identify points not different from chance (p>0.05). Curves were fit with a logistic function.

The tactile capture of taste may also explain the interesting finding of patients with lateralized ageusia who are not aware of their taste deficit.

4.8.4. Spatial Summation

The spatial dimension of taste also plays an important role in taste intensity. Intensity is represented as a function of the concentration of the compound and the area being stimulated (Bujas and Ostajcic, 1941; Hara, 1955; Smith, 1971). This relationship is described by

$$C \times A^m = k$$

for equal intensity functions where C is concentration, A is area, and m is proportional to the exponent of the concentration–intensity power function exponent (Bujas and Ostojcic, 1941; Smith, 1971). The two parameters (C and A) trade off differently for different compounds as a function of their power function exponents. For some compounds, however, such as quinine and saccharin, concentration and area trade off quite well (m ≈ 0.8) over a range of 4 to 126 mm^2 (Smith, 1971). That is, a constant intensity could be maintained when concentration was decreased by an order of ten by increasing area by an order of ten. This linearity of spatial summation is similar to that found for thermal intensity in the skin trading thermal energy with area of stimulation (Stevens, 1991).

4.9. Interactions with Other Modalities

Most people intuitively recognize that tastes interact with one another and with other sensory modalities. In particular, people enjoy multimodal heterogeneity in their foods. In other words, we like our condiments layered on top of our burgers and not mixed in with the meat prior to cooking. But why? Both have all of the same sapid compounds in them, assuming both conditions provide equal access of stimuli to receptors. If the taste system is a purely noninteractive labeled-line system, then it should not matter. The fact that people do care tells us that tastes interact with the other sensory modalities. People prefer foods with distinction; they like the individuality of different tastes and flavors in the diet, and dislike having all flavors mixed together into one jumble.

There are clear interactions of taste with other modalities. Certain odors and colors and textures interact with taste. For example, the sweet taste of sucrose is reported as greater either when the solution is given a fruit flavor, like strawberry (Hyman, 1983; Frank and Byram, 1988; Cliff and Noble, 1990), a fruit color, such as red (Clydesdale, 1993), or a texture that may indicate higher concentrations, such as with a thickener (Christensen, 1980). The interaction of these attributes by combining red and strawberry with a slight thickening of the solution can enhance sweet taste further still (Dubose et al., 1980). The magnitude of these effects, however, is dependent upon the testing methodology. For example, Strawberry odor will not contribute to sweetness intensity if subjects rate the intensity of fruitiness in addition to sweetness (Clark and Lawless, 1994; Frank et al., 1993). In general, there is clear evidence that tastes and smells do, in fact, interact in a usually suppressive manner at suprathreshold concentrations (Commetto-Muñiz, 1981; Garcia-Medina, 1981).

Cruz and Green (2000) have provided evidence for interactions between the taste system and the thermal sensory systems. Heating and cooling a small area of the tongue induces sensations of taste. This is consistent with electrophysio-logical recordings that show thermal sensitivity in peripheral taste neurons. On anterior tongue, warming cavses sweetness in some subjects and cooling can evoke sourness and/or saltiness. Thermal taste may be found on posterior tongue, but the qualities of taste elicited may be different. Whether chemically induced thermal sensations, such as from capsaicin and menthol, would interact similarly with tastes remains to be determined. More recently, Dalton et al. (2000) have shown that tastes and smells interact positively, even when their respective intensities are below absolute detection. Thus, a sub-threshold taste and a sub-threshold odor were detected when presented together at approximately 63% of their individual detection thresholds. These data suggest direct neural integration of the two modalities, rather than the attentional or congnitive mechanisms engaged with supra-threshold stimuli.

There also have been a few reports of interactions between irritants and taste. In the case of capsaicin, the principal irritant from chilies, taste has been observed to interact with burn sensations when presented sequentially (e.g., burn followed by taste) (Lawless and Stevens, 1984; Lawless et al., 1985) and not to interact by other authors when presented simultaneously (Cowart, 1987). Carbonation interacts with tastes to decrease sweetness and saltiness and to increase sourness and to a small degree bitterness (Cowart, 1998).

5. CLINICAL EFFECTS ON TASTE FROM DISEASE AND AGING

There are several recent reviews on taste pathology and the effects of aging, so I will only briefly discuss them here. For clinical issues I refer the reader to Cowart et al. (1997) and Getchell et al. (1991) and for aging issues to Bartoshuk (1989), Cowart (1981), and Murphy and Gilmore (1989).

5.1. Clinical Etiologies

Of the chemosensory disorders people experience, taste problems are in the clear minority and olfactory problems in the majority. Only 4% of patients who presented chemosensory complaints to the University of Pennsylvania Smell and Taste Center had taste deficits (N=750) (Deems et al., 1991; see also Goodspeed, 1987). Of these, taste disorders are most frequently quality (or compound) specific, not involving all taste sensation, and usually involve taste losses, although taste phantoms also occur (Cowart et al., 1997). The primary causes of taste dysfunction can be broken into two main categories: (1) drug and toxin effects, and (2) disease effects, including (i) infections/periodontal disease and other local effects, (ii) nervous disorders/herpes zoster, (iii) nutritional disorders, and (iv) endocrine disorders (Schiffman and Gatlin, 1993). The most common of the two are drug and toxin effects on taste. Drugs may impact taste by direct systemic stimulation of taste receptors by the drug, altering normal function of transduction processes or cellular function, altering salivary function/ flow, or perhaps altering central neural processes (Ackerman and Kasbekar, 1997; Cowart et al., 1997). At this time, we know little of how drugs impact taste. Phantoms occur for several reasons, including oral yeast infections (Brightman et al., 1968; Osaki et al., 1996), nerve damage (both mechanical and infectious) (Blackburn and Bramley, 1989; Grant et al., 1989; Kveton and Bartoshuk, 1994; Yanagisawa et al., 1998), and head trauma (Costanzo and Zassler, 1991).

5.2. Aging and Taste

Although taste function decreases with age, the loss of taste is much less pronounced than for olfaction (Stevens et al., 1984; Cowart, 1989). Taste declines specifically for certain qualities or representative compounds with age. In particular, sensitivity to bitter and salty stimuli may decrease (Weiffenbach et al., 1982, 1986; Murphy and Gilmore, 1989; Schiffman and Gatlin, 1993; Cowart et al., 1994; Stevens, 1996), although, large losses in sensitivity to compounds such as citric acid can be shown for localized gustatory areas, such as the tongue tip (Bartoshuk, 1989). Studies of aging have even provided further evidence for differences in transduction mechanisms for different bitter compounds. For example, urea appears to be detected via an age insensitive mechanism while quinine detection exhibits decreasing function with age (Cowart et al., 1994).

Young and elderly subjects may differ greatly in detection threshold for standard test stimuli, where elderly require two to nine times greater concentrations to detect the compounds (Stevens and Cain, 1993). Usually, much smaller differences exist between the age groups for suprathreshold whole-mouth concentrations (with notable exceptions for Quinine (Cowart et al., 1994). This could result in the false impression that taste losses are of little consequence among the elderly, but when detecting target stimuli in the presence of a background masking tastes, elderly are two to three times less sensitive than young subjects (Stevens et al., 1991; Stevens and Cain, 1993; Stevens, 1996; Stevens and Traverzo, 1997). Despite often not being able to detect the presence of key ingredients, like NaCl, in everyday foods, like soups (Stevens et al., 1991), elderly often seem able to enjoy food and derive pleasure from eating (Stevens, 1989). The chronic overconsumption of NaCl or of bitter toxins, however, could pose a health risk among the elderly.

6. THE FUTURE

6.1. Have the Old Questions Been Answered?
What Attempts Have Been Made to Answer Them?
Do We Need To Revisit These Questions?

Surprisingly, the big questions raised in Section 3.1.1 still have not been answered, that is, whether all sweeteners elicit the same quality of sweetness, all bitter agents the same bitterness, etc. Although, there has been recent work published on the problem (Breslin et al., 1994, 1996) (Fig. 16.4). Such information has implications for how the taste system is organized physiologically, since it would be unusual (though not impossible) to have several redundant perceptual channels that code for only four or five tastes. More likely there is one (or a few) perceptual channels for each distinct quality of taste. More precisely, there should be N perceptual channels corresponding to the number of taste primaries necessary to match all tastes (assuming one can control for spatial, temporal, and somatosensory effects of stimuli). Perceptual channel is defined here not as a labeled line but as a unique physiological informational channel that transmits gustatory information to the brain which when activated either

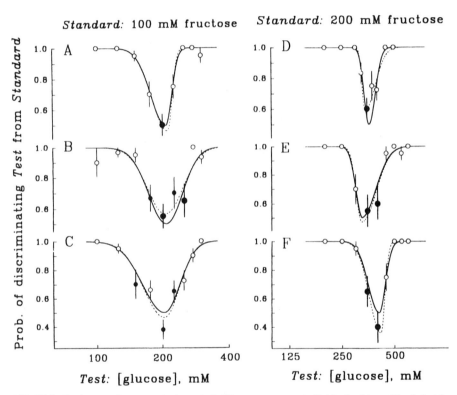

FIG. 16.4. Each row of two panels (e.g., A & D) represents an individual subject. The left side column of panels (A–C) represent functions in which subjects discriminated between 100 mM fructose and a range of glucose concentrations, and on the right side subjects discriminated 200 mM fructose from glucose. Symbols represent empirical data values ±s.e.m. Large filled circles are not significantly different from chance, (p ≥ 0.10); small filled symbols depict points that are weakly different from chance, (0.05 ≤ p < 0.10); open circles are points that are significantly different from chance. The dashed lines are maximum likelihood estimates (MLEs) of an inverted Gaussian curve fit to the data. The solid lines are MLEs of an inverted Gaussian that has been forced to chance performance.

alone or in combination with all other gustatory perceptual channels give rise to all gustatory sensations (c.f., color coding channel).

Another critical question that the older literature attempted to address was the very issue of the number of perceptual channels, that is, the coding dimensionality of taste. How many primary stimuli are necessary to match the taste of any other taste stimulus? This is arguably the single most important issue in human taste perception for understanding how taste is constructed psychologically or coded. The parallel work in color vision that was begun by Helmholtz and later Young and was pursued at the beginning of this century has led: (a) to a firm understanding of the psychological space of color and the linear laws that primary color stimuli follow, (b) to inferences about physiological coding mechanisms that have almost all been born out by physiological research, and (c) industrial engineering innovations that include the development of color television and color film. Until similar methods have been attempted in taste, we won't know whether the same benefits can be obtained for the human gustatory system. The single largest difficulty in this work is control over the physical stimuli, since every concentration and mixture solution must be prepared, as opposed to the color stimuli which were generated with the simple turn of a dial. Also, taste stimuli have the added temporal and spatial dimensions that are either nonexistent in color vision or are much easier to control.

6.2. Future Research Directions

6.2.1. Metameric Matching

If there are classes of compounds that elicit indistinguishable tastes, then according to the *N-geusia hypothesis* (mentioned above) there is either a small number of receptor/transduction sequences underlying all taste experiences, or a large number of several receptor/transduction sequences with converging signals (Bartoshuk and Cleveland, 1977). Psychophysical techniques can determine whether physiological action by two stimuli converge distally or more centrally. The most distal location where the stimuli activate the sensory system in the same manner is referred to as the critical locus. A set of psychophysical laws first described to determine the color vision matching equations can be applied to taste to find the critical locus (Grassmann, 1853). For example, if two stimuli, A and B, are indistinguishable, and they accomplish this by stimulating the same receptors to the same degree, then anything that can be done to these receptors should affect A and B equally, and their match should be maintained. Therefore, most pretreatments, adaptational states (self- or cross-), blocking agents, or additives should affect the two stimuli identically, and their match should be maintained.

In concurrence with von Skramlik, I believe that the number of taste coding channels is relatively small, although almost certainly greater than four. This is in agreement with the views of many other investigators (von Skramlik, 1926; McBurney et al., 1972; Birch, 1987; Shallenberger, 1993; Breslin et al., 1994,

1996). Others, however, believe the number is large as evidenced by the many reportedly unique ("unmatchable") tastes that accompany stimulation with different taste stimuli (Schiffman et al., 1981; Erickson, 1982; Lawless and Stevens, 1983; Faurion, 1987b). If the number is small, then it should be possible to match "complex" tastes (e.g., inorganic salts) as von Skramlik claimed.

6.2.2. Threshold Testing Under Conditions of Selective Cross-Adaptation

The work of Stiles in color vision focused on the changes in detection threshold as a function of varying color adaptation states of the observer. He described several Pi (π) mechanisms that are today believed to describe the properties of early level signal processing in retinal ganglion cells (Pugh and Kirk, 1986). This type of work has rarely (if ever) been attempted in taste (c.f., Stevens, 1996; Stevens and Traverzo, 1997).

6.2.3. Individual Difference Studies

Individual differences shows promise as an area for uncovering potential physiological and qualitative links among different compounds that elicit similar qualities. In particular, the study of subjects from the extreme ends of a sensitivity-frequency histogram may yield interesting similarities and differences to their sensitivities to many other compounds both within the same quality of taste and across qualities (if not also modalities). Thus far this type of analysis has only been attempted on the compounds PTC and PROP because of their obvious genetic advantage, however, similar analyses of other compounds could prove useful as well.

6.2.4. Selective Blockade of Qualities of Taste

The complete blockade of a quality of taste, such as with sweetness by lactisole, yields interesting information. The discovery of blockers that inhibit all sensations of sourness or saltiness in humans would be exciting. Furthermore, a compound that would block umami taste or conversely the combination of salty, sweet, bitter, and sour taste blockers that did not block umami taste would provide conclusive evidence that umami can not be matched by mixtures of sucrose, quinine, citric acid, and NaCl—rather umami sensations arise from their own sensory coding channel (see Crocker and Henderson, 1932, for contradictory evidence). In addition, the identification of blockers that clearly work on some but not all compounds in a category may provide evidence in favor of multiple transduction mechanisms within a taste quality.

6.2.5. Mixture Interactions

Mixtures are the most common taste stimuli. The evolution of the human gustatory system has surely been shaped by our exposure to mixtures of compounds. The continuing study of taste mixtures can only shed more valuable information about how the taste system is organized and shed light on how the system is coded. The term "mixtures" includes not only mixtures within the same quality, but also heteroquality mixtures and mixtures across modalities.

6.2.6. Fischer Hypothesis: Taste and Metabolic Physiology

Fischer and Griffin (1963) hypothesized that the taste system provides a window of conscious access to internal metabolic physiology for toxic bitter tasting compounds. The more bitter a compound tastes to an individual, relative to others tasting the same concentration of the same compound, the more toxic that substance will be to that person. This is an intriguing idea that remains to be fully tested (see Joyce et al., 1968).

6.2.7. Contributions of Undissociated Protons to Sour Taste?

One mystery of human taste not likely to be resolved using metameric matching (Section 6.3.1), cross-adaptation (Section 6.3.2), or pharmacological blockade techniques (Section 6.3.4) involves the role that undissociated protons play in sourness perception. Weak organic acids (i.e., malic, tartaric, citric) do not fully dissociate in solution, yet it is clear that they taste more sour than their pH (negative log of dissociated proton concentration) suggests, compared to strong, inorganic fully dissociating acids at the same pH (i.e., HCl, HNO_3). The undissociated protons clearly contribute to the sourness, however, the exact role that they play remains unknown (Sowalsky and Noble, 1998).

7. CONCLUSIONS

In this chapter I have attempted to describe the power of psychophysical techniques to help us understand the orderly laws that perceptual experience follows as a function of variation in the physical world, as well as the potential for psychophysics to describe functional properties of the underlying physiological mechanisms. Although the recent taste literature has made great strides toward understanding some of the laws that guide our perception, I conclude that the excellent questions asked around the turn of the century have largely gone unanswered. Thus, I see a need to return to the older, fundamental questions of taste coding that may be answered psychophysically.

Qualitative experiences cannot be directly measured with common physiological techniques such as calcium imaging, electrophysiological recordings from cells or of membrane potentials, or with the measurement of humoral agents. Since it is impossible to define subjective experience in terms other than by discussion of the physics and physiology that normally give rise to the sensations, one must rely on the fact that the terms sweet, sour, salt, and bitter are "in-principle-intersubjectively confirmable" (Feigl, 1958). Someone who is ageusic may understand the physics, the physiology, the psychophysics, and the bridges that link these three together, but will never know what is meant by the term sweet. Since individuals show very large differences in gustatory sensitivity (particularly for bitterness, see Section 4.5), intersubjective confirmation of taste sensations is not always as simple as one might initially believe. Nevertheless, physiological measures are useful tools in understanding our psychological experiences; but we must not lose sight of the psychological experience itself. Without direct measure of perception we would have no means to obtain psychophysical or psychophysiological relationships, we would be reduced to understanding only physiophysical (or biophysical) relationships (removing our experiences from the field of query). On the other hand, any theory of taste psychophysics must take into account the relevant electrophysiological and biophysical data, otherwise psychophysics is reduced to linear-systems black-box input/output analysis. Unless taste psychophysics and electrophysiology take account of each other, the interpretation of data will likely prove misguided. The physical and physiological realms are not linked in any logical manner to phenomenal mental life, so relations between the two must be made with formal, empirically based bridges (Feigl, 1958).

REFERENCES

Ackerman, B. H. and N. Kasbekar (1997). Disturbances of taste and smell induced by drugs. *Pharmacotherapy* **17**:482–496.

Anand, K. K. and J. R. Zuniga (1997). Effect of amiloride on suprathreshold NaCl, LiCl, and KCl salt taste in humans. *Physiol Behav* **62**:925–929.

Bartoshuk, L. M. (1968). Water taste in man. *Percept Psychophys* **3**:69–72.

Bartoshuk, L. M. (1975). Taste mixtures: is mixture suppression related to compression? *Physiol Behav* **14**:643–649.

Bartoshuk, L. M. (1989). Taste: robust across the age span? *Ann N Y Acad Sci* **561**: 65–75.

Bartoshuk, L. M. and C. T. Cleveland (1977). Mixtures of substances with similar tastes: a test of a new model of taste mixture interactions. *Sen Proc* **1**:177–186.

Bartoshuk, L. M. and J. F. Gent (1984). Taste mixtures: an analysis of synthesis. In: D. W. Pfaff (ed). *Taste, Olfaction and the Central Nervous System: A Festschrift in Honor of Carl Pfaffmann*. New York: Rockefeller University Press, pp 210–232.

Bartoshuk, L. M., D. H. McBurney, and C. Pfaffmann (1964). Taste of sodium chloride solutions after adaptation to sodium chloride: implications for the "water taste." *Science* **143**:967–968.

Bartoshuk, L. M., V. B. Duffy, and I. J. Miller (1994). PTC/PROP tasting: anatomy, psychophysics, and sex effects. *Physiol Behav* **56**:1165–1171.

Bernstein, I. L. and C. J. Hennessy (1987). Amiloride-sensitive sodium channels and expression of sodium appetite in rats. *Am J Physiol* **253**:R371–R374.

Birch, G. G. (1987). Chemical aspects of sweetness. In: J. Dobbing (ed). *Sweetness*. New York: Springer-Verlag, pp 3–14.

Blackburn, C. W. and P. A. Bramley (1989). Lingual nerve damage associated with the removal of lower third molars. *Br Dental J* **167**:103–107.

Breslin, P. A. S. (1996). Interactions among salty, sour and bitter compounds. *Trends Food Sci Technol* **7**:390–399.

Breslin, P. A. S. and G. K. Beauchamp (1995). Suppression of bitterness by sodium: variation among bitter taste stimuli. *Chem Senses* **20**:609–623.

Breslin, P. A. S. and G. K. Beauchamp (1997). Salt enhances flavour by suppressing bitterness. *Nature* **387**:563.

Breslin, P. A. S. and C. D. Tharp (2000). Reduction of saltiness and bitterness after a chlordexidine rinse. *Chem Senses* (in press).

Breslin, P. A. S., S. Kemp, and G. K. Beauchamp (1994). Single sweetness signal. *Nature* **369**:447–448.

Breslin, P. A. S., G. K. Beauchamp, and E. N. Pugh Jr. (1996). Monogeusia for fructose, glucose, sucrose, and maltose. *Percept Psychophys* **58**:327–341.

Brightman, V. J., J. Guggenheimer, and I. Ship (1968). Changes in the oral microbial flora during treatment of recurrent aphthous ulcers. (Abstract) *J Dent Res* **47**(suppl):126.

Bujas, Z. and A. Ostojcic (1941). La sensibilité gustative en fonction de la surface exciteé. *Acta Inst Psychol Univ Zagreb* **13**:1–19.

Christensen, C. M. (1980). Effects of solution viscosity on perceived saltiness and sweetness. *Percept Psychophys* **28**:347–353.

Clark, C. C. and H. T. Lawless (1994). Limiting response alternatives in time-intensity scaling: an examination of the halo-dumping effect. *Chem Senses* **19**:583–594.

Cliff, M. A. and B. G. Green (1994). Sensory irritation and coolness produced by menthol: evidence for selective desensitization of irritation. *Physiol Behav* **56**:1021–1029.

Cliff, M. and A. C. Noble (1990). Time-intensity evaluation of sweetness and fruitiness and their interaction in a model solution. *J Food Sci* **55**:450–454.

Clydesdale, F. M. (1993). Color as a factor in food choice. *Crit Rev Food Sci Nutr* **33**: 83–101.

Collings, V. B. (1974). Human taste response as a function of locus of stimulation on the tongue and soft palate. *Percept Psychophys* **16**:169–174.

Commeto-Muñiz, J. E. (1981). Odor, taste, and flavor perception of some flavoring agents. *Chem Senses* **6**:215–223.

Costanzo, R. M. and N. D. Zassler (1991). Head trauma. In: T. V. Getchell, R. L. Doty, L. M. Bartoshuk, J. B. Snow Jr. (eds). *Smell and Taste in Health and Disease*. New York: Raven Press, pp 711–730.

Cowart, B. J. (1981). Development of taste perception in humans: sensitivity and preference throughout the life span. *Psychol Bull* **90**:43–73.

Cowart, B. J. (1987). Oral chemical irritation: does it reduce perceived taste intensity? *Chem Senses* **12**:467–479.

Cowart, B. J. (1989). Relationships between taste and smell across the adult life span. *Ann N Y Acad Sci* **561**:39–55.

Cowart, B. J. (1998). The addition of CO_2 to traditional taste solutions alters taste quality. *Chem Senses* **23**:397–402.

Cowart, B. J., Y. Yokomukai, and G. K. Beauchamp (1994). Bitter taste in aging: compound-specific decline in sensitivity. *Physiol Behav* **6**:1237–1241.

Cowart, B. J., I. M. Young, R. S. Feldman, and L. D. Lowry (1997). Clinical disorders of smell and taste. In: G. K. Beauchamp and L. M. Bartoshuk (eds). *Tasting and Smelling.* New York: Academic Press, pp 175–198.

Crocker, E. C. and L. F. Henderson (1932). The glutamic taste: a study of the flavor of meat and its possible duplication by chemical and vegetable means. *Am Perfum Essent Oil Rev* **27**:156–158.

Cruz, A. and B. G. Green (2000). Thermal stimulation of taste. *Nature* **403**:889–892.

Dalton, P., N. Doolittle, H. Nagata, and P. A. S. Breslin (2000). The merging of the senses: Integration of subthreshold taste and smell. *Nature Neurosci* **3**:431–432.

Deems, D. A., R. L. Doty, R. G. Settle, V. Moore-Gillon, P. Shaman, A. F. Mester, C. P. Kimmelman, V. J. Brightman and J. B. Snow Jr. (1991). Smell and taste disorders: a study of 750 patients from the University of Pennsylvania Smell and Taste Center. *Arch Otolaryng Head Neck Surg* **117**:519–528.

De Graaf, C. and J. E. Frijters (1988). Assessment of the taste interaction between two qualitatively similar-tasting substances: a comparison between comparison rules. *J Exp Psychol Hum Percept Perform* **14**:526–538.

Delwiche, J. F., M. F. Lera, and P. A. S. Breslin (2000). Selective removal of a target stimulus localized by taste in humans. *Chem Senses* **25**:181–187.

Dubose, C. N., A. V. Cardello, and O. Maller (1980). Effects of colorants, and flavorants on identification, perceived flavor intensity, and hedonic quality of fruit-flavored beverages and cake. *J Food Sci* **45**:1393–1399.

Duffy, D. B., L. M. Bartoshuk, R. Striegel-Moore, and J. Rodin (1998). Taste changes across pregnancy. *Ann N Y Acad Sci* **855**:805–809.

Dzendolet, E. and H. L. Meiselman (1967). Gustatory quality changes as a function of solution concentration. *Percept Psychophys* **2**:29–33.

Erickson, R. P. (1963). Sensory neural patterns and gustation. In: Y. Zotterman (ed). *Olfaction and Taste I.* New York: Macmillan Co., pp 205–213.

Erickson, R. P. (1982). Studies on the perception of taste: do primaries exist? *Physiol Behav* **28**:57–62.

Faurion, A. (1987a). MSG as one of the sensitivities within a continuous taste space: electrophysiological and psychophysical studies. In: Y. Kawamura and M. R. Kare (eds). *Umami: A Basic Taste,* New York: Marcel Dekker, pp 387–408.

Faurion, A. (1987b). Physiology of the sweet taste. In: D. Ottoson (ed). *Progress in Sensory Neurobiology Heidelberg*: Springer-Verlag, pp. 129–201.

Feigl, H. (1958). The mental and the physical. In: H. Feigl, M. Scriven, and G. Maxwell (eds). *Concepts, Theories, and the Mind-Body Problem, Minnesota Studies in the Philosophy of Science, Vol. II.* Minneapolis: University of Minnesota Press, pp 370–497.

Fischer, R. and F. Griffin (1963). Quinine dimorphism: a cardinal determinant of taste sensitivity. *Nature* **200**:343–347.

Frank, M. E. (1991). Taste-responsive neurons of the glossopharyngeal nerve of the rat. *J Neurophysiol* **65**:1452–1463.

Frank, R. A. and J. Byram (1988). Taste-smell interactions are tastant and odorant dependent. *Chem Senses* **13**:445–455.

Frank, R. A., N. J. van der Klaauw, and H. N. Schifferstein (1993). Both perceptual and conceptual factors influence taste-odor interactions. *Percept Psychophys* **54**:343–354.

Frijters, J. E. and C. De Graaf (1987). The equiratio taste mixture model successfully predicts the sensory response to the sweetness intensity of complex mixtures of sugars and sugar alcohols. *Perception* **16**:615–628.

Garcia-Medina, M. R. (1981). Flavor-odor taste interactions in solutions of acetic acid and coffee. *Chem Senses* **6**:13–22.

Gerber, L. M., G. C. Williams, and S. J. Gray (1999). The nutrient-toxin dosage continuum in human evolution and modern health. *Quart Rev Biol* **74**:273–289.

Getchell, T. V., R. L. Doty, L. M. Bartoshuk, and J. B. Snow Jr. (1991). *Smell and Taste in Health and Disease*. New York: Raven Press.

Glanville, E. V. and A. R. Kaplan (1965). The menstrual cycle and sensitivity of taste perception. *Am J Obstet Gyn* **92**:189–194.

Goodspeed, R. B., J. F. Gent, and F. A. Catalanotto (1987). Chemosensory dysfunction: clinical evaluation results from a taste and smell clinic. *Postgrad Med* **81**:251–260.

Grant, R., S. Miller, D. Simpson, P. J. Lamey, and I. Bone (1989). The effect of chorda tympani section on ipsilateral and contralateral salivary secretion and taste in man. *J Neurol Neurosurg Psychol* **52**:1058–1062.

Grassmann, H. (1853). Zur Theorie der Farbenmischung. *Poggendorfs Annalen der Physik* **84**:69–84.

Green, B. G. (1992). The sensory effects of l-menthol on human skin. *Somatosens Mot Res* **9**:235–244.

Green, B. G., G. S. Shaffer, and M. M. Gilmore (1993). Derivation and evaluation of a semantic scale of oral sensation magnitude with apparent ratio properties. *Chem Senses* **18**:683–702.

Green, B. G., P. Dalton, B. Cowart, G. Shaffer, K. Rankin, and J. Higgins (1996). Evaluating the 'Labeled Magnitude Scale' for measuring sensations of taste and smell. *Chem Senses* **21**:323–334.

Green, D. M. (1988). Audition: psychophysics and perception. In: R. C. Atkinson, R. J. Hernstein, G. Lindzey, and R. D. Luce (eds). *Steven's Handbook of Experimental Psychology Vol. I: Perception and Motivation*. New York: John Wiley and Sons, pp 327–376.

Guinard, J.-X, D. Y. Hong, C. Zoumas-Morse, C. Budwig, and G. F. Russel (1994). Chemoreception and perception of the bitterness of isohumulones. *Physiol Behav* **56**:1257–1263.

Halpern, B. P. (1998). Amiloride and vertebrate gustatory responses to NaCl. *Neurosci Biobehav Rev* **23**:5–47.

Halpern, B. P. and R. B. Darlington (1998). Effects of amiloride on gustatory quality descriptions and temporal patterns produced by NaCl. *Chem Senses* **23**:501–511.

Hänig, D. P. (1901). Zur Psychophysik des Geschmackssinnes. *Philosoph Stud* **17**:576–623.

Hara, S. (1955). Interrelationship among stimulus intensity, stimulated area and reaction time in the human gustatory sensation. *Bull Tokyo Med Dent Univ* **2**:147–158.

Harada, S., T. Yamamoto, K. Yamaguchi, and Y. Kasahara (1997). Different characteristics of gustatory responses between the greater superficial petrosal and chorda tympani nerves in the rat. *Chem Senses* **22**:133–140.

Hellekant, G. and Y. Ninomiya (1991). On the taste of umami in chimpanzee. *Physiol Behav* **49**:927–934.

Hellekant, G. and Y. Ninomiya (1994). Bitter taste in signle chorda tympani taste fibers from chimpanzee. *Physiol Behav* **56**:1185–1188.

Hellekant, G., Y. Ninomiya, and V. Danilova (1997). Taste in chimpanzees II: Single chorda tympani fibers. *Physiol Behav* **61**:829–841.

Hellekant, G., Y. Ninomiya, and V. Danilova (1998). Taste in chimpanzees III: Labeled-line coding in sweet taste. *Physiol Behav* **65**:191–200.

Helms, J. A., M. A. Della-Fera, A. E. Mott, and M. E. Frank (1995). Effects of chlorhexidine on human taste perception. *Arch Oral Biol* **40**:913–920.

Hensel, H. (1982). *Thermal Sensation and Thermoreceptors in Man*. Springfield, IL: C. C. Thomas Pub.

Hill, D. L., B. K. Formaker, and K. S. White (1990). Perceptual characteristics of the amiloride-suppressed sodium chloride taste response in the rat. *Behav Neurosci* **104**:734–741.

Holway, A. H. and L. M. Hurvich (1938). On the psychophysics of taste. *J Exp Psychol* **23**:191–198.

Hurvich, L. (1981). *Color Vision*. Sunderland, MA: Sinauer Associates Inc.

Hurvich, L. M. and D. Jameson (1955). Some quantitative aspects of an opponent-colors theory. II. Brightness, saturation, and hue in normal and dichromatic vision. *J Opt Soc Am* **45**:602–616.

Hyman, A. (1983). The influence of color on taste perception of carbonated water preparations. *Bull Psychonom Soc* **21**:145–148.

Johnson, C., G. G. Birch, and D. B. MacDougall (1994). The effect of the sweetness inhibitor 2(-4-methoxyphenoxy)propanoic acid (sodium salt) (Na-PMP) on the taste of bitter-sweet stimuli. *Chem Senses* **19**:349–358.

Joyce, C. R. B., P. Lynn, and D. D. Varanos (1968). Taste sensitivity may be used to predict pharmacological effects. *Life Sci* **7**:533–537.

Kalmus, H. (1971). The genetics of taste. In: L. M. Beidler (ed). *Handbook of Sensory Physiology: Vol. IV, Chemical Senses, (2): Taste*. New York: Springer-Verlag, pp 165–179.

Kamen, J. M., F. J. Pilgrim, N. J. Gutman, and B. J. Kroll (1961). Interactions of suprathreshold taste stimuli. *J Exp Psychol* **62**:348–356.

Katsuragi, Y. and K. Kurihara (1997). Specific inhibitor for bitter taste. In: G. Roy (ed). *Modifying Bitterness: Mechanism, Ingredients, and Applications*. Lancaster: Technomic Pub Co, pp 255–281.

Katsuragi, Y., M. Kashiwayanagi, and K. Kurihara (1997). Specific inhibitor for bitter taste: inhibition of frog taste nerve responses and human taste sensation to bitter stimuli. *Brain Res Protoc* **3**:292–298.

Kelling, S. T. and B. P. Halpern (1988). Taste judgements and gustatory stimulus duration: taste quality, taste intensity, and reaction time. *Chem Senses* **13**:559–586.

Kobal, G., S. Van Toller, and T. Hummel (1989). Is there directional smelling? *Experientia* **45**:130–131.

Kroeze, J. H. and L. M. Bartoshuk (1985). Bitterness suppression as revealed by split-tongue taste stimulation in humans. *Physiol Behav* **35**:779–783.

Kveton, J. F. and L. M. Bartoshuk (1994). The effect of unilateral chorda tympani damage on taste. *Laryngoscope* **104**:25–29.

Lang, N. P., F. A. Catalanotto, R. U. Knöpfli, and A. A. A. Antczak (1988). Quality-specific taste impairment following the application of chlorhexidine digluconate mouthrinses. *J Clin Periodontol* **15**:43–48.

Lawless, H. (1982). Paradoxical adaptation to taste mixtures. *Physiol Behav* **29**:149–152.

Lawless, H. T. (1979). Evidence for neural inhibition in bittersweet taste mixtures. *J Comp Physiol Psychol* **93**:538–547.

Lawless, H. T. (1998). Theoretical note: tests of synergy in sweetener mixtures. *Chem Senses* **23**:447–451.

Lawless, H. T. and D. A. Stevens (1983). Cross adaptation of sucrose and intensive sweeteners. *Chem Sens* **7**:309–315.

Lawless, H. and D. A. Stevens (1984). Effect of oral chemical irritation on taste. *Physiol Behav* **32**:995–998.

Lawless, H., P. Rozin, and J. Shenker (1985). Effects of oral capsaicin on gustatory, olfactory, and irritant sensations and flavor identification in humans who regularly or rarely consume chili pepper. *Chem Senses* **10**:579–589.

Leach, E. J. and A. C. Noble (1986). Comparison of bitterness of caffeine and quinine by a time-intensity procedure. *Chem Senses* **11**:339–345.

Lindley, M. G. (1991). Phenoxyalkanoic acid sweetness inhibitors. In: D. E. Walters, F. T. Orthoefer, and G. E. DuBois (eds). *Sweeteners: Discovery, Molecular Design, and Chemoreception*. ACS Symposium Series 450. Washington: *Amer Chem Soc* pp 251–260.

Marks, L. E., G. Borg, and J. Westerlund (1992). Differences in taste perception assessed by magnitude matching and by category-ratio scaling. *Chem Senses* **17**:493–506.

McBride, R. L. (1989). Three models for taste mixtures. In: D. G. Laing (ed). *Perception of Complex Smells and Tastes*. Sydney, Australia: Academic Press, pp 265–282.

McBride, R. L. (1983). A JND-scale/category-scale convergence in taste. *Percept Psychophys* **34**:77–83.

McBurney, D. H. (1976). Temporal properties of the human taste system. *Sens Proces* **1**:150–162.

McBurney, D. H. and T. R. Shick (1971). Taste and water taste of twenty six compounds for man. *Percept Psychophys* **10**:249–252.

McBurney, D. H., D. V. Smith, and T. R. Shick (1972). Gustatory cross-adaptation: sourness and bitterness. *Percept Psychophys* **11**:228–232.

McCutcheon, N. B. (1991). Sodium deficient rats are unmotivated by sodium chloride solutions mixed with the sodium channel blocker amiloride. *Behav Neurosci* **105**:764–766.

McCutcheon, N. B. (1992). Human psychophysical studies of saltiness suppression by amiloride. *Physiol Behav* **51**:1069–1074.

Meiselman, H. L. and B. P. Halpern (1970). Human judgements of *Gymnema sylvestre* and sucrose mixtures. *Physiol Behav* **8**:945–948.

Miller, I. J. and L. M. Bartoshuk (1991). Taste perception, taste bud distribution, and spatial relationships. In: T. V. Getchell, R. L. Doty, L. M. Bartoshuk, and J. B. Snow Jr. (eds). *Smell and Taste in Health and Disease*. New York: Raven Press, pp 205–233.

Ming, D., Y. Ninomiya, and R. F. Margolskee (1999). Blocking taste receptor activation of gustducin inhibits gustatory responses to bitter compounds. *Proc Natl Acad Sci U S A* **96**:9903–9908.

Miyamoto, T., R. Fujiyama, Y. Okada, and T. Sato (1998). Salty and sour transduction: multiple mechanisms and strain differences. *Ann N Y Acad Sci* **855**:128–133.

Murphy, C. and M. M. Gilmore (1989). Quality-specific effects of aging on the human taste system. *Percept Psychophys* **45**:121–128.

O'Mahony M. (1986). *Sensory Evaluation of Food: Statistical Methods and Procedures*. New York: Marcel Dekker.

O'Mahony, M., S. Atassi-Sheldon, L. Rothman, and T. Murphy-Ellison (1983). Relative singularity/mixedness judgements for selected taste stimuli. *Physiol Behav* 31:749–755.

Osaki, T., M. Ohshima, Y. Tomita, N. Matsugi, and Y. Nomura (1996). Clinical and physiological investigations in patients with taste abnormality. *J Oral Pathol Med* 25:38–43.

Ossebaard, C. A. and D. V. Smith (1996). Amiloride suppresses the sourness of NaCl and LiCl. *Physiol Behav* 60:1317–1322.

Ossebaard, C. A. and D. V. Smith (1997). Effect of amiloride on the taste of NaCl, Na-gluconate and KCl in humans: implications for Na$^+$ receptor mechanism. *Chem Senses* 20:37–46.

Ossebaard, C. A., I. A. Polet, and D. V. Smith (1997). Amiloride effects on taste quality: comparison of single and multiple response category preocedures. *Chem Senses* 22:267–275.

Pelchat, M. L. and P. Rozin (1982). The special role of nausea in the acquisition of food dislikes by humans. *Appet* 3:341–351.

Pfaffmann, C., L. M. Bartoshuk, and D. H. McBurney (1971). Taste psychophysics. In: L. M. Beidler (ed). *Handbook of Sensory Physiology: Vol. IV, Chemical Senses. Part 2. Taste*. New York: Springer-Verlag, pp 75–101.

Powley, T. L. and H. R. Berthoud (1985). Diet and cephalic phase insulin release. *Am J Clin Nutr* 42(5 suppl.):991–1002.

Pugh, E. N. Jr. (1988). Vision: physics and retinal physiology. In: R. C. Atkinson, R. J. Hernstein, G. Lindzey, and R. D. Luce (eds). *Steven's Handbook of Experimental Psychology Vol. I: Perception and Motivation*. New York: John Wiley and Sons, pp 75–164.

Pugh, E. N. Jr and D. B. Kirk (1986). The π mechanisms of WS Stiles: a historical review. *Perception* 15:705–728.

Purdy, D. (1931). Spectral hue as a function of intensity. *Am J Psychol* 43:541–559.

Reed, D. R., E. Nanthakumar, M. North, C. Bell, L. M. Bartoshuk, and R. A. Price (1999). Localization of a gene for bitter-taste perception to human chromosome 5p15. *Am J Hum Genet* 64:1478–1480.

Reed, D. R., L. M. Bartoshuk, V. Duffy, S. Marino, and R. A. Price (1995). Propylthiouracil tasting: Determination of underlying threshold distributions using maximum likelihood. *Chem Senses* 20:529–533.

Riha, W. E. III, J. G. Brand, and P. A. S. Breslin (1997). Salty taste enhancement with amino acids. *ISOT XII* (Abstract).

Rivkin, B. A. and L. M. Bartoshuk (1980). Taste synergism between monosodium glutamate and disodium 5'-guanylate. *Physiol Behav* 24:1169–1172.

Roitman, M. F. and I. L. Bernstein (1999). Amiloride-sensitive sodium signals and salt appetite: multiple gustatory pathways. *Am J Physiol* 276:R1732–R1738.

Schifferstein, H. N. J. (1996). An equiratio mixture model for non-additive components: a case study for aspartame/acesulfame-K mixtures. *Chem Senses* 21:1–11.

Schifferstein, H. N. J. and J. E. R. Frijters (1992). Two-stimulus versus one-stimulus procedure in the framework of functional measurement: A comparative investigation using quinineHCl/NaCl mixtures. *Chem Senses* 17:127–150.

Schifferstein, H. N. J. and R. W. Kleykers (1996). An empirical taste of Olsson's interaction model using mixtures of tastants. *Chem Senses* **21**:283–291.

Schiffman, S. S. and C. A. Gatlin (1993). Clinical physiology of taste and smell. *Annu Rev Nutr* **13**:405–436.

Schiffman, S. S., D. A. Reilly, and T. B. Clark (1979). Qualitative differences among sweeteners. *Physiol Behav* **23**:1–9.

Schiffman, S. S., H. Cahn, and M. G. Lindley (1981). Multiple receptor sites mediate sweetness: evidence from cross adaptation. *Pharm Biochem Behav* **15**:377–388.

Schiffman, S. S., E. Lockhead, and F. W. Maes (1983). Amiloride reduces the taste intensity of Na^+ and Li^+ salts and sweeteners. *Proc Nat Acad Sci U S A* **80**:6136–6140.

Schiffman, S. S., B. J. Booth, B. T. Carr, M. L. Losee, E. A. Sattely-Miller, and B. G. Graham (1995). Investigation of synergism in binary mixtures of sweeteners. *Brain Res Bull* **38**:105–120.

Schiffman, S. S., B. J. Booth, E. A. Sattely-Miller, B. G. Graham, and K. M. Gibes (1999). Selective inhibition of sweetness by the sodium salt of $+/-$ 2-(4-methoxyphenoxy)propanoic acid. *Chem Senses* **24**:439–447.

Sclafani, A. and C. Perez (1997). Cypha [propionic acid, 2-(4-methoxyphenol)salt] inhibits sweet taste in humans, but not in rats. *Physiol Behav* **61**:25–29.

Seitz, J. A. (1998). A kinetic analysis of 2-(4-methoxyphenoxy)propanoic acid and its inhibitory effect on the time-intensity functions of sweeteners. Unpublished Honor's Thesis. University of Pennsylvania.

Shallenberger, R. S. (1993). *Taste Chemistry*. New York: Blackie Academic and Professional, pp 153–188.

Shikata, H., D. B. T. McMahon, and P. A. S. Breslin (1999). Psychophysics of taste lateralization on anterior tongue. *Percept Psychophys* **62**:684–694.

Shimizu, K., M. Ozeki, K. Tanaka, K. Itoh, S. Nakajyo, N. Urakawa, and M. Atsuchi (1997). Suppression of glucose absorption by extracts from leaves of *Gymnema inodorum*. *J Vet Med Sci* **59**:753–757.

Smith, D. V. (1971). Taste intensity as a function of area and concentration: differentiation between compounds. *J Exp Psychol* **87**:163–171.

Smith, D. V. and S. L. Bealer (1975). Sensitivity of the rat gustatory system to the rate of stimulus onset. *Physiol Behav* **15**:303–314.

Smith, D. V. and D. H. McBurney (1969). Gustatory cross-adaptation: does a single mechanism code the salty taste? *J Exp Psychol* **80**:101–105.

Smith, D. V. and C. A. Ossebaard (1995). Amiloride suppression of the taste intensity of sodium chloride: evidence from direct magnitude scaling. *Physiol Behav* **57**:773–777.

Smith, D. V. and N. J. van der Klaauw (1995). The perception of saltiness is eliminated by NaCl adaptation: implications for gustatory transduction and coding. *Chem Senses* **20**:545–557.

Sowalsky, R. A. and A. C. Noble (1998). Comparison of the effects of concentration, pH andanion species on astringency and sourness of organic acids. *Chem Senses* **23**:343–349.

Spector, A. C., N. A. Guagliardo, S. J. St. John (1996). Amiloride disrupts NaCl versus KCl discrimination performance: Implications for salt taste coding in rats. *J Neurosci* **16**:8115–8122.

St. John S. J. and A. C. Spector (1998). Behavioral discrimination between quinine and KCl is dependent on input from the seventh cranial nerve: Implications for the functional roles of the gustatory nerves in rats. *J Neurosci* **18**:4353–4362.

Steiner, J. E. (1973). The gustofacial response: observation on normal and anencephalic newborn infants. *Symp Oral Sens Percept* **4**:254–278.

Steiner, J. E. (1974). Discussion paper: Innate, discriminative human facial expression to taste and smell stimulation. *Ann N Y Acad Sci* **237**:229–233.

Stevens, J. C. (1989). Food quality reports from noninstitutionalized aged. *Ann N Y Acad Sci* **561**:87–83.

Stevens, J. C. (1991). Thermal sensibility. In: M. A. Heller, and W. Schiff (eds). *The Psychology of Touch.* Hillsdale, NJ: Lawrence Erlbaum Associates, pp 61–90.

Stevens, J. C. (1995). Detection of heteroquality taste mixtures. *Percept Psychophys* **57**:18–26.

Stevens, J. C. (1996). Detection of tastes in a mixture with other tastes: issues of masking and aging. *Chem Senses* **21**:211–221.

Stevens, J. C. (1997). Detection of very complex taste mixtures: generous integration across constituent compounds. *Physiol Behav* **62**:1137–1143.

Stevens, J. C. and W. S. Cain (1993). Changes in taste and flavor in aging. *Crit Rev Food Sci Nutri* **33**:27–37.

Stevens, J. C. and A. Traverzo (1997). Detection of a target taste in a complex masker. *Chem Senses* **22**:529–534.

Stevens, J. C., L. M. Bartoshuk, and W. S. Cain (1984). Chemical senses and aging: taste versus smell. *Chem Senses* **9**:167–179.

Stevens, J. C., W. S. Cain, A. Demarque, and A. M. Ruthruff (1991). On the discrimination of missing ingredients: aging and salt flavor. *Appetite* **16**:129–140.

Stevens, J. C., L. A. Cruz, J. M. Hoffman, and M. Q. Patterson (1995). Taste sensitivity and aging: high incidence of decline revealed by repeated threshold measures. *Chem Senses* **20**:451–459.

Stevens, S. S. (1969). Sensory scales of taste intensity. *Percept Psychophys* **6**:302–308.

Stevens, S. S. (1975). *Psychophysics: Introduction to Its Perceptual Neural and Social Prospects.* New York: Wiley and Sons.

Teff, K. L., and K. Engelman (1996). Oral sensory stimulation improves glucose tolerance in humans: effects on insulin, C-peptide, and glucagon. *Am J Physiol* **270**:R1371–R1379.

Tennissen, A. M. (1992). Amiloride reduces intensity responses of human fungiform papillae. *Physiol Behav* **51**:1061–1068.

Tennissen, A. M. and N. B. McCutcheon (1996). Anterior tongue stimulation with amiloride suppresses NaCl saltiness, but not citric acid sourness in humans. *Chem Senses* **21**:113–120.

Tichener, E. B. (1909). *A Textbook of Psychology.* New York: Macmillan.

Todrank, J. and L. M. Bartoshuk (1991). A taste illusion: taste sensation localized by touch. *Physiol Behav* **50**:1027–1031.

von Fick, A. (1864). *Lehrbuch der Anatomie und Physiologie der Sinnesorgane.* Lahr: Schauenberg.

von Skramlik, E. (1921a). Mischungsgleichungen im gebiete des geschmacksinns. *Zeit Psychol Physiol Sinnesorgane* **53**(2 Abt.):36–78.

von Skramlik, E. (1921b). Mischungsgleichungen im gebiete des geschmacksinns.II. *Zeit Psychol Physiol Sinnesorgane* **53**(2 Abt.):219–233.

von Skramlik, E. (1924). Uber die Lokalisation der Empfindungen bei den niederen Sinnen. *Zeitsch Psychol Physiol Sinnesorgane* **56**:69–88.

von Skramlik, E. (1926). Die Physiologie des Geruch- und Geschmackssinnes. In: *Handbuch der Physiologie der niederen Sinne, Vol. I.* Leipzig: Thieme, p 532.

Weiffenbach, J. M., B. J. Baum, and R. Burghauser (1982). Taste thresholds: quality specific variation with human aging. *J Gerontol* **37**:372–377.

Weiffenbach, J. M., B. J. Cowart, and B. J. Baum (1986). Taste intensity perception in aging. *J Gerontol* **41**:460–468.

Weinstein, S. (1968). Intensive and extensive aspects of tactile sensitivity as a function of body part, sex, and laterality. In: D. R. Kenshalo (ed). *The Skin Senses.* Springfield, IL: C. C. Thomas Pub. pp 195–222.

Yamaguchi, S. (1987). Fundamental properties of umami in human taste sensation. In: Y. Kawamura and M. R. Kare (eds). *Umami: A Basic Taste.* New York: Marcel Dekker, Inc., pp 41–73.

Yanagisawa, K., L. M. Bartoshuk, F. A. Catalanotto, T. A. Karrer, and J. F. Kveton (1998). Anesthesia of the chorda tympani nerve and taste phantoms. *Physiol Behav* **63**:329–335.

Yokomukai, Y., B. J. Cowart, and G. K. Beauchamp (1993). Individual differences in sensitivity to bitter-tasting substances. *Chem Senses* **18**:669–681.

INDEX